Induction Machines Handbook

Electric Power Engineering Series

Series Editor:
Leonard L. Grigsby

Electromechanical Systems, Electric Machines, and Applied Mechatronics
Sergey E. Lyshevski

Power Quality
C. Sankaran

Power System Operations and Electricity Markets
Fred I. Denny and David E. Dismukes

Electric Machines
Charles A. Gross

Electric Energy Systems
Analysis and Operation
Antonio Gomez-Exposito, Antonio J. Conejo, and Claudio Canizares

The Induction Machines Design Handbook, Second Edition
Ion Boldea and Syed A. Nasar

Linear Synchronous Motors
Transportation and Automation Systems, Second Edition
Jacek F. Gieras, Zbigniew J. Piech, and Bronislaw Tomczuk

Electric Power Generation, Transmission, and Distribution, Third Edition
Leonard L. Grigsby

Computational Methods for Electric Power Systems, Third Edition
Mariesa L. Crow

Electric Energy Systems
Analysis and Operation, Second Edition
Antonio Gomez-Exposito, Antonio J. Conejo, and Claudio Canizares

Induction Machines Handbook, Third Edition (Two-Volume Set)
Ion Boldea

Induction Machines Handbook, Third Edition
Steady State Modeling and Performance
Ion Boldea

Induction Machines Handbook, Third Edition
Transients, Control Principles, Design and Testing
Ion Boldea

For more information about this series, please visit: https://www.crcpress.com/Electric-Power-Engineering-Series/book-series/CRCELEPOWENG

Induction Machines Handbook

Steady State Modeling and Performance

Third Edition

Ion Boldea

CRC Press
Taylor & Francis Group
Boca Raton London New York

CRC Press is an imprint of the
Taylor & Francis Group, an **informa** business

Third edition published 2020
by CRC Press
6000 Broken Sound Parkway NW, Suite 300, Boca Raton, FL 33487-2742

and by CRC Press
4 Park Square, Milton Park, Abingdon, Oxon OX14 4RN

First issued in paperback 2023

First edition published by CRC Press 2001
Second edition published by CRC Press 2009

CRC Press is an imprint of Taylor & Francis Group, an Informa business

Visit the Taylor & Francis Web site at
http://www.taylorandfrancis.com

and the CRC Press Web site at
http://www.crcpress.com

Library of Congress Cataloging-in-Publication Data
Names: Boldea, I., author.
Title: Induction machines handbook: steady state modeling and performance / Ion Boldea.
Description: Third edition. | Boca Raton: CRC Press, 2020. |
Series: Electric power engineering | Includes bibliographical references and index. |
Contents: v. 1. Induction machines handbook: steady stat — v.
2. Induction machines handbook: transients
Identifiers: LCCN 2020000304 (print) | LCCN 2020000305 (ebook) |
ISBN 9780367466121 (v. 1 ; hbk) | ISBN 9780367466183 (v. 2 ; hbk) |
ISBN 9781003033417 (v. 1 ; ebk) | ISBN 9781003033424 (v. 2 ; ebk)
Subjects: LCSH: Electric machinery, Induction—Handbooks, manuals, etc.
Classification: LCC TK2711 .B65 2020 (print) | LCC TK2711 (ebook) |
DDC 621.34—dc23
LC record available at https://lccn.loc.gov/2020000304
LC ebook record available at https://lccn.loc.gov/2020000305

ISBN: 978-1-03-258268-9 (pbk)
ISBN: 978-0-367-46612-1 (hbk)
ISBN: 978-1-003-03341-7 (ebk)

DOI: 10.1201/9781003033417

Typeset in Times
by codeMantra

Publisher's Note
The publisher has gone to great lengths to ensure the quality of this reprint but points out that some imperfections in the original copies may be apparent.

A humble, late, tribute to:

Nikola Tesla

Galileo Ferraris

Dolivo-Dobrovolski

Contents

Preface

MOTIVATION

The 2010–2020 decade has seen notable progress in induction machines (IMs) technology such as

- Extension of analytical and finite element modelling (FEM) for better precision and performance
- Advanced FEM-assisted optimal design methodologies with multi-physics character
- Introduction of upgraded premium efficiency IM international standards
- Development and fabrication of copper cage rotor IM drives for traction on electric vehicles
- Extension of wound rotor induction generators (WRIG or DFIG) with partial rating A.C.–D.C.–A.C. converters in wind energy conversion and to pump storage reversible power plants (up to 400 MVA/unit)
- Extension of cage rotor induction generators with full power pulse width modulation (PWM) converters for wind energy conversion (up to 5 MVA/unit)
- Development of cage (or nested cage) rotor dual stator winding induction generators/motors with partial rating power electronics for wind energy and vehicular technologies (autonomous operation)
- Development of line-start premium efficiency IMs with cage rotor, provided with PMs and/or magnetic saliency for self-synchronisation and operation at synchronism (three phase and one phase), for residential applications, etc.
- Introduction of multiphase (m > 3) IMs for higher torque density and more fault-tolerant electric drives.

All the above, reflected in a strong increase of line-start IMs and variable speed IM motor and generator drives markets, have prompted us to prepare a new (third) edition of the book.

VOLUME 1: *INDUCTION MACHINE: STEADY STATE MODELING AND PERFORMANCE,* THIRD EDITION/437 PP.

Short Description

In essence, as a compromise between not confusing the readers of the second edition, but bringing new knowledge in too, it was decided to introduce the new knowledge towards the end of each chapter mainly, but not only, as new paragraphs.

Also the text, equations, figures, and numbers in the numerical examples have been checked carefully with local corrections/improvements/additions wherever necessary.

Finally, new/recent representative references have been added and assessed in the text (sometimes in the existing paragraphs) of each chapter.

The new paragraphs are

Chapter 4/4.14 – Multiphase and multilayer tooth-wound coil windings
Chapter 6/6.14 – The brushless doubly fed induction machine (BDFIM)
Chapter 7/7.19 – Equivalent circuits for brushless doubly fed IMs
Chapter 8/8.6 – Control basics of DFIMs
Chapter 9/9.9 – Magnetic saturation effects or current/slip and torque/slip curves
Chapter 9/9.10 – Rotor slot leakage reactance saturation effects
Chapter 9/9.11 – Closed slot IM saturation effects

Chapter 10/10.11 – The origin of electromagnetic vibration: by practical experience
Chapter 13/13.10 – PM-assisted split-phase cage-rotor IMs.

All efforts have been made to keep the mathematics extension under control **but introduce ready-to-use expressions of industry parameters and performance computation**. In compensation, numerical examples have been generously spread all over the book to facilitate a strong feeling of magnitudes which is, in our view, indispensable in engineering.

There are in total 14 chapters (1–14) in Volume I, which totalise about 437 pages.

Contents

As already mentioned, Volume I refers to steady-state modelling and performance of induction machines, covering 14 chapters. Here is a short presentation of the 14 chapters by content.

Chapter 1: "Induction Machines: An Introduction"/16 pages.

Induction machines (IMs) are related to mechanical into electrical energy conversion via magnetic energy storage in an ensemble of coupled electric and magnetic circuits. Energy is paramount to prosperity, while electric energy conversion in electric power plants and in motion digital control by power electronics is the key to higher industrial productivity with reasonable impact on the environment. A historical touch is followed by IMs' application examples in a myriad of industries, with IM not only the workhorse but also one important racehorse of the industry (at variable speed).

Chapter 2: "Construction Aspects and Operation Principles"/18 pages.

This chapter introduces the main parts of IMs and the principles of electromagnetic induction (Faraday law) in producing electromagnetic force (torque) on the rotor (mover) made of a cage or two- (three-) phase A.C. winding (coil) placed in a slotted laminated silicon-iron core. The principles explained this way apply to both linear motion and rotary electric machines (where the torque concept replaces the tangential force concept).

The recent premium IMs with PM and magnetic saliency added beneath the rotor cage are also referred to.

Chapter 3: "Magnetic, Electric, and Insulation Materials for IM"/16 pages.

Main active materials – magnetic, electric, and insulation types – used in the IM fabrication are presented with their characteristic performance, with additional data on recently developed such materials (and pertinent source literature). Core loss basic formulae have been derived starting from Maxwell equations, and tables illustrate core losses/kg of silicon steel at various flux densities and a few frequencies of interest.

Chapter 4: "Induction Machine Windings and Their mmfs"/40 pages.

The fundamental concept of magnetomotive force (mmf) as the source of a travelling magnetic field in the airgap of IMs is introduced, first produced with three-phase alternative current (A.C.) windings (coils), then with two-phase A.C. windings (for single-phase A.C. source supplies), and finally (new) with multiphase (5, 7, 2×3, 3×3 phases) A.C. windings, introduced recently. The placement of A.C. windings in slots is discussed in detail. The winding factor's three components for distributed windings (integer or fractionary slot/pole/phase, $q \geq 1$) are derived both for the fundamental component (average torque producing) and for the m.m.f. space harmonics. Finally, the "skewing" m.m.f. concept is introduced, only to be, later in this book, used to prove the axially non-uniform magnetic saturation of the stator core, especially at high currents (above 2–3 p.u.; p.u. means relative values to rated current).

Chapter 5: "The Magnetisation Curve and Inductance"/32 pages.

The magnetisation curve (magnetisation fundamental airgap) flux linkage Ψ_{m1} versus stator fundamental phase current i_{m10} (with zero rotor current and/or the magnetisation inductance $L_m = \Psi_{m1}/i_{m10}$) are crucial to IM performance and are calculated first analytically by a simplified and then by analytical iterative model (AIM) both accounting for magnetic saturation and for the slotting influences, with results proved experimentally. The magnetic saturation airgap, teeth, and back-iron flux density harmonics are illustrated as a basis to calculate iron losses in a later chapter.

The electromagnetic force (emf) induced in an A.C. winding is calculated illustrating the same three components of the winding factors already derived for mmfs and the emf time harmonics, produced essentially by the mmf and magnetic saturation-caused space harmonics.

Finally, finite element modelling (FEM) – numerical – computation of airgap flux density for zero rotor currents (ideal no-load currents: at ideal (synchronous) no-load speed n_1 of mmf wave fundamental: $n_1 = f_1/p_1$; f_1 – A.C. stator currents frequency, p_1 – number of pole pairs (periods) of travelling field along one revolution) is performed to illustrate the space fundamental and harmonics produced by the slotting (and magnetic saturation). A numerical detailed example for the magnetisation curve calculation gives a stronger feeling of magnitudes for its components in air and iron cores.

Chapter 6: "Leakage Inductances and Resistances"/25 pages.

The part of the magnetic field that does not embrace both stator and rotor windings, called generically leakage field, being mostly in air, is decomposed in differential (space harmonics), slot (distinct for different practical slot geometries), zig-zag – airgap and end – connection components to give rise to respective leakage inductance components in the stator and in the rotor (L_{sl} and L_{rl}). This important aspect of IM technology is presented quantitatively in ample detail. Numerical examples illustrate the magnitudes, and very recent refined calculation methodologies of leakage inductances – with FEM validation – are synthesised (**new**). The stator- and rotor-phase resistances (R_s and R_r) are calculated too.

Chapter 7: "Steady-State Equivalent Circuit and Performance"/47 pages.

Based on the main phase circuit parameters, extracted from magnetic field distribution (fluxes, energy), – L_m, L_{sl}, L_{rl}, R_s, R_r – the standard phase equivalent circuit of the IM with three phases is introduced and then, particularised for: ideal no-load speed ($n_0 = n_1 = n$; $S_{0i} = (n_1 - n)/n_1 = 0$), no-load motoring ($n = n_0$, $S_0 = (n_1 - n_0)/n_1$; $S_0 > S_{0i}$), on load ($S = (n_1 - n)/n_1 > s_0$) but in the interval of 0.05–0.005, with increasing power, generator to the grid, capacitor (self-excited) autonomous generator, mechanical characteristic, efficiency, power factor, unbalanced stator or rotor operation, voltage sags, swells, and time harmonics, with numerical examples. Finally, the steady-state equivalent circuit of nested-cage dual stator winding IM and of cascaded – dual stator winding IM are presented in view of their recent proposition for wind energy conversion (**new**).

Chapter 8: "Starting and Speed Control Methods"/31 pages.

Starting and close loop control of IMs is an art of itself – electric drives – but the principles of it have to be derived first by a deep knowledge of IM operation modes. This chapter presents first the starting and speed control methods of line-start (constant stator frequency) cage and wound rotor IMs. Then the two main categories of close loop speed control: V/f control (with stabilizing loops recently) and field-oriented control (FOC), illustrated by exemplary mechanical characteristics and block (structural) control diagrams are unfolded and complimented by self-explanatory numerical examples.

Chapter 9: "Skin and On-Load Saturation Effects"/51 pages.

Skin (frequency) and on-load magnetic saturation influence on IM resistances, inductances, characteristics, and performance are investigated in detail in this chapter as they are key issues in optimal design of IMs for various applications from industrial drives to wind generators, electric vehicle propulsion, or deep underground (or underwater) fluid pump motors, etc., or home appliance split-phase motors.

Both these phenomena are presented quantitatively, with practical methodologies of industrial value. A comprehensive analytical non-linear approach for linkage saturation and skin effect in IMs is unfolded to shed light on its various aspects, with experimental validation.

Finally, FEM is illustrated as a suitable approach to calculate skin effects in slots with multiple conductors. New paragraphs deal with such subtle phenomena in very recent treatments (torque/slip curves as influenced by saturation harmonics eddy currents in high-speed IM closed slot (saturation effects)).

Chapter 10: "Airgap Field Space Harmonics Parasitic Torques, Radial Forces, and Noise Basics"/32 pages.

The airgap field space harmonics produced by mmf space harmonics, airgap magnetic permeance harmonics, and magnetic saturation give rise to parasitic torques (time pulsations in the torque), strong local variations in time of radial forces, and thus, additional vibration and more.

These phenomena are all treated quantitatively for uniform airgap and for the influence of static and dynamic eccentricity, via many numerical examples. The main result is founding a way to choose proper combinations of stator and rotor slot numbers for various number of pole pairs, which is crucial in a good industrial IM design.

A **new** section (10.11) dealing with the origin of electromagnetic vibration by practical experience is introduced.

Chapter 11: "Losses in Induction Machines"/41 pages.

Losses defined as the difference between input and output power of IM have many components. There is, however, a notable difference between the losses measured under direct load and the ones calculated from the separation of losses methods (in no-load and short-circuit tests). This difference is called stray load losses, and they vary from 0.5% to 2.5% of full power; they occur even on motor no-load operation.

This chapter attempts a rather exhaustive approach to calculate the additional (stray load) losses produced by the magnetic field space harmonics in the presence of slotting and finally of current (and magnetic flux) time harmonics so common in variable speed drives fed from PWM static power converters. Numerical results illustrate the concepts step by step and offer a strong feeling of magnitudes. Loss computation by FEM, which lumps in all aspects is also illustrated at the end of this chapter. A comparison between sinusoidal source and PWM source inverter IM losses via an experimental case study is added as **new**.

Chapter 12: "Thermal Modelling and Cooling"/24 pages.

A highly non-linear system, the thermal model of IM may be approached by a lump equivalent circuit method or by FEM.

This chapter develops such an equivalent thermal circuit with expressions for its parameters, to calculate steady-state temperatures in a few points (nodes). It also develops on temperature variation in time (thermal transients) and on FEM computation of temperatures in the machine with lumped thermal parameter estimation. As this subject represents an art of itself, recent/representative literature is added for the diligent reader to explore it further on his own.

Chapter 13: "Single-Phase Induction Machines: The Basics"/22 pages.

Single-phase A.C. sources are common in residential/building supply power technologies.

Line-start split-phase IMs have been developed for worldwide spread applications; so efficiency is paramount.

A classification of them is offered, with the shaded-pole IM also added. The principle of operation with two A.C. windings at start (at least) to create a strong forward travelling field is described together with the symmetrical components' general model and its complex-variable steady-state equivalent circuit. The d-q model is introduced as a tool to investigate more complex topologies (Steinmetz connection) and transients.

The **new** section (13.10) describes the model of the split-phase IM when PMs and (or) a magnetic saliency are added to the cage rotor, to improve efficiency.

Chapter 14: "Single-Phase Induction Motors: Steady State"/41 pages.

This chapter starts with steady-state operation with open auxiliary winding: equivalent circuits, mechanical characteristics, efficiency, and power factor, via a numerical example.

The split-phase capacitor IM steady state is then investigated by the complex equivalent circuit defined in the previous chapter via another numerical example. Then, the symmetrisation simetrization conditions, starting torque and current inquires, typical motor characteristics, non-orthogonal winding modelling, via extended numerical graphical delta from a dedicated

MATLAB® computer code, mmf space harmonics parasitic torques, interbar rotor currents, voltage harmonics effects, the doubly tapped winding capacitor IM, a 2/4 pole split-phase capacitor motor are all subjects treated quantitatively as issues of industrial importance (new).

Timisoara, 2019

Author

Ion Boldea, IEEE Life Fellow and Professor Emeritus at University Politehnica Timisoara, Romania, has taught, did research, and published extensively papers and books (monographs and textbooks) over more than 45 years, related to rotary and linear electric motor/generator variable speed drives, and maglevs. He was a visiting professor in the USA and UK for more than 5 years since 1973 to present.

He was granted four IEEE Best Paper Awards, has been a member of IEEE IAS, IE MEC, and IDC since 1992, was the guest editor of numerous special sections in IEEE Trans, vol. IE, IA, delivered keynote addresses at quite a few IEEE-sponsored International Conferences, participated in IEEE Conference tutorials, and is an IEEE IAS distinguished lecturer since 2008 (with lecture in the USA, Brasil, South Korea, Denmark, Italy, etc.). He held periodic intensive graduate courses for Academia and Industry in the USA and Denmark in the last 20 years.

He was a general chair of ten biannual IEEE-sponsored OPTIM International Conferences (www.info-optim.ro) and is the founding and current chief editor, since 2000, of the Internet-only *Journal of Electrical Engineering*, "www.jee.ro".

As a full member of Romanian Academy, he received the IEEE-2015 "Nikola Tesla Award" for his contributions to the development of rotary and linear electric motor/generator drives and maglevs modelling, design, testing, and control in industrial applications.

1 Induction Machines
An Introduction

1.1 ELECTRIC ENERGY AND INDUCTION MOTORS

The level of prosperity of a community is related to its capability to produce goods and services. However, producing goods and services is strongly related to the use of energy in an intelligent way.

Motion and temperature (heat) control are paramount in energy usage. Energy comes into use in a few forms such as thermal, mechanical, and electrical.

Electrical energy, measured in kWh, represents more than 30% of all used energy, and it is on the rise. Part of electrical energy is used directly to produce heat or light (in electrolysis, metallurgical arch furnaces, industrial space heating, lighting, etc.).

The larger part of electrical energy is converted into mechanical energy in electric motors. Amongst electric motors, induction motors are most used both for home appliances and in various industries [1–11].

This is so because they have been traditionally fed directly from the three-phase A.C. electric power grid through electromagnetic power switches with adequate protection. It is so convenient.

Small-power induction motors, in most home appliances, are fed from the local single-phase A.C. power grids. Induction motors are rugged and have moderate costs, explaining their popularity.

In developed countries today, there are more than 3 kW of electric motors per person and most of them are induction motors.

While most induction motors are still fed from the three- or single-phase power grids, some are supplied through frequency changers (or power electronics converters) to provide variable speeds.

In developed countries, already 20% of all induction motor power is converted in variable speed drive applications. The annual growth rate of variable speed drives has been 9% in the past decade, while the electric motor markets showed an average annual growth rate of 4% in the same time.

Variable speed drives with induction motors are used in transportation, pumps, compressors, ventilators, machine tools, robotics, hybrid or electric vehicles, washing machines, etc.

The forecast is that in the next decade, up to 50% of all electric motors will be fed through power electronics with induction motors covering 60%–70% of these new markets.

The ratings of induction motors vary from a few tens of watts to 33,120 kW (45,000 HP). The distribution of ratings in variable speed drives is shown in Table 1.1 [1].

Intelligent use of energy means higher productivity with lower active energy and lower losses at moderate costs. Reducing losses leads to lower environmental impact where the motor works and lower thermal and chemical impacts at an electric power plant that produces the required electrical energy.

Variable speed through variable frequency is paramount in achieving such goals. As a side effect, the use of variable speed drives leads to current harmonics pollution in the power grid and

TABLE 1.1
Variable Speed A.C. Drives Ratings

Power (kW)	1–4	5–40	40–200	200–600	>600
Percentage	21	26	26	16	11

electromagnetic interference (EMI) with the environment. So power quality and EMI have become new constraints on electric induction motor drives.

Digital control is now standard in variable speed drives, while autonomous intelligent drives to be controlled and repaired via the Internet are on the horizon. And new application opportunities abound: from digital appliances to hybrid and electric vehicles and more electric aircraft.

So much on the future, let us now go back to the first two invented induction motors.

1.2 A HISTORICAL TOUCH

Faraday discovered the electromagnetic induction law around 1831, and Maxwell formulated the laws of electricity (or Maxwell's equations) around 1860. The knowledge was ripe for the invention of the induction machine (IM) which has two fathers: Galileo Ferraris (1885) and Nikola Tesla (1886). Their IMs are shown in Figures 1.1 and 1.2.

Both motors have been supplied from a two-phase A.C. power source and thus contained two-phase concentrated coil windings 1-1′ and 2-2′ on the ferromagnetic stator core.

In Ferrari's patent, the rotor was made of a copper cylinder, while in Tesla's patent, the rotor was made of a ferromagnetic cylinder provided with a short-circuited winding.

Though the contemporary induction motors have more elaborated topologies (Figure 1.3) and their performance is much better, the principle has remained basically the same.

That is, a multiphase A.C. stator winding produces a travelling field that induces voltages that produce currents in the short-circuited (or closed) windings of the rotor. The interaction between the stator-produced field and the rotor-induced currents produces torque and thus operates the induction motor. As the torque at zero rotor speed is nonzero, the induction motor is self-starting. The three-phase A.C. power grid capable of delivering energy at a distance to induction motors and other consumers has been put forward by Dolivo-Dobrovolsky around 1880.

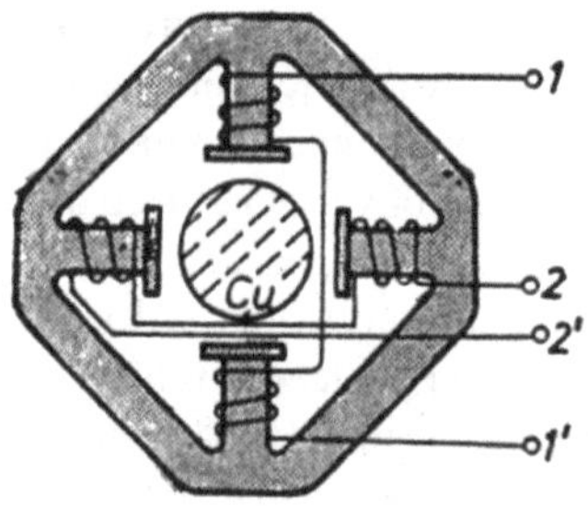

FIGURE 1.1 Ferrari's induction motor (1885).

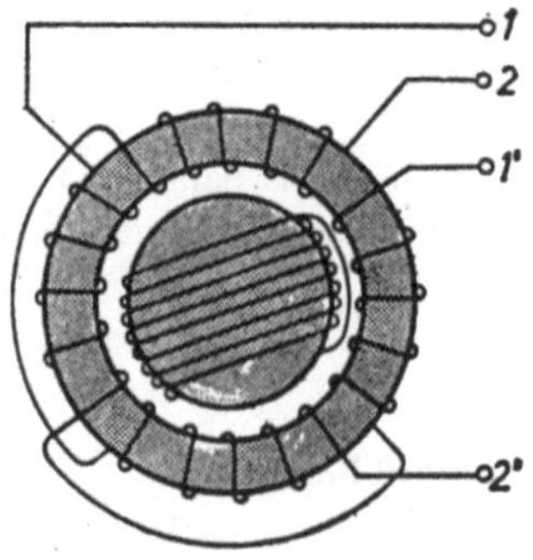

FIGURE 1.2 Tesla's induction motor (1886).

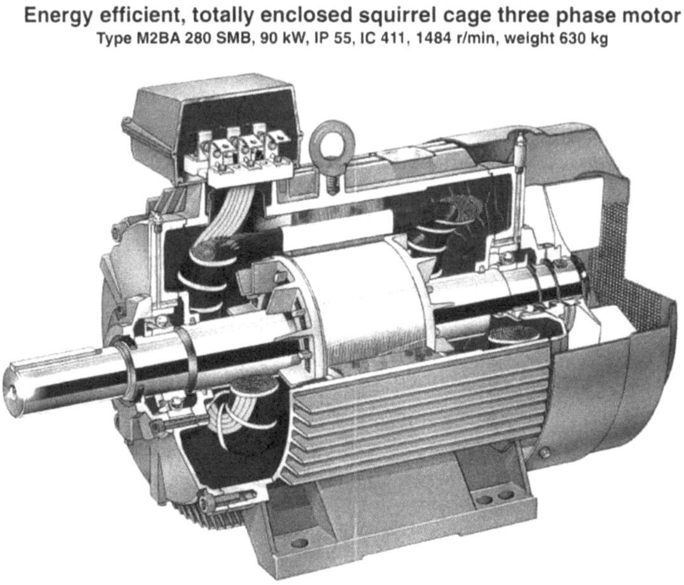
Energy efficient, totally enclosed squirrel cage three phase motor
Type M2BA 280 SMB, 90 kW, IP 55, IC 411, 1484 r/min, weight 630 kg

FIGURE 1.3 A state-of-the-art three-phase induction motor. (Source: ABB motors.)

In 1889, Dolivo-Dobrovolsky invented the three-phase induction motor with the wound rotor and subsequently the cage rotor in a topology very similar to that used today. He also, apparently, invented the double-cage rotor.

Thus, around 1900, the induction motor was ready for wide industrial use. No wonder that before 1910, in Europe, locomotives provided with induction motor propulsion were capable of delivering 200 km/h.

However, at least for transportation, the D.C. motor took over all markets until around 1985 when the insulated gate bipolar transistor pulse width modulation (IGBT PWM) inverter has provided for efficient frequency changers. This promoted the induction motor spectacular comeback in variable speed drives with applications in all industries.

Mainly due to power electronics and digital control, the induction motor may add to its old nickname of "the workhorse of industry" and the label of "the racehorse of high-tech".

A more complete list of events that marked the induction motor history follows:

- Better and better analytical models for steady-state and design purposes
- The orthogonal (circuit) and space phasor models for transients
- Better and better magnetic and insulation materials and cooling systems
- Design optimisation deterministic and stochastic methods
- IGBT PWM frequency changers with low losses and high power density (kW/m^3) for moderate costs
- Finite element modellings (FEMs) for field distribution analysis and coupled circuit-FEM models for comprehensive exploration of IMs with critical (high) magnetic and electric loading
- Development of induction motors for super-high speeds and high powers
- A parallel history of linear induction motors with applications in linear motion control has unfolded
- New and better methods of manufacturing and testing for IMs
- Integral induction motors: induction motors with the PWM converter integrated into one piece.

1.3 INDUCTION MACHINES IN APPLICATIONS

Induction motors are, in general, supplied from single- or three-phase A.C. power grids.

Single-phase supply motors, which have two-phase stator windings to provide self-starting, are used mainly for home applications (fans, washing machines, etc.): up to 2.2–3 kW. A typical contemporary single-phase induction motor with dual (start and run) capacitor in the auxiliary phase is shown in Figure 1.4.

Three-phase induction motors are sometimes built with aluminium frames for general-purpose applications below 55 kW (Figure 1.5).

Standard efficiency (IE1), high efficiency (IE2), premium efficiency (IE3 and NEMA premium), and super-premium efficiency (IE4) have been defined in the second edition of the IEC 60034-30. Standard induction motors have been introduced to promote further energy savings both at constant and variable speeds (Figure 1.6). For IE4 (IE5), line-start permanent magnet motors have been considered (Table 1.2), but too large starting/rated current, the lower peak, rated and starting/rated torque, size, and cost are issues to be dealt with (notice the cost of line start synchronous permanent motor (LSPM) is 230% of IM cost in Table 1.2). However, even at this high cost, the payback time of LSPM, for 6000 hours/year, is less than 3 years.

FIGURE 1.4 Start-run capacitor single-phase induction motor. (Source: ABB.)

FIGURE 1.5 Aluminium frame induction motor. (Source: ABB.)

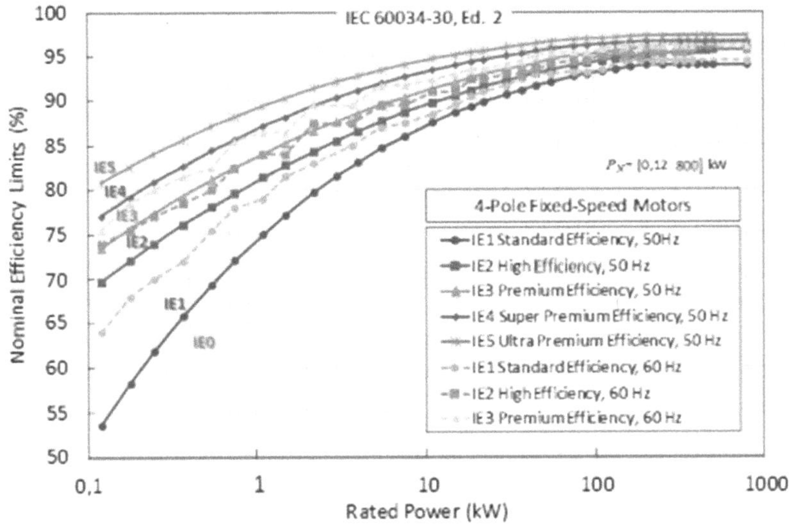

FIGURE 1.6 Rated efficiency class limits proposed in IEC60034-30 for four-pole motors (0.12–800kW) [12].

TABLE 1.2
Commercial IE-2, IE-3, IE-4 Class 7.5kW Four-Pole Motors [12]

Standard and the Year Published	State
IEC 60034-1, Ed. 12, 2010, Rating and performance.	Active.
Application: Rotating electrical machines.	
IEC 60034-2-1, Ed. 1, 2007, Standard method for determining losses and efficiency from tests (excluding machines for traction vehicles). Establishes methods of determining efficiencies from tests and also specifies methods of obtaining specific losses.	Active but under revision.
Application: D.C. machines and A.C. synchronous and IMs of all sizes within the scope of 1EC 60034-1.	
IEC 60034-2-2, Ed. 1, 2010, Specific methods for determining separate losses of large machines from tests – supplement to IFC60034-2-I. Establishes additional methods of determining separate losses and to define an efficiency supplementing IEC 60034-2-1. These methods apply when full-load testing is not practical and result in a greater uncertainty.	Active.
Application: Special and large rotating electrical machines.	
IEC 60034-2-3, Ed. 1, 2011, Specific test methods for determining losses and efficiency of convener-fed A.C. motors.	Not active. Draft.
Application: Convener-fed motors.	
IEC 60034-30, Ed. 1, 2008, Efficiency classes of single-speed, three-phase, cage induction motors (IEC code).	Active but under revision.
Application: 0.75–375kW, 2,4, and 6 poles, 50 and 60Hz.	
IEC60034-31, Ed. 1, 2010, Selection of energy-efficient motors including variable speed applications – application guide.	Active.
Provides a guideline of technical aspects for the application of energy – efficient, three-phase, electric motors. It not only applies to motor manufacturers, original equipment manufacturers, end users, regulators, and legislators but also to all other interested parties.	
Application: Motors covered by IEC 60034-30 and variable frequency/speed drives.	
IEC 60034-17, Ed 4, 2006, Cage induction motors when fed from conveners – application guide.	Active.
Deals with the steady-state operation of cage induction motors within the scope of IEC 60034-12, when fed from converters. Covers the operation over the whole speed setting range but does not deal with starting or transient phenomena.	
Application: Cage induction motors fed from converters.	

Cast iron finned frame efficient motors up to 2000kW are built today with axial exterior air cooling. The stator and the rotor have laminated single stacks.

Typical values of efficiency and sound pressure for such motors built for voltages of 3800–11,500V and 50–60Hz are shown in Table 1.3 (source: ABB). For large starting torque, dual-cage rotor induction motors are built (Figure 1.7).

There are applications (such as overhead cranes) where for safety reasons, the induction motor should be braked quickly when the motor is turned off. Such an induction motor with an integrated brake is shown in Figure 1.8.

Induction motors used in pulp and paper industry need to be kept clean from excess pulp fibres. Rated to IP55 protection class, such induction motors prevent the influence of ingress, dust, dirt, and damp (Figure 1.9).

Aluminium frames offer special corrosion protection. Bearing grease relief allows for greasing the motor while it is running.

IMs are extensively used for wind turbines up to 2000kW per unit and more [13]. A typical dual winding (speed) such induction generator with cage rotor is shown in Figure 1.10.

Wind power conversion to electricity has shown a steady growth since 1985 [2].

50 GW of wind power were in operation, with 25% wind power penetration in Denmark, in 2005 and 180 GW were predicted for 2010 (source Windforce10). About 600 GW were expected to be installed worldwide by 2019.

TABLE 1.3
Typical Values of Efficiency and Sound Pressure for High-Voltage Induction Machines

Output	**Efficiency (%)**		
	Typical Values of High-Voltage Four-Pole Machines		
kW	4/4 load	3/4 load	1/2 load
500	96.7	96.7	96.1
630	97.0	97.0	96.4
710	97.1	97.1	96.5
800	97.3	97.2	96.8
900	97.4	97.4	96.9
1000	97.4	97.4	97.1
1250	97.6	97.7	97.5
1400	97.8	97.8	97.5
2000	97.9	97.8	97.5

Frame/rpm	**3000**	**1500**	**1000**	**≤750**
	Typical Sound Pressure Levels in dB (A) at 1 m Distance			
315	79	78	76	–
355	79	78	76	–
400	79	78	76	75
450	80	78	76	75
500	80	78	76	75
560	80	78	76	75

The variation and measuring tolerance of the figures is 3dB (A).

FIGURE 1.7 Dual-cage rotor induction motors for large starting torque. (Source: ABB.)

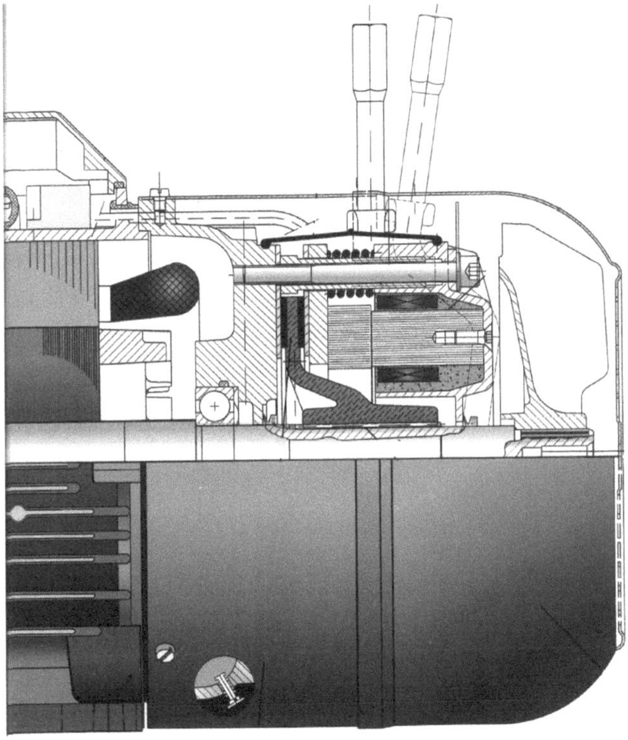

FIGURE 1.8 Induction motor with integrated electromagnetic brake. (Source: ABB.)

The environmentally clean solutions to energy conversion are likely to grow in the near future. A 10% coverage of electrical energy needs in many countries of the world seems within reach in the next 20 years. Also, small power hydropower plants with induction generators may produce twice as much that amount.

Induction motors are used more and more for variable speed applications in association with PWM converters.

FIGURE 1.9 Induction motor in pulp and paper industries. (Source: ABB.)

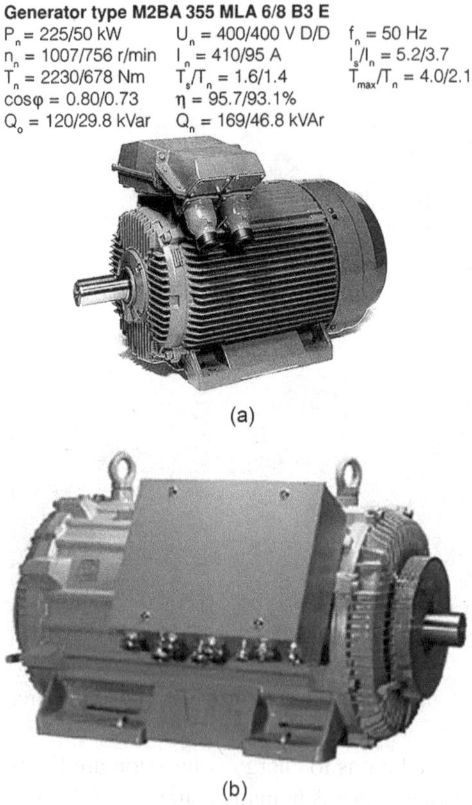

FIGURE 1.10 (a) Dual-stator winding induction generator for wind turbines. (b) Wound rotor induction generator. (Source: ABB.) 750/200 kW, cast iron frame, liquid-cooled generator. Output power: kW and MW range; shaft height: 280–560. Features: air or liquid cooled; cast iron or steel housing. Single, two-speed, doubly-fed design or full variable speed generator.

Up to 5000 kW at 690 V (line voltage, RMS), PWM voltage-source IGBT converters are used to produce variable speed drives with induction motors. A typical frequency converter with a special induction motor series is shown in Figure 1.11.

Constant ventilator speed cooling by integrated forced ventilation independent of motor speed provides high continuous torque capability at low speed in servo drive applications (machine tools, etc.).

Roller tables use several low-speed ($2p_1 = 6$–12 poles) induction motors with or without mechanical gears, supplied from one or more frequency converters for variable speeds.

The high torque load and high ambient temperature, humidity, and dust may cause damage to induction motors unless they are properly designed and built.

Totally enclosed induction motors are fit for such demanding applications (Figure 1.12). Mining applications (hoists, trains, conveyors, etc.) are somewhat similar.

Induction motors are extensively used in marine environments for pumps, fans, compressors, etc. for power up to 700 kW or more. Due to the aggressive environment, they are totally enclosed and may have aluminium (at low power), steel, or cast iron frames (Figure 1.13).

Aboard ship, energy consumption reduction is essential, especially money-wise, as electric energy on board is produced through a diesel engine electrical generator system.

Suppose that electric motors aboard a ship amount to 2000 kW running 8000 hours/year. With an energy cost of US\$0.15/kWh, the energy bill difference per year between two induction motor supplies with a 2% difference in motor efficiency is: $0.02 \times 2000 \times 8000\,\text{hours} \times 0.15 = \text{US\$}55{,}200$ per year.

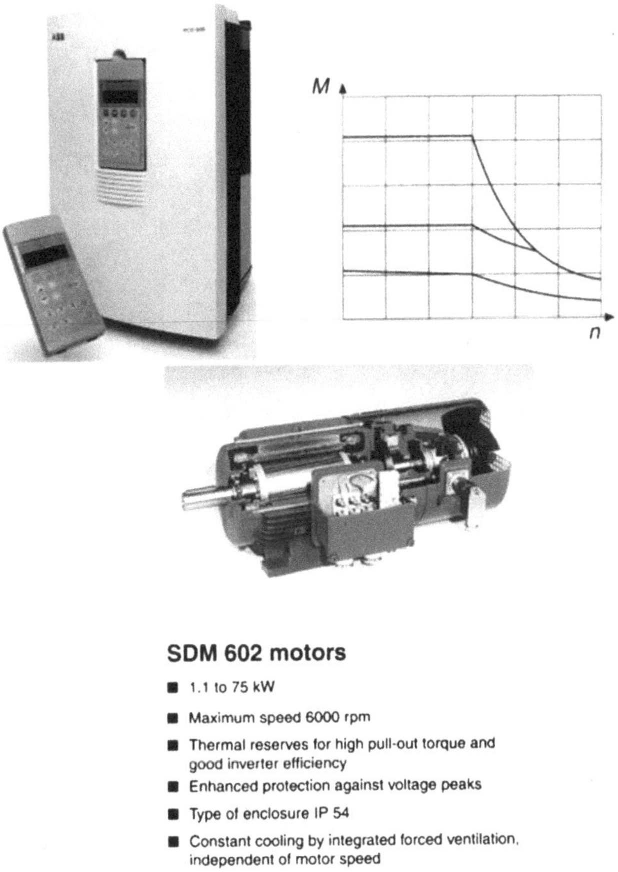
M
n
SDM 602 motors
1.1 to 75 kW
Maximum speed 6000 rpm
Thermal reserves for high pull-out torque and good inverter efficiency
Enhanced protection against voltage peaks
Type of enclosure IP 54
Constant cooling by integrated forced ventilation, independent of motor speed

FIGURE 1.11 Frequency converter with induction motor for variable speed applications. (Source: ABB.)

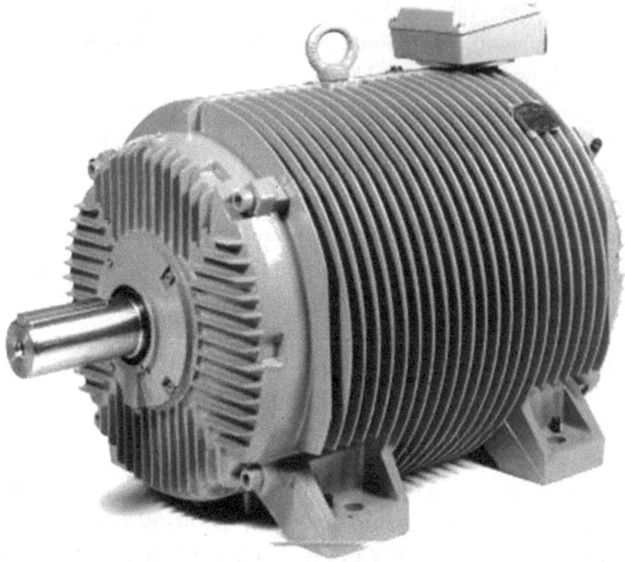

(a)

(b)

FIGURE 1.12 Roller table induction motors without (a) and with (b) a gear. (Source: ABB.)

Electric trains, light rail people movers in or around town, or trolleybuses of the last generation are propelled by variable speed induction motor drives.

Most pumps, fans, conveyors, or compressors in various industries are driven by constant or variable speed induction motor drives.

The rotor of a 2500 kW, 3 kV, 400 Hz, two-pole (24,000 rpm) induction motor in different stages of production as shown in Figure 1.14 proves the suitability of induction motors to high-speed and high-power applications.

Figure 1.15a shows a 3.68 kW (5 HP), 3200 Hz (62,000 rpm) induction motor, with direct water stator cooling, which weighs only 2.268 Kg (5 Pds). A high-speed gyroscope dual IM with high

FIGURE 1.13 Induction motor driving a pump aboard a ship. (Source: ABB.)

FIGURE 1.14 A 2500 kW, 3 kV, 24,000 rpm induction motor. (Source: ABB.)

inertia external rotor is shown in Figure 1.15b. This is to show that it is rather the torque than the power that determines the electric motor size.

Copper-cage induction motor drives have been introduced to electric vehicles (Figure 1.16) due to their ruggedness and simpler control: the copper in the rotor cage leads to a better efficiency at high starting/accelerating torques and cruising speed.

At the other end of the applications scale, four- or two-pole changing IMs have been investigated for better efficiency line-start small refrigerator compressor drives (Figure 1.17).

It starts and operates mostly on four-pole configuration (with permanent magnets (PMs) under the rotor cage for an 88% efficiency at 50 W, 1500 rpm), but for short duty, it switches to two-pole operation (with a single-phase winding) for special compressor operation conditions.

In parallel with the development of rotary induction motor, power electronics driven linear motion induction motors have witnessed intense studies with quite a few applications [9,10,15]. Amongst them, Figure 1.18 shows the UTDC-built linear induction motor people mover (BC transit) in Vancouver now in use for more than two decades.

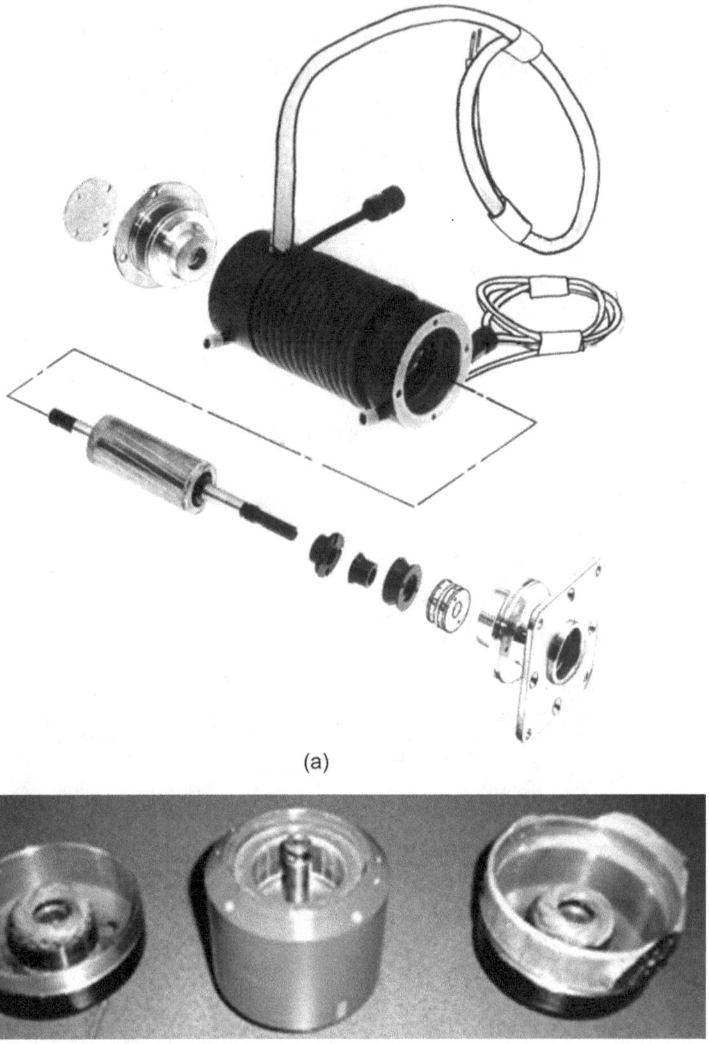

(a)

(b)

FIGURE 1.15 A 3.68 kW (5 HP), 3200 Hz (62,000 rpm) induction motor with forced liquid cooling (a); high-speed gyroscope dual IM with outer rotor (b).

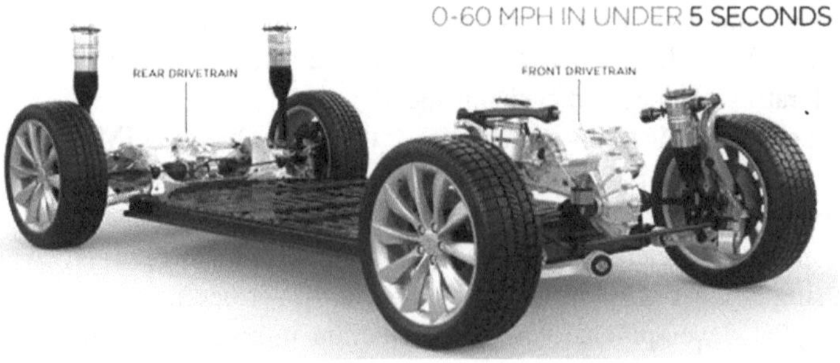

FIGURE 1.16 Electric power train of "Tesla" electric vehicle, with copper cage rotor induction motor drives. (www.pinterest.com/pin/270145677620776613/).

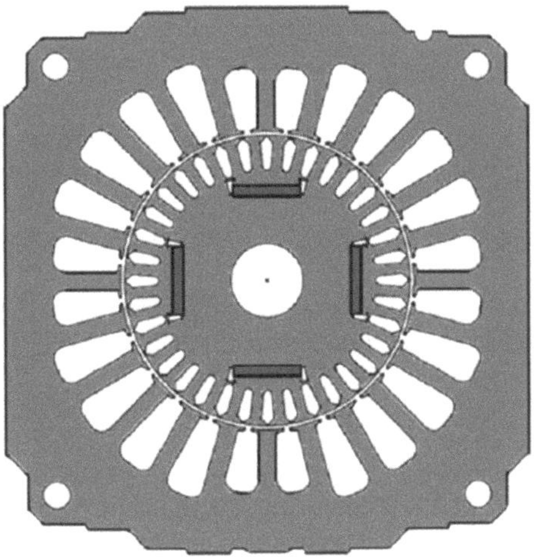

FIGURE 1.17 Four- or two-pole changing split phase capacitor IM for small compressor drives [14].

FIGURE 1.18 The BC transit system in Vancouver: with linear motion induction motor propulsion. (Source: UTDC.)

The panoramic view of induction motor applications sketched above is only to demonstrate the extraordinary breadth of IM speed and power ratings and of its applications both for constant and variable speeds.

1.4 CONCLUSION

After 1885, more than one century from its invention, the induction motor steps into the 21st century with a vigour hardly paralleled by any other motor.

Power electronics, digital control, computer-aided design, and new and better materials have earned the induction motor the new sobriquet of "the racehorse of industry" in addition to the earlier one of "the workhorse of industry".

Present in all industries and in-home appliances in constant and variable speed applications, the induction motor seems now ready to make the X by wire and even the electric starter/generator systems aboard of the hybrid electric vehicles of the near future [16].

The new challenges in modelling, optimisation design in the era of FEMs, its control as a motor and generator for even better performance when supplied from PWM converters, and its enormous application potential hopefully justify this rather comprehensive book on IMs at the beginning of the 21st century.

REFERENCES

1. R. J. Kerkman, G. L. Skibinski, D. W. Schlegel, AC drives; Year 2000 and beyond, *Record of IEEE – APEC '99*, March, 1999.
2. P. Gipe, *Wind Energy Comes of Age*, Wiley & Sons Inc., New York, 1995.
3. H. Sequenz, *The Windings of Electric Machines, Vol. 3: A.C. Machines*, Springer Verlag, Vienna, 1950 (in German).
4. R. Richter, *Electric Machines-Vol. 4-Induction Machines*, Verlag Birkhauser, Bassel/Stuttgart, 1954 (in German).
5. Ph. Alger, *The Nature of Induction Machines*, 2nd edition, Gordon & Breach, New York, 1970.
6. C. Veinott, *Theory and Design of Small Induction Motors*, McGraw-Hill, New York, 1959.
7. J. Stepina, *Single Phase Induction Motors*, Springer Verlag, Vienna, 1981 (in German).
8. B. Heller, V. Hamata, *Harmonic Field Effects in Induction Machines*, Elsevier Scientific Publishing Company, Amsterdam, 1977.
9. E. Laithwaite, *Induction Machines for Special Purposes*, Newness, London, UK, 1966.
10. I. Boldea, S. A. Nasar, *Linear Motion Electromagnetic Systems*, Wiley Interscience, London, UK, 1985.
11. EURODEEM by European Commission on Internet: http://iamest.jrc.it/projects/eem/eurodeem.htm
12. T. De Almeida, F.J.T.E. Ferreira, A. Q. Duarte, Technical and economical considerations on superhigh – Efficiency three phase motors, *IEEE Transactions on Industry Applications*, Vol. IA-50, No. 2, 2014, pp. 1274–1275.
13. I. Boldea, *Electric Generators, Part 2 Variable Speed Generators*, CRC Press, Taylor and Francis, New York, 2005.
14. F. J. Kalluf, L. N. Tutelea, I. Boldea, A. Espindola, 2/4-pole split-phase capacitor motor for small compressors: A comprehensive motor characterization, *IEEE Transactions on Industry Applications*, Vol. IA-50, No. 1, 2014, pp. 356–363.
15. I. Boldea, S. A. Nasar, *Linear Motion Electromagnetic Devices*, Taylor and Francis, New York, 2001.
16. M. Ehsani, Y. Gao, S. E. Gay, A. Emadi, *Modern Electric, Hybrid Electric, and Fuel Cell Vehicles*, CRC Press, Taylor & Francis, New York, 2004.

2 Construction Aspects and Operation Principles

The induction machine (IM) is basically an A.C. polyphase machine connected to an A.C. power grid, either in the stator or in the rotor. The A.C. power source is, in general, three phase, but it may also be single phase. In both cases, the winding arrangement on the part of the machine – the primary – connected to the grid (the stator in general) should produce a travelling field in the machine airgap. This travelling field will induce voltages in conductors on the part of the machine not connected to the grid (the rotor or the mover in general) – the secondary. If the windings on the secondary (rotor) are closed, A.C. currents occur in the rotor.

The interaction between primary field and secondary currents produces torque from zero rotor speed onwards. The rotor speed at which the rotor currents are zero is called the ideal no-load (or synchronous) speed. The rotor winding may be multiphase (wound rotors) or made of bars short-circuited by end rings (cage rotors).

All the windings on primary and secondary are placed in uniform slots stamped into thin silicon steel sheets called laminations.

The IM has a rather uniform airgap of 0.2–3 mm. The largest values correspond to large powers, 1 MW or more. The secondary windings may be short-circuited or connected to an external impedance or to a power source of variable voltage and frequency. In the latter case, the IM works, however, as a synchronous machine as it is doubly fed and both stator and rotor slip frequencies are imposed (the latter is speed dependent).

Though historically double-stator and double-rotor machines have also been proposed to produce variable speed more conveniently, they did not make it to the markets. Today's power electronics seem to move such solutions even further into forgetfulness.

In this chapter, we discuss construction aspects and operation principles of IMs. A classification is implicit.

The main parts of any IM are as follows:

- The stator slotted magnetic core
- The stator electric winding
- The rotor slotted magnetic core
- The rotor electric winding
- The rotor shaft
- The stator frame with bearings
- The cooling system
- The terminal box.

The IMs may be classified in many ways. Here are some of them.

- With rotary or linear motion
- Three-phase or single-phase supply
- With wound or cage rotor.

In very rare cases, the internal primary is the mover and the external secondary is at standstill. In most rotary IMs, the primary is the stator and the secondary is the rotor. It is not so for linear IMs. Practically all IMs have a cylindrical rotor, and thus, a radial airgap between stator and rotor, though, in principle, axial airgap IMs with disk-shaped rotor may be built to reduce volume and weight in special applications.

First, we discuss construction aspects of the abovementioned types of IMs and then essentials of operation principles and modes.

2.1 CONSTRUCTION ASPECTS OF ROTARY IMS

Let us start with the laminated cores.

2.1.1 The Magnetic Cores

The stator and rotor magnetic cores are made of thin silicon steel laminations with unoriented grain to reduce hysteresis and eddy current losses. The stator and rotor laminations are packed into a single (Figure 2.1) or multiple stacks (Figure 2.2). The latter has radial channels (5–15 mm wide) between elementary stacks (50–150 mm long) for radial ventilation.

Single stacks are adequate for axial ventilation.

Single-stack IMs have been traditionally used below 100 kW, but recently, they have been introduced up to 2 MW as axial ventilation has been improved drastically. The multistack concept is, however, necessary for large power (torque) with long stacks.

The multiple stacks lead to additional winding losses, up to 10%, in the stator and rotor as the coils (bars) lead through the radial channels without producing torque. Also, the electromagnetic field energy produced by the coil (bar) currents in the channels translates into additional leakage inductances which tend to reduce the breakdown torque and the power factor. They also reduce the starting current and torque. Typical multistack IMs are shown in Figure 2.2.

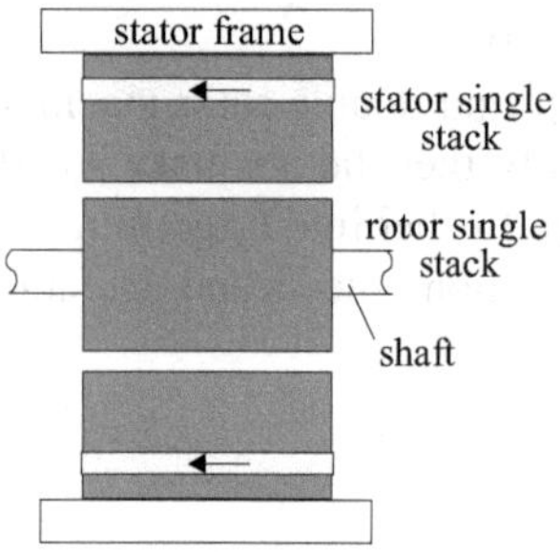

FIGURE 2.1 Single-stack magnetic core.

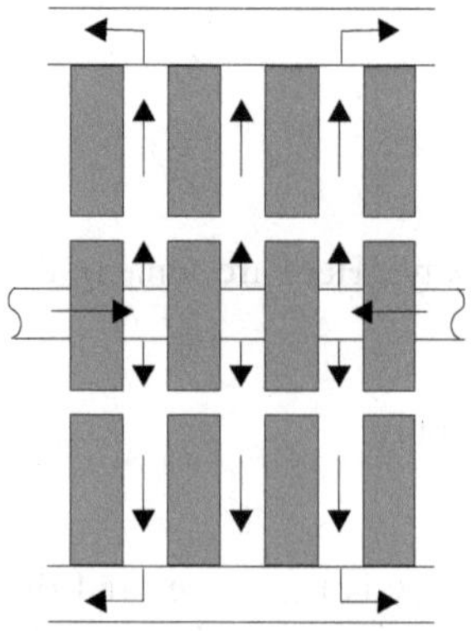

FIGURE 2.2 Multiple-stack IM.

For IMs of fundamental frequency up to 300 Hz, 0.5 mm thick silicon steel laminations lead to reasonable core losses 2–4 W/Kg at 1 T and 50 Hz.

For higher fundamental frequency, thinner laminations are required. Alternatively, anisotropic magnetic powder materials may be used to cut down the core losses at high fundamental frequencies, above 500 Hz, at lower power factor, however (see Chapter 3 on magnetic materials).

2.1.2 Slot Geometry

The airgap, or the air space between the stator and the rotor, has to be travelled by the magnetic field produced by the stator. This in turn will induce voltages and produce currents in the rotor windings. Magnetising air requires large magnetomotive forces (mmfs) or ampere-turns. The smaller the air (nonmagnetic) gap, the smaller the magnetisation mmf. The lower limit of airgap g is determined by mechanical constraints and by the ratio of the stator and rotor slot openings b_{os} and b_{or} to airgap g in order to keep surface core and tooth flux pulsation additional losses within limits. The tooth is the lamination radial sector between two neighbouring slots.

Putting the windings (coils) in slots has the main merit of reducing the magnetisation current. Second, the winding manufacturing and placing in slots becomes easier. Third, the winding in slots is better off in terms of mechanical rigidity and heat transmission (to the cores). Finally, the total mmf per unit length of periphery (the coil height) could be increased, and thus, large power IMs could be built efficiently. What is lost is the possibility to build windings (coils) that can produce purely sinusoidal distribution ampere-turns (mmfs) along the periphery of the machine airgap. But this is a small price to pay for the incumbent benefits.

The slot geometry depends mainly on IM power (torque) level and thus on the type of magnetic wire – with round or rectangular cross section – from which the coils of windings are made. With round wire (random wound) coils for small power IMs (below 100 kW in general), the coils may be introduced in slots wire by wire, and thus, the slot openings may be small (Figure 2.3a). For preformed coils (in large IMs), made, in general, of rectangular cross-section wire, open or semiopen slots are used (Figure 2.3b and c).

In general, the slots may be rectangular, straight trapezoidal, or rounded trapezoidal. Open and semiopen slots tend to be rectangular (Figure 2.3b and c) in shape and the semiclosed ones are trapezoidal or rounded trapezoidal (Figure 2.3a).

In an IM, only slots on one side are open, while on the other side, they are semiclosed or semiopen.

The reason is that a large slot opening, b_{os}, per gap, g, ratio ($b_{os}/g > 6$) leads to lower average flux density, for given stator mmf and to large flux pulsation in the rotor tooth, whose pulsations (harmonics) will produce large additional core losses. In the airgap, flux density harmonics lead to parasitic torques, noise, and vibration as presented in subsequent, dedicated chapters. For semiopen and semiclosed slots, $b_{os}/g \cong (4–6)$ in general. For the same reasons, the rotor slot opening per airgap $b_{or}/g \cong 3–4$ wherever possible. Too a small slot opening per gap ratio leads to a higher magnetic field in the slot neck (Figure 2.3) and thus to a higher slot leakage inductance, which causes lower starting torque and current and lower breakdown torque.

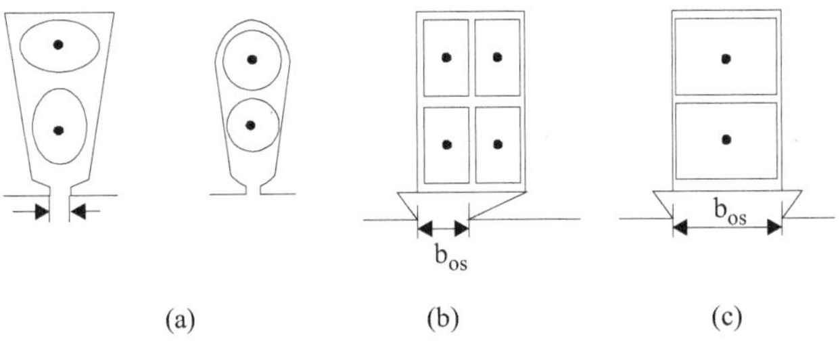

FIGURE 2.3 Slot geometrics to locate coil windings (a) semiclosed, (b) semiopen, and (c) open.

Slots as in Figure 2.3 are used both for stator and wound rotors. Rotor slot geometry for cage rotors is much more diversified depending upon

- Starting and rated load constraints (specifications)
- Constant voltage/frequency (V/f) or variable voltage/frequency supply operation
- Torque range.

Less than rated starting torque, high-efficiency IMs for low power at constant V/f or for variable V/f may use round semiclosed slots (Figure 2.4a). Rounded trapezoidal slots with rectangular teeth are typical for medium starting torque (around rated value) in small power IMs (Figure 2.4b).

Closed rotor slots may be used to reduce noise and torque pulsations for low power circulating fluid pumps for homes at the expense of large rotor leakage inductance: that is, lower breakdown torque. In essence, the iron bridge (0.5–1 mm thick), above the closed rotor slot, already saturates at 10%–15% of rated current at a relative permeability of 50 or less that drops further to 15–20 for starting conditions (zero speed, full voltage).

For high starting torque, high rated slip (lower rated speed with respect to ideal no-load speed), rectangular deep-bar rotor slots are used (Figure 2.5a). Inverse trapezoidal or double-cage slots are used for low starting current and moderate and large starting torque (Figure 2.5b and c). In all these cases, the rotor slot leakage inductance increases, and thus, the breakdown torque is reduced to as low as 150%–200% rated torque.

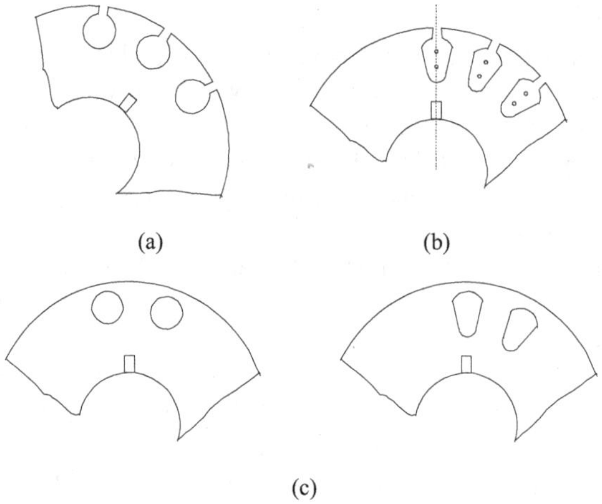

FIGURE 2.4 Rotor slots for cage rotors: (a) semiclosed and round, (b) semiclosed and round trapezoidal, and (c) closed slots.

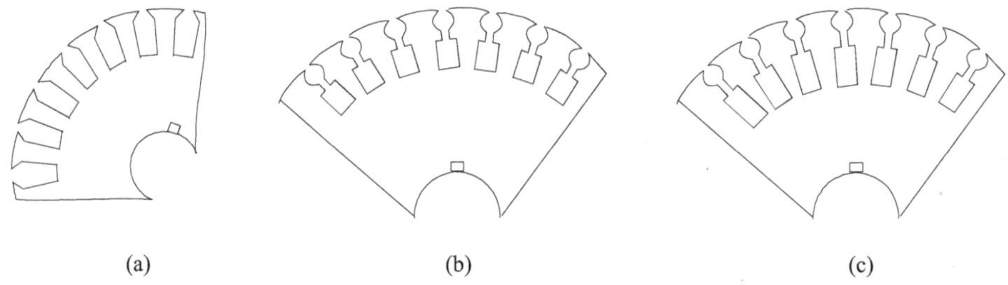

FIGURE 2.5 Rotor slots for low starting current IMs: (a) high slip, high starting torque and (b) moderate starting torque, and (c) very high.

More general, optimal shape cage rotor slots may today be generated through direct FEM optimisation techniques to meet the desired performance constraints for special applications. In general, low stator current and moderate and high starting torque rely on the rotor slip frequency effect on rotor resistance and leakage inductance.

At the start, the frequency of rotor currents is equal to stator (power grid) frequency f_1, while at full load, $f_{sr} = S_n f_1$; S_n, the rated slip, is below 0.08 and less than 0.01 in large IMs:

$$S = \frac{f_1 - np_1}{f_1}; \quad \text{n-speed in rps} \tag{2.1}$$

p_1 is the number of spatial periods of airgap travelling field wave per revolution produced by the stator windings:

$$B_{g0}(x,t) = B_{gm0}\cos(p_1\theta_1 - \omega_1 t) \tag{2.2}$$

θ_1 – mechanical position angle; $\omega_1 = 2\pi f_1$.

Remember that, for variable voltage and frequency supply (variable speed), the starting torque and current constraints are eliminated as the rotor slip frequency Sf_1 is always kept below that corresponding to breakdown torque.

Very important in variable speed drives is efficiency, power factor, breakdown torque, and motor initial or total costs (with capitalised loss costs included).

2.1.3 IM Windings

The IM is provided with windings both on the stator and on the rotor. Stator and rotor windings are treated in detail in Chapter 4.

Here we refer only to a primitive stator winding with six slots for two poles (Figure 2.6).

Each phase is made of a single coil whose pitch spans half of the rotor periphery. The three phases (coils) are space shifted by 120°. For our case, there are 120 mechanical degrees between phase axes as $p_1 = 1$ pole pair. For $p_1 = 2, 3, 4, 5, 6$, there will be $120°/p_1$ mechanical degrees between phase axes.

The airgap field produced by each phase has its maximum in the middle of the phase coil (Figure 2.6), and with the slot opening eliminated, it has a rectangular spatial distribution whose fundamental varies sinusoidally in time with frequency f_1 (Figure 2.7).

It is evident from Figure 2.7 that when the time angle θ_t electrically varies by $\pi/6$, so does the fundamental maximum of airgap flux density with space harmonics neglected, and a travelling wave field in the airgap is produced. Its direction of motion is from phase a to phase b axis, if the current in phase a leads (in time) the current in phase b and phase b leads phase c. The angular speed of this field is simply ω_1, in electrical terms, or ω_1/p_1 in mechanical terms (see also Equation (2.2)).

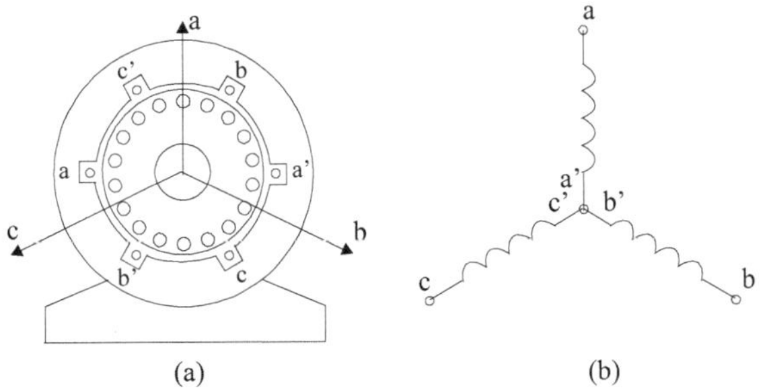

FIGURE 2.6 Primitive IM with (a) six stator slots and (b) phase connection.

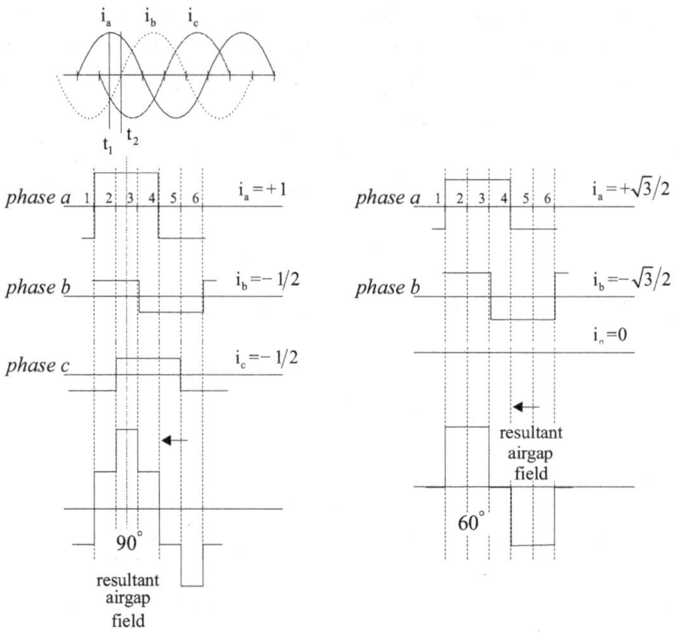

FIGURE 2.7 Stator currents and airgap field at times t_1 and t_2.

$$\Omega_1 = 2\pi n_1 = \omega_1 / p_1 \tag{2.3}$$

So n_1, the travelling field speed in rps (rotations per second), is

$$n_1 = f_1 / p_1 \tag{2.4}$$

This is how for 50(60) Hz, the no-load ideal speed is 3000/3600 rpm for $p_1 = 1$, 1500/1800 rpm for $p_1 = 2$, and so on.

As the rated slip S_n is small (less than 8% for most IMs), the rated speed is only slightly lower than a submultiple of f_1 in rps. The crude configuration in Figure 2.7 may be improved by increasing the number of slots and by using two layers of coils in each slot. This way the harmonic content of airgap flux density diminishes, approaching a better pure travelling field, despite the inherently discontinuous placement of conductors in slots.

A wound stator is shown in Figure 2.8. The three phases may be star or delta connected. Sometimes, during starting, the connection is changed from star to delta (for delta-designed IMs) to reduce starting currents in weak local power grids.

Wound rotors are built in a similar way (Figure 2.9). The slip rings are visible to the right. The stator-placed brush system is not visible. Single-phase-supply IMs have, on the other hand, in general, two windings on the stator.

The main winding (m) and the auxiliary (or starting) one (a) are used to produce a travelling field in the airgap. Similar to the case of three phases, the two windings are spatially phase shifted by 90° (electrical) in general. Also to phase shift the current in the auxiliary winding, a capacitor is used.

In reversible motion applications, the two windings are identical. The capacitor is switched from one phase to the other to change the direction of travelling field.

When auxiliary winding works continuously, each of the two windings uses half the number of slots. Only the number of turns and the cross section of wire differ.

The presence of auxiliary winding with capacitance increases the torque, efficiency, and power factor. In capacitor-start low-power (below 250 W) IMs, the main winding occupies 2/3 of stator slots

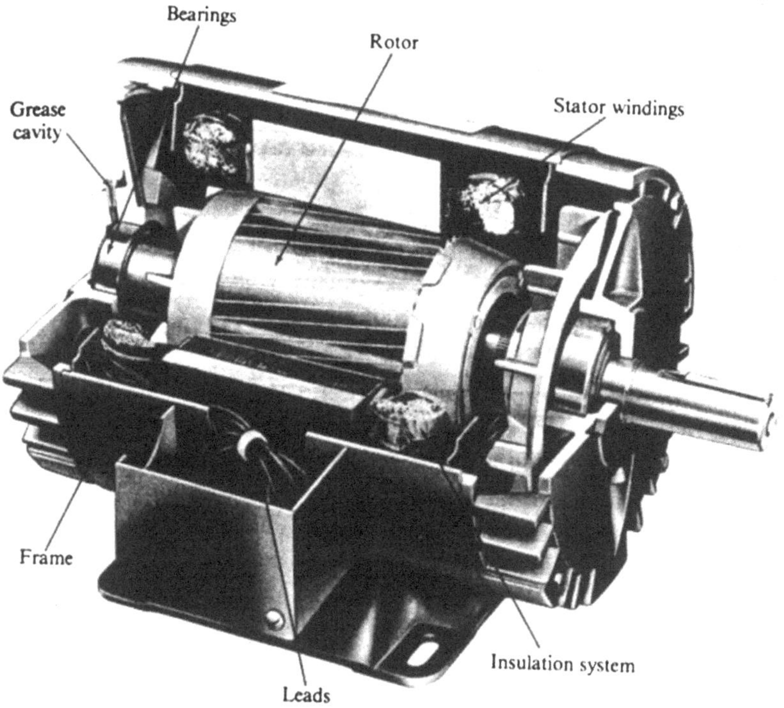

FIGURE 2.8 IM wound three-phase stator winding with cage rotor.

FIGURE 2.9 Three-phase wound rotor.

and the auxiliary (starting) winding only 1/3. The capacitor winding is turned off in such motors by a centrifugal or time relay at a certain speed (time) during starting. In all cases, a cage winding is used in the rotor (Figure 2.10). For very low power levels (below 100 W in general), the capacitor may be replaced by a resistance to cut cost at the expense of lower efficiency and power factor.

Finally, it is possible to produce travelling field with a single-phase concentrated coil winding with shaded poles (Figure 2.11). The short-circuiting ring is retarding the magnetic flux of the stator in the shaded pole area with respect to the unshaded pole area.

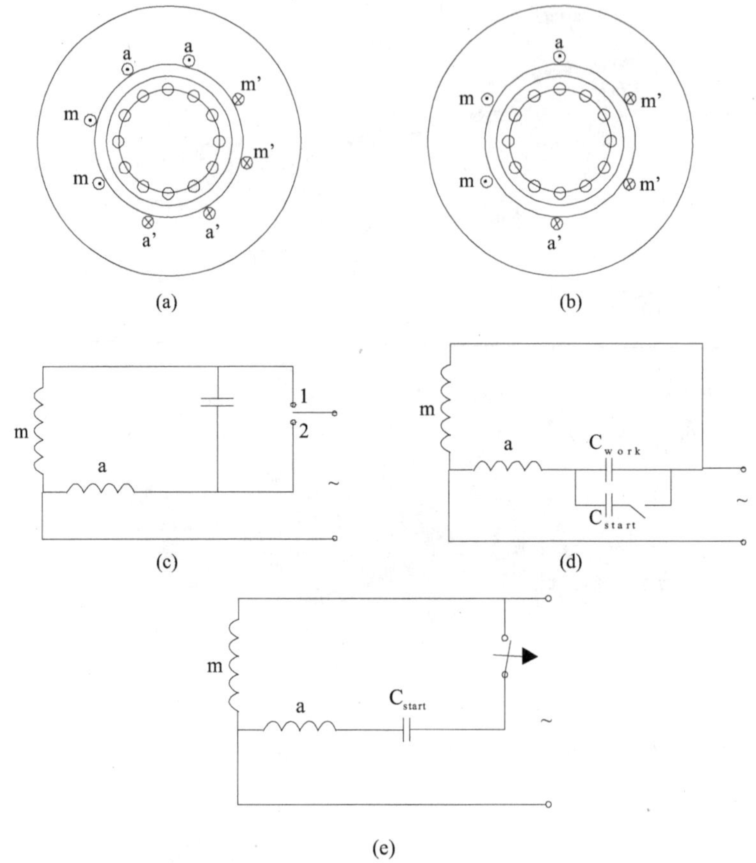

FIGURE 2.10 Single-phase supply capacitor IMs: (a) primitive configuration with equally strong windings, (b) primitive configuration with 2/3, 1/3 occupancy windings, (c) reversible motor, (d) dual capacitor connection, and (e) capacitor start-only connection.

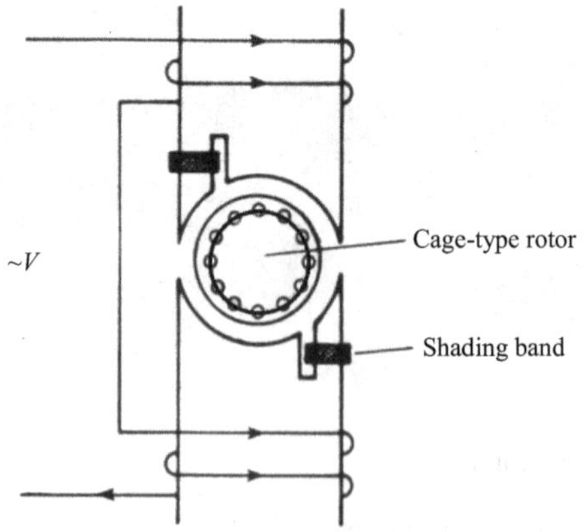

FIGURE 2.11 Single-phase (shaded pole) IM.

The airgap field has a travelling component and the motor starts rotating from the unshaded to the shaded pole zone. The rotor has a cage winding. The low cost and superior ruggedness of shaded pole single-phase IM is paid for by lower efficiency and power factor. This motor is more of historical importance and it is seldom used, mostly below 100 W, where cost is the prime concern.

2.1.4 Cage Rotor Windings

As mentioned above, the rotor of IMs is provided with single- or double-cage windings (Figure 2.12), besides typical three-phase windings on wound rotors.

The cage bars and end rings are made of die-cast aluminium for low and medium power and from brass or copper for large powers. For medium and high powers, the bars are silver rings – welded to end to provide low resistance contact.

For double cages, we may use brass (higher resistivity) for the upper cage and copper for the lower cage. In this case, each cage has its own end ring, mainly due to thermal expansion constraints.

For high-efficiency IMs, copper tends to be preferred due to higher conductivity, larger allowable current density, and working temperatures.

As die-cast copper cage manufacturing is maturing, even for small motors (in home appliances), the copper cage will be used more and more, besides today's welded copper-cage IM's hot water pump of home furnaces.

The die casting of aluminium at rather low temperatures results in low rotor mass production costs for low power IMs.

The debate over aluminium or copper is not decided yet, and both materials are likely to be used depending on the application and power (torque) level.

Although some construction parts such as frames, cooling systems, shafts, bearings, and terminal boxes have not been described yet here, we will not dwell on them at this time as they will "surface" again in subsequent chapters. Instead, Figure 2.13 presents a rather complete cutaway view of a fairly modern induction motor. It has a single-stack magnetic core; thus, axial ventilation is used by a fan located on the shaft beyond the bearings. The heat evacuation area is increased by the finned stator frame. This technology has proved practical up to 2 MW in low-voltage IMs.

The IM in Figure 2.13 has a single-cage rotor winding. The stator winding is built in two layers out of round magnetic wire. The coils are random wound. The stator and rotor slots are of the semiclosed type. Configuration in Figure 2.13 is dubbed as totally enclosed fan cooled (TEFC), as the ventilator is placed outside bearings on the shaft.

It is a low-voltage IM (below 690 V RMS – line voltage).

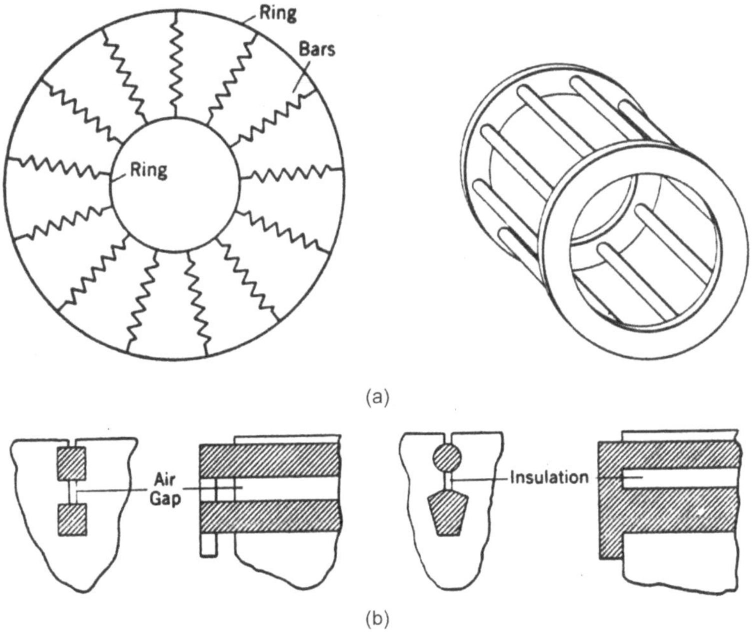

FIGURE 2.12 Cage rotor windings: (a) single cage and (b) double cage.

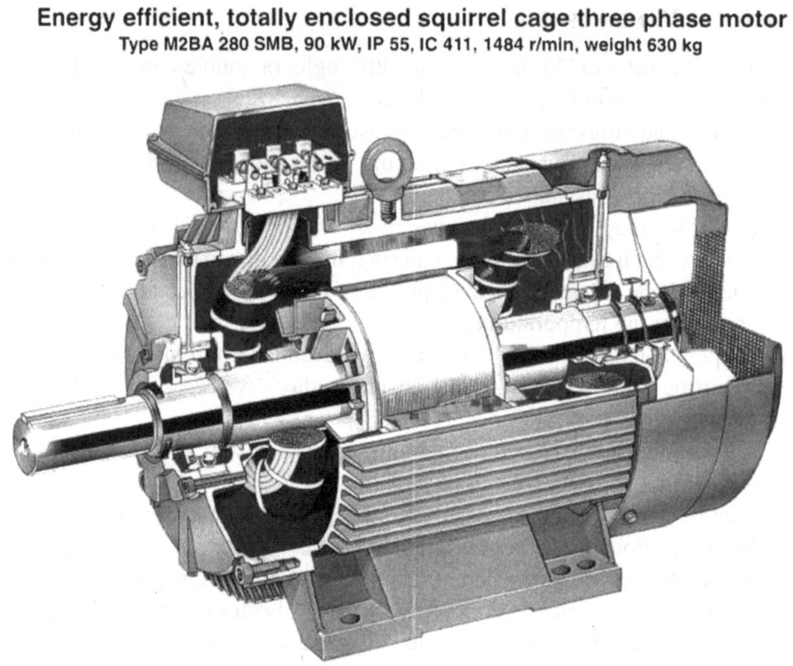
Energy efficient, totally enclosed squirrel cage three phase motor
Type M2BA 280 SMB, 90 kW, IP 55, IC 411, 1484 r/min, weight 630 kg

FIGURE 2.13 Cutaway view of a modern induction motor.

2.2 CONSTRUCTION ASPECTS OF LINEAR INDUCTION MOTORS

In principle, for each rotary IM, there is a linear motion counterpart. The imaginary process of cutting and unrolling the rotary machine to obtain the linear induction motor (LIM) is by now classic (Figure 2.14) [1].

The primary may now be shorter or larger than the secondary. The shorter component will be the mover. In Figure 2.14, the short primary is the mover. The primary may be double- (Figure 2.14d) or single-sided (Figure 2.14c and e).

The secondary material is copper or aluminium for the double-sided LIM, and it may be aluminium (copper) on solid iron for the single-sided LIM. Alternatively, a ladder conductor secondary placed in the slots of a laminated core may be used, as for cage rotor rotary IMs (Figure 2.14c). This latter case is typical for short travel (up to a few metres), low-speed (below 3 m/s) applications.

Finally, the secondary solid material may be replaced by a conducting fluid (liquid metal), when a double-sided linear induction pump is obtained [2].

All configurations shown in Figure 2.14 may be dubbed as flat ones. The primary winding produces an airgap field with a strong travelling component at the linear speed u_s, for the pole pitch τ and frequency f_1

$$u_s = \tau \cdot \frac{\omega_1}{\pi} = 2\tau f_1 \tag{2.5}$$

The number of pole pairs does not influence the ideal no-load linear speed u_s. Incidentally, the peripheral ideal no-load speed in rotary IMs has the same formula (2.5) where τ is also the pole pitch (the spatial semiperiod of the travelling field).

In contrast to rotary IMs, the LIM has an open magnetic structure along the direction of motion. Additional phenomena called longitudinal effects occur due to this. They tend to increase with speed,

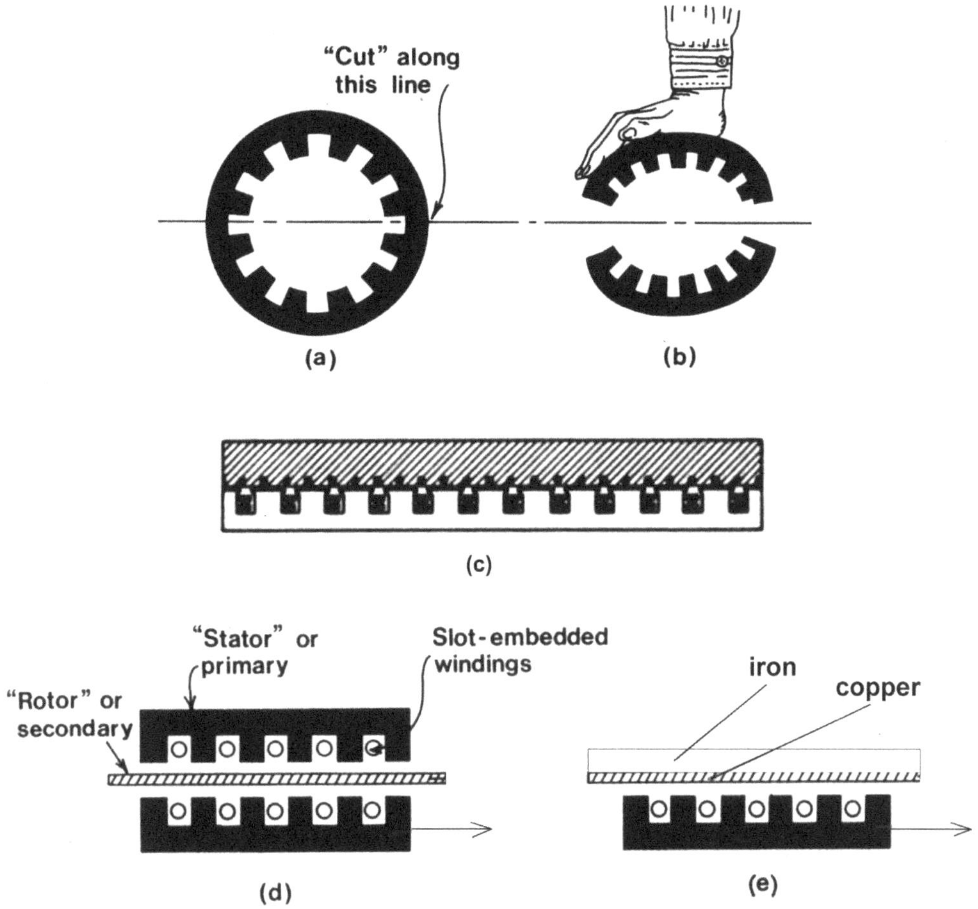

FIGURE 2.14 Cutting and unrolling process to obtain an LIM.

deteriorating the thrust, efficiency, and power factor performance. Also, above 3–5 m/s speed, the airgap has to be large (due to mechanical clearance constraints): in general, 3–12 mm. This leads to high magnetisation currents and, thus, lower power factor than in rotary IMs.

However, the LIM produces electromagnetic thrust directly and, thus, eliminates any mechanical transmission in linear motion applications (transportation).

Apart from flat LIM, tubular LIMs may be obtained by rerolling the flat LIM around the direction of motion (Figure 2.15).

The coils of primary winding are now of circular shape. The rotor may be an aluminium cylinder on solid iron. Alternatively, a secondary cage may be built. In this case, the cage is made of ring-shaped bars. Transverse laminations of disc shape may be used to make the magnetic circuit easy to manufacture, but care must be exercised to reduce the core losses in the primary. The blessing of circularity renders this LIM more compact and proper for short travels (1 m or less).

In general, LIMs are characterised by a continuous thrust density (N/cm^2 of primary) of up to 2 (2.5) N/cm^2 without forced cooling. The large values correspond to larger LIMs. The current LIM used for a few urban transportation systems in North America, Middle East, and East Asia has proved that they are rugged and almost maintenance free. More on LIMs are presented in Chapter 12 (Volume 2).

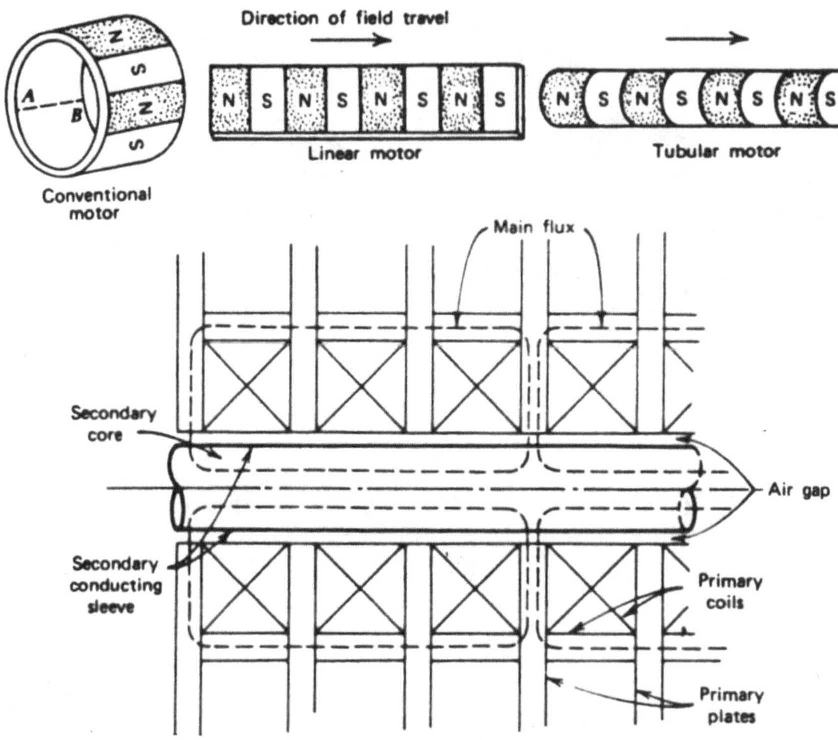

FIGURE 2.15 The tubular LIM.

2.3 OPERATION PRINCIPLES OF IMS

The operation principles are basically related to torque (for rotary IMs) and, respectively, thrust (for LIMs) production. In other words, it is about forces in travelling electromagnetic fields. Or even simpler, why the IM spins and the LIM moves linearly. Basically the torque (force) production in IMs and LIMs may be approached via

- Forces on conductors in a travelling field
- The Maxwell stress tensor [3]
- The energy (coenergy) derivative
- Variational principles (Lagrange equations) [4].

The electromagnetic travelling field produced by the stator currents exists in the airgap and crosses the rotor teeth to embrace the rotor winding (say, rotor cage) – Figure 2.16. Only a small fraction of it traverses radially the top of the rotor slot which contains conductor material.

It is thus evident that, with rotor and stator conductors in slots, there are no main forces experienced by the conductors themselves. Therefore, the method of forces experienced by conductors in fields does not apply directly to rotary IMs with conductors in slots.

The current occurs in the rotor cage (in slots) because the magnetic travelling flux produced by the stator in any rotor cage loop varies in time even at zero speed (Figure 2.17). If the cage rotor rotates at speed n (in rps), the stator-produced travelling flux density in the airgap (2.2) moves with respect to the rotor with the relative speed n_{sr}.

$$n_{sr} = \frac{f_1}{p_1} - n = S \cdot \frac{f_1}{p_1} \tag{2.6}$$

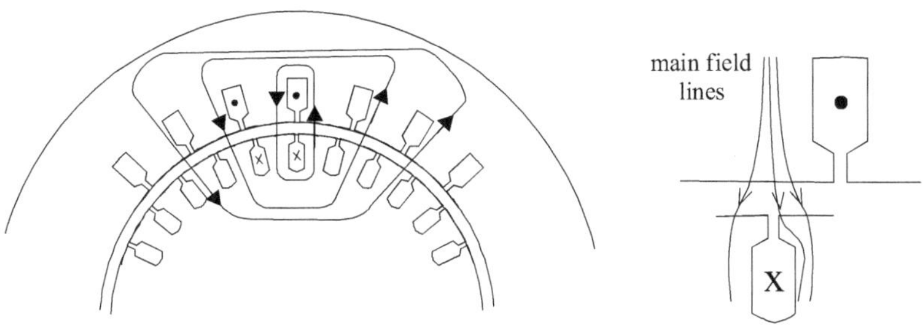

FIGURE 2.16 Flux paths in IMs.

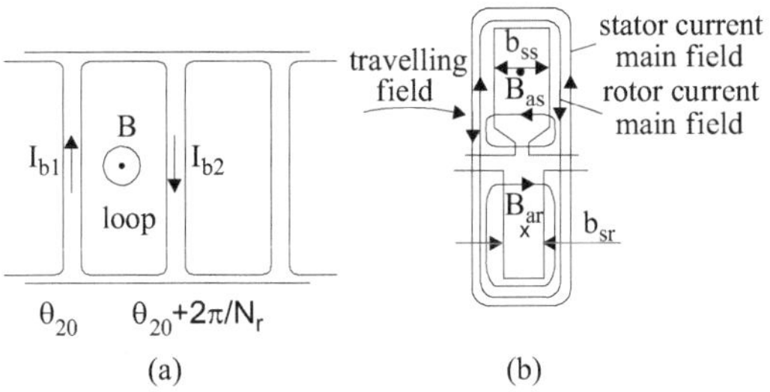

FIGURE 2.17 Travelling flux crossing the rotor cage loops (a), leakage and main fields (b).

S is called IM slip.

So in rotor coordinates, (2.2) may be written as

$$B(\theta_r,t)=B_m\cos(p_1\theta_2-S\omega_1 t) \tag{2.7}$$

Consequently, with the cage bars in slots, according to electromagnetic induction law, a voltage is induced in loop 1 of the rotor cage, and thus, a current occurs in it such that its reaction flux opposes the initial flux (visible in Figure 2.17).

The current which occurs in the rotor cage, at rotor slip frequency Sf_1 (see Equation (2.7)), produces a reaction field that crosses the airgap. This is the main reaction field. Thus, the resultant airgap field is the "product" of both stator and rotor currents. As the two currents tend to be shifted more than $2\pi/3$, the resultant (magnetisation) current is reasonably low; in fact, it is 25%–80% of the rated current, depending on the machine airgap g to pole pitch τ ratio. The higher the ratio τ/g, the smaller the magnetisation current in p.u. (per unit or relative values).

The stator and rotor currents also produce leakage flux paths crossing the slots: B_{as} and B_{ar} (Figure 2.17).

According to the Maxwell stress tensor theory, at the surface border between mediums with different magnetic materials and permeabilities (μ_0 in air, $\mu\neq\mu_0$ in the core), the magnetic field produces forces. The interaction force component perpendicular to the rotor slot wall is [3]

$$F_{tn}=\frac{B_{ar}(\theta_r,t)B_{tr}(\theta_r,t)}{\mu_0}\bar{n}_{tooth} \tag{2.8}$$

The magnetic field has a radial (B_{tr}) and a tangential (B_{ar}) component only.

Now, for the rotor slot, $B_{ar}(\theta_r)$ – tangential flux density – is the slot leakage flux density.

$$B_{ar}(\theta_r)=\frac{\mu_0 I_b(\theta_r,t)}{b_{sr}} \tag{2.9}$$

The radial flux density that counts in the torque production is that produced by the stator currents, $B_{tr}(\theta_r)$, in the rotor tooth.

$$B_{tr}(\theta_r,t) = B(\theta_r,t)\cdot\frac{b_{tr}+b_{sr}}{b_{sr}} \tag{2.10}$$

In (2.10), b_{tr} is the mean rotor tooth width while $B(\theta_r,t)$ is the airgap flux density produced by the stator currents in the airgap.

When we add the specific Maxwell stress tensors [1] on the left- and right-side walls of the rotor slot, we should note that the normal direction changes sign on the two surfaces. Thus, the addition becomes a subtraction.

$$f_{tooth}\left(N/m^2\right) = -\left(\frac{B_{ar}(\theta_r+\Delta\theta,t)B_{tr}(\theta_r+\Delta\theta,t)}{\mu_0} - \frac{B_{ar}(\theta_r,t)B_{tr}(\theta_r,t)}{\mu_0}\right) \tag{2.11}$$

Essentially, the slot leakage field B_{ar} does not change with $\Delta\theta$ – the radial angle that corresponds to a slot width.

$$f_{tooth}\left(N/m^2\right) = -\frac{B_{ar}(\theta_r,t)}{\mu_0}\left(B_{tr}(\theta_r+\Delta\theta,t) - B_{tr}(\theta_r,t)\right) \tag{2.12}$$

The approximate difference may be replaced by a differential when the number of slots is large. Also from (2.9):

$$f_{tooth}\left(N/m^2\right) = -\frac{-I_b(\theta_r,t)}{b_{sr}}\cdot\frac{\Delta B(\theta_r,t)}{\Delta\theta_{slot}}\cdot\frac{(b_{tr}+b_{sr})}{b_{sr}}\cdot\Delta\theta_{slot} \tag{2.13}$$

Therefore, it is the change of stator-produced field with θ_r, the travelling field existence, which produces the tangential force experienced by the walls of each slot. The total force for one slot may be obtained by multiplying the specific force in (2.13) by the rotor slot height and the stack length.

It may be demonstrated that with a pure travelling field and rotor current travelling wave $I_b(\theta_r,t)$, the tangential forces on each slot pair of walls add up to produce finally a smooth torque. Not so if the field is not purely travelling.

Based on the same rationale, the opposite direction tangential force on stator slot walls may be calculated. It is produced by the interaction of stator leakage field with rotor main reaction field. This is to be expected as action equals reaction according to Newton's third law. So the tangential forces that produce the torque occur on the tooth radial walls. Despite this reality, the principle of IM is traditionally explained by forces on currents in a magnetic field.

It may be demonstrated that, mathematically, it is correct to "move" the rotor currents from rotor slots, eliminate the slots, and place them in an infinitely thin conductor sheet on the rotor surface to replace the actual slot-cage rotor configuration (Figure 2.18). This way the tangential force will be exerted directly on the "rotor" conductors. Let us use this concept to further explain the operation modes of IM.

The relative speed between rotor conductors and stator travelling field is $\overline{U}-\overline{U}_s$, so the induced electrical field in the rotor conductors is

$$\overline{E} = \left(\overline{U}-\overline{U}_s\right)\times\overline{B} \tag{2.14}$$

As the rotor cage is short-circuited, no external electric field is applied to it, so the current density in the rotor conductor $\overline{J}$ is

$$\overline{J} = \sigma_{Al}\overline{E} \tag{2.15}$$

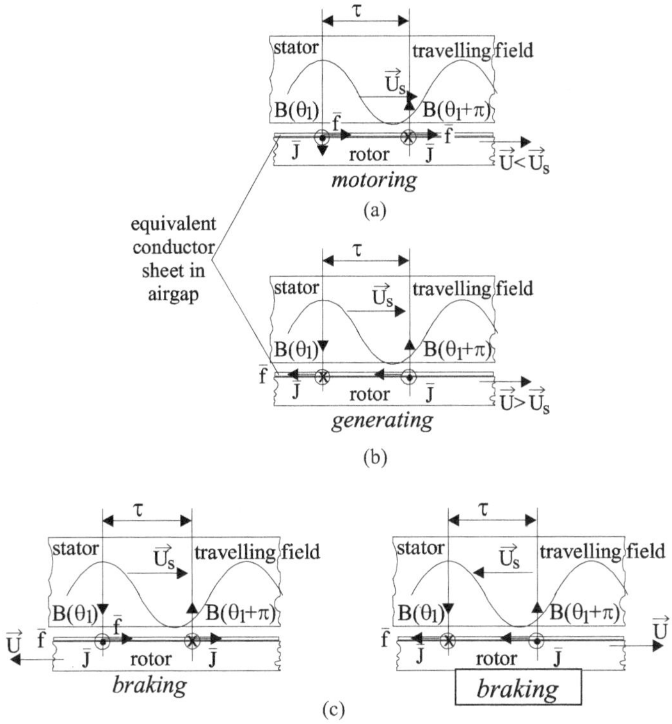

FIGURE 2.18 Operation modes of IMs: (a) motoring: $\vec{U} < \vec{U}_s$ both in the same direction, (b) generating: $\vec{U} > \vec{U}_s$; both in the same direction, (c) braking: $\vec{U}$ opposite to $\vec{U}_s$; either $\vec{U}$ or $\vec{U}_s$ changes direction.

Finally, the force (per unit volume) exerted on the rotor conductor by the travelling field, f_t, is

$$f_t = \bar{J} \times \bar{B} \tag{2.16}$$

Applying these fundamental equations for rotor speed $\bar{U}$ and field speed $\bar{U}_s$ in the same direction, we obtain the motoring mode for $\vec{U} < \vec{U}_s$, and, respectively, the generating mode for $\vec{U} > \vec{U}_s$, as shown in Figure 2.17a and b. In the motoring mode, the force on rotor conductors acts along the direction of motion, while, in the generating mode, it acts against it. At the same time, the electromagnetic (airgap) power P_e is negative.

$$\begin{aligned} P_e &= \bar{f}_t \cdot \bar{U}_s > 0; \text{ for motoring } (ft > 0, U > 0) \\ P_e &= \bar{f}_t \cdot \bar{U}_s < 0; \text{ for generating } (ft < 0, U > 0) \end{aligned} \tag{2.17}$$

This simply means that in the generating mode, the active power travels from rotor to stator to be sent back to the grid after losses are covered. In the braking mode ($U < 0$ and $U_s > 0$ or $U > 0$ and $U_s < 0$), as seen in Figure 2.18c, the torque acts against motion again but the electromagnetic power P_e remains positive ($\bar{U}_s > 0$ and $\bar{f}_t > 0$ or $\bar{U}_s < 0$ and $\bar{f}_t < 0$). Consequently, active power is drawn from the power source. It is also drawn from the shaft. The summation of the two is converted into IM losses.

The braking mode is, thus, energy-intensive and should be used only at low frequencies and speeds (low U_s and U), in variable speed drives, to “lock” the variable speed drive at standstill under load.

The LIM operation principles and operation modes are quite similar to those presented for rotary IMs.

In wound rotor IMs, the rotor may be fed from a separate source at slip frequency $f_2 = S \cdot f_1$, variable with speed. The principle is, however, the same as for the cage rotor IM (interaction between stator field and rotor currents (or their leakage field)).

Recently, permanent magnets and (or) magnetic saliency have been added on the IM cage rotor to enhance their efficiency in constant speed drives, while dual stator windings with nested cage rotors have been introduced for variable speed generators with fractional kVA pulse width modulation (PWM) converters. They will be dedicated special paragraphs in this book (see the Contents).

2.4 SUMMARY

- The IM is an A.C. machine. It may be energised directly from a three- or single-phase A.C. power grids. Alternatively, it may be energised through a PWM converter at variable voltage (V) and frequency (f).
- The IM is essentially a travelling field machine. It has an ideal no-load speed $n_1 = f_1/p_1$; p_1 is the number of travelling field periods per one revolution.
- The main parts of IM are the stator and rotor slotted magnetic cores and windings. The magnetic cores are, in general, made of thin silicon steel sheets (laminations) to reduce the core losses to values such as 2–4 W/Kg at 60 Hz and 1 T.
- Three- or two-phase windings are placed in the primary (stator) slots. Windings are coil systems connected to produce a travelling mmf (ampere-turns) in the airgap between the stator and the rotor cores.
- The slot geometry depends on power (torque) level and performance constraints.
- Starting torque and current, breakdown torque, rated efficiency, and power factor are typical constraints (specifications) for power grid directly energised IMs.
- Two-phase windings are used for capacitor IMs energised from a single-phase A.C. supply to produce travelling field. Single-phase A.C. supply is typical for home appliances.
- Cage windings made of solid bars (of aluminum, brass or copper [5]) in slots with end rings are used on most IM rotors.
- The rotor bar cross section is tightly related to all starting and running performance. Deep-bar or double-cage windings are used for high starting torque, low starting current IMs fed from the power grid (constant V and f).
- LIMs are obtained from rotary IMs by the cut-and-unroll process. Flat and tubular configurations are feasible with single- or double-sided primary. Either primary or secondary may be the mover in LIMs. Ladder or aluminium sheet or iron are typical for single-sided LIM secondaries. Continuous thrust densities up to 2–2.5 N/cm^2 are feasible with air cooling LIMs.
- In general, the airgap g per pole pitch τ ratio is larger than for rotary IM, and thus, the power factor and efficiency are lower. However, the absence of mechanical transmission in linear motion applications leads to virtually maintenance-free propulsion systems. Urban transportation systems with LIM propulsion are now in use in a few cities from three continents.
- The principle of operation of IMs is related to torque production. By using the Maxwell stress tensor concept it has been shown that, with windings in slots, the torque (due to tangential forces) is exerted mainly on slot walls and not on the conductors themselves.
- Stress analysis during severe transients should illustrate this reality. It may be demonstrated that the rotor winding in slots can be "mathematically" moved in the airgap and transformed into an equivalent infinitely thin current sheet. The same torque is now exerted directly on the rotor conductors in the airgap. The LIM with conductor sheet on iron resembles this situation naturally.

- Based on the $\bar{J}\times\bar{B}$ force principle, three operation modes of IM are easily identified (with U_s – ideal no-load speed, U – mover speed):
  - Motoring: $|U| < |U_s|$; U and U_s either positive or negative
  - Generating: $|U| > |U_s|$; U and U_s either positive or negative
  - Braking: $(U > 0\ \&\ U_s < 0)$ or $(U < 0\ \&\ U_s > 0)$.
- For the motoring mode, the torque acts along the direction of motion, while, for the generator mode, it acts against it as it does during braking mode.
- However, during generating, the IM returns some power to the grid, after covering the losses, while for braking, it draws active power also from the power grid.
- Generating is energy conversion advantageous while braking is energy-intensive. Braking is recommended only at low frequency and speed, with variable V/f PWM converter supply, to stall the IM drive on load (like in overhead cranes). The energy consumption is moderate in such cases anyway (as the frequency is small).

REFERENCES

1. I. Boldea, S. A. Nasar, *Linear Motion Electric Machines*, Wiley Interscience, New York, 1976.
2. I. Boldea, S. A. Nasar, *Linear Motion Electromagnetic Systems*, Wiley Interscience, New York, 1985, Chapter 5.
3. M. Schwartz, *Principles of Electrodynamics*, Dover Publications Inc., New York, 1972, pp.180.
4. D. C. White, H. H. Woodson, *Electromechanical Energy Conversion*, John Wiley & Sons Inc. London, 1959.
5. R. Kimmich, M. Doppelbauer, D. T. Peters, J. G. Cowie, E. F. Brush, Jr., *Die-cast Copper Rotor Motor Via Simple Substitution and Motor Redesign for Copper*, Record of ICEM-2006, Ref. No. 358.

3 Magnetic, Electric, and Insulation Materials for IM

3.1 INTRODUCTION

Induction machines (IMs) contain magnetic circuits travelled by A.C. and travelling magnetic fields and electric circuits flowed by alternative currents. The electric circuits are insulated from the magnetic circuits (cores). The insulation system comprises the conductor insulation, slot, and interphase insulation.

Magnetic, electrical, and insulation materials are characterised by their characteristics (B–H curve, electrical resistivity, dielectric constant, and breakdown electric field (V/m)) and their losses.

At frequencies encountered in IMs (up to tens of kHz, when pulse width modulation (PWM) inverter fed), the insulation losses are neglected. Soft magnetic materials are used in IM as the magnetic field is currently produced. The flux density (B)/magnetic field (H) curve and cycle depend on the soft material composition and fabrication process. Their losses in W/kg depend on the B–H hysteresis cycle, frequency, electrical resistivity, and the A.C. (or) travelling field penetration into the soft magnetic material.

Silicon steel sheets are standard soft magnetic materials for IMs. Amorphous soft powder materials have been introduced recently with some potential for high-frequency (high-speed) IMs. The pure copper is the favourite material for the stator electric circuit (windings), while aluminium or brass is used mainly for rotor squirrel-cage windings.

Insulation materials are getting thinner and better and are ranked into a few classes: A (105°C), B (130°C), F (155°C), and H (180°C).

3.2 SOFT MAGNETIC MATERIALS

In free space, the flux density B and the magnetic field H are related by the permeability of free space $\mu_0 = 4\pi 10^{-7}$ H/m (S.I.)

$$B\left[\frac{Wb}{m^2}\right] = \mu_0\left[\frac{H}{m}\right] \cdot H\left[\frac{A}{m}\right] \tag{3.1}$$

Within a certain material, a different magnetisation process occurs.

$$B = \mu \cdot H; \quad \mu = \mu_0 \mu_R \tag{3.2}$$

In (3.2), μ is termed as permeability and μ_R as relative permeability (non-dimensional).

Permeability is defined for homogenous (uniform quality) and isotropic (same properties in all directions) materials. In non-homogeneous or (and) non-isotropic materials, μ becomes a tensor. Most common materials are non-linear: μ varies with B.

A material is classified according to the value of its relative permeability, μ_R, which is related to its atomic structure.

Most non-magnetic materials are either paramagnetic – with μ_R slightly greater than 1.0 – or diamagnetic with μ_R slightly less than 1.0. Superconductors are perfect diamagnetic materials. In such materials when $B \to 0$, $\mu_R \to 0$.

Magnetic properties are related to the existence of permanent magnetic dipoles within the matter.

There are quite a few classes of more magnetic materials ($\mu_R \gg 1$). Amongst them, we will deal here with soft ferromagnetic materials. Soft magnetic materials include alloys made of iron, nickel, cobalt, and one rare earth element and/or soft steels with silicon.

There is also a class of magnetic materials made of powdered iron particles (or other magnetic material) suspended in an epoxy or plastic (non-ferrous) matrix. These soft powder magnetic materials are formed by compression or injection, moulding, or other techniques.

There are a number of properties of interest in a soft magnetic material such as permeability versus B, saturation flux density, H(B), temperature variation of permeability, hysteresis characteristics, electric conductivity, Curie temperature, and loss coefficients.

The graphical representation of non-linear B(H) curve (besides the pertinent table) is of high interest (Figure 3.1). Also of high interest is the hysteresis loop (Figure 3.2).

There are quite a few standard laboratory methods to obtain these two characteristics. The B–H curve can be obtained in two ways: the virgin (initial) B–H curve, obtained from a totally demagnetised sample; and the normal (average) B–H curve, obtained as the tips of hysteresis loops of increasing magnitude. There is only a small difference between the two methods.

The B–H curve is the result of domain changes within the magnetic material. The domains of soft magnetic materials are 10^{-4}–10^{-7}m in size. When completely demagnetised, these domains have random magnetisation with zero flux in all finite samples.

When an external magnetic field H is applied, the domains aligned to H tend to grow when B is increased (region I on Figure 3.1). In region II, H is further increased and the domain walls move rapidly until each crystal of the material becomes a single domain. In region III, the domains rotate towards alignment with H. This results in magnetic saturation B_s. Beyond this condition, the small increase in B is basically due to the increase in the space occupied by the material for $B = \mu_0 H_{r0}$.

This "free-space" flux density may be subtracted to obtain the intrinsic magnetisation curve. The non-linear character of B–H curve (Figure 3.1) leads to two different definitions of relative permeability:

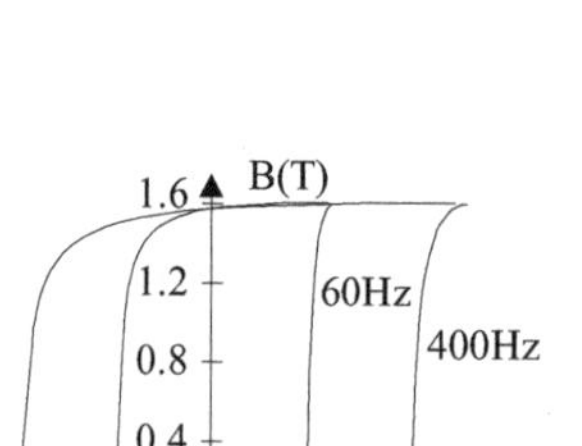

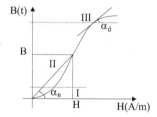

FIGURE 3.1 Typical B–H curve.

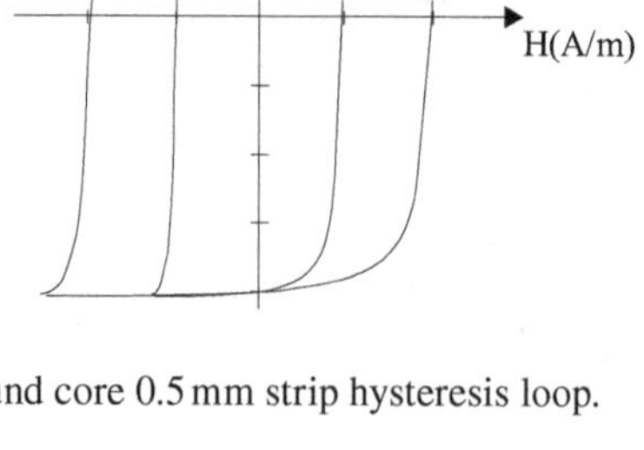

FIGURE 3.2 Deltamax tape-wound core 0.5 mm strip hysteresis loop.

- The normal permeability μ_{Rn}:

$$\mu_{Rn} = \frac{B}{\mu_0 H} = \frac{\tan\alpha_n}{\mu_0} \tag{3.3}$$

- The differential relative permeability μ_{Rd}:

$$\mu_{Rd} = \frac{dB}{\mu_0 dH} = \frac{\tan\alpha_d}{\mu_0} \tag{3.4}$$

Only in region II, $\mu_{Rn} = \mu_{Rd}$. In regions I and III, in general, $\mu_{Rn} > \mu_{Rd}$ (Figure 3.3). The permeability is maximum in region II. For M19 silicon steel sheets, $B_s = 2$ T, $H_s = 40{,}000$ A/m, and $\mu_{Rmax} = 10{,}000$).

So the minimum relative permeability is

$$(\mu_{Rn})_{B_s=2.0\,T} = \frac{2.0}{4\pi 10^{-7} \cdot 40{,}000} = 39.8! \tag{3.5}$$

The second graphical characteristic of interest is the hysteresis loop (Figure 3.2). This is a symmetrical hysteresis loop obtained after a number of reversals of magnetic field (force) between $\pm H_c$. The area within the loop is related to the energy required to reverse the magnetic domain walls as H is reversed. This nonreversible energy is called hysteresis loss and varies with temperature and frequency of H reversals in a given material (Figure 3.2). A typical magnetisation curve B–H for silicon steel nonoriented grain is given in Table 3.1.

It has been shown experimentally that the magnetisation curve varies with frequency as shown in Table 3.2. This time the magnetic field is kept in original data (1Oe = 79.5 A/m) [1].

In essence for the same flux density B, the magnetic field increases with frequency. It is recommended to reduce the design flux density when the frequency increases above 200 Hz as the core losses grow markedly with frequency.

Special materials such as Hiperco 50 show saturation flux densities $B_s = 2.3$–2.4 T at $H_s < 10{,}000$ A/m, which allows for higher airgap flux density in IMs and, thus, lower weight designs (Tables 3.3 and 3.4).

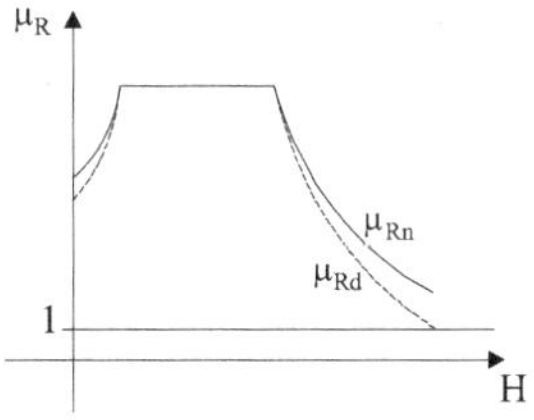

FIGURE 3.3 Relative permeability versus H.

TABLE 3.1
B–H Curve for Silicon (3.5%) Steel (0.5 mm thick) at 50 Hz

B (T)	0.05	0.1	0.15	0.2	0.25	0.3	0.35	0.4	0.45	0.5
H (A/m)	22.8	35	45	49	57	65	70	76	83	90
B (T)	0.55	0.6	0.65	0.7	0.75	0.8	0.85	0.9	0.95	1
H (A/m)	98	106	115	124	135	148	162	177	198	220
B (T)	1.05	1.1	1.15	1.2	1.25	1.3	1.35	1.4	1.45	1.5
H (A/m)	237	273	310	356	417	482	585	760	1050	1340
B (T)	1.55	1.6	1.65	1.7	1.75	1.8	1.85	1.9	1.95	2.0
H (A/m)	1760	2460	3460	4800	6160	8270	11,170	15,220	22,000	34,000

TABLE 3.2
1Oe = 79.5 A/m

Induction	**Typical D.C. and Derived A.C. Magnetising Force (Oe) of As-Sheared 29-Cage M19 Fully Processed Cold Rolled Nonoriented (CRNO) at Various Frequencies**											
(kG)	**D.C.**	**50 Hz**	**60 Hz**	**100 Hz**	**150 Hz**	**200 Hz**	**300 Hz**	**400 Hz**	**600 Hz**	**1000 Hz**	**1500 Hz**	**2000 Hz**
1.0		0.333	0.334	0.341	0.349	0.356	0.372	0.385	0.412	0.485	0.564	0.642
2.0	0.401	0.475	0.480	0.495	0.513	0.533	0.567	0.599	0.661	0.808	0.955	1.092
4.0	0.564	0.659	0.669	0.700	0.39	0.777	0.846	0.911	1.040	1.298	1.557	1.800
7.0	0.845	0.904	0.916	0.968	1.030	1.094	1.211	1.325	1.553	2.000	2.483	2.954
10.0	1.335	1.248	1.263	1.324	1.403	1.481	1.648	1.822	2.169	2.867	3.697	4.534
12.0	2.058	1.705	1.718	1.777	1.859	1.942	2.129	2.326	2.736	3.657	4.769	5.889
13.0	2.951	2.211	2.223	2.273	2.342	2.424	2.609	2.815	3.244	4.268	5.499	
14.0	5.470	3.508	3.510	3.571	3.633	3.691	3.855	4.132				
15.0	13.928	8.276	8.313	8.366	8.366	8.478	8.651	9.737				
15.5	22.784	13.615	13.587	13.754	13.725	13.776	14.102	16.496				
16.0	35.201	21.589	21.715	21.800	21.842	21.884						
16.5	50.940	32.383	32.506	32.629	32.547	32.588						
17.0	70.260	46.115	46.234	46.392	46.644	46.630						
18.0	122.01											
19.0	201.58											
20.0	393.50											
21.0	1111.84											

3.3 CORE (MAGNETIC) LOSSES

Energy loss in the magnetic material itself is a very significant characteristic in the energy efficiency of IMs. This loss is termed core loss or magnetic loss.

Traditionally, core loss has been divided into two components: hysteresis loss and eddy current loss. The hysteresis loss is equal to the product between the hysteresis loop area and the frequency of the magnetic field in sinusoidal systems.

$$P_h \approx k_h f B_m^2 \,[W/kg]; \quad B_m \text{- maximum flux density} \tag{3.6}$$

Hysteresis losses are 10%–30% higher in travelling fields than in A.C. fields for $B_m < 1.5(1.6)$ T. However, in a travelling field, they have a maximum, in general, between 1.5 and 1.6 T and then decrease to low values for B > 2.0 T. The computation of hysteresis losses is still an open issue due to the hysteresis cycle's complex shape, its dependence on frequency and on the character of the magnetic field (travelling or A.C.) [2].

Preisach modelling of hysteresis cycle is very popular [3], but neural network models have proved much less computation time consuming [4].

Eddy current losses are caused by induced electric currents in the magnetic material by an external A.C. or travelling magnetic field.

TABLE 3.3
Hiperco 50 Magnetisation Curve B/H – T/(A/m)

B (T)	0.017	0.7	1.5	1.9	2	2.1	2.15	2.2	2.25	2.267	2.275	2.283
H (A/m)	0	39.75	79.5	159	318	477	715.5	1431	3975	7950	11,925	15,900

TABLE 3.4
Hiperco 50 Losses W/Kg

f = 60 Hz											
Core loss (W/Kg)	0.8866	1.0087	1.2317	1.3946	1.5671	1.7589	1.9410	2.1614	2.4202	2.6215	2.8612
B (T)	1	1.1	1.2	1.3	1.4	1.5	1.6	1.7	1.8	1.9	2
f = 400 Hz											
Core loss (W/Kg)	8.5429	10.196	11.849	13.734	15.432	17.636	19.290	21.770	23.975	25.904	27.282
B (T)	1	1.1	1.2	1.3	1.4	1.5	1.6	1.7	1.8	1.9	2
f = 800 Hz											
Core loss (W/Kg)	23.589	27.282	31.416	35.274	40.786	45.072	51.768	58.137	65.485	72.671	76.917
B (T)	1	1.1	1.2	1.3	1.4	1.5	1.6	1.7	1.8	1.9	2
f = 1200 Hz											
Core loss (W/Kg)	40.738	47.242	58.659	65.745	77.162	84.642	92.909	104.32	113.38	128.34	135.43
B (T)	1	1.1	1.2	1.3	1.4	1.5	1.6	1.7	1.8	1.9	2

$$P_e \approx k_e f^2 B_m^{\,2}\left[W/kg\right] \tag{3.7}$$

Finite elements are used to determine the magnetic distribution with zero electrical conductivity, and then the core losses may be calculated by some analytical approximations as (3.6) and (3.7) or [5]

$$P_{core} \approx k_h f B_m^{\,\alpha} K(B_m) + \frac{\sigma_{Fe}}{12}\frac{d^2 f}{\gamma_{Fe}}\int_{1/f}\left(\frac{dB}{dt}\right)^2 dt + K_{ex} f \int_{1/f}\left(\frac{dB}{dt}\right)^{1.5} \tag{3.8}$$

where $K = 1 + \frac{0.65}{B_m}\sum^{n}\Delta B_i$

B_m – maximum flux density
f – frequency
γ_{Fe} – material density
d – lamination thickness
K_h – hysteresis loss constant
K_{ex} – excess loss constant
ΔB_i – change of flux density during a time step
n – total number of time steps.

Equation (3.8) is a generalisation of Equations (3.6) and (3.7) for nonsinusoidal time-varying magnetic fields as produced in PWM inverter IM drives. Recently, better fits of power losses formula for cold-rolled motor laminations for a wide frequency and magnetisation range have been proposed [6].

For sinusoidal systems, the eddy currents in a thin lamination may be calculated rather easily by assuming the external magnetic field $H_0 e^{j\omega_1 t}$ acting parallel to the lamination plane (Figure 3.4).

Maxwell's equations yield

$$\frac{\partial H_y}{\partial x} = J_z; \quad H_{0y} = H_0 e^{j\omega_1 t}$$
$$-\frac{\partial E_z}{\partial x} = -j\omega_1\mu\left(H_{0y} + H_y\right); \quad \sigma_{Fe} E_z = J_z \tag{3.9}$$

where J is current density and E is electric field.

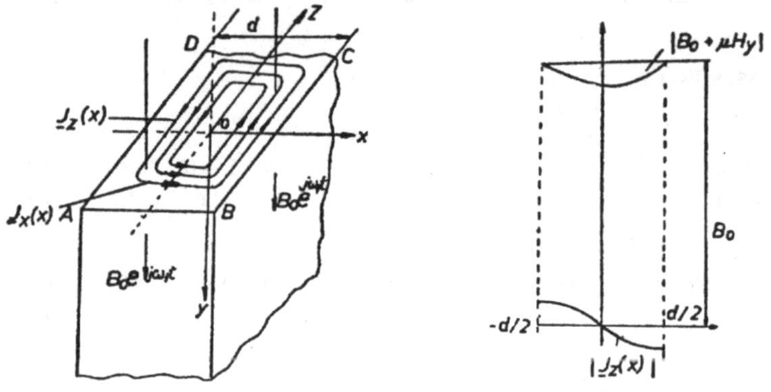

FIGURE 3.4 Eddy current paths in a soft material lamination.

As the lamination thickness is small in comparison with its length and width, J_x contribution is neglected. Consequently, (3.9) is reduced to

$$\frac{\partial^2 H_y}{\partial x^2} - j\omega_1\mu\sigma_{Fe}H_y = j\omega_1\sigma_{Fe}B_0 \tag{3.10}$$

$B_0 = \mu_0 H_0$ is the initial flux density on the lamination surface.

The solution of (3.10) is

$$H_y(x) = A_1e^{\gamma x} + A_2e^{-\gamma x} + \frac{B_0}{\mu_0} \tag{3.11}$$

$$\gamma = \beta(1+j); \quad B = \sqrt{\frac{\omega_1\mu\sigma_{Fe}}{2}} \tag{3.12}$$

The current density $J_z(x)$ is

$$J_z(x) = \frac{\partial H_y}{\partial x} = \gamma\left(A_1e^{\gamma x} + A_2e^{-\gamma x}\right) \tag{3.13}$$

The boundary conditions are

$$H_y\left(\frac{d}{2}\right) = H_y\left(-\frac{d}{2}\right) = 0 \tag{3.14}$$

Finally,

$$A_1 = A_2 = \frac{B_0}{2\mu\cosh\beta\frac{d}{2}(1+j)} \tag{3.15}$$

$$J_z(x) = -\frac{\beta(1+j)}{\mu}\frac{B_0\sinh(1+j)\beta x}{\cosh\beta\frac{d}{2}(1+j)} \tag{3.16}$$

The eddy current loss per unit weight P_e is

$$P_e = \frac{2\gamma_{Fe}}{d\sigma_{Fe}}\frac{1}{2}\int_0^{d/2}(J_z(x))^2\,dx = \frac{\beta\gamma_{Fe}d\omega_1}{\mu}B_0^2\left[\frac{\sinh(\beta d) - \sin(\beta d)}{\cosh(\beta d) + \cos(\beta d)}\right]\left[\frac{W}{kg}\right] \tag{3.17}$$

The iron permeability has been considered constant within the lamination thickness though the flux density slightly decreases.

For a good utilisation of the material, the flux density reduction along lamination thickness has to be small. In other words, $\beta d \ll 1$. In such conditions, the eddy current losses increase with the lamination thickness.

The electrical conductivity σ_{Fe} is also influential, and silicon added to soft steel reduces σ_{Fe} to $(2–2.5) \times 10^6\ (\Omega m)^{-1}$. This is why 0.5–0.6 mm thick laminations are used at 50(60) Hz and, in general, up to 200–300 Hz IMs.

For such laminations, eddy current losses may be approximated to

$$P_e \approx K_w B_m^{\ 2}\left[\frac{W}{kg}\right]; \quad K_w = \frac{\omega_1^{\ 2}\sigma_{Fe} d^2}{24}\gamma_{Fe} \tag{3.18}$$

The above loss formula derivation process is valid for A.C. magnetic field excitation. For pure travelling field, the eddy current losses are twice as much for same laminations, frequency, and peak flux density.

Given the complexity of eddy current and hysteresis losses, it is recommended that tests be run to measure them in conditions very similar to those encountered in the particular IM.

Soft magnetic material producers manufacture laminations for many purposes. They run their own tests and provide data on core losses for practical values of frequency and flux density.

Besides Epstein's traditional method, made with rectangular lamination samples, the wound toroidal cores method has also been introduced [7] for A.C. field losses. For travelling field loss measurement, a rotational loss tester may be used [8].

Typical core loss data for M15 – 3% silicon 0.5 mm thick lamination material – used in small IMs, is given in Figure 3.5a [9].

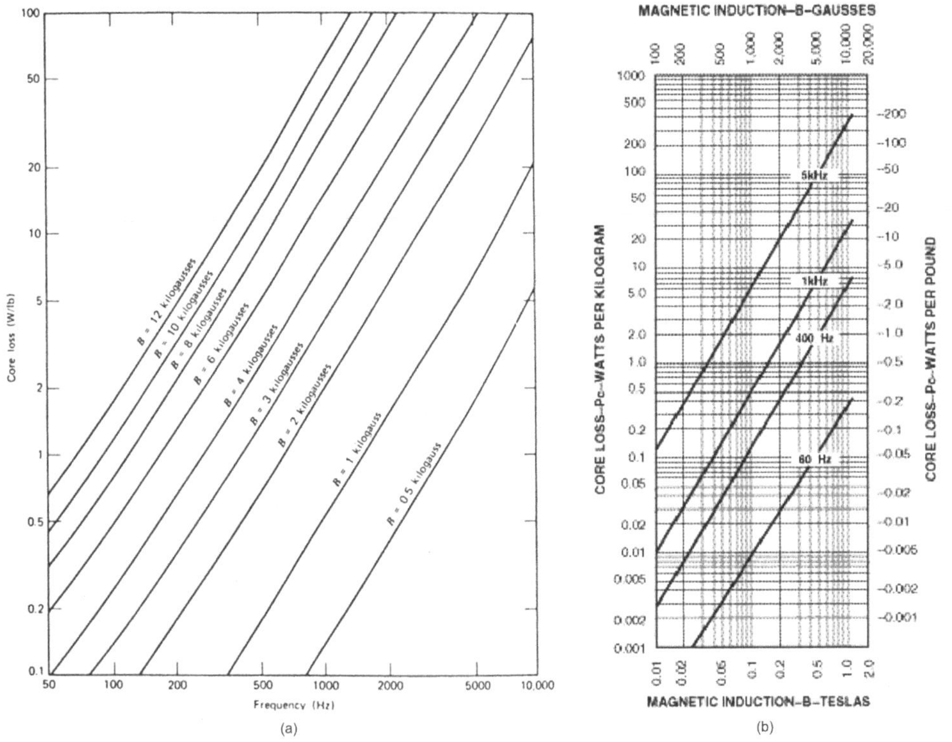

FIGURE 3.5 Core losses for M15 – 3% silicon 0.5 mm thick laminations [9], (a) and 0.1 mm thick, 3% silicon Metglass (AMM) core losses, (b).

TABLE 3.5
Typical Core Loss (W/lb) As-Sheared 29 Cage M19 Fully Processed CRNO at Various Frequencies [1]

Induction (kG)	50 Hz	60 Hz	100 Hz	150 Hz	200 Hz	300 Hz	400 Hz	600 Hz	1000 Hz	1500 Hz	2000 Hz
1.0	0.008	0.009	0.017	0.029	0.042	0.074	0.112	0.205	0.465	0.900	1.451
2.0	0.031	0.039	0.072	0.0119	0.173	0.300	0.451	0.812	1.786	3.370	5.318
4.0	0.109	0.134	0.252	0.424	0.621	1.085	1.635	2.960	6.340	11.834	18.523
7.0	0.273	0.340	0.647	1.106	1.640	2.920	4.450	8.180	17.753	33.720	53.971
10.0	0.494	0.617	1.182	2.040	3.060	5.530	8.590	16.180	36.303	71.529	116.702
12.0	0.687	0.858	1.648	2.860	4.290	7.830	12.203	23.500	54.258	108.995	179.321
13.0	0.812	1.014	1.942	3.360	5.060	9.230	14.409	27.810	65.100	131.98	
14.0	0.969	1.209	2.310	4.000	6.000	10.920	17.000				
15.0	1.161	1.447	2.770	4.760	7.150	13.000	20.144				
15.5	1.256	1.559	2.990	5.150	7.710	13.942	21.619				
16.0	1.342	1.667	3.179	5.466	8.189						
16.5	1.420	1.763	3.375	5.788	8.674						
17.0	1.492	1.852	3.540	6.089	9.129						

As expected, core losses increase with frequency and flux density. A similar situation occurs, with a superior but still common material: steel M19 FP (0.4 mm) 29 gauge (Table 3.5) [1].

A rather complete up-to-date data source on soft magnetic materials' characteristics and losses may be found in Ref. [1].

Core loss represents 25%–35% of all losses in low-power 50 (60) Hz IMs and slightly more in medium- and large-power IMs at 50(60) Hz. The development of high-speed IMs, up to more than 45,000 rpm at 20 kW [10], has caused a new momentum in the research for better magnetic materials as core losses are even larger than winding losses in such applications.

Thinner (0.35 mm or less) laminations of special materials (3.25% silicon) with special thermal treatment are used to strike a better compromise between low 60 Hz and moderate 800/1000 Hz core losses (1.2 W/kg at 60 Hz, 1 T; 28 W/kg at 800 Hz, 1 T).

6.5% silicon steel nonoriented steel laminations for low-power IMs at 60 Hz have shown capable of a 40% reduction in core losses [11]. The noise level has also been reduced this way [11]. Similar improvements have been reported with 0.35 mm thick oriented grain laminations by alternating laminations with perpendicular magnetisation orientation or crossed magnetic structure (CMS) [12].

Recent Progress on Core Loss Assessment

As the fundamental frequency in electric machine tends to increase (for high-speed variable speed drives) formulae as (3.8), with constant coefficients do not fit experimental results.

Also, high-temperature annealing core laminations have been proven to reduce the hysteresis losses but increase the eddy current losses.

Finally, the magnetic core properties degrade due to manufacturing process (punching, pressing, welding, packaging), and variability due to manufacturing has to be observed. To treat the above problems and thus provide reliable data on hysteresis cycle and losses and on eddy current losses, recent R & D effort concentrated on the following:

- Proving that the variation of hysteresis cycle with frequency is due to skin effect in the lamination especially when the frequency increases [13].
- Investigating experimentally, by a modified electromagnetic Halbach core test rig, the rotational core losses as they tend to have a maximum of around 1.5–1.6 T monotonously with flux density [14].

- Measuring the stator core loss degradation by the manufacturing process by a stator toroidal model that uses the stator core and investigates mainly the stator back core losses [15,16].

Soft magnetic composites (SMCs) have been produced by powder metallurgy technologies. The magnetic powder particles are coated by insulation layers and a binder which are compressed such that to provide

- Large enough magnetic permeability
- Low-enough core losses
- Densities above 7.1 g/cm^3 (for high-enough permeability).

The eddy current loss tends to be constant with frequency, while hysteresis loss increases almost linearly with frequency (up to 1 kHz or so).

At 400–500 Hz and above, the losses in SMC become smaller than for 0.5 mm thick silicon steels. However, the relative permeability is still low: 100–200. Only for recent materials, fabricated by cold compression, the relative permeability has been increased above 500 for flux densities in the 1 T range [17,18]. On the other hand, amorphous magnetic materials with plastic fill between magnetic particles, such as Metglass (Figure 3.5b), show not only lower losses at 300–500 Hz but also allow for higher flux densities.

Added advantages such as more freedom in choosing the stator core geometry and the increase of slot-filling factor by coil in slot magnetic compression-embedded windings [19] may lead to wide use of SMCs in induction motors. The electric loading may thus be increased. The heat transmissivity also increases [17].

In the near future, better silicon 0.5 mm (0.35 mm) thick steel laminations with nonoriented grain seem to remain the basic soft magnetic materials for IM fabrication. For high speed (frequency above 300 Hz), thinner laminations are to be used. The insulation coating layer of each lamination is getting thinner and thinner to retain a good stacking factor (above 85%).

3.4 ELECTRICAL CONDUCTORS

Electrical copper conductors are used to produce the stator three (two)-phase windings. The same is true for wound-rotor windings.

The electrical copper has a high purity and is fabricated by involved electrolysis process. The purity is well above 99%. The cross section of copper conductors (wires) to be introduced in stator slots is either circular or rectangular (Figure 3.6). The electrical resistivity of magnetic wire (electric conductor) $\rho_{Co} = (1.65–1.8) \times 10^{-8}\ \Omega$m at 20°C and varies with temperature as

$$\rho_{Co}(T) = (\rho_{Co})_{20^\circ}\left[1 + (T - 20)/273\right] \tag{3.19}$$

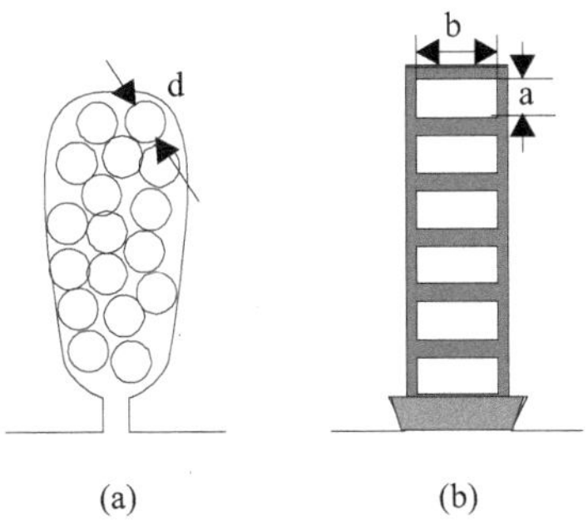

FIGURE 3.6 Stator slot with round (a) and rectangular (b) conductors.

Round magnetic wires come into standardised gauges up to a bare copper diameter of about 2.5 mm (3 mm) (or 0.12 inch), in general (Tables 3.6 and 3.7).

The total cross section A_{con} of the coil conductor depends on the rated phase current I_{1n} and the design current density J_{con}:

$$A_{con} = I_{1n} / J_{con} \tag{3.20}$$

TABLE 3.6
Round Magnetic Wire Gauges in Inches

	Bare Wire Diameter	Film Additions (inches)		Overall Diameter (inches)			Weight at 20°C–68°F		Resistance at 20°C–68°F			
Awg Size	Nominal (inches)	Min.	Max.	Min.	Nom.	Max.	Lbs./M Ft. Nom.	Ft./Lb. Nom.	Ohms./M Ft. Nom.	Ohms./ Lb. Nom.	Wires/ In. Nom.	Awg Size
8	0.1285	0.0016	0.0026	0.1288	0.1306	0.1324	50.20	19.92	0.6281	0.01251	7.66	8
9	0.1144	0.0016	0.0026	0.1149	0.1165	0.1181	39.81	25.12	0.7925	0.01991	8.58	9
10	0.1019	0.0015	0.0025	0.1024	0.1039	0.1054	31.59	31.66	0.9988	0.03162	9.62	10
11	0.0907	0.0015	0.0025	0.0913	0.0927	0.0941	25.04	39.94	1.26	0.05032	10.8	11
12	0.0808	0.0014	0.0024	0.0814	0.0827	0.0840	19.92	50.20	1.59	0.07982	12.1	12
13	0.0720	0.0014	0.0023	0.0727	0.0738	0.0750	15.81	63.25	2.00	0.1265	13.5	13
14	0.0641	0.0014	0.0023	0.0649	0.0659	0.0659	12.49	80.06	2.52	0.2018	15.2	14
15	0.0571	0.0013	0.0022	0.0578	0.0588	0.0599	9.948	100.5	3.18	0.3196	17.0	15
16	0.0508	0.0012	0.0021	0.0515	0.0525	0.0534	7.880	126.9	4.02	0.5101	19.0	16
17	0.0453	0.0012	0.0020	0.0460	0.0469	0.0478	6.269	159.5	5.05	0.8055	21.3	17
18	0.0403	0.0011	0.0019	0.0410	0.0418	0.0426	4.970	201.2	6.39	1.286	23.9	18
19	0.0359	0.0011	0.0019	0.0366	0.0374	0.0382	3.943	253.6	8.05	2.041	26.7	19
20	0.0320	0.0010	0.0018	0.0327	0.0334	0.0341	3.138	318.7	10.1	3.219	29.9	20
21	0.0285	0.0010	0.0018	0.0292	0.0299	0.0306	2.492	401.2	12.8	5.135	33.4	21
22	0.0253	0.0010	0.0017	0.0260	0.0267	0.0273	1.969	507.9	16.2	8.228	37.5	22
23	0.0226	0.0009	0.0016	0.0233	0.0238	0.0244	1.572	636.1	20.3	12.91	42.0	23
24	0.0201	0.0009	0.0015	0.0208	0.0213	0.0218	1.240	806.5	25.7	20.73	46.9	24
25	0.0179	0.0009	0.0014	0.0186	0.0191	0.0195	988	1012	32.4	32.79	52.4	25
26	0.0159	0.0008	0.0013	0.0165	0.0169	0.0174	779	1284	41.0	52.64	59.2	26
27	0.0142	0.0008	0.0013	0.0149	0.0153	0.0156	0.623	1605	51.4	82.50	65.4	27
28	0.0126	0.0007	0.0012	0.0132	0.0136	0.0139	0.491	2037	65.3	133.0	73.5	28
29	0.0113	0.0007	0.0012	0.0119	0.0122	0.0126	0.395	2532	81.2	205.6	82.0	29
30	0.0100	0.0006	0.0011	0.0105	0.0109	0.112	0.310	3226	104	335.5	91.7	30
31	0.0089	0.0006	0.0011	0.0094	0.0097	0.0100	0.246	4065	131	532.5	103	31
32	0.0080	0.0006	0.0010	0.0085	0.0088	0.0091	0.199	5025	162	814.1	114	32
33	0.0071	0.0005	0.0009	0.0075	0.0078	0.0081	0.157	6394	206	1317	128	33
34	0.0063	0.0005	0.0008	0.0067	0.0070	0.0072	0.123	8130	261	2122	143	34
35	0.0056	0.0004	0.0007	0.0059	0.0062	0.0064	0.0977	10,235	331	3388	161	35
36	0.0050	0.0004	0.0007	0.0053	0.0056	0.0058	0.0783	12,771	415	5300	179	36
37	0.0045	0.0003	0.0006	0.0047	0.0050	0.0052	0.0632	15,823	512	8101	200	37
38	0.0040	0.0003	0.0006	0.0042	0.0045	0.0047	0.0501	19,960	648	12,934	222	38
39	0.0035	0.0002	0.0005	0.0036	0.0039	0.0041	0.0383	26,110	847	22,115	256	39
40	0.0031	0.0002	0.0005	0.0032	0.0035	0.0037	0.0301	33,222	1080	35,880	286	40
41	0.0028	0.0002	0.0004	0.0029	0.0031	0.0033	0.0244	40,984	1320	54,099	323	41
42	0.0025	0.0002	0.0004	0.0026	0.0028	0.0030	0.0195	51,282	1660	85,128	357	42
43	0.0022	0.0002	0.0003	0.0023	0.0025	0.0026	0.0153	65,360	2140	139,870	400	43
44	0.0020	0.0001	0.0003	0.0020	0.0022	0.0024	0.0124	80,645	2590	208,870	455	44

TABLE 3.7
Typical Round Magnetic Wire Gauges in mm

Rated Diameter (mm)	Insulated Wire Diameter (mm)	Rated Diameter (mm)	Insulated Wire Diameter (mm)
0.3	0.327	0.75	0.7949
0.32	0.348	0.80	0.8455
0.33	0.359	0.85	0.897
0.35	0.3795	0.90	0.948
0.38	0.4105	0.95	1.0
0.40	0.4315	1.00	1.051
0.42	0.4625	1.05	1.102
0.45	0.4835	1.10	1.153
0.48	0.515	1.12	1.173
0.50	0.536	1.15	1.2035
0.53	0.567	1.18	1.2345
0.55	0.5875	1.20	1.305
0.58	0.6185	1.25	1.325
0.60	0.639	1.30	1.356
0.63	0.6705	1.32	1.3765
0.65	0.691	1.35	1.407
0.67	0.7145	1.40	1.4575
0.70	0.742	1.45	1.508
0.71	0.7525	1.50	1.559

The design current density varies between 3.5 and 15 A/mm^2 depending on the cooling system, service duty cycle, and the targeted efficiency of the IM. High-efficiency IMs are characterised by lower current density (3.5–6 A/mm^2). If the A_{con} in (3.19) is larger than the cross section of the largest round wire gauge available, a few conductors of lower diameter are connected in parallel and wound together. Up to 6–8 elementary conductors may be connected together.

If A_{con} is larger than 30–40 mm^2 (that is, 6–8, 2.5 mm diameter wires in parallel), rectangular conductors are recommended.

In many countries, rectangular conductor cross sections are also standardised. In some cases, small cross sections such as (0.8–2)·2 mm × mm or (0.8–6) × 6 mm × mm are used for rectangular conductors.

In general, the rectangular conductor height a is kept low ($a < 3.55$ mm) to reduce the skin effect; that is, to keep the A.C. resistance low. A large cross-section area of 3.55 × 50 mm × mm would be typical for large-power IMs.

The rotor cage is in general made of aluminium: die-casted aluminium in low-power IMs (up to 300 kW or so) or of aluminium bars attached through brazing or welding processes to end rings.

Fabricated rotor cages are made of aluminium or copper alloys and of brass (the upper cage of a double cage) for powers above 300 kW, in general. The casting process of aluminium uses the rotor lamination stack as a partial mould because the melting point of silicon steel is much higher than that of aluminium. The electrical resistivity of aluminium $\rho_{Al} \cong (2.7\text{–}3.0) \times 10^{-8}\ \Omega\text{m}$ and varies with temperature as shown in (3.19).

Though the rotor cage bars are in general uninsulated from the magnetic core, most of the current flow through the cage bars as their resistivity is more than 20–30 times smaller than that of the laminated core.

Insulated cage bars would be ideal, but this would severely limit the rotor temperature unless a special high-temperature (high-cost) insulation coating is used.

3.5 INSULATION MATERIALS

The primary purpose of stator insulation is to withstand turn-to-turn, phase-to-phase, and phase-to-ground voltage such that to direct the stator phase currents through the desired paths of stator windings.

Insulation serves a similar purpose in phase-wound rotors whose phase leads are connected to insulated copper rings and then through brushes to stationary devices (resistances or/and special power electronic converters). Insulation is required to withstand voltages associated to: brush rigging (if any), winding connections, winding leads, auxiliaries such as temperature probes, and bearings (especially for PWM inverter drives).

The stator laminations are insulated from each other by special coatings (0.013 mm thick) to reduce eddy current core losses.

In standard IMs, the rotor (slip) frequency is rather small, and thus, inter-lamination insulation may not be necessary, unless the IM is to work for prolonged intervals at large slip values.

For all wound-rotor motors, the rotor laminations are insulated from each other. The bearing sitting is insulated from the stator to reduce the bearing (shaft) voltage (current), especially for large-power IMs whose stator laminations are made of a few segments, thus allowing a notable A.C. axial flux linkage. This way, premature bearing damage may be prevented and even more so in PWM inverter-fed IMs, where additional common voltage mode superhigh frequency capacitor currents through the bearings occur.

Stator winding insulation systems may be divided into two types, related to power and voltage levels.

- Random-wound conductor IMs with small and round conductors
- Form-wound conductor IMs with relatively large area rectangular conductors.

Insulation systems for IMs are characterised by voltage and temperature requirements. The IM insulation has to withstand the expected operating voltages between conductors, conductors (phase) and ground, and phase to phase.

The American National Standards Institute (ANSI) specifies that the insulation test voltage shall be twice the rated voltage plus 1000 V applied to the stator winding for 1 minute.

The heat produced by the winding currents and the core losses causes hot-spot temperatures that have to be limited in accordance with the thermal capability of the organic (resin) insulation used in the machine and to its chemical stability and capability to prevent conductor-to-conductor and conductor-to-ground short-circuits during IM operation.

There is continuous, but slow deterioration of the organic (resin) insulation by internal chemical reaction, contamination, and chemical interactions. Thermal degradation develops cracks in the enamel, varnish, or resin, reducing the dielectric strength of insulation.

Insulation materials for electric machines have been organised in stable temperature classes at which they can perform satisfactorily for the expected service lifetime.

The temperature classes are (again):

class A: 105°C	class F: 155°C
class B: 130°C	class G: 180°C

The main insulation components for the random-wound coil windings are the enamel insulation on the wire, the insulation between coils and ground/slot walls, and slot liner insulation and between phases (Figure 3.7).

The connections between the coils of a phase and the leads to terminal box have to be insulated. Also, the binding cord used to tie down end windings to reduce their vibration is made of insulation materials.

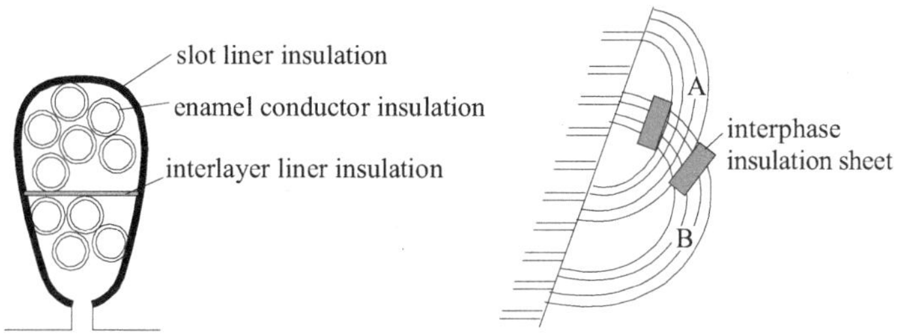

FIGURE 3.7 Random-wound coils insulation.

Random-wound IMs are built for voltages below 1 kV. The moderate currents involved can be handled by wound conductors (eventually a few in parallel) where enamel insulation is the critical component. To apply the enamel, the wire is passed through a solution of polymerisable resin and into the high-curing temperature tower where it turns into a thin, solid, and flexible coating.

3.5.1 Random-Wound IM Insulation

Several passes are required for the desired thickness (0.025 mm thick or so). There are dedicated standards that mention the tests on enamel conductors (ASTMD-1676); American Society for Testing and Materials (ASTM) standards part 39 for electric insulation test methods: solids and solidifying liquids should be considered for the scope.

Enamel wire, stretched and scraped when the coils are introduced in slots, should survive this operation without notable damage to the enamel. Some insulation varnish is applied over the enamel wire after the stator winding is completed. The varnish provides additional enamel protection against moisture, dirt, and chemical contamination and also provides mechanical support for the windings.

Slot and phase-to-phase insulation for class A temperatures is a somewhat flexible sheet material (such as cellulose paper), 0.125–0.25 mm thick, or a polyester film. In some cases, fused resin coatings are applied to stator slot walls by electrostatic attraction of polymerisable resin powder. The stator is heated to fuse and cure the resin to a smooth coating.

For high-temperature IMs (class F, H), glass cloth mica paper or asbestos treated with special varnishes are used for slot and phase-to-phase insulation. Varnishes may interact with the enamel to reduce thermal stability. Enamels and varnishes are tested separately according to ASTM (D2307, D1973, and D3145) or International Electrotechnical Commission (IEC) standards.

Model motor insulation systems (motorettes) are tested according to Institute of Electrical and Electronics Engineers (IEEE) standards for small motors.

All these insulation-accelerated life tests involve the ageing of insulation test specimens until they fail at temperatures higher than the operating temperature of the respective motor. The logarithms of the accelerated ageing times are then graphed against their reciprocal Kelvin test temperatures (Arrhenius graph). The graph is then extrapolated to the planed (reduced) temperature to predict the actual lifetime of insulation.

3.5.2 Form-Wound Windings

Form-wound windings are employed in high-power IMs. The slots are rectangular and so are the conductors. The slot filling factor increases due to this combination.

The insulation of the coil conductors (turns) is applied before inserting the coils in slots. The coils are also vacuum impregnated outside the machine. The slot insulation is made of resin-bonded mica applied as a wrapper or tape with a fibrous sheet for support (in high-voltage IMs above 1–2 kV).

Vacuum impregnation is done with polymerisable resins which are then cured to solids by heating. During the cure, the conductors may be constraint to size to enter the slot as the epoxy-type resins are sufficiently elastic for the scope.

Voltage, through partial discharges, may cause insulation failure in higher voltage IMs. Incorporating mica in the major insulation schemes solves this problem to a large degree.

A conducting paint may be applied over the slot portion of the coils to fill the space between the insulated preformed coil and slot wall, to avoid partial discharges. Lower and medium voltage coil insulation is measured in accelerated higher temperature tests (IEEE standard 275) by using the model system called formette. Formette testing is similar to motorette testing for random-wound IMs [20].

Diagnostic non-destructive tests to check the integrity and capability of large IM insulation are also standardised [20–23].

3.6 SUMMARY

- The three main materials used to build IMs are of magnetic, electric, and insulation type.
- As IM is an A.C. machine, reducing eddy current losses in its magnetic core is paramount.
- It is shown that these losses increase with the soft magnetic sheet thickness parallel to the external A.C. field.
- Soft magnetic materials (silicon steel) used in thin laminations (0.5 mm thick up to 200 Hz) have low hysteresis and eddy current losses (about or less 2 W/kg at 1 T and 60 Hz).
- Besides losses, the B–H (magnetisation) curve characterises a soft magnetic material [13].
- The magnetic permeability $\mu = B/H$ varies from (5000–8000) μ_0 at 1 T to (40–60) μ_0 at 2.0 T in modern silicon steel laminations. High permeability is essential to low magnetisation (no load) current and losses.
- High-speed IMs require frequencies above 300 Hz (and up to 800 Hz and more). Thinner silicon lamination steels with special thermal treatments are required to secure core losses in the order of 30–50 W/kg at 800 Hz and 1 T.
- 6.5% silicon steel lamination for small IMs have been proved adequate to reduce core losses by as much as 40% at 50 (60) Hz.
- Also, interspersing oriented grain (transformer) laminations (0.35 mm thick) with orthogonal orientation laminations has been shown to produce a 30%–40% reduction in core losses at 50 (60) Hz and 1 T in comparison with 0.5 mm thick nonoriented grain silicon steel used in most IMs.
- SMCs have been introduced and shown to produce lower losses than silicon steel laminations only above 300 Hz but at the expense of lower permeability ((100–200) μ_0 in general). Cold compression methods are expected to increase slot filling factor notably and thus increase the current loading. Size reduction is obtained also due to the increase of heat transmissivity through SMCs. Amorphous magnetic materials (Metglass) have been recently introduced for lower losses and high permeability (0.1 mm thick), 400 Hz, 1.6 T: 50% of core losses of silicon steel at 0.1 mm thickness.
- Electric conductors for stator windings and wound rotors are made of pure (electrical) copper.
- Cast aluminium is used for rotor cage windings up to 300 kW.
- Fabricated aluminium or copper bars and rings are used for higher power IM cage rotors. Die-cast methods for copper cage have been proposed recently [24]. Lightly ferromagnetic aluminium rotor bars have also been proposed to increase starting torque at lower starting current, while preserving rated speed performance [25].
- The rotor cage bars are not, in general, insulated from the rotor lamination core. Interbar currents may thus occur.
- The windings are made out of random-wound coils with round wire and form-wound coils for large IMs with rectangular wire.

- The windings are insulated from the magnetic core through insulation materials. Also, the conductors are enamelled to insulate one conductor from another.
- Insulation systems are classified according to temperature limits in four classes: Class A-105°C, Class B-130°C, Class F-155°C, and Class G-180°C.
- Insulation testing is thoroughly standardised as the insulation breakdown diminishes the operation life of an IM through short-circuit.
- Thinner and better insulation materials keep "surfacing" as they are crucial to better performance IMs fed from the power grid and PWM inverters.
- Better finite element (FE) and analytical methods and test procedures to appraise existing and novel materials for IMs are being proposed.

REFERENCES

1. S. Sprague, D. Jones, Using the new lamination steels database in motor design, *Proceedings of SMMA-2000 Fall Conference in Chicago*, pp. 1–12.
2. M. Birkfeld, K. A. Hempel, Calculation of the magnetic behaviour of electrical steel sheet under two dimensional excitation by means of the reluctance tensor, *IEEE Transactions on Magnetics*, Vol. MAG-33, No. 5, 1997, pp. 3757–3759.
3. I. D. Mayergoyz, *Mathematical Models for Hysteresis*, Springer Verlag, New York, 1991.
4. H. H. Saliah, D. A. Lowther, B. Forghani, A neural network model of magnetic hysteresis for computational magnetics, *IEEE Transactions on Magnetics*, Vol. MAG-33, No. 5, 1997, pp. 4146–4148.
5. M. A. Mueller et al., Calculation of iron losses from time – Stepped finite element models of cage induction machines, *Seventh International Conference on EMD*, IEE Conference Publication No. 412, Durham, UK.
6. M. Popescu, D. M. Ionel, A best-fit model of power losses in cold rolled motor lamination steel operating in a wide range of frequency and magnetisation, *IEEE Transactions on Magnetics*, Vol. MAG-43, No. 4, 2007, pp. 1753–1756.
7. A. J. Moses, N. Tutkun, Investigation of power losses in wound toroidal cores under PWM excitation, *IEEE Transactions on Magnetics*, Vol. MAG-33, No. 5, 1997, pp. 3763–3765.
8. M. Enokizono, T. Tanabe, Studies on a new simplified rotational loss tester, *IEEE Transactions on Magnetics*, Vol. 33, No. 5, 1997, pp. 4020–4022.
9. S. A. Nasar, *Handbook of Electrical Machines*, Chapter 2, p. 211, McGraw Hill Inc., New York, 1987.
10. W. L. Soong, G. B. Kliman, R. N. Johnson, R. White, J. Miller, Novel high speed induction motor for a commercial centrifugal compressor, *IEEE Transactions on Industry Applications*, Vol. IA-36, No. 3, 2000, pp. 706–713.
11. M. Machizuki, S. Hibino, F. Ishibashi, Application of 6.5% silicon steel sheet to induction motor and to magnetic properties, *EMPS*, Vol. 22, No. 1, 1994, pp. 17–29.
12. A. Boglietti, P. Ferraris, M. Lazzari, F. Profumo, Preliminary consideration about the adoption of unconventional magnetic materials and structures for motors, *IBID*, Vol. 21, No. 4, 1993, pp. 427–436.
13. M. Ibrahim, P. Pillay, Advanced testing and modelling of magnetic materials including a new method of core loss separation for electrical machines, *IEEE Transactions on Industry Applications*, Vol. IA-48, no. 5, 2012, pp. 1507–1515.
14. N. Alatawneh, P. Pillay, Design of a novel test fixture to measure rotational core losses in machine laminations, *IEEE Transactions on Industry Applications*, Vol. IA-48, no. 5, 2012, pp. 1467–1477.
15. A. J. Clerc, A. Muetze, Measurement of stator core magnetic degradation during the manufacturing process, *IEEE Transactions on Industry Applications*, Vol. IA-48, no. 4, 2012, pp. 1344–1352.
16. M. Enokizono, S. Takahashi, A. Ikariga, A measurement system for two-dimensional D.C. biased magnetic property, *KIEE Transactions on ENECS*, Vol. 2-B, No. 4, 2002, pp. 143–148.
17. D. Gay, Composite iron powder for A.C. electromagnetic applications: History and use, *Record of SMMA-2000 Fall Conference*, Chicago, Oct. 4–6, 2000.
18. M. Persson, P. Jansson, A. G. Jack, B. C. Mecrow, Soft magnetic materials–Use for electric machines, *IEEE 7th International Conference on EMD*, 1995, pp. 242–246.
19. E. A. Knoth, Motors for the 21st century, *Record of SMMA-2000 Fall Conference Chicago*, Oct. 4–6, 2000.
20. T. W. Dakin, Electric machine insulations, Chapter 13 in *Electric Machine Handbook*, Edited by S.A. Nasar, McGraw-Hill, New York, 1987.

21. P. L. Cochran, *Polyphase Induction Motors*, CRC Press, Boca Raton, FL, 1989, Chapter 11.
22. R. M. Engelmann, W. H. Middendorf (Eds.), *Handbook of Electric Machines*, Marcel Dekker Inc., New York, 1995.
23. R. Morin, R. Bartnikas, P. Menard, A three phase multistress accelerated electrical aging test facility for stator bars, *IEEE Transactions on Energy Conversion*, Vol. EC-15, No. 2, 2000, pp. 149–156.
24. R. Kimmich, M. Doppelbauer, D. T. Pefers, J. G. Cowie, E. F. Bruvn Jr., Die cast copper rotor motors via simple substitution and motor redesign for copper, *Record of ICEM-2006*, Ref. No. 358.
25. M. R. Khan, M. F. Momen, Lightly ferromagnetic rotor bars for three phase squirrel-cage induction machines, *IEEE Transactions on Industry Applications*, Vol. IA-40, No. 6, 2004, pp. 1536–1540.

4 Induction Machine Windings and Their mmfs

4.1 INTRODUCTION

As shown in Chapter 2, the slots of the stator and rotor cores of induction machines (IMs) are filled with electric conductors, insulated (in the stator) from cores, and connected in a certain way. This ensemble constitutes the windings. The primary (or the stator) slots contain a polyphase (three- or two-phase) A.C. winding. The rotor may have either a three- or two-phase winding or a squirrel cage. Here we will discuss the polyphase windings.

Designing A.C. windings means, in fact, assigning coils in the slots to various phases, establishing the direction of currents in coil sides and coil connections per phase and between phases, and finally calculating the number of turns for various coils and the conductor sizing.

We start with single pole number three-phase windings as they are most commonly used in induction motors. Then pole-changing windings are also treated in some detail. Such windings are used in wind generators or in doubly fed variable speed configurations. Two-phase windings are given then special attention. Finally, squirrel cage winding magnetomotive forces (mmfs) are analysed.

Keeping in mind that A.C. windings are a complex subject having books dedicated to it [1,2], we will treat here first its basics. Then we introduce new topics such as "pole amplitude modulation", "polyphase symmetrisation" [3,4], "intersperse windings" [5], "simulated annealing" [6], and "the three equation principle" [7] for pole changing. These are new ways to produce A.C. windings for special applications (for pole changing or mmf chosen harmonics elimination). Finally, fractional multilayer three-phase windings with reduced harmonics content are treated in some detail [8,9]. The present chapter is structured to cover both the theory and case studies of A.C. winding design, classifications, and mmf harmonic analysis.

4.2 THE IDEAL TRAVELLING MMF OF A.C. WINDINGS

The primary (A.C. fed) winding is formed by interconnecting various conductors in slots around the circumferential periphery of the machine. As shown in Chapter 2, we may have a polyphase winding on the so-called wound rotor. Otherwise the rotor may have a squirrel cage in its slots. The objective with polyphase A.C. windings is to produce a pure travelling mmf, through proper feeding of various phases with sinusoidal symmetrical currents. And all this mmf (F_{s1} (x, t)) in order to produce constant (rippleless) torque under steady state is

$$F_{s1}(x,t) = F_{s1m}\cos\left(\frac{\pi}{\tau}x - \omega_1 t - \theta_0\right) \tag{4.1}$$

where

x – coordinate along stator bore periphery
τ – spatial half period of mmf ideal wave
ω_1 – angular frequency of phase currents
θ_0 – angular position at $t = 0$

We may decompose (4.1) into two terms

$$F_{s1}(x,t) = F_{s1m}\left[\cos\left(\frac{\pi}{\tau}x - \theta_0\right)\cos\omega_1 t + \sin\left(\frac{\pi}{\tau}x - \theta_0\right)\sin\omega_1 t\right] \tag{4.2}$$

Equation (4.2) has a special physical meaning. In essence, there are now two mmfs at standstill (fixed) with sinusoidal spatial distribution and sinusoidal currents. The space angle lag and the time angle between the two mmfs is $\pi/2$. This suggests that a pure travelling mmf may be produced with two symmetrical windings $\pi/2$ shifted in time (Figure 4.1a). This is how the two-phase IM evolved.

Similarly, we may decompose (4.1) into three terms

$$F_{s1}(x,t) = \frac{2}{3}F_{s1m}\left[\cos\left(\frac{\pi}{\tau}x - \theta_0\right)\cos\omega_1 t + \cos\left(\frac{\pi}{\tau}x - \theta_0 - \frac{2\pi}{3}\right)\cos\left(\omega_1 t - \frac{2\pi}{3}\right)\right.$$
$$\left. + \cos\left(\frac{\pi}{\tau}x - \theta_0 + \frac{2\pi}{3}\right)\cos\left(\omega_1 t + \frac{2\pi}{3}\right)\right] \tag{4.3}$$

Consequently, three mmfs (single-phase windings) at standstill (fixed) with sinusoidal spatial (x) distribution and departured in space by $2\pi/m$ radians, with sinusoidal symmetrical currents – equal amplitude, $2\pi/3$ radians time lag angle – are also able to produce a travelling mmf (Figure 4.1b).

In general, m phases with a phase lag (in time and space) of $2\pi/m$ can produce a travelling wave. Six phases (m = 6) would be a rather practical case besides m = 3 phases. Recently, five and seven phases have been introduced for fault-tolerant IM drives. The number of mmf electrical periods per one revolution is called the number of pole pairs p_1.

$$p_1 = \frac{\pi D}{2\tau}; \quad 2p_1 = 2,4,6,8,... \tag{4.4}$$

where D is the stator bore diameter.

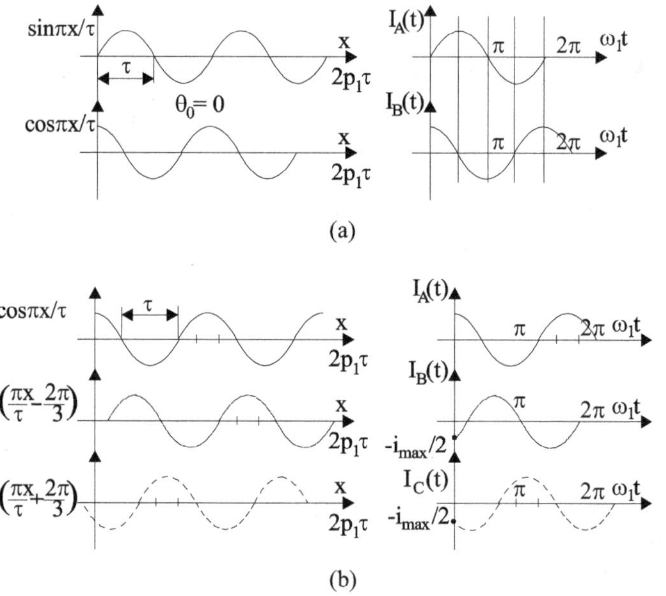

FIGURE 4.1 Ideal multiphase mmfs: (a) two-phase and (b) three-phase machines.

It should be noted that, for $p_1 > 1$, according to (4.4), the electrical angle α_e is p_1 times larger than the mechanical angle α_g

$$\alpha_e = p_1 \alpha_g \tag{4.5}$$

A sinusoidal distribution of mmfs (ampere-turns) would be feasible only with the slotless machine and windings placed in the airgap. Such a solution is hardly practical for IMs because the magnetisation of a large total airgap would mean very large magnetisation mmf and, consequently, low power factor and efficiency. It would also mean problems with severe mechanical stress acting directly on the electrical conductors of the windings.

In practical IMs, the coils of the windings are always placed in slots of various shapes (Chapter 2).

The total number of slots per stator N_s should be divisible to the number of phases m so that

$$N_s / m = \text{integer} \tag{4.6}$$

A parameter of great importance is the number of slots per pole per phase q:

$$q = \frac{N_s}{2p_1 m} \tag{4.7}$$

The number of slots/pole/phase q may be an integer ($q = 1, 2, \ldots 12$) or fraction.

In most IMs, q is an integer to provide complete (pole to pole) symmetry for the winding.

The windings are made of coils. Lap and wave coils are used for IMs (Figure 4.2).

The coils may be placed in slots in one (Figure 4.3a) or two layers (Figure 4.3b).

Single-layer windings imply full pitch ($y = \tau$) coils to produce an mmf fundamental with pole pitch τ.

Double-layer windings also allow chorded (or fractional pitch) coils ($y < \tau$) such that the end connections of coils are shortened, and thus, copper loss is reduced. Moreover, as shown later in this chapter, the space harmonics content of mmf may be reduced by chorded coils. Unfortunately, so is the fundamental mmf.

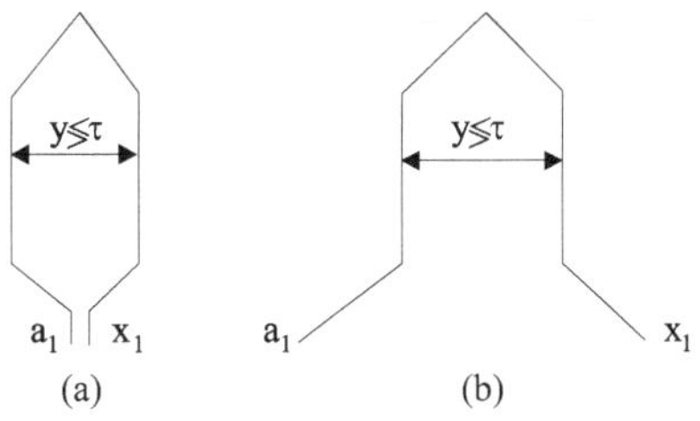

FIGURE 4.2 Lap (a) and wave (b) single turn (bar) coils.

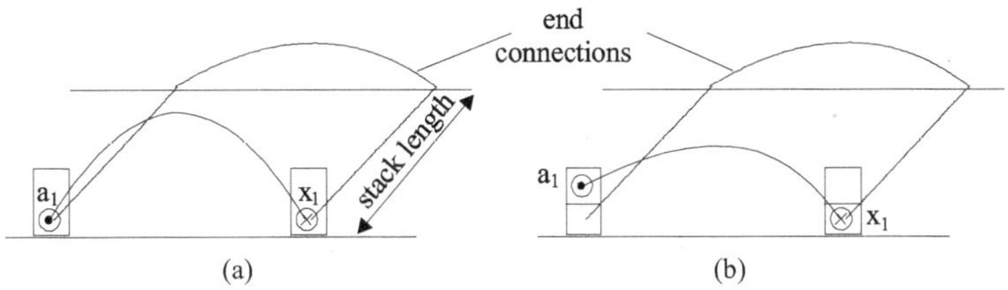

FIGURE 4.3 Single-layer (a) and double-layer (b) coils (windings).

4.3 A PRIMITIVE SINGLE-LAYER WINDING

Let us design a four-pole ($2p_1 = 4$) three-phase single-layer winding with q = 1 slots/pole/phase. $N_s = 2p_1qm = 2{\cdot}2{\cdot}1{\cdot}3 = 12$ slots in all.

From the previous paragraph, we infer that for each phase, we have to produce an mmf with $2p_1 = 4$ poles (semiperiods). To do so, for a single-layer winding, the coil pitch $y = \tau = N_s/2p_1 = 12/4 = 3$ slot pitches.

For 12 slots, there are 6 coils in all. That is, two coils per phase to produce four poles. It is now obvious that the four-phase A slots are $y = \tau = 3$ slot pitches apart. We may start in slot 1 and continue with slots 4, 7, and 10 for phase A (Figure 4.4a).

Phases B and C are placed in slots by moving 2/3 of a pole pitch (two-slot pitches in our case) to the right. All coils/phases may be connected in series to form one current path (a = 1) or they may be connected in parallel to form two current paths in parallel (a = 2). The number of current paths a is obtained in general by connecting part of coils in series and then the current paths in parallel such that all the current paths are symmetric. Current paths in parallel serve to reduce wire gauge (for given output phase current) and, as shown later, to reduce uncompensated magnetic pull between rotor and stator in the presence of rotor eccentricity.

If the slot is considered infinitely thin (or the slot opening $b_{os} \approx 0$), the mmf (ampere-turns) jumps up, as expected, by $n_{cs}{\cdot}i_{A,B,C}$, along the middle of each slot.

For the time being, let us consider $b_{os} = 0$ (a virtual closed slot).

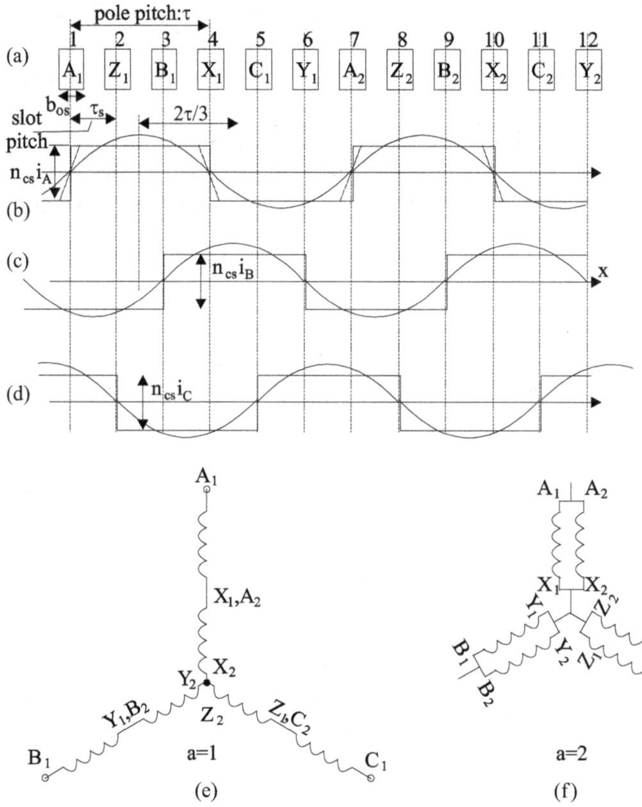

FIGURE 4.4 Single-layer three-phase winding for $2p_1 = 4$ poles and q = 1 slots/pole/phase: (a) slot/phase allocation; (b–d) ideal mmf distribution for the three phases when their currents are maximum; (e) star series connection of coils/phase; (f) parallel connection of coils/phase.

The rectangular mmf distribution may be decomposed into harmonics for each phase. For phase A, we simply obtain

$$F_{A1}(x,t) = \frac{2}{\pi} \cdot \frac{n_{cs} I\sqrt{2} \cos \omega_1 t}{\nu} \cos \frac{\nu \pi x}{\tau} \quad (4.8)$$

For the fundamental, $\nu = 1$, we obtain the maximum amplitude. The higher the order of the harmonic, the lower its amplitude in (4.8).

While in principle such a primitive machine works, the harmonics content is too rich.

It is only intuitive that if the number of steps in the rectangular ideal distribution would be increased, the harmonics content would be reduced. This goal could be met by increasing q or (and) via chording the coils in a two-layer winding. Let us then present such a case.

4.4 A PRIMITIVE TWO-LAYER CHORDED WINDING

Let us still consider $2p_1 = 4$ poles, m = 3 phases but increase q from 1 to 2. Thus, the total number of slots $N_s = 2p_1qm = 2{\cdot}2{\cdot}2{\cdot}3 = 24$.

The pole pitch τ measured in slot pitches is $\tau = N_s/2p_1 = 24/4 = 6$. Let us reduce the coil throw (span) y such that $y = 5\tau/6$.

We still have to produce four poles. Let us proceed as in the previous paragraph but only for one layer, disregarding the coil throw.

In a two-layer winding, the total number of coils is equal to the number of slots. So in our case, there are $N_s/m = 24/3$ coils per phase. Also, there are eight slots occupied by one phase in each layer, four with inward and four with outward current direction. With each layer, each phase has to produce four poles in our case. So slots 1,2; 7′,8′; 13,14; 19′,20′ in layer one belong to phase A. The superscript prime refers to outward current direction in the coils. The distance between neighbouring slot groups of each phase in one layer is always equal to the pole pitch to preserve the mmf distribution half period (Figure 4.5).

Notice that in Figure 4.5, for each phase, the second layer is displaced to the left by $\tau - y = 6 - 5 = 1$ slot pitch with respect to the first layer. Also, after two poles, the situation repeats itself. This is typical for a fully symmetrical winding.

Each coil has one side in one layer, say in slot 1, and the second one in slot $y + 1 = 5 + 1 = 6$. In this case, all coils are identical, and thus, the end connections occupy less axial room and are shorter due to chording. Such a winding is typical with random wound coils made of round magnetic wire.

For this case, we explore the mmf ideal resultant distribution for the situation when the current in phase A is maximum ($i_A = i_{max}$). For symmetrical currents, $i_B = i_C = -i_{max}/2$ (Figure 4.1b).

Each coil has n_c conductors and, with zero slot opening, the mmf jumps up at every slot location by the total number of ampere-turns. Notice that half the slots have coils of same phase while the other half accommodate coils of different phases.

The mmf of phase A, for maximum current value (Figure 4.5b), has two steps per polarity as q = 2. It had only one step for q = 1 (Figure 4.4). Also, the resultant mmf has three unequal steps per polarity ($q + \tau - y = 2 + 6 - 5 = 3$). It is indeed closer to a sinusoidal distribution. Increasing q and using chorded coils reduces the space harmonics content of the mmf.

Also shown in Figure 4.5 is the movement by $2\tau/3$ (or $2\pi/3$ electrical radians) of the mmf maximum when the time advances with $2\pi/3$ electrical (time) radians or T/3 (T is the time period of sinusoidal currents).

4.5 THE MMF HARMONICS FOR INTEGER q

Using the geometrical representation in Figure 4.5, it becomes fairly easy to decompose the resultant mmf in space harmonics noticing the step form of the distributions.

Proceeding with phase A, we obtain (by some extrapolation for integer q),

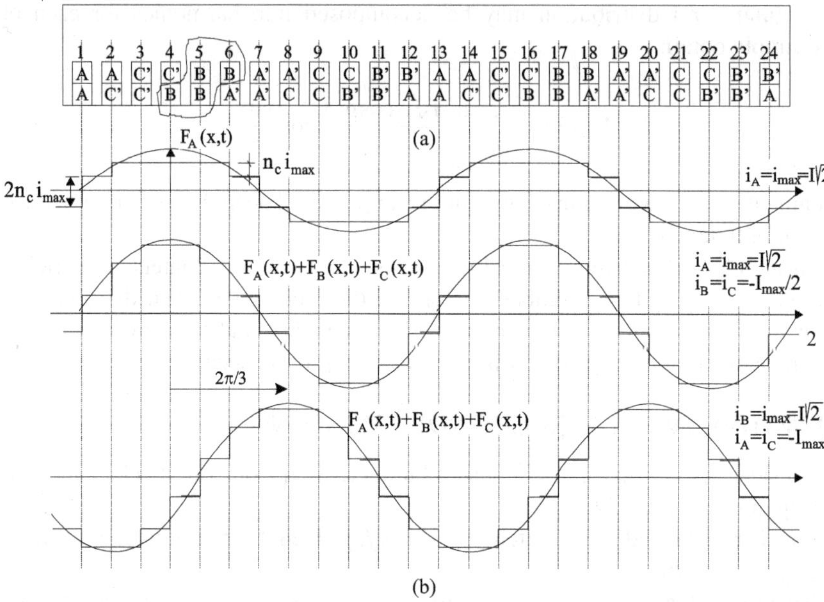

FIGURE 4.5 Two-layer winding for Ns = 24 slots, 2 p_1 = 4 poles, $y/\tau = 5/6$ (a) slot/phase allocation and (b) mmfs distribution.

$$F_{A1}(x,t) = \frac{2}{\pi} n_c q I\sqrt{2} K_{q1} K_{y1} \cos\frac{\pi}{\tau} x \cos\omega_1 t \tag{4.9}$$

with

$$K_{q1} = \sin\pi/6/\left(q\sin\pi/6q\right) \le 1; \quad K_{y1} = \sin\frac{\pi}{2} y/\tau \le 1 \tag{4.10}$$

K_{q1} is known as the zone (or spread) factor and K_{y1} the chording factor. For q = 1, $K_{q1} = 1$, and for full pitch coils, $y/\tau = 1$, $K_{y1} = 1$, as expected.

To keep the winding fully symmetric, $y/\tau \ge 2/3$. This way all poles have similar slot/phase allocation.

Assuming now that all coils per phase are in series, the number of turns per phase W_1 is

$$W_1 = 2p_1 q n_c \tag{4.11}$$

With (4.11), Equation (4.9) becomes

$$F_{A1}(x,t) = \frac{2}{\pi P_1} W_1 I\sqrt{2} K_{q1} K_{y1} \cos\frac{\pi}{\tau} x \cos\omega_1 t \tag{4.12}$$

For three phases, we obtain

$$F_1(x,t) = F_{1m}\cos\left(\frac{\pi}{\tau}x - \omega_1 t\right) \tag{4.13}$$

with

$$F_{1m} = \frac{3W_1 I\sqrt{2}K_{q1}K_{y1}}{\pi p_1} \quad \text{(amperturns per pole)} \tag{4.14}$$

The derivative of pole mmf with respect to position x is called linear current density (or current sheet) A (in A/m)

$$A_1(x,t) = \frac{\partial F_1(x,t)}{\partial x} = A_{1m}\sin\left(-\frac{\pi}{\tau}x + \omega_1 t\right) \tag{4.15}$$

$$A_{1m} = \frac{3\sqrt{2}W_1 I\sqrt{2}K_{q1}K_{y1}}{P_1\tau} = \frac{\pi}{\tau}F_{1m} \tag{4.16}$$

A_{1m} is the maximum value of the current sheet and is also identified as current loading. The current loading is a design parameter (constant) $A_{1m} \approx 5000–50{,}000$ A/m, in general, for IMs in the power range of kilowatts to megawatts. It is limited by the temperature rise mainly and increases with machine torque (size).

The harmonics content of the mmf is treated in a similar manner to obtain

$$F(x,t) = \frac{3W_1 I\sqrt{2}K_{q\nu}K_{y\nu}}{\pi p_1 \nu}\cdot$$

$$\cdot\left[K_{BI}\cos\left(\frac{\nu\pi}{\tau}x - \omega_1 t - (\nu-1)\frac{2\pi}{3}\right) - K_{BII}\cos\left(\frac{\nu\pi}{\tau}x + \omega_1 t - (\nu+1)\frac{2\pi}{3}\right)\right] \tag{4.17}$$

with

$$K_{q\nu} = \frac{\sin \nu\pi/6}{q\sin \nu\pi/6q}; \quad K_{y\nu} = \sin\left(\frac{\nu\pi y}{2\tau}\right) \tag{4.18}$$

$$K_{BI} = \frac{\sin(\nu-1)\pi}{3\sin(\nu-1)\pi/3}; \quad K_{BII} = \frac{\sin(\nu+1)\pi}{3\sin(\nu+1)\pi/3} \tag{4.19}$$

Due to mmf full symmetry (with q = integer), only odd harmonics occur. For three-phase star connection, 3K harmonics may not occur as the current sum is zero and their phase shift angle is $3K \cdot 2\pi/3 = 2\pi K$.

We are left with harmonics $\nu = 3K \pm 1$; that is, $\nu = 5, 7, 11, 13, 17, \ldots$

We should notice in (4.19) that for $\nu_d = 3K + 1$, $K_{BI} = 1$ and $K_{BII} = 0$. The first term in (4.17) represents, however, a direct (forward) travelling wave as, for a constant argument under cosines, we do obtain

$$\left(\frac{dx}{dt}\right) = \frac{\omega_1\tau}{\pi\nu} = \frac{2\tau f_1}{\nu}; \quad \omega_1 = 2\pi f_1 \tag{4.20}$$

On the contrary, for $\nu = 3K - 1$, $K_{BI} = 0$ and $K_{BII} = 1$. The second term in (4.17) represents a backward travelling wave. For a constant argument under cosine, after a time derivative, we have

$$\left(\frac{dx}{dt}\right)_{\nu=3K-1} = \frac{-\omega_1\tau}{\pi\nu} = \frac{-2\tau f_1}{\nu} \quad (4.21)$$

We should also notice that the travelling speed of mmf space harmonics, due to the placement of conductors in slots, is ν times smaller than that of the fundamental (ν = 1).

The space harmonics of the mmf just investigated are due both to the placement of conductors in slots and to the placement of various phases as phase belts under each pole. In our case, the phase belt spread is π/3 (or one-third of a pole). There are also two-layer windings with 2π/3 phase belts, but the π/3 (60°) phase belt windings are more practical.

So far, the slot opening influences on the mmf stepwise distribution have not been considered. It will be done later in this chapter.

Notice that the product of zone (spread or distribution) factor $K_{q\nu}$ and the chording factor $K_{y\nu}$ is called the stator winding factor $K_{w\nu}$.

$$K_{w\nu} = K_{q\nu}K_{y\nu} \quad (4.22)$$

As in most cases, only the mmf fundamental (ν = 1) is useful, reducing most harmonics and cancelling some is a good design attribute. Chording the coils to cancel $K_{y\nu}$ leads to

$$\sin\left(\frac{\nu\pi y}{2\tau}\right) = 0; \quad \frac{\nu\pi y}{2\tau} = n\pi; \quad \frac{y}{\tau} > \frac{2}{3} \quad (4.23)$$

As the mmf harmonic amplitude (4.17) is inversely proportional to the harmonic order, it is almost standard to reduce (cancel) the fifth harmonic (ν = 5) by making n = 2 in (4.23).

$$\frac{y}{\tau} = \frac{4}{5} \quad (4.23a)$$

In reality, this ratio may not be realised with an integer q (q = 2) and thus y/τ = 5/6 or 7/9 is the practical solution which keeps the fifth mmf harmonic low. Chording the coils also reduces K_{y1}. For y/τ = 5/6, $\sin\frac{\pi}{2}\frac{5}{6} = 0.966 < 1.0$, but a 4% reduction in the mmf fundamental is worth the advantages of reducing the coil end-connection length (lower copper losses) and a drastic reduction of fifth mmf harmonic.

As will be shown later in the book, mmf harmonics produce parasitic torques, radial forces, additional core and winding losses, noise, and vibration.

Example 4.1

Let us consider an IM with the following data: stator core diameter D = 0.15 m, number of stator slots $N_s = 24$, number of poles $2p_1 = 4$, y/τ = 5/6, two-layer winding; slot area $A_{slot} = 100\,mm^2$, total slot fill factor $K_{fill} = 0.5$, current density $j_{Co} = 5$ A/mm², number of turns per coil $n_c = 25$. Let us calculate

a. The rated current (RMS value), wire gauge
b. The pole pitch τ
c. K_{q1} and K_{y1}, K_{w1}
d. The amplitude of the mmf F_{1m} and current sheet A_{1m}
e. K_{q7}, K_{y7}, and F_{7m} (ν = 7)

Solution

Part of the slot is filled with insulation (conductor insulation, slot wall insulation, and layer insulation) while there is some room between round wires. The total filling factor of slot takes care of all these aspects. The mmf per slot is

$$2n_cI = A_{slot} \cdot K_{fill} \cdot J_{Co} = 100 \cdot 0.5 \cdot 5 = 250 \text{ Aturns}$$

As nc = 25; I = 250/(2·25) = 5A (rms).

The wire gauge d_{Co} is

$$d_{Co} = \sqrt{\frac{4}{\pi}\frac{I}{J_{Co}}} = \sqrt{\frac{4}{\pi}\frac{5}{5}} = 1.128\,\text{mm}$$

The pole pitch τ is

$$\tau = \frac{\pi D}{2p_1} = \frac{\pi \cdot 0.15}{2 \cdot 2} = 0.11775 \text{ m}$$

From (4.10),

$$K_{q1} = \frac{\sin\frac{\pi}{6}}{2\sin\frac{\pi}{6 \cdot 2}} = 0.9659$$

$$K_{y1} = \sin\frac{\pi}{2}\cdot\frac{5}{6} = 0.966; \quad K_{w1} = K_{q1}K_{y1} = 0.9659 \cdot 0.966 = 0.933$$

The mmf fundamental amplitude, (from 4.14), is

$$W_1 = 2P_1qn_c = 2 \cdot 2 \cdot 2 \cdot 25 = 200\,\text{turns/phase}$$

$$F_{1m} = \frac{3W_1I\sqrt{2}K_{w1}}{\pi p} = \frac{3 \cdot 200 \cdot 5\sqrt{2} \cdot 0.933}{\pi \cdot 2} = 628\,\text{turns/pole}$$

From (4.16), the current sheet (loading) A_{1m} is

$$A_{1m} = F_{1m}\frac{\pi}{\tau} = 628 \cdot \frac{\pi}{0.15} = 13155.3\,\text{Aturns/m}$$

From (4.18),

$$K_{q7} = \frac{\sin(7\pi/6)}{2\sin(7\pi/6 \cdot 2)} = -0.2588; \qquad K_{y7} = \sin\frac{7\pi}{2}\cdot\frac{5}{6} = 0.2588$$

$$K_{w7} = -0.2588 \cdot 0.2588 = -0.066987$$

From (4.18),

$$F_{7m} = \frac{3W_1I\sqrt{2}K_{q7}K_{y7}}{\pi p_1 7} = \frac{3 \cdot 200 \cdot 5\sqrt{2} \cdot 0.066987}{\pi \cdot 2 \cdot 7} = 6.445\,\text{Aturns/pole}$$

This is less than 1% of the fundamental F_{1m} = 628 Aturns/pole.

Note. It may be shown that for 120° phase belts [10], the distribution (spread) factor K_{qv} is

$$K_{qv} = \frac{\sin \nu(\pi/3)}{q\sin(\nu\pi/3\cdot q)} \tag{4.24}$$

For some cases, q = 2 and ν = 1, we find $K_{q1} = \sin \pi/3 = 0.867$.

This is much smaller than 0.9659, the value obtained for the 60° phase belt, which explains in part why the latter case is preferred in practice.

Now that we introduced ourselves to A.C. windings through two case studies, let us proceed and develop general rules to design practical A.C. windings.

4.6 RULES FOR DESIGNING PRACTICAL A.C. WINDINGS

The A.C. windings for induction motors are usually built in one or two layers.

The basic structural element is represented by coils. We already pointed out (Figure 4.2) that there may be lap and wave coils. This is the case for single turn (bar) coils. Such coils are made of continuous bars (Figure 4.6a) for open slots or from semibars bent and welded together after insertion in semiclosed slots (Figure 4.6b).

These are preformed coils generally suitable for large machines.

Continuous bar coils may also be made from a few elementary conductors in parallel to reduce the skin effect to acceptable levels.

On the other hand, round-wire, mechanically flexible coils forced "wire by wire" into semiclosed slots are typical for low-power IMs.

Such coils may have various shapes as shown in Figure 4.7.

A few remarks are in order.

- Wire coils for single-layer windings, typical for low-power induction motors (kW range and $2p_1 = 2$ pole), have in general wave shape
- Coils for single-layer windings are always full pitch as an average
- The coils may be concentrated or identical
- The main concern should be to produce equal resistance and leakage inductance per phase
- From this point of view, rounded concentrated or chain-shape identical coils are to be preferred for single-layer windings.

Double-layer winding coils for low-power IMs are trapezoidal and round-shaped wire types (Figure 4.8a and b).

For large power motors, preformed multibar (rectangular wire) (Figure 4.8c) or unibar coils (Figure 4.6) are used.

Now, to return to the basic rules for A.C. winding design, let us first remember that they may be integer q or fractional q (q = a + b/c) windings with the total number of slots $N_s = 2p_1qm$. The number

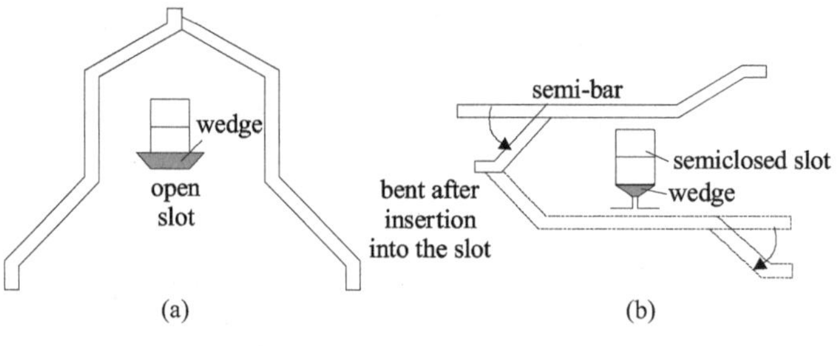

FIGURE 4.6 Bar coils: (a) continuous bar and (b) semibar.

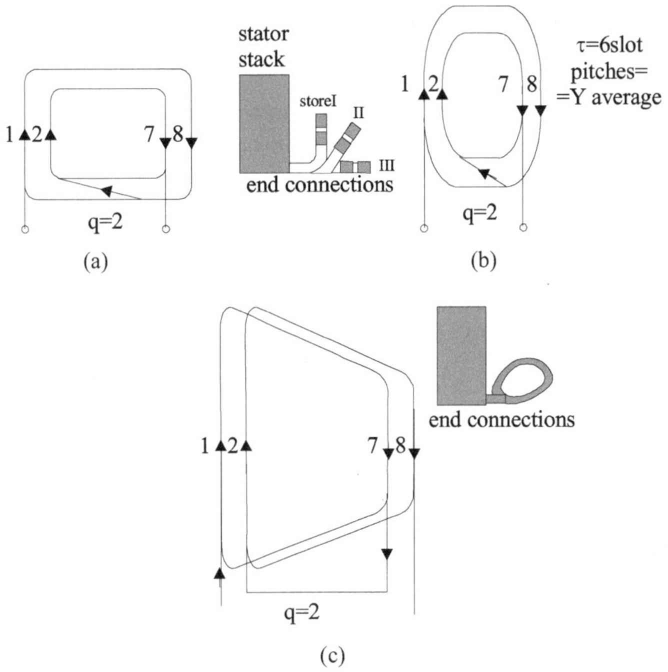

FIGURE 4.7 Full pitch coil groups/phase/pole – for q = 2 – for single-layer A.C. windings: (a) with concentrated rectangular shape coils and 2 (3) store end connections; (b) with concentrated rounded coils; and (c) with chain shape coils.

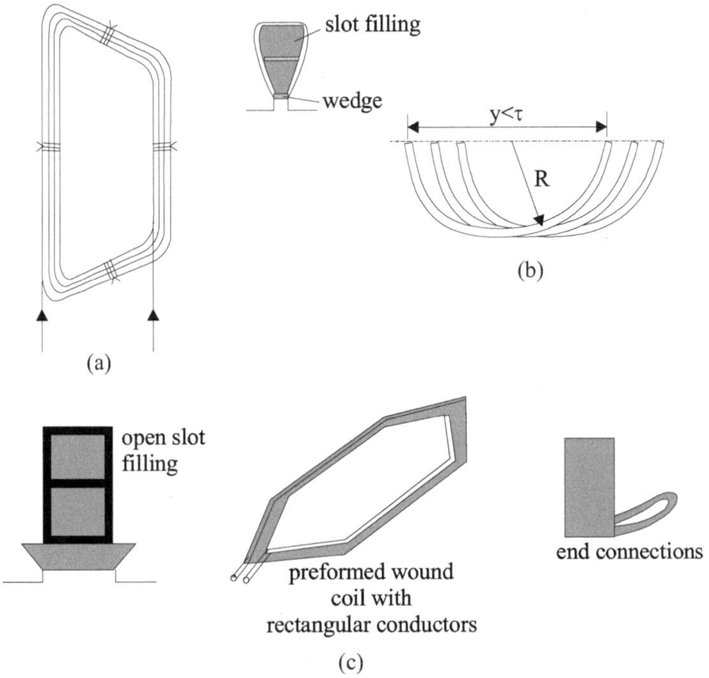

FIGURE 4.8 Typical coils for two-layer A.C. windings: (a) trapezoidal flexible coil (round wire); (b) rounded flexible coil (rounded wire); and (c) preformed wound coil (of rectangular wire) for open slots.

of slots per pole could be only an integer. Consequently, for a fractional q, the latter is different and integer for a phase under different poles. Only the average q is fractional. Single-layer windings are built only with an integer q.

As one coil sides occupy 2 slots, it means that $N_s/2m$ = an integer (m – number of phases; m = 3 in our case) for single-layer windings. The number of inward current coil sides is evidently equal to the number of outward current coil sides.

For two-layer windings, the allocation of slots per phase is performed in one (say upper) layer. The second layer is occupied "automatically" by observing the coil pitch whose first side is in one layer and the second one in the second layer. In this case, it is sufficient to have N_s/m = an integer.

A pure travelling stator mmf (4.13), with an open rotor winding and a constant airgap (slot opening effects are neglected), when the stator and iron core permeability is infinite, will produce a no-load ideal flux density in the airgap as

$$B_{g10}(x,t) = \frac{\mu_0 F_{1m}}{g}\cos\left(\frac{\pi}{\tau}x - \omega_1 t\right) \tag{4.25}$$

according to Biot–Savart law.

This flux density will self-induce sinusoidal emfs in the stator windings.

The emfs induced in coil sides placed in neighbouring slots are thus phase shifted by α_{es}

$$\alpha_{es} = \frac{2\pi P_1}{N_s} \tag{4.26}$$

The number of slots with emfs in phase, t, is

$$t = \text{greatest common divisor}\ (N_s, p_1) = \text{g.c.d.}\ (N_s, p_1) \le p_1 \tag{4.27}$$

Thus, the number of slots with emfs of distinct phase is N_s/t. Finally, the phase shift between neighbouring distinct slot emfs α_{et} is

$$\alpha_{et} = \frac{2\pi t}{N_s} \tag{4.28}$$

If $\alpha_{es} = \alpha_{et}$, that is $t = p_1$, the counting of slots in the emf phasor star diagram is the real one in the machine.

Now let us reconsider the case of a single-layer winding with $N_s = 24$, $2p_1 = 4$. In this situation,

$$\alpha_{es} = \frac{2\pi P_1}{N_s} = \frac{2\pi \cdot 2}{24} = \frac{\pi}{6} \tag{4.29}$$

$$t = \text{g.c.d.}\ (N_s, p_1) = \text{g.c.d.}(24,2) = 2 = p_1 \tag{4.30}$$

So the number of distinct emfs in slots is $N_s/t = 24/2 = 12$ and their phase shift $\alpha_{et} = \alpha_{es} = \pi/6$. So their counting (order) is the natural one (Figure 4.9).

The allocation of slots to phases to produce a symmetric winding is to be done as follows for

- Single-layer windings
  - Build up the slot emf phasor star based on calculating α_{et}, α_{es}, and N_s/t distinct arrows counting them in natural order after α_{es}.

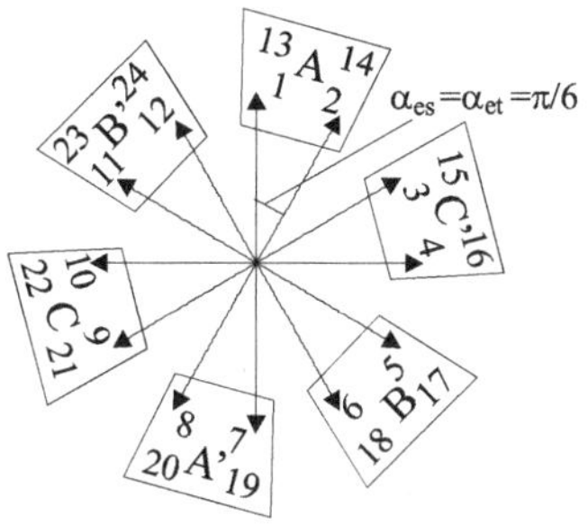

FIGURE 4.9 The star of slot emf phasors for a single-layer winding with $q=2$, $2p_1=3$, $m=3$, and $N_s=24$ slots.

- Choose randomly $N_s/2m$ successive arrows to make up the inward current slots of phase A (Figure 4.9).
- The outward current arrows of phase A are phase shifted by π radians with respect to the inward current ones.
- By skipping $N_s/2m$ slots from phase A, we find the slots of phase B.
- Skipping further $N_s/2m$ slots from phase B, we find the slots of phase C.

- Double-layer windings
  - Build up the slot emf phasor star as for single-layer windings.
  - For each phase choose N_s/m arrows and divide them into two groups (one for inward and one for outward current sides) such that they are as opposite as possible.
  - The same routine is repeated for the other phases providing a phase shift of $2\pi/3$ radians between phases.

It is well understood that the above rules are also valid for the case of fractional q. Fractional q windings are built only in two layers and small q to reduce the order of first-slot harmonic.

- Placing the coils in slots

For single-layer full-pitch windings, the inward and outward side coils occupy entirely the allocated slots from left to right for each phase. There will be $N_s/2m$ coils/phase.

The chorded coils of double-layer windings, with a pitch y ($2\tau/3 \le y < \tau$ for integer q and single-pole count windings), are placed from left to right for each phase, with one side in one layer and the other side in the second layer. They are connected observing the inward (A, B, C) and outward (A′, B′, C′) directions of currents on their sides.

- Connecting the coils per phase

The $N_s/2m$ coils per phase for single-layer windings and the N_s/m coils per phase for double-layer windings are connected in series (or series/parallel) such that for the first layer, the inward/outward directions are observed. With all coils/phase in series, we obtain a single current path ($a = 1$). We may obtain "a" current paths if the coils from $2p_1/a$ poles are connected in series and, then, the "a " chains in parallel.

Example 4.2

Let us design a single-layer winding with $2p_1 = 2$ poles, $q = 4$, $m = 3$ phases

Solution

The angle α_{es} (4.26), t (4.27), and α_{et} (4.28) are

$$N_s = 2p_1qm = 24; \quad \alpha_{es} = \frac{2\pi p_1}{N_s} = \frac{2\pi \cdot 1}{24} = \frac{\pi}{12}$$

$t = \text{g.c.d.}(N_s, p_1) = \text{g.c.d.}(24,1) = 1$

$$\alpha_{et} = \frac{2\pi t}{N_s} = \frac{2\pi \cdot 1}{24} = \frac{\pi}{12}$$

Also the count of distinct arrows of slot emf star $N_s/t = 24/1 = 24$.

Consequently, the number of arrows in the slot emf star is 24 and their order is the real (geometrical) one (1, 2, …, 24) – Figure 4.10.

Making use of Figure 4.10, we may, thus, allocate the slots to phases as shown in Figure 4.11.

Example 4.3

Let us consider a double-layer three-phase winding with $q = 3$, $2p_1 = 4$, $m = 3, (N_s = 2p_1qm = 36\,\text{slots})$, chorded coils $y/\tau = 7/9$ with $a = 2$ current paths

Solution

Proceeding as explained above, we may calculate α_{es}, t, and α_{et}

$$\alpha_{es} = \frac{2\pi p_1}{N_s} = \frac{2\pi \cdot 2}{36} = \frac{\pi}{9}$$

$t = \text{g.c.d.}(36, 2) = 2$

$$\alpha_{et} = \frac{2\pi t}{N_s} = \frac{2\pi \cdot 2}{36} = \frac{\pi}{9} = \alpha_{es} \quad N_s/t = 36/2 = 18$$

There are 18 distinct arrows in the slot emf star as shown in Figure 4.12.

The winding layout is shown in Figure 4.13. We should notice that the second-layer slot allocation is lagging by $\tau - y = 9 - 7 = 2$ slots from the first-layer allocation.

Phase A produces four fully symmetric poles. Also, the current paths are fully symmetric. Equipotential points of two current paths U – U′, V – V′, and W – W′ could be connected to each other to handle circulating currents due to, say, rotor eccentricity.

Having two current paths, the current in the coils is half the current at the terminals. Consequently, the wire gauge of conductors in the coils is smaller, and thus, the coils are more flexible and easier to handle.

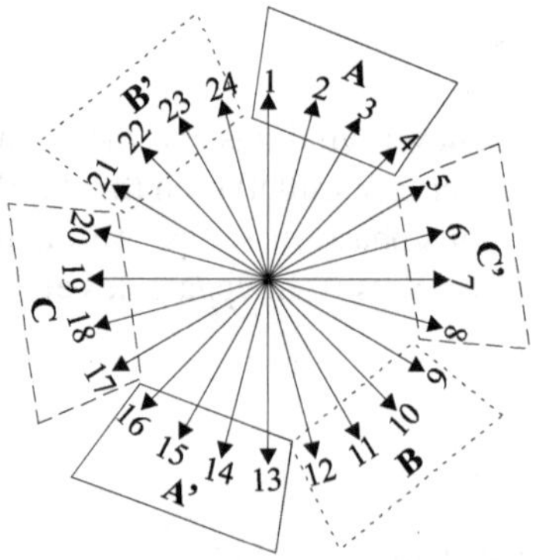

FIGURE 4.10 The star of slot emf phasors for a single-layer winding: $q = 1$, $2p_1 = 2$, $m = 3$, and $N_s = 24$.

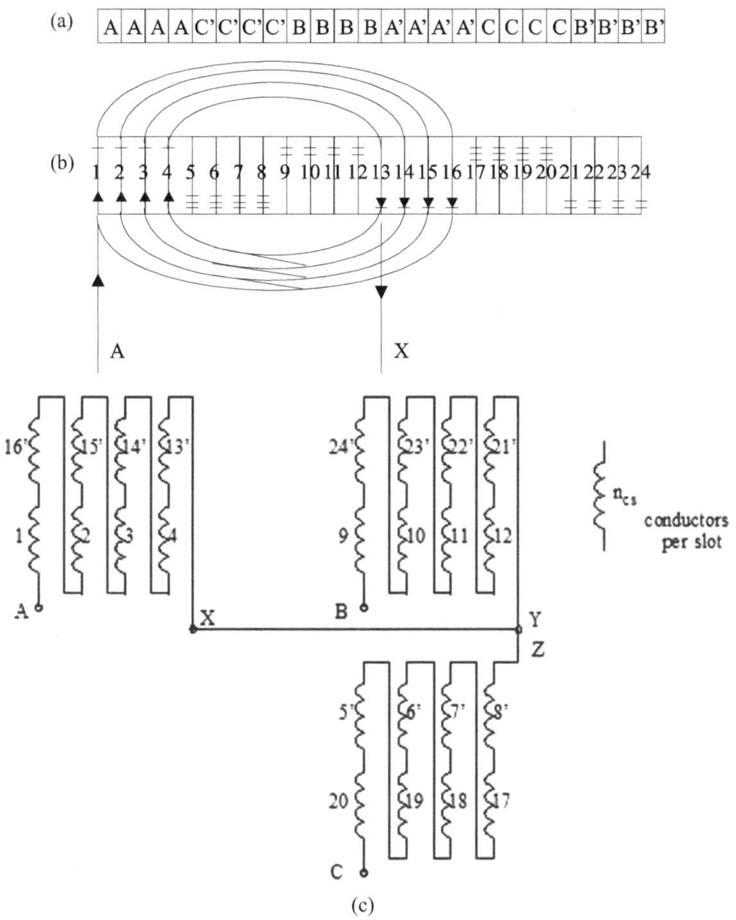

FIGURE 4.11 Single-layer winding layout: (a) slot/phase allocation; (b) rounded coils of phase A; and (c) coils per phase.

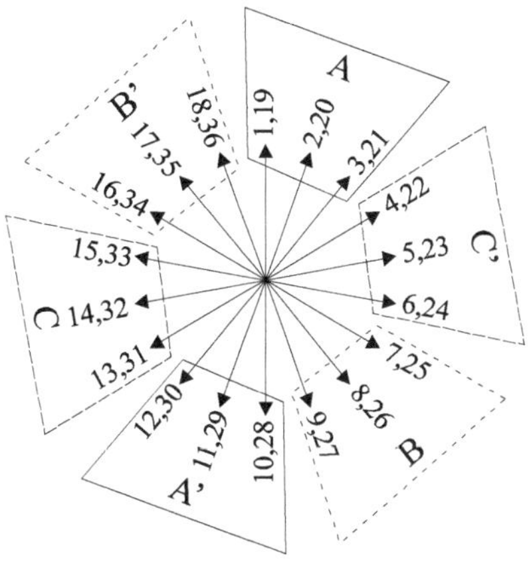

FIGURE 4.12 The star of slot emf phasors for a double-layer winding (one layer shown) with $2p_1 = 4$ poles, $q = 3$ slots/pole/phase, $m = 3$, $N_s = 36$.

	1	2	3	4	5	6	7	8	9	10	11	12	13	14	15	16	17	18	19	20	21	22	23	24	25	26	27	28	29	30	31	32	33	34	35	36
layer1	A	A	A	C'	C'	C'	B	B	B	A'	A'	A'	C	C	C	B'	B'	B'	A	A	A	C'	C'	C'	B	B	B	A'	A'	A'	C	C	C	B'	B'	B'
layer2	A	C'	C'	C'	B	B	B	A'	A'	A'	C	C	C	B'	B'	B'	A	A	A	C'	C'	C'	B	B	B	A'	A'	A'	C	C	C	B'	B'	B'	A	A

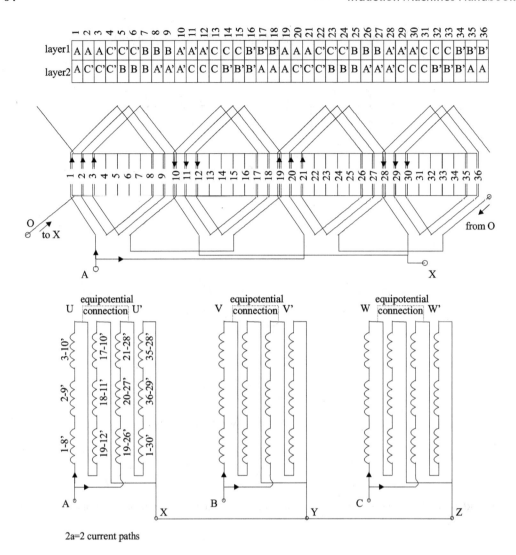

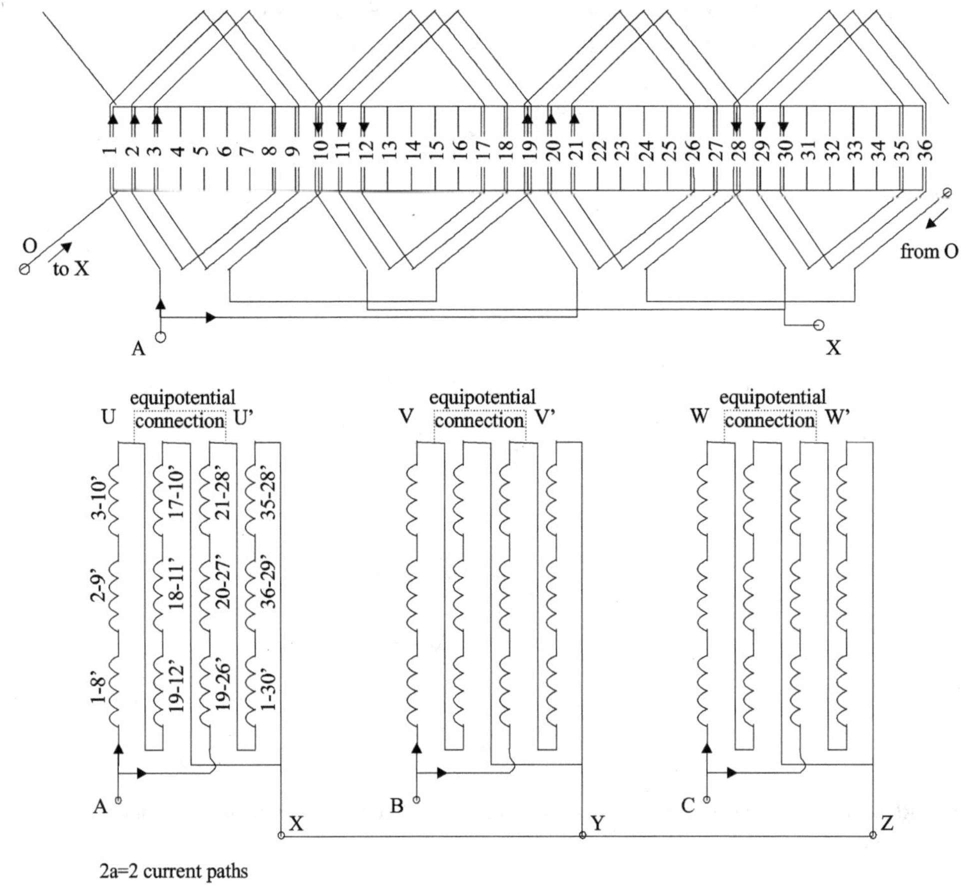

FIGURE 4.13 Double-layer winding: $2p_1 = 4$ poles, $q = 3$, $y/\tau = 7/9$, $N_s = 36$ slots, $a = 2$ current paths.

Note that using wave coils is justified in single-bar coils to reduce the external leads (to one) by which the coils are connected to each other in series. Copper, labour, and space savings are the advantages of this solution.

4.7 BASIC FRACTIONAL q THREE-PHASE A.C. WINDINGS

Fractional q A.C. windings are not typical for induction motors due to their inherent pole asymmetry as slot/phase allocation under adjacent poles is not the same in contrast to integer q three-phase windings. However, with a small q ($q \le 3$) to reduce the harmonics content of airgap flux density, by increasing the order of the first-slot harmonic from $6q \pm 1$ for integer q to $6(ac + b) \pm 1$ for $q = (ca + b)/c$ = fractional, two-layer such windings are favoured to single-layer versions. To set the rules to design such a standard winding – with identical coils – we may proceed with an example.

Let us consider a small induction motor with $2p_1 = 8$ and $q = 3/2$, $m = 3$. The total number of slots $N_s = 2p_1qm = 2{\cdot}4{\cdot}3/2{\cdot}3 = 36$. The coil span y is

$$y = \text{integer}\left(N_s/2p_1\right) = \text{integer}(36/8) = 4 \text{ slot pitches} \tag{4.31}$$

The parameters t, α_{es}, α_{et} are

$$t = \text{g.c.d.}\left(N_s, p_1\right) = \text{g.c.d.}(36, 4) = 4 = p_1 \tag{4.32}$$

$$\alpha_{es} = \frac{2\pi p_1}{N_s} = \frac{\pi \cdot 8}{36} = \frac{2\pi}{9} = \alpha_{et} \tag{4.33}$$

The count of distinct arrows in the star of slot emf phasors is $N_s/t = 36/4 = 9$. This shows that the slot/phase allocation repeats itself after each pole pair (for an integer q it repeats after each pole). Thus, mmf subharmonics, or fractional space harmonics, are still absent in this case of fractional q. This property holds for any q = (2l + 1)/2 for two-layer configurations.

The star of slot emf phasors has q arrows and the counting of them is the natural one ($\alpha_{es} = \alpha_{et}$) (Figure 4.14a).

A few remarks are in order:

- The actual value of q for each phase under neighbouring poles is 2 and 1, respectively, to give an average of 3/2
- Due to the periodicity of two poles (2τ), the mmf distribution does not show fractional harmonics ($\nu < 1$)
- There are both odd and even harmonics as the positive and negative polarities of mmf (Figure 4.14c) are not fully symmetric
- Due to a two-pole periodicity, we may have a = 1 (Figure 4.14d) or a = 2, 4
- The chording and distribution (spread) factors (K_{y1}, K_{q1}) for the fundamental may be determined from Figure 4.14e using simple phasor composition operations.

$$K_{y1} = \sin\left[\frac{\pi p_1}{N_s}\text{integer}\left(N_s/2p_1\right)\right] \tag{4.34}$$

$$K_{q1} = \frac{1 + 2\cos\left(\frac{\pi t}{N_s}\right)}{3} \tag{4.35}$$

(a)

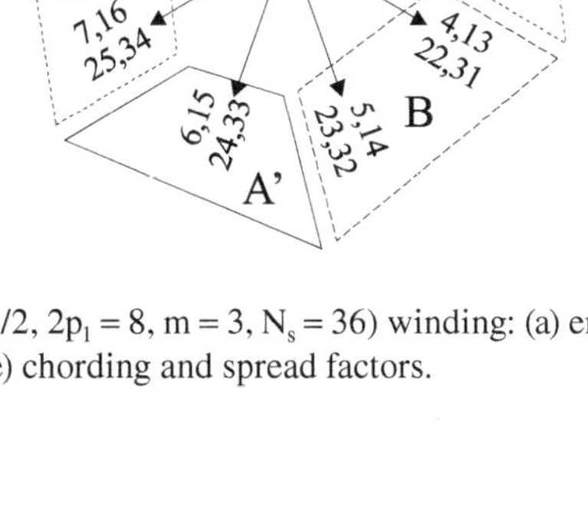

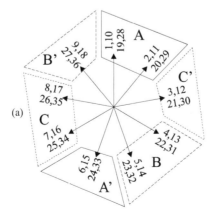

FIGURE 4.14 Fractionary q (q = 3/2, $2p_1 = 8$, m = 3, $N_s = 36$) winding: (a) emf star; (b) slot/phase allocation; (c) mmf; (d) coils of phase A; and (e) chording and spread factors.

(Continued)

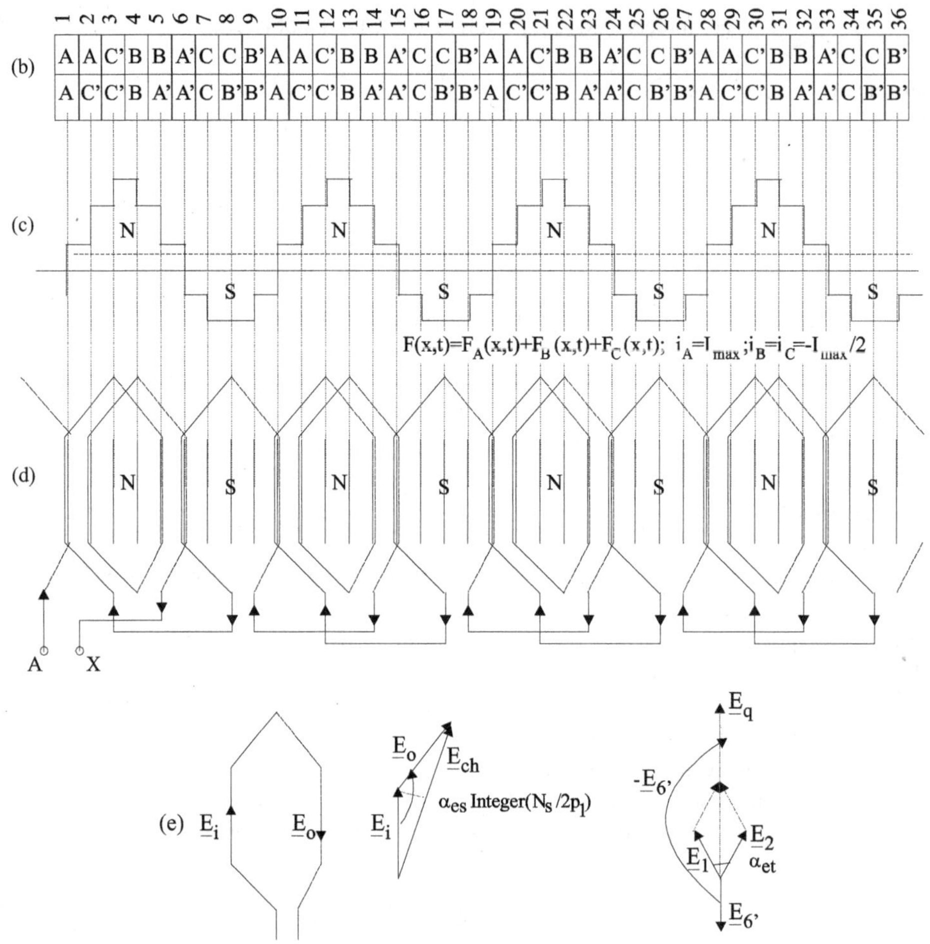

FIGURE 4.14 (CONTINUED) Fractionary q ($q = 3/2$, $2p_1 = 8$, $m = 3$, $N_s = 36$) winding: (a) emf star; (b) slot/phase allocation; (c) mmf; (d) coils of phase A; and (e) chording and spread factors.

This is a kind of general method valid both for integer and fractional q.

Extracting the fundamental and the space harmonics of the mmf distribution (Figure 4.14c) takes care implicitly of these factors both for the fundamental and for the harmonics.

4.8 BASIC POLE-CHANGING THREE-PHASE A.C. WINDINGS

From (4.20), the speed of the mmf fundamental dx/dt is

$$\left(\frac{dx}{dt}\right)_{\nu=1} = 2\tau f_1 \tag{4.36}$$

The corresponding angular speed is

$$\Omega_1 = \frac{dx}{dt}\frac{\pi}{D} = \frac{2\pi f_1}{p_1}; \quad n_1 = \frac{f_1}{p_1} \tag{4.37}$$

The mmf fundamental wave travels at a speed $n_1 = f_1/p_1$. This is the ideal speed of the motor with cage rotor.

Changing the speed may be accomplished either by changing the frequency (through a static power converter) or by changing the number of poles.

Changing the number of poles to produce a two-speed motor is a traditional method. Its appeal is still strong today due to low hardware costs where continuous speed variation is not required. In any case, the rotor should have a squirrel cage to accommodate both pole pitches. Even in variable speed drives with variable frequency static converters, when a very large constant power speed range (over 2(3) to 1) is required, such a solution is to be considered to avoid a notable increase in motor weight (and cost).

Two-speed induction generators are also used for wind energy conversion to allow for a notable speed variation in order to extract more energy from the wind speed.

There are two possibilities to produce a two-speed motor. The most obvious one is to place two distinct windings in the slots. The number of poles would be $2p_1 > 2p_2$. However, the machine becomes very large and costly, while for the winding placed on the bottom of the slots, the slot leakage inductance will be very large with all due consequences.

Using a pole-changing winding seems thus a more practical solution. However, standard pole-changing windings have been produced mainly for $2p_1/2p_2 = 1/2$ or $2p_1/2p_2 = 1/3$.

The most acclaimed winding has been invented by Dahlander and bears his name.

In essence, the current direction (polarity) in half of a $2p_2$ pole winding is changed to produce only $2p_1 = 2p_2/2$ poles. The two halves may be reconnected in series or parallel and Y or Δ connections of phases are applied. Thus, for a given line voltage and frequency supply, with various such connections, constant power or torque or a certain ratio of powers for the two speeds may be obtained.

Let us now proceed with an example and consider a two-layer three-phase winding with $q = 2$, $2p_2 = 4$, $m = 3$, $N_s = 24$ slots, and $y/\tau = 5/6$ and investigate the connection changes to switch it to a two-pole ($2p_1 = 2$) machine.

The design of such a winding is shown in Figure 4.15. The variables are $t = \text{g.c.d}\,(N_s, p_2) = 2 = p_2$, $\alpha_{es} = \alpha_{et} = 2\pi p_2/N_s = \pi/6$, and $N_s/t = 12$. The star of slot emf phasors is shown in Figure 4.15a.

Figure 4.15c illustrates the fact that only the current direction in the section A2 – X2 of phase A is changed to produce a $2p_1 = 2$ pole winding. A similar operation is done for phases B and C.

A possible connection of phase halves called Δ/2Y is shown in Figure 4.16.

It may be demonstrated that for the Δ ($2p_2 = 4$)/2Y ($2p_1 = 2$) connection, the power obtained for the two speeds is about the same.

We should also notice that with chorded coils for $2p_2 = 4$ ($y/\tau = 5/6$), the mmf distribution for $2p_1 = 2$ has a rather small fundamental and is rich in harmonics.

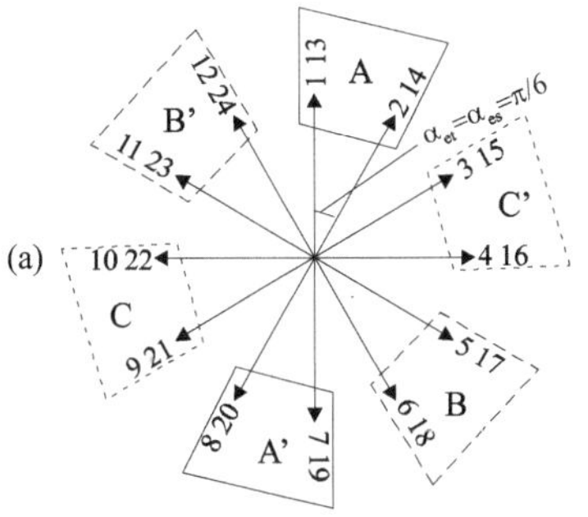

FIGURE 4.15 Two- or four-pole winding ($N_s = 24$) (a) emf star, (b) slot/phase allocation, (c) coils of phase A, (d), (e) mmf for $2p_2 = 4$ and $2p_1 = 2$.

(Continued)

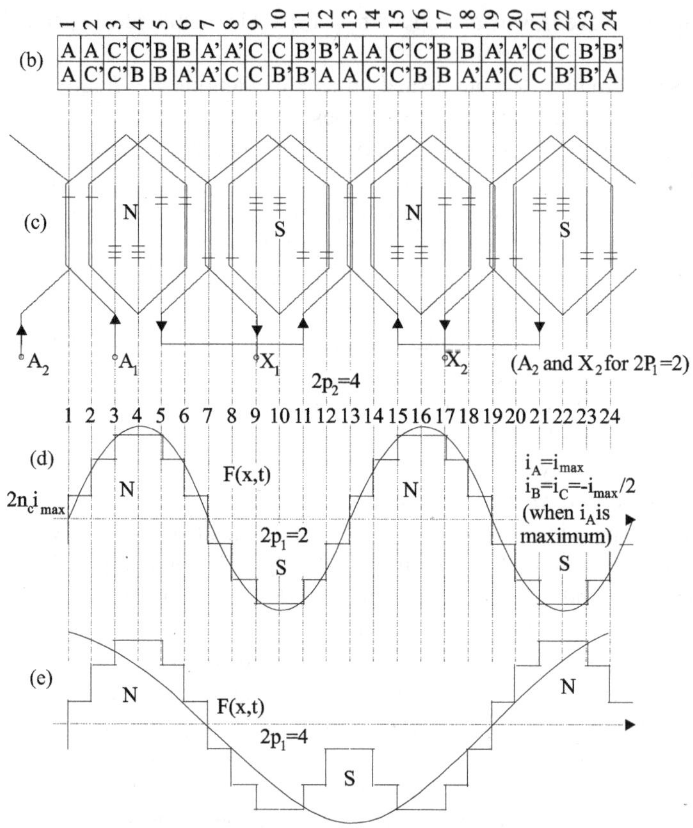

FIGURE 4.15 (CONTINUED) Two- or four-pole winding ($N_s = 24$) (a) emf star, (b) slot/phase allocation, (c) coils of phase A, (d), (e) mmf for $2p_2 = 4$ and $2p_1 = 2$.

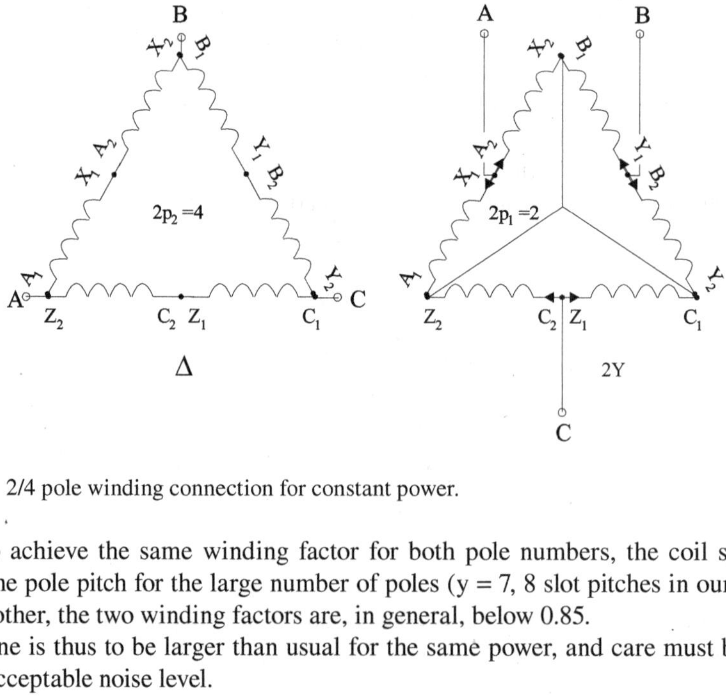

FIGURE 4.16 2/4 pole winding connection for constant power.

In order to achieve the same winding factor for both pole numbers, the coil span should be greater than the pole pitch for the large number of poles (y = 7, 8 slot pitches in our case). Even if close to each other, the two winding factors are, in general, below 0.85.

The machine is thus to be larger than usual for the same power, and care must be exercised to maintain an acceptable noise level.

Various connections of phases may produce designs with constant (same rated torque for both speeds) or variable torque. For example, a Y–YY parallel connection for $2p_2/2p_1 = 2/1$ is producing a ratio of power $P_2/P_1 = 0.35 - 0.4$ as needed for fan driving. Also, in general, when switching the pole number, we may need to modify the phase sequence to keep the same direction of rotation.

One may check if this operation is necessary by representing the stator mmf for the two cases at two instants in time. If the positive maximum of the mmfs advances in time in opposite directions, then the phase sequence has to be changed.

4.9 TWO-PHASE A.C. WINDINGS

When only a single-phase supply is available, two-phase windings are used. One is called the main winding (M) and the other, connected in series with a capacitor, is called the auxiliary winding (Aux).

The two windings are displaced from each other, as shown earlier in this chapter, by 90° (electrical) and are symmetrised for a certain speed (slip) by choosing the correct value of the capacitance. Symmetrisation for start (slip = 0) with a capacitance C_{start} and again for rated slip with a capacitance C_{run} is typical ($C_{start} >> C_{run}$) (Figure 4.17).

When a single capacitor is used, as a compromise between acceptable starting and running, the two windings may even be shifted in space by more than 90° (105°–110°).

Single capacitor configurations for good start are characterised by the disconnection of the auxiliary winding (and capacitor) after starting. In this case, the machine is called capacitor start and the main winding occupies 66% of stator periphery. On the other hand, if bi-directional motion is required, the two windings should each occupy 50% of the stator periphery and should be identical (Figure 4.17b).

Two-phase windings are used for low power (0.1–2 kW) and thus have rather low efficiency. As these motors are made in large numbers due to their use in home appliances, any improvement in the design may be implemented at competitive costs. For example, to reduce the mmf harmonics content, both phases may be placed in all (most) slots in different proportions. Notice that in two-phase windings, multiple of three harmonics may exist and they may deteriorate performance notably.

Let us first consider the capacitor-start motor windings.

As the main winding (M) occupies 2/3 of all slots (Figure 4.18a and b), its mmf distribution (Figure 4.18c) is notably different from that of the auxiliary winding (Figure 4.18d). Also the distribution factors for the two phases (K_{q1M} and K_{q1Aux}) are expected to be different as $q_M = 4$ and $q_{Aux} = 2$.

For $N_s = 16$, $2p_1 = 2$, the above winding could be redesigned for reversible motion where both windings occupy the same number of slots. In that case, the windings look as those shown in Figure 4.19.

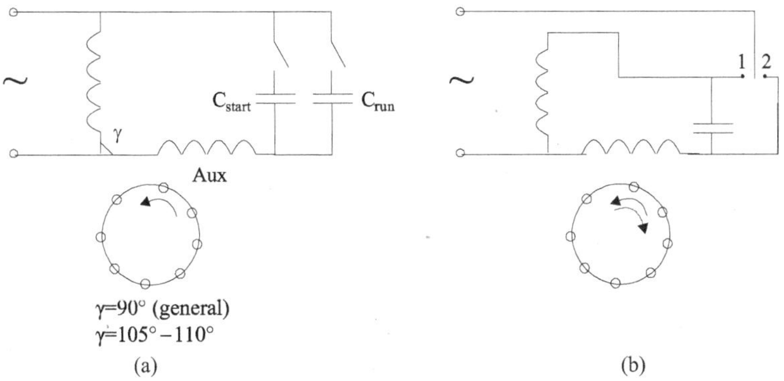

FIGURE 4.17 Two-phase induction motor for (a) unidirectional motion and (b) bidirectional motion (1 – closed for forward motion; 2 – closed for backward motion).

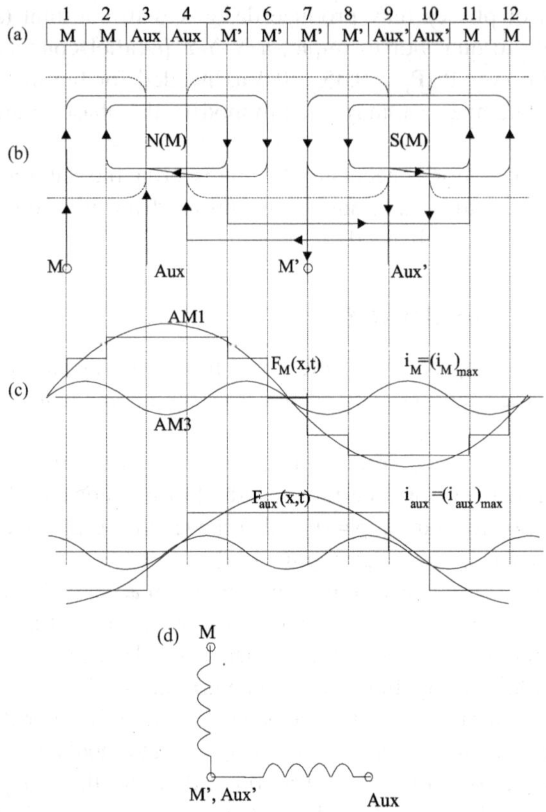

FIGURE 4.18 Capacitor start motor two-phase single-layer winding; $2p_1 = 2$ poles, $N_s = 12$ slots: (a) slot/phase allocation: M/Aux = 2/1; (b) coils per phase connections; (c) mmf distribution for main phase (M); and (d) mmf distribution for auxiliary phase (Aux).

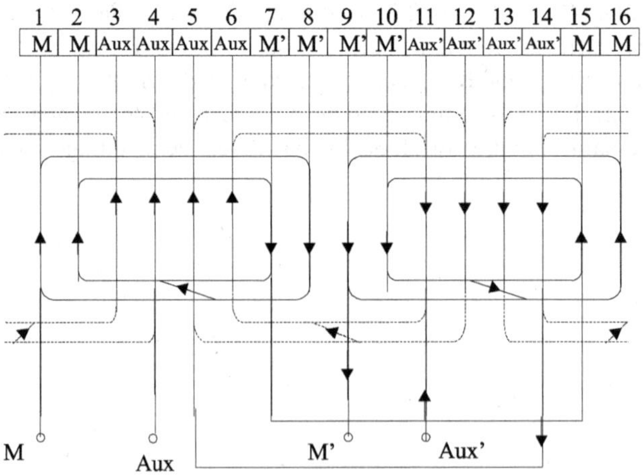

FIGURE 4.19 Reversible motion (capacitor run) two-phase winding with $2p_1 = 2$, $N_s = 16$, single layer.

The two windings have the same number of slots and same distribution factors and have the same number of turns per coil and same wire gauge only for reversible motion when the capacitor is connected to either of the two phases. The quantity of copper is not the same since the end connections may have slightly different lengths for the two phases.

Double-layer windings, for this case ($q_M = q_{Aux}$ – slot/pole/phase), may be built as done for three-phase windings with identical chorded coils. In this case, even for different number of turns/coil and wire gauge – capacitor run motors – the quantity of copper is the same and the ratio of resistances and self-inductances is proportional to the number of turns/coil squared.

As mentioned above, and in Chapter 2, capacitor induction motors of high performance need almost sinusoidal mmf distribution to cancel the various space harmonic bad influences on performance.

One way to do it is to build almost sinusoidal windings; that is, all slots contain coils of both phases but in different proportions (number of turns) so as to obtain a sinusoidal mmf. An example for the two-pole 12-slot case is given in Figure 4.20.

Now if the two windings are equally strong, having the same amount of copper, the slots could be identical. If not, in case of a capacitor start-only or capacitor run-only, part of the slots of auxiliary winding may have a smaller area. This way, the motor weight may be reduced. It is evident from Figure 4.20c that the mmf distribution is now much closer to a sinusoid.

Satisfactory results may be obtained for small power resistor started induction motors if the distribution of conductor/slot/phase runs linear rather than sinusoidal with a flat (open) zone of one-third of a pole pitch. In this case, however, the number of slots N_s should be a multiple of $6p_1$.

An example of an ideal two one-layer sinusoidal winding is shown in Figure 4.21a, for $2p_1 = 2$, $N_s = 24$ slots. A practical one is shown in Figure 4.21b.

Similar to Figure 4.20b, the number of turns/coils for the main winding in slots 1–5 is

$$\begin{aligned}
n_{c1}^{m} &= K_M \cos\alpha_{es}/2 = n_{c12}^{m} = n_{c13}^{m} = n_{c24}^{m} = 0.991 \cdot K_M \\
n_{c2}^{m} &= K_M \cos\frac{3}{2}\alpha_{es}/2 = n_{c11}^{m} = n_{c14}^{m} = n_{c23}^{m} = 0.924 \cdot K_M \\
n_{c3}^{m} &= K_M \cos\frac{5}{2}\alpha_{es}/2 = n_{c10}^{m} = n_{c15}^{m} = n_{c22}^{m} = 0.7933 \cdot K_M \\
n_{c4}^{m} &= K_M \cos\frac{7}{2}\alpha_{es}/2 = n_{c9}^{m} = n_{c16}^{m} = n_{c21}^{m} = 0.6087 \cdot K_M \\
n_{c5}^{m} &= K_M \cos\frac{9}{2}\alpha_{es}/2 = n_{c8}^{m} = n_{c17}^{m} = n_{c20}^{m} = 0.38268 \cdot K_M
\end{aligned} \tag{4.38}$$

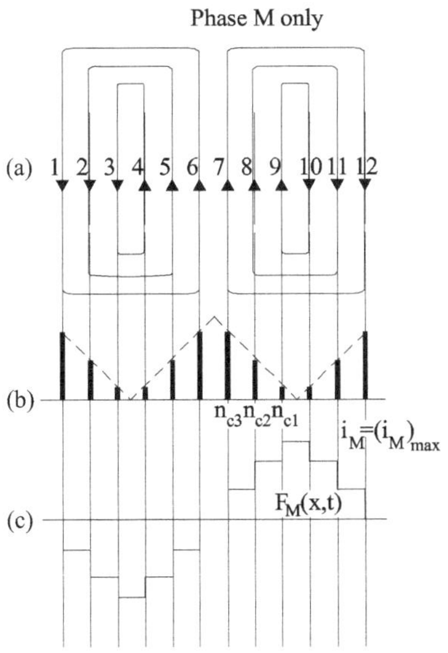

FIGURE 4.20 Quasi-sinusoidal winding (main winding M shown): (a) coils in slot/one layer; (b) turns/coil/slot; and (c) mmf distribution.

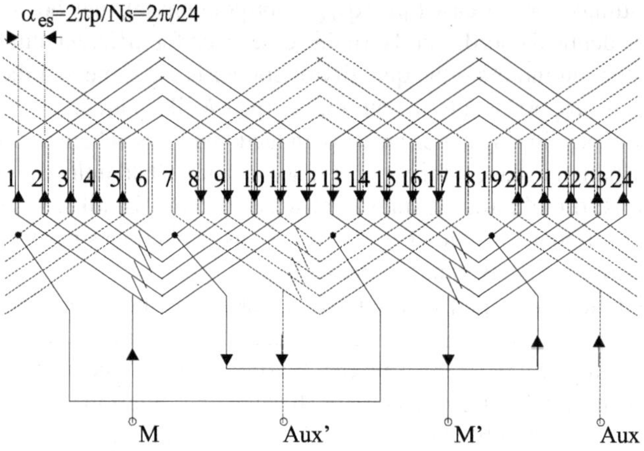

(a)

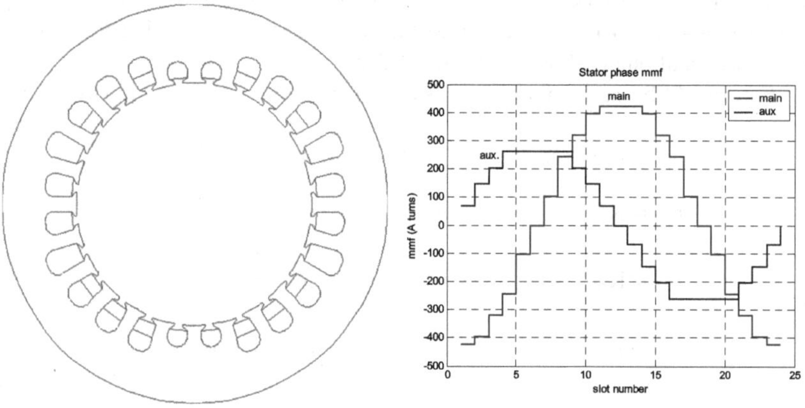

(b)

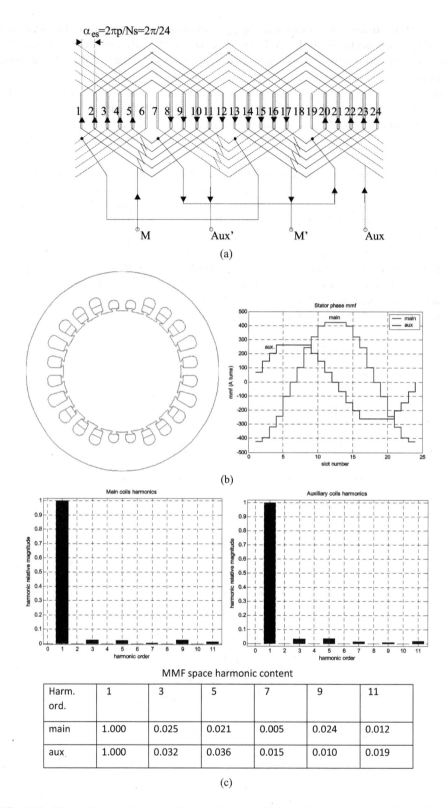

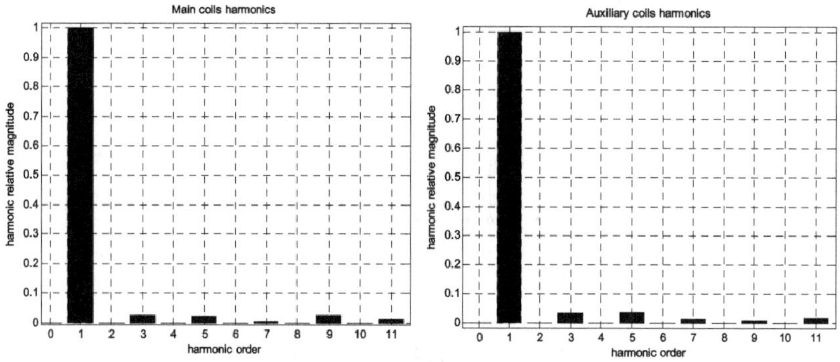

MMF space harmonic content

Harm. ord.	1	3	5	7	9	11
main	1.000	0.025	0.021	0.005	0.024	0.012
aux	1.000	0.032	0.036	0.015	0.010	0.019

(c)

FIGURE 4.21 Two-pole two-phase winding with sinusoidal conductor count/slot/phase distribution: (a) all slots are filled; (b) practical case; and (c) mmf harmonics content.

A similar division of conductor counts is valid for the auxiliary winding.

$$\begin{aligned}
n_{c6}^{a} &= K_A \cos\alpha_{es}/2 = n_{c7}^{a} = n_{c18}^{a} = n_{c19}^{a} = 0.991 \cdot K_A \\
n_{c5}^{a} &= K_A \cos\frac{3}{2}\alpha_{es}/2 = n_{c8}^{a} = n_{c17}^{a} = n_{c20}^{a} = 0.924 \cdot K_A \\
n_{c4}^{a} &= K_A \cos\frac{5}{2}\alpha_{es}/2 = n_{c9}^{a} = n_{c16}^{a} = n_{c21}^{a} = 0.7933 \cdot K_A \\
n_{c3}^{a} &= K_A \cos\frac{7}{2}\alpha_{es}/2 = n_{c10}^{a} = n_{c15}^{a} = n_{c22}^{a} = 0.6087 \cdot K_A \\
n_{c2}^{a} &= K_A \cos\frac{9}{2}\alpha_{es}/2 = n_{c11}^{a} = n_{c14}^{a} = n_{c23}^{a} = 0.38268 \cdot K_A
\end{aligned} \tag{4.39}$$

For the practical winding (Figure 4.21b), the number of turns/slot/phase is a bit different from (4.39).

4.10 POLE CHANGING WITH SINGLE-PHASE SUPPLY INDUCTION MOTORS

A $2/1 = 2p_2/2p_1$ pole-changing winding for single-phase supply can be approached with a three-phase pole-changing winding. The Dahlander winding may be used (see Figure 4.16) as shown in Figure 4.22.

4.11 SPECIAL TOPICS ON A.C. WINDINGS

Most of the windings treated so far in this chapter are in industrial use today. At the same time, new knowledge has surfaced recently. Here we review the most significant part of it.

- **A new general formula for mmf distribution**

Let us consider the slot opening angle β, n_{cK} – conductors/slot/phase, i_K (t) current in slot K, and θ_K – slot position per periphery.

A linear current density in a slot $I_K(\theta_K,t)$ may be defined as

$$\begin{cases} I_K(\theta,t) = \dfrac{2n_{cK} \cdot i_K(t)}{\beta D} & \text{for } \theta \in \left(\theta_K - \beta/2, \theta_K + \beta/2\right) \\ I_K(\theta,t) = 0 & \text{for } \theta \notin \left(\theta_K - \beta/2, \theta_K + \beta/2\right) \end{cases} \tag{4.40}$$

For a single slot, the linear current density may be decomposed in mechanical harmonics as

$$I_K(\theta,t) = \frac{2n_{cK} \cdot i_K(t)}{\pi D}\left[\frac{1}{2} + \sum_{\nu}^{\infty} K_\nu \cos(\theta - \theta_\nu)\right] \tag{4.41}$$

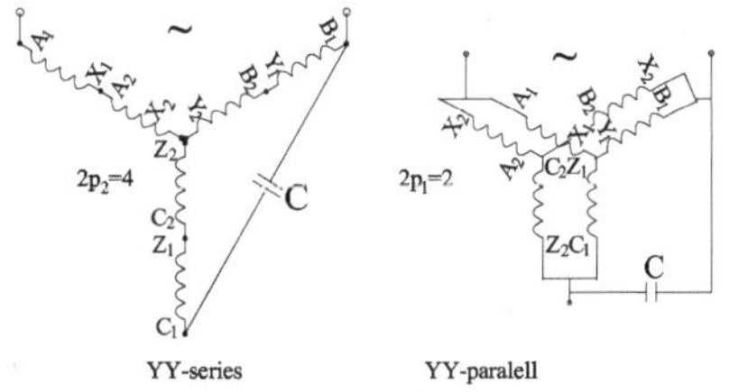

FIGURE 4.22 4/2 pole two-phase winding for single-phase supply made from a three-phase winding.

with

$$K_{\nu} = \frac{2}{\nu\beta}\sin\frac{\nu\beta}{2} \tag{4.42}$$

The number n_{cK} may be positive or negative depending on inward/outward sense of connections and it is zero if no conductors belong to that particular phase. By adding the contributions of all slots per phase A, we obtain its mmf [10,11].

$$F_A(\theta,t) = \frac{2}{\pi}W_1 i_A(t)\sum_{\nu}^{\infty}\frac{K_{\nu}K_{w\nu}}{\nu}\sin(\theta-\theta_{A\nu}) \tag{4.43}$$

with

$$K_{w\nu} = \frac{1}{2W_1}\sqrt{\left(\sum_{K=1}^{N_s} n_{cK}\cos\nu\theta_K\right)^2 + \left(\sum_{K=1}^{N_s} n_{cK}\sin\nu\theta_K\right)^2} \tag{4.44}$$

$$\theta_{A\nu} = \frac{1}{\nu}\tan^{-1}\left[\left(\sum_{K=1}^{N_s} n_{cK}\sin\nu\theta_K\right)\Big/\left(\sum_{K=1}^{N_s} n_{cK}\cos\nu\theta_K\right)\right] \tag{4.45}$$

where W_1 is the total number of turns/phase. When there is more than one current path ($a > 1$), W_1 is replaced by W_{1a} (turns/phase/path) and only the slots/phase/path are considered.

$K_{w\nu}$ is the total winding factor: spread and chording factors multiplied. This formula is valid for the general case (integer q or fractional q windings, included). The effect of slot opening on the phase mmf is considered through the factor $K_{\nu} < 1$(which is equal to unity for zero slot opening – ideal case). In general, the mmf harmonics amplitude is reduced by slot opening but so is its fundamental.

- **Low-space harmonic content windings**

Fractional pitch windings have been, in general, avoided for induction motor, especially due to low-order (sub or fractional) harmonics and their consequences: parasitic torques, noise, vibration, and losses.

However, as basic resources for IM optimisation are exhausted, new ways to reduce copper weight and losses are investigated. Among them, fractional windings with two-pole symmetry ($q = (2l + 1)/2$) are investigated thoroughly [9,12]. They are characterised by the absence of sub-(fractional) harmonics. The case of $q = 3/2$ is interesting low-power motors for low-speed high-pole count induction generators and motors. In essence, the number of turns of the coils in the two groups/phase belonging to neighbouring poles (one with $q_1 = l$ and the other with $q_2 = l + 1$) varies so that designated harmonics could be reduced or destroyed. Also, using coils of different number of conductors in various slots of a phase with standard windings (say $q = 2$) should lead to the cancellation of low-order mmf harmonics and thus render the motors less noisy.

A reduction of copper weight and (or) an increase in efficiency may thus be obtained. The price is using non-identical coils.

The general formula of the winding factor (4.44) may be simplified for the case of $q = 3/2, 5/2, 7/2 \ldots$ where two-pole symmetry is secured and thus the summation in (4.44) extends only along the two poles.

With γ_i the coil pitch, the winding factor $K_{w\nu}$, for two poles, is [9]

$$K_{w\nu} = \sum_{i}^{q_1=l} n'_{ci} \sin\frac{\nu\gamma_i}{2} - (-1)^{\nu} \sum_{K}^{q_2=l+1} n'_{cK} \sin\frac{\nu\gamma_K}{2} \tag{4.46}$$

n_{ci}' and n_{cK}' are relative number of turns with respect to total turns/pole pair/phase; therefore, the coil pitches may be different under the two poles. Also, the number of such coils for the first pole is $q_1 = l$ and for the second is $q_2 = l + 1$ ($q = (2l + 1)/2$).

As $q = (2l + 1)/2$, the coil pitch angles γ_i will refer to an odd number of slot pitches while γ_K refers to an even number of slot pitches. So it could be shown that the winding factor $K_{w\nu}$ of some odd and even harmonic pairs are equal to each other. For example, for $q = 3/2$, $K_{w4} = K_{w5}$, $K_{w2} = K_{w7}$, $K_{w1} = K_{w8}$. These relationships may be used when designing the windings (choosing γ_i, γ_K, and n_{ci}', n_{cK}') to cancel some harmonics.

- **A quasi-sinusoidal two-layer winding** [8]

Consider a winding with $2l + 1$ coils/phase/pole pairs. Consequently, $q = (2l + 1)/2$ slot/pole/phase.

There are two groups of coils per phase. One for the first pole containing $K + 1$ coils of span $3l + 1$, $3l - 1$ … and, for the second pole, K coils of span $3l$, $3l - 2$. The problem consists of building a winding in two layers to cancel all harmonics except those multiples of 3 (which will be inactive in star connection of phases anyway) and the $3(2l + 1) \pm 1$ or slot harmonics.

$$\sum K_{w\nu}\left(n'_{ci}+n'_{cK}\right)=0; \quad \nu = 2,4,5,\ldots\text{(in all 2l values)} \tag{4.47}$$

with

$$\sum_{i=1}^{2l+1}\left(n'_{ci}+n'_{cK}\right)=1 \tag{4.48}$$

This linear system has a unique solution, which, in the order of increased number of turns, yields

$$n'_{ci} = 4\sin\left(\frac{\pi}{6(2l+1)}\right)\sin\frac{(2i-1)\pi}{6(2l+1)}; \quad i = 1,2,3\ldots 2l+1 \tag{4.49}$$

For the fundamental mmf wave, the winding factor K_{w1} is

$$K_{w1} = (2l+1)\sqrt{3}\sin\frac{\pi}{6(2l+1)} \tag{4.50}$$

For $l = 1$, $2l + 1 = 3$ coils/phase/pole pair. In this case, the relative number of turns per coil for phase A is (from 4.50): $n_{c1}' = 0.1206$, $n_{c2}' = 0.3473$, and $n_{c3}' = 0.5321$.

As we can see, $n_{c1}' + n_{c3}' \cong 2\, n_{c2}'$, so the filling of all slots is rather uniform though not identical. The winding layout is shown in Figure 4.23.

The fundamental winding factor of this winding is $K_{w1} = 0.9023$, which is satisfactory (both distribution and chording factors are included). To further cancel the multiples of 3 ($3l$) harmonics, two-/three-layer windings based on the same principle have been successfully tested [9] though their fundamental winding factor is below 0.8.

The same methodology could be applied for q = integer, by using concentrated coils with various numbers of turns to reduce mmf harmonics. Very smooth operation has been obtained with such a 2.2 kW motor with $q = 2$, $2p_1 = 4$, $N_s = 24$ slots up to 6000 rpm [12].

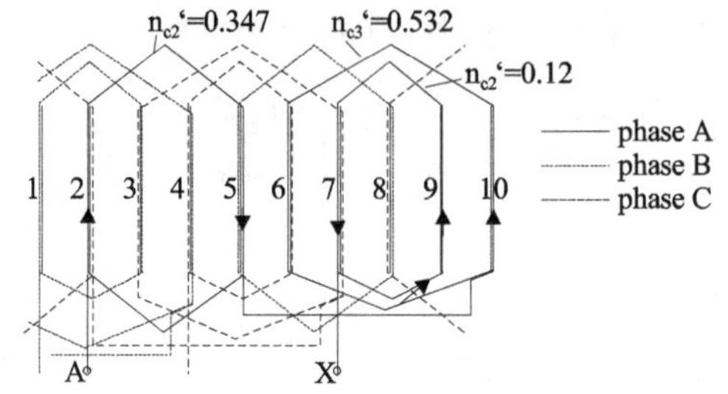

FIGURE 4.23 Sinusoidal winding with $q = 3/2$ in two layers.

In general, for same copper weight, such sinusoidal windings are claimed to produce a 20% increase in starting torque and 5–8 dB noise reduction, at about the same efficiency.

- **Better pole-changing windings**

Pole amplitude modulation [3–5] and symmetrisation [13,14] techniques have been introduced in the 1960s and revisited in the 1990s [15]. Like interspersing [10, pp. 37–39], these methods have produced mixed results so far. On the other hand, the so-called "3 equation principle" for better pole-changing winding has been presented in Ref. [6] for various pole count combinations. This is also a kind of symmetrisation method but with a well-defined methodology. The connections of phases are Δ/Δ and (3Y + 2Y)/3Y for $2p_2$ and $2p_1$ pole counts, respectively. Higher fundamental winding factor, lower harmonics content, better winding and core utilisation, and simpler switching devices from $2p_2$ to $2p_1$ poles than with Dahlander connections are all merits of such a methodology. Two single-throw switches (soft starters) suffice (Figure 4.24) to change speed.

The principle consists of dividing the slot periphery into three equivalent parts by three slots counted n, $n + N_s/3$, and $n + 2N_s/3$. Now the electrical angle between these slots mmf is $2\pi/3p_i$ as expected ($p_i = p_1, p_2$). If $2p_i$ is not a multiple of three, the coils in slots n, $n + N_s/3$, and $n + 2N_s/3$ belong to all three phases and the winding is to be symmetric. The Δ connection is used (Figure 4.24a). In contrast, if $2p_i$ is a multiple of 3, the emfs in slots n, $n + N_s/3$, $n + N_s + 2N_s/3$ are in phase and thus

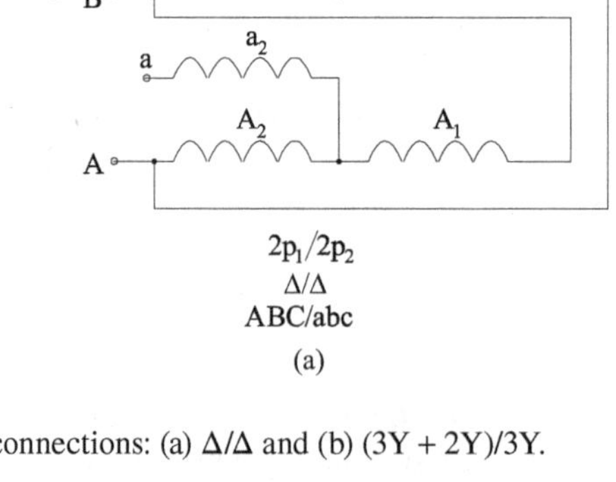

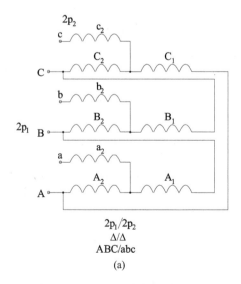

FIGURE 4.24 $2p_1/2p_2$ pole connections: (a) Δ/Δ and (b) (3Y + 2Y)/3Y.

(Continued)

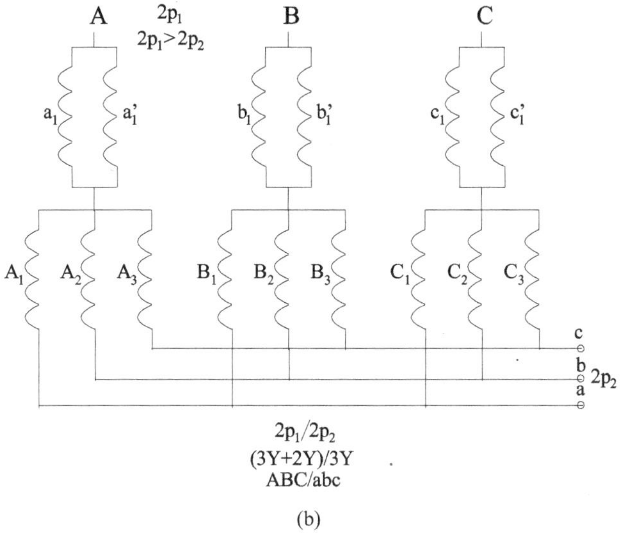

FIGURE 4.24 (CONTINUED) $2p_1/2p_2$ pole connections: (a) Δ/Δ and (b) (3Y + 2Y)/3Y.

they belong to the same phase and we may build three parallel branches (paths) per phase in a 3Y connection (Figure 4.24b). In general, n = 1, 2, … $N_s/3$.

As already mentioned, the Δ/Δ connection (Figure 4.24a) is typical for nonmultiples of three pole counts ($2p_1/2p_2 = 8/10$, for example).

Suppose every coil group $A_{1,2,3}$, $B_{1,2,3}$, $C_{1,2,3}$ is made of t_1 coils ($t_1 < N_s/6$). The coil group A_1 is made of t_1 coils in slots $n_1, n_2, \ldots, n_{t1}$ while groups B_1 and C_1 are displaced by $N_s/3$ and $2N_s/3$, respectively. The sections A_1, B_1, and C_1, which belong to the three phases, do not change phase when the number of poles is changed from $2p_1$ to $2p_2$.

Coil groups A_2, B_2, and C_2 are composed of $t_2 < N_s/6$ coils, and thus, coil group A_2 refers to slots $n_1', n_2', \ldots, n_{t2}'$ while coil groups B_2 and C_2 are displaced by $N_s/3$ and $2N_s/3$ with respect to group A_2. Again, sections A_2, B_2, and C_2 are symmetric.

It may be shown that for $2p_1$ and $2p_2$ equal to $2p_1 = 6K_1 + 2(4)$ and $2p_2 = 6K_2 + 1$, respectively, the mmf waves for the two pole counts travel in the same direction.

On the other hand, if $2p_1 = 6K_1 + 2(4)$ and $2p_2 = 6K_2 + 4(2)$, the mmf waves for the two pole counts move in opposite direction. Swapping two phases is required to keep the same direction of motion for both pole counts (speeds).

In general, t_1 or t_2 are equal to $N_s/6$. For voltage adjustment for higher pole counts, $t_3 = N_s/3 - \Sigma t_i$ slots/phase are left out and distributed later, also with symmetry in mind.

Adjusting groups of t_3 slots/phase should not produce emfs phase shifted by more than 30° with respect to groups $A_iB_iC_i$ in order to secure a high spread (distribution) factor.

An example of such an 8/6 pole Δ/Δ winding is shown in Table 4.1 [6]; the coil connections from Table 4.1 are shown in Figure 4.25 [6].

Such a pole-changing winding seems suitable for fan or windmill applications as the power delivered is reduced to almost half when switching from 8 to 10 poles, that is, with a 20% reduction in speed. The procedure is similar for the (3Y + 2Y)/3Y connection [7].

This type (Figure 4.24b) for $a_1 = a_1' = b_1 = b_1' = c_1 = c_1' = 0$ and pole count ratio of say $2p_2/2p_1 = 6/2$ has been known for long and recently proposed for dual winding (doubly fed) stator, nest cage rotor IMs [16].

Once the slot/phase allocation is done, the general formula for the winding factor (4.44) or the mmf step-shape distribution with harmonics decomposition may be used to determine the mmf fundamental and harmonics content for both pole counts. These results are part of any IM design process.

TABLE 4.1
(8/6 Pole, Δ/Δ, N_s = 72 Slots)

Coil Group	Slot (Coil) Distribution Principle	Slot (Coil) Number
A_1	$n\ (t_1 = 10)$	72,1,2,3,–10,–11,–65,–66,–67,–68
B_1	$n + N_s/3$	24,25,26,27,–34,–35,–17,–18,–19,–20
C_1	$n + 2N_s/3$	48,49,50,51,–58,–59,–41,–42,–43,–44
A_2	$n'(t_2 = 10)$	21,22,23,–29,–30,–31,–32,37,38,39
B_2	$n' + N_s/3$	45,46,47,–53,–54,–55,–56,61,62,63
C_2	$n' + 2N_s/3$	69,70,71,–5,–6,–7,–8,13,14,15
a_2	$n''\ (t_3 = 4)$	–52,60,–9,16
b_2	$n'' + N_s/3$	–4,12,33,40
c_2	$n'' + 2N_s/3$	–28,26,–57,64

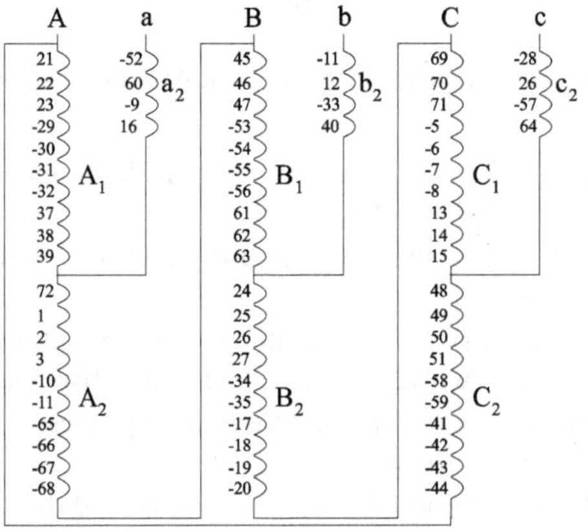

FIGURE 4.25 8/6 pole winding with 72 slots.

- **A pure mathematical approach to A.C. winding design**

As we have seen so far, A.C. winding design is more an art than a science. Intuition and symmetry serve as tools to this aim.

Since we have a general expression for the harmonics winding factor valid for all situations when we know the number of turns per coil and what current flows through those coils in each slot, we may develop an objective function to define an optimum performance.

Maximum airgap flux energy for the fundamental mmf, or minimum of some harmonics squared winding factors summation, or the maximum fundamental winding factor, may constitute practical objective functions. Now copper weight may be considered a constraint. So is the pole count $2p_1$. Some of the above objectives may become constraints. What then would be the variables for this problem? It seems that the slot/phase allocation and, ultimately, the number of turns/coil are essential variables to play with, together with the number of slots.

In a pole-changing ($2p_2/2p_1$) winding, the optimisation process has to be done twice. This way the problem becomes typical: non-linear, with constraints.

All optimisation methods used now in electric machine design also could apparently be applied here.

The first attempt has been made in [6] by using the "simulated annealing method" – a kind of direct (random) search method, for windings with identical coils. The process starts with fixing

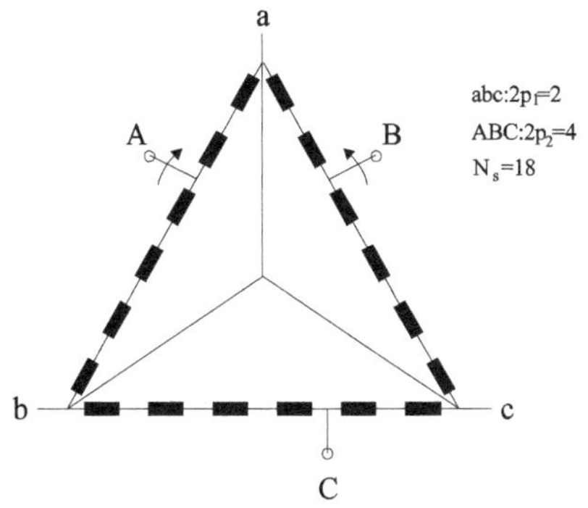

FIGURE 4.26 Initial (random) winding on the way to "rediscover" the Dahlander connection.

the phase connection: Y or Δ with a random allocation of slots to phases A, B, and C. Then we take the instant when the current in phase A is maximum i_{Max} and $i_B = i_C = -i_{Max}/2$. Allowing for a given number of turns per coil n_c and 1 or 2 layers, we may start with given (random) mmfs in each phase.

Then we calculate the objective function for this initial situation and proceed to modify the slot/phase allocation either by changing the connection (beginning or end) of the coil or its position (see Figure 4.26). Various rules of change may be adopted here.

The objective function is calculated again. The computation cycle is redone until sufficient convergence is obtained for a limited computation time.

In [6], by using such an approach, the 4/2 pole Dahlander connection for an 18 slot IM is rediscovered after 2 hours CPU time on a large mainframe computer. The computing time is still large, so the direct (random) search approach has to be refined or changed, but the method becomes feasible, especially for some particular applications.

The method was pursued in Ref. [17], with harmonics distortion factors as fitness functions and unity ratio for airgap flux density fundamentals, for integer (Dahlander) or fractional pole-changing windings [17].

4.12 THE MMF OF ROTOR WINDINGS

IM rotors have either three-phase windings star-connected to three-phase slip rings or have squirrel cage-type windings. The rotor three-phase windings are made of full pitch coils and one or two layers located in semiclosed or semiopen slots. Their end connections are braced against centrifugal forces. Wound rotors are typical for IMs in the hundreds and thousands of kW ratings and more recently in a few hundreds of MW as motor/generators in pumped storage hydro power plants.

The voltage rating of rotor windings is of the same order as that in the stator of the same machine. The mmf of such windings is similar to that of stator windings treated earlier in this chapter.

Squirrel-cage windings – single cage, double cage, deep bar, solid rotor, and nest cage types – have been introduced in Chapter 2. Here we deal with the mmf of a single cage winding. We consider a fully symmetrical cage (no broken bars or end rings) with straight slots (no skewing) – Figure 4.27a.

By replacing the cage with an N_r/p_1 phase winding ($m = N_r/p_1$) with one conductor per slot, $q = 1$, and full pitch coils [17], we may use the formula for the three winding (4.43) to obtain

$$F_2(\theta,t) = \frac{N_r}{\pi P_1} I_b \sqrt{2} \sum \frac{K_\mu}{\mu} \sin\frac{\nu P_1}{N_r} \cdot \left[K_{f\nu}\cos(\omega_1 t - \mu\theta) + K_{b\nu}\cos(\omega_1 t + \mu\theta)\right] \quad (4.51)$$

where I_b is the RMS bar current and

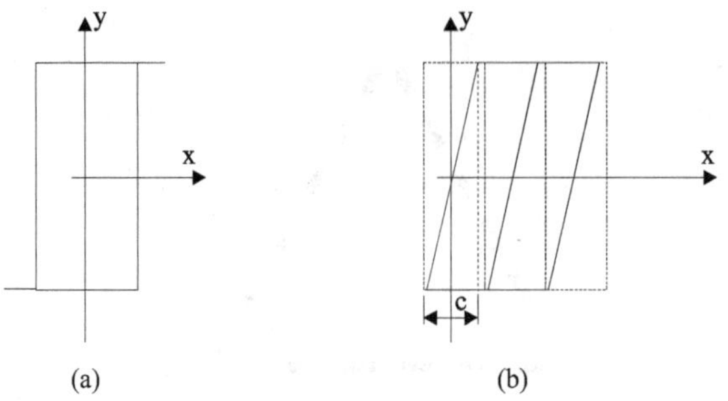

FIGURE 4.27 Rotor cage with (a) straight rotor slots and (b) with skewed rotor slots.

$$K_{fv,bv} = \frac{P_1}{N_s}\left[1+\left(\frac{N_r}{P_1}-1\right)\cos\left(\mu \pm P_1\right)\frac{2\pi}{N_r}\right] \tag{4.52}$$

Let us remember that μ is a spatial harmonic and θ is a geometrical angle: $\theta_e = p_1\,\theta$. Also, $\mu = p_1$ means the fundamental of mmf.

There are forward (f) and backward (b) harmonics. Only those of the order $\mu = KN_r + p_1$ have $K_{f\nu} = 1$ while for $\mu = KN_r - p_1$, they have $K_{b\nu} = 1$. All the other harmonics have $K_{f\nu,\,b\nu}$ as zero. But the above harmonics are, in fact, caused by the rotor slotting. The most important are the ones first obtained for K = 1: $\mu_{fmin} = N_r - p_1$ and $\mu_{bmin} = N_r + p_1$.

For many sub-MW-power IMs, the rotor slots are skewed, in general, by about one stator slot pitch or more, to destroy the first stator-slot-caused stator mmf harmonic ($\nu = N_s - p_1$). Again ν is a spatial harmonic; that is, $\nu_1 = p_1$ for the fundamental.

In this situation, let us take the case of the fundamental only. The situation is as if the rotor position in (4.53) varies with y (Figure 4.27b).

$$F_2(\theta,t)_{\nu=p_1} \approx \frac{N_r}{\pi p_1} I_b\sqrt{2}\cdot\cos\left[\omega_1 t-\left(\theta_e-\frac{\pi y}{\tau}\right)-\left(\pi-\gamma_{1,2}\right)\right]; \quad \theta_e = \frac{\pi x}{\tau} \tag{4.53}$$

for

$$y \in \left[0.0, \pm c/2\right] \tag{4.54}$$

Now as ω_1, the stator frequency, was used in the rotor mmf, it means that (4.53) is written in a stator reference system,

$$\omega_1 = \omega_2 + \omega_r; \quad \omega_r = \Omega_r p_1 \tag{4.55}$$

with ω_r as the rotor speed in electrical terms and ω_2 the frequency of the currents in the rotor. The angle $\gamma_{1,2}$ is a reference angle (at zero time) between the stator $F_1(\theta,t)$ and rotor $F_2(\theta,t)$ fundamental mmfs.

4.13 THE "SKEWING" MMF CONCEPT

In an induction motor on load, there are currents both in the stator and the rotor. Let us consider here only the fundamentals for the sake of simplicity. With $\theta_e = \pi x/\tau$, (4.13) becomes

$$F_1(\theta_e,t)=F_{1m}\cos(\omega_1 t-\theta_e) \tag{4.56}$$

with F_{1m} from (4.14).

The resultant mmf is the sum of stator and rotor mmfs, $F_1(\theta_e,t)$, and $F_2(\theta_e,t)$. It is found by using (4.53) and (4.56).

$$F_{r1}(\theta_e,t)=F_{1m}\cos(\omega_1 t-\theta_e)+F_{2m}\cos\left(\omega_1 t-\theta_e-\frac{\pi y}{\tau}-(\pi-\gamma_{1,2})\right) \tag{4.57}$$

with

$$F_{1m}=\frac{3W_1 I\sqrt{2}K_{w1}}{\pi P_1};\quad F_{2m}=\frac{N_r}{\pi P_1}I_b\sqrt{2} \tag{4.58}$$

The two mmf fundamental amplitudes, F_{1m} and F_{2m}, are almost equal to each other at standstill. In general,

$$\frac{F_{2m}}{F_{1m}}\approx\sqrt{1-\left(\frac{I_0}{I}\right)^2} \tag{4.59}$$

where I_0 is the no-load current.

For rated current, with $I_0 = (0.7 \text{ to } 0.3)I_{Rated}$, F_{2m}/F_{1m} is about (0.7 to 0.95) at rated load. As at start (zero speed) $I_s = (4.5 \text{ to } 7)I_n$, it is now obvious that in this case,

$$\left(\frac{F_{2m}}{F_{1m}}\right)_{s=1}\approx 1.0 \tag{4.60}$$

The angle between the two mmfs $\gamma_{1,2}$ varies from a few degrees at standstill (S = 1) to 45° at peak (breakdown) slip S_K and then goes to zero at S = 0. The angle $\gamma_{1,2} > 0$ for motoring and $\gamma_{1,2} < 0$ for generating.

The resultant mmf $F_r(\theta_e,t)$, with (4.53) and (4.56), is

$$F_{r1}(\theta_e,t)=\left[F_{1m}\cos(\omega_1 t-\theta_e)-F_{2m}\cos(\omega_1 t-\theta_e+\gamma_{1,2})\cos\frac{\pi y}{\tau}\right]$$
$$+F_{2m}\sin(\omega_1 t-\theta_e+\gamma_{1,2})\,\sin\frac{\pi y}{\tau};\ y\in(-c/2,+c/2) \tag{4.61}$$

The term outside the square brackets in (4.61) is zero if there is no skewing. Consequently, the term inside the square brackets is the compensated or the magnetising mmf which tends to be small as $\cos \pi y/\tau \cong 1$, and thus, the skewing $c = (0.5 \text{ to } 2.0)\tau/mq$ is small and the angle $\gamma_{1,2} \in (\pi/4, -\pi/4)$ for all slip values (motoring and generating).

Therefore, we define F_{2skew} by

$$F_{2skew}=F_{2m}\sin(\omega_1 t-\theta_e+\gamma_{1,2})\sin\frac{\pi}{\tau};\quad y\in(-c/2,+c/2) \tag{4.62}$$

the uncompensated rotor-produced mmf, which is likely to cause large airgap flux densities varying along the stator stack length at standstill (and large slip) when the rotor currents (and F2m) are large.

Thus, heavy saturation levels are expected along main flux paths in the stator and rotor core at standstill as a form of leakage flux, which strongly influences the leakage inductances as shown later in this book.

4.14 MULTIPHASE AND MULTILAYER TOOTH-WOUND COIL WINDINGS

Though in general two-phase (in split-phase capacitor motors) and three-phase windings ($q \geq 1$) are typically placed in one or two layers in slots, multiphase windings have also been proposed for IMs to:

- Provide for increased fault tolerance in safety critical applications
- At large powers to reduce the ratings of silicon-controlled rectifiers (SCRs) and finally also increase the efficiency by providing lower harmonics loss in static power converter supplies.

A typical five-phase winding (with, say one-layer $q = 1$ slots/pole/phase and pole span coils) is shown in Figure 4.28, for three connection schemes with increasing torque-speed envelope in variable speed drives [18].

Though intuitively sinusoidal current control may be targeted in special applications, a third harmonic current may be added.

So far, the efficiency of distributed ($q \geq 1$) multiphase IMs has not been proven superior to the conventional three-phase one, but the fault tolerance is better. Also, in an attempt to simplify the stator windings (which dominates the fabrication costs), tooth-coil ($q \leq 0.5$) windings have been recently tried (Figure 4.29) [19].

The four-layer tooth-coil winding in Figure 4.29 is meant to reduce the mmf space harmonics and thus to reduce torque pulsations, but they are still high (30%) and the efficiency, power factor, and torque density increase with respect to distributed ($q > 1$) windings is still uncertain [20].

Dual three-phase stator windings have been proposed both with same number of poles (and standard cage rotors) or with different pole pairs, p_p, p_c (and nested cage rotors) and one of them connected to the grid (or load) and the other connected to a variable frequency PWM converter designed at fractional kVA ratings [21–26].

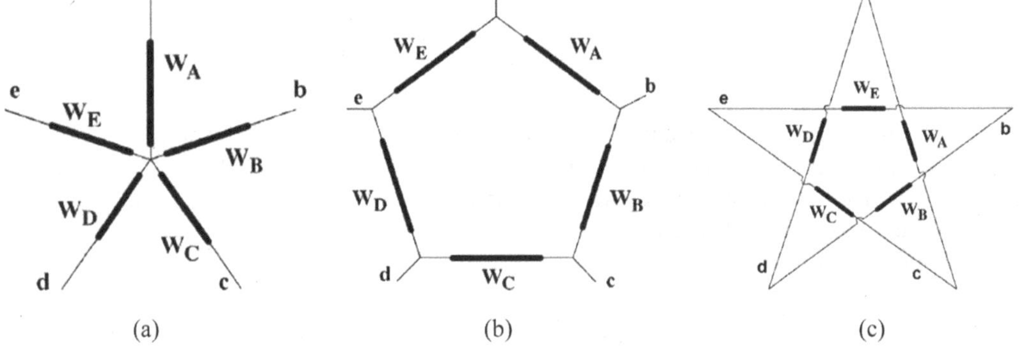

FIGURE 4.28 Five-phase A.C. winding with three connection schemes: (a) star; (b) pentagon; and (c) pentacle.

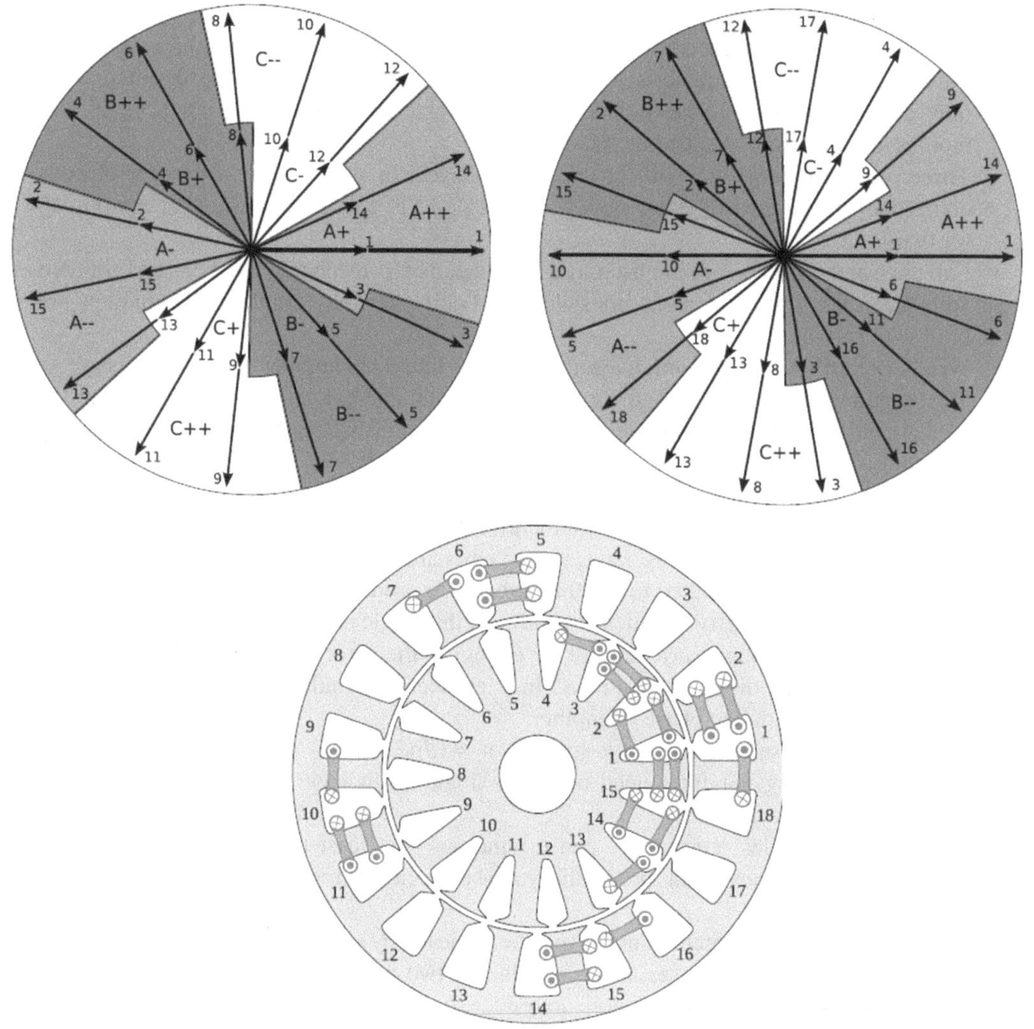

FIGURE 4.29 Four-layer tooth-coil stator and rotor windings in an 18 slot stator, 15 slot rotor, and 14 pole IM.

4.15 SUMMARY

- Alternate current winding design deals with assigning coils to slots and phases, establishing coil connections inside and between phases, and calculating the number of turns/coil and the wire gauge.
- The mmf of a winding means the spatial/time variation of ampere-turns in slots along the stator periphery.
- A pure travelling mmf is the ideal; it may be approximately realised through intelligent placing of coils of phases in their slots.
- Travelling wave mmfs are capable of producing ripple-free torque at steady state.
- The practical mmf wave has a spatial period of two pole pitches 2τ.
- There are p_1 electrical periods for one mechanical revolution.
- The speed of the travelling wave (fundamental) $n_1 = f_1/p_1$; f_1 – current frequency.

- Two phases with stepwise mmfs, phase shifted by $\pi/2p_1$ mechanical degrees or m phases (three in particular), phase shifted by π/mp_1 geometrical degrees, may produce a practical travelling mmf wave characterised by a large fundamental and small space harmonics for integer q. Harmonics of order $\nu = 5, 11, 17$ are reverse in motion, while $\nu = 7, 13, 19$ harmonics move forward at speed $n_\nu = f_1/(\nu p_1)$.
- Three-phase windings are built in one or two layers in slots; the total number of coils equals half the number of stator slots N_s for single-layer configurations and is equal to N_s for two-layer windings.
- Full pitch and chorded coils are used; full pitch means π/p_1 mechanical radians and chorded coils mean less than that; sometimes elongated coils are used; single-layer windings are built with full pitch coils.
- Windings for IMs are built with integer and, rarely, fractional number of slots/pole/phase, q.
- The windings are characterised by their mmf fundamental amplitude (the higher the better) and the space harmonic contents (the lower, the better); the winding factor $K_{w\nu}$ characterises their influence.
- The star of slot emf phasors refers to the emf phasors in every slot conductors drawn with a common origin and is based on the fact that the airgap field produced by the mmf fundamental is also a travelling wave; so, the emfs are sinusoidal in time.
- The star phasors are allocated to the three phases to produce three resultant phasors 120° (electrical) apart; in partly symmetric windings, this angle is only close to 120°.
- Pole-changing windings may be built by changing the direction of connections in half of each phase or by dividing each phase in a few sections (multiples of 3, in general) and switching them from one phase to another.
- Pole changing is used to modify the speed as $n_1 = f_1/p_1$.
- New pole-changing windings need only two single-throw switches while standard ones need more costly switches.
- Single-phase supplies require two-phase windings – at least to start; a capacitor (or a resistor at very low power) in the auxiliary phase provides a resultant travelling mmf for a certain slip (speed).
- Reducing harmonics content in mmfs may be achieved by varying the number of turns per coil in various slots and phases; this is standard in two-phase (single-phase supply) motors and rather new in three-phase motors, to reduce the noise level.
- Two-phase (for single-phase supply) pole-changing windings may be obtained from such three-phase windings by special connections.
- Rotor three-phase windings use full pitch coils in general.
- Rotor-cage mmfs have a fundamental ($\mu = p_1$) and harmonics (in mechanical angles) $\mu = K * N_r \pm p_1$, N_r – number of rotor slots.
- The rotor cage may be replaced by an N_r/p_1 phase winding with one conductor per slot.
- The rotor mmf fundamental amplitude F_{2m} varies with slip (speed) up to F_{1m} (of the stator) and so does the phase shift angle between them $\gamma_{1,2}$. The angle $\gamma_{1,2}$ varies in the interval $\gamma_{1,2} \in (0, \pm \pi/4)$.
- For skewed rotor slots, a part of rotor mmf, variable along stator stack length (shaft direction), remains uncompensated; this mmf is called here the "skewing" mmf which may produce heavy saturation levels at low speed (and high rotor currents).
- The sum of the stator and rotor mmfs, which solely exist in the absence of skewing, constitutes the so-called "magnetisation" mmf and produces the main (useful) flux in the machine.
- Even this magnetising mmf varies with slip (speed) – for constant voltage and frequency; however, it decreases with slip (reduction in speed); this knowledge is to be used in Chapters 5, 6, and 8.

- New pole-changing single-phase windings are still proposed [27].
- Multiple (5, 2×3, 3×3) phase IM drives with rather trapezoidal shape currents have been recently proposed for higher fault tolerance but did not yet reach commercial stage [25,26].
- A recent strong tutorial view on combined star delta-connected windings' winding factors to improve efficiency is available in [28].

REFERENCES

1. H. Sequenz, *The Windings of Electrical Machines*, Vol.3, A.C. Machines, Springer Verlag, Vienna, 1950 (in German).
2. M. M. Liwschitz-Garik, *Winding Alternating Current Machines*, Van Nostrand Publications, 1950.
3. M. Bhattacharyya, Pole amplitude modulation applied to salient pole synchronous machines, *Ph.D. Thesis*, University of Bristol, Bristol, UK, 1966.
4. G. H. Rawcliffe, G. H. Burbidge, W. Fong, Induction motor speed changing by pole amplitude modulation, *Proceedings of IEE - Part A: Power Engineering*, Vol. 105A, No. 3, August 1958, pp. 411–419.
5. B. J. Chalmers, Interspersed A.C. windings, *Proceedings of IEE- Part A: Power Engineering*, Vol. 111, 1964, p. 1859.
6. M. Poloujadoff, J. C. Mipo, Designing 2 pole and 2/4 pole windings by simulated annealing method, *EMPS*, Vol. 26, No. 10, 1998, pp. 1059–1066.
7. C. Zhang, J. Yang, X. Chen, Y. Guo, A new design principle for pole changing winding–The three equation principle, *EMPS*, Vol. 22, No. 2, 1994, pp. 187–189.
8. M. V. Cistelecan, E. Demeter, A new approach to the three-phase monoaxial nonconventional windings for A.C., *Machines, Record of IEEE – IEMDC, 99*, Seattle, WA, USA, 1999, pp. 323–325.
9. M. V. Cistelecan, E. Demeter, A closer approach of the fractional multilayer three-phase A.C. windings, *Record of Electromotion – 1999,* Patros, Greece, May 1999, Vol. 1, pp. 51–54.
10. B. Chalmers, A. Williamson, *A.C. Machines: Electromagnetics and Design*, Research Studies Press Ltd., Somerset, England, 1991, p. 28.
11. D. M. Ionel, M. V. Cistelecan, T. J. E. Miller, M. I. McGilp, A new analytical method for the computation of air-gap reactances in 3-phase induction motors, *Conference Record of 1998 IEEE Industry Applications Conference Thirty-Third IAS Annual Meeting* (Cat. No.98CH36242), St. Louis, MO, USA, 12–15 October, 1998, Vol. 1, pp. 65–72.
12. M. Cistelecan, E. Demeter, New 3 phase A.C. windings with low spatial harmonic content, *Record of Electromotion Symposium*, Cluj – Napoca, Romania, 1997, pp. 98–100.
13. W. Fong, Polyphase symmetrisation, UK Patent, 1964.
14. A. R. W. Broadway, Part–symmetrisation of 3-phase windings, *Proceedings of the IEE*, Vol. 122, July 1975, p. 125.
15. M. Bhattacharyya, G. K. Singh, Observations on the design of distributed armature windings by poly–phase symmetrisation, *EMPS*, Vol. 19, No. 3, 1991, pp. 363–379.
16. R. Li, A Wallace, R. Spee, Two-axis model development of cage-rotor brushless doubly-fed machines, *IEEE Transactions on Energy Conversion*, Vol. EC – 6, No. 3, 1991, pp. 453–460.
17. J. C. Mippo, M. Poloujadoff, M. M. Radulescu, Simulated annealing approach to the design optimisation of two–speed induction–motor windings, *Electromotion* (ISSN-057X), Vol. 12, No. 1, 2005, pp. 19–25.
18. S. Sadeghi, L. Guo, H. A. Toliyat, L. Parsa, Wide operational speed range of five-phase permanent magnet machines by using different stator winding configurations, *IEEE Transactions on Industrial Electronics*, Vol. IE-59, No. 6, 2012, pp. 2621–2631.
19. L. Alberti, N. Bianchi, Design and tests on a fractional – slot induction machine, *Record of IEEE-ECCE*, 2012, pp. 166–172.
20. A. S. Abdel-Khalik, S. Ahmed, Performance evaluation of five – phase modular winding induction machine, *IEEE Transactions on Industrial Electronics*, Vol. IE-59, No. 6, 2012, pp. 2654–2669.
21. F. Bu, W. Huang, Y. Hu, K. Shi, An excitation-capacitor-optimized dual stator – winding with static excitation controller for wind application, *IEEE Transactions on Energy Conversion*, Vol. EC-26, No. 1, 2011, pp. 122–131.
22. L. N. Tutelea, I. Boldea, N. Muntean, S. I. Deaconu, Modeling and performance of novel scheme dual winding cage-rotor variable speed induction generator with D.C. link power delivery, *Record of IEEE-ECCE*, 2014, pp.271–278.

23. A. Oraee, E. Abdi, S. Abdi, R. McMahon, P. J. Tavner, Effects of rotor windings structure on the BDFM equivalent circuit parameters, *IEEE Transactions on Energy Conversion*, Vol. EC-30, No. 4, 2015, pp. 1660–1669.
24. Y. Park, J.-M. Yoo, S.-K. Sul, Vector control of double-delta sourced winding for a dual-winding induction machine, *IEEE Transactions on Industry Applications*, Vol. IA-53, No. 1, 2017, pp. 171–180.
25. E. Levi, Multiphase electric machines for variable speed applications, *IEEE Transactions on Industry Electronics,* Vol. IE-55, No. 5, 2008, pp. 1893–1909.
26. H. Xu, F. Bu, W. Huang, Y. Hu, H. Liu, Control and performance of five – phase dual stator winding induction generator DC generating system, *IEEE Transactions on Industrial Electronics*, Vol. IE-64, No. 7, 2017, pp. 5276–5285.
27. H. Nam, S.-K. Jung, G.-H. Kang, J.-P. Hong, T.-U. Jung, S.-M. Baek, Design of pole-change single-phase induction motor for household appliances, *IEEE Transactions on Industry Applications*, Vol. IA-40, No. 3, May–June 2004, pp. 780–788.
28. S. M. Raziee, O. Misir, B. Ponick, Combined star-delta winding analysis, *IEEE Transactions on Energy Conversion*, Vol. EC-33, No. 1, 2018, pp. 383–394.

5 The Magnetisation Curve and Inductance

5.1 INTRODUCTION

As shown in Chapters 2 and 4, the induction machine configuration is quite complex. So far, we elucidated the subject of windings and their mmfs. With windings in slots, the mmf has (in three- or two-phase symmetric windings) a dominant wave and harmonics. The presence of slot openings on both sides of the airgap is bound to amplify (influence at least) the mmf stepwise shape harmonics. Many of them will be attenuated by rotor cage-induced currents. To further complicate the picture, the magnetic saturation of the stator (rotor) teeth and back irons (cores or yokes) also influences the airgap flux distribution producing new harmonics.

Finally, the rotor eccentricity (static and/or dynamic) introduces new space harmonics in the airgap field distribution.

In general, both stator and rotor currents produce a resultant field in the machine airgap and iron parts.

However, with respect to fundamental, torque-producing, airgap flux density, the situation does not change notably from zero rotor currents to rated rotor currents (rated torque) in most induction machines, as experience shows.

Thus, it is only natural and practical to investigate, first, the airgap field fundamental with uniform equivalent airgap (slotting accounted through correction factors) as influenced by the magnetic saturation of stator and rotor teeth and back cores for zero rotor currents.

This situation occurs in practice with the wound rotor winding kept open at standstill or with the squirrel cage rotor machine fed with symmetrical A.C. voltages in the stator and driven at mmf wave fundamental speed ($n_1 = f_1/p_1$).

As in this case, the pure travelling mmf wave runs at rotor speed, no induced voltages occur in the rotor bars. The mmf space harmonics (step harmonics due to the slot placement of coils and slot opening harmonics, etc.) produce some losses in the rotor core and windings. They do not influence notably the fundamental airgap flux density and thus, for this investigation, they may be neglected, only to be revisited in Chapter 11.

To calculate the airgap flux density distribution in the airgap, for zero rotor currents, a rather precise approach is the finite element modelling (FEM). With FEM, the slot openings could be easily accounted for; however, the computation time is prohibitive for routine calculations or optimisation design algorithms.

In what follows, we first introduce the Carter coefficient K_c to account for the slotting (slot openings) and the equivalent stack length in the presence of radial ventilation channels. Then, based on magnetic circuit and flux laws, we calculate the dependence of stator mmf per pole F_{1m} on airgap flux density, accounting for magnetic saturation in the stator and rotor teeth and back cores, while accepting a pure sinusoidal distribution of both stator mmf F_{1m} and airgap flux density, B_{1g}.

The obtained dependence of $B_{1g}(F_{1m})$ is called the magnetisation curve. Industrial experience shows that such standard methods, in modern, rather heavily saturated magnetic cores, produce notable errors in the magnetising curves, at 100%–130% rated voltage at ideal no load (zero rotor currents). The presence of heavy magnetic saturation effects, such as teeth or back core flux density flattening (or peaking), and the rough approximation of mmf calculations in the back irons are the main causes for such discrepancies.

Improved analytical methods have been proposed to produce satisfactory magnetisation curves. One of them is presented here in extenso with some experimental validation.

Based on the magnetisation curve, the magnetisation inductance is defined and calculated.

Later, the emf induced in the stator and rotor windings and the mutual stator/rotor inductances are calculated for the fundamental airgap flux density [1]. This information prepares the ground to define the parameters of the equivalent circuit of the induction machine, that is, for the computation of performance for any voltage, frequency, and speed conditions.

5.2 EQUIVALENT AIRGAP TO ACCOUNT FOR SLOTTING

The actual flux path for zero rotor currents when current in phase A is maximum $i_A = I\sqrt{2}$ and $i_B = i_C = -I\sqrt{2}/2$, obtained through FEM, is shown in Figure 5.1a [4].

The corresponding radial airgap flux density is shown in Figure 5.1b. In the absence of slotting and stator mmf harmonics, the airgap field is sinusoidal, with an amplitude of B_{g1max}.

In the presence of slot openings, the fundamental of airgap flux density is B_{g1}. The ratio of the two amplitudes is called the Carter coefficient.

$$K_C = \frac{B_{g1\,max}}{B_{g1}} \tag{5.1}$$

When the magnetic circuit is not heavily saturated, K_C may also be written as the ratio between smooth and slotted airgap magnetic permeances or between a larger equivalent airgap g_e and the actual airgap g

$$K_C = \frac{g_e}{g} \geq 1 \tag{5.2}$$

FEM allows for the calculation of Carter coefficient from (5.1) when it is applied to smooth and double slotted structure (Figure 5.1).

On the other hand, easy-to-handle analytical expressions of K_C, based on conformal transformation or flux tube methods, have been traditionally used in the absence of saturation, though. First, the airgap is split in the middle and the two slottings are treated separately. Although many other formulas have been proposed, we still present Carter's formula as it is one of the best.

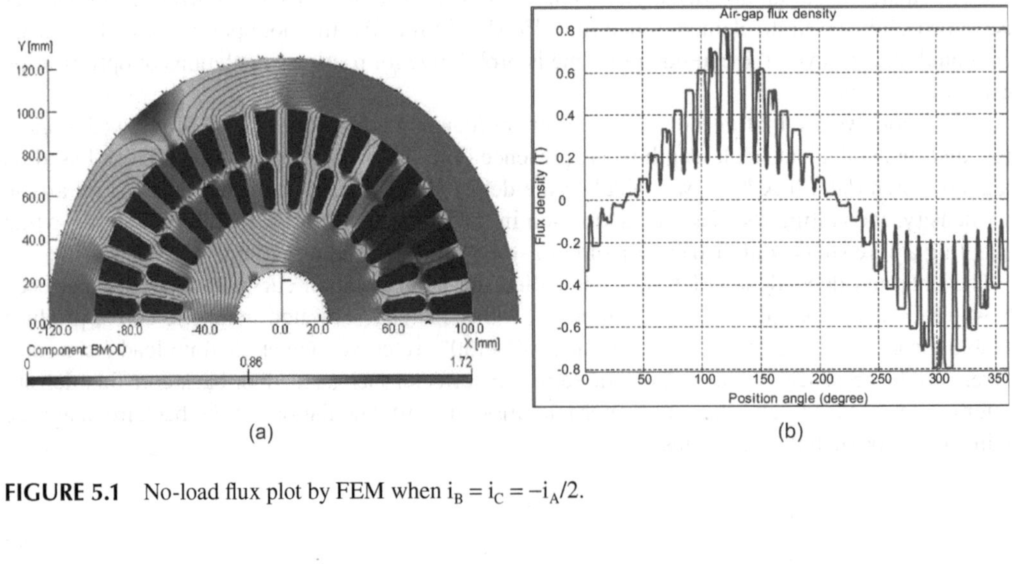

(a) (b)

FIGURE 5.1 No-load flux plot by FEM when $i_B = i_C = -i_A/2$.

$$K_{C1,2} = \frac{\tau_{s,r}}{\tau_{s,r} - \gamma_{1,2} \cdot g/2} \tag{5.3}$$

$\tau_{s,r}$ – stator/rotor slot pitch, g – the actual airgap, and

$$\gamma_{1,2} = \frac{4}{\pi}\left[\frac{b_{os,r}}{g} \Big/ \tan\left(\frac{b_{os,r}}{g}\right) - \ln\sqrt{1+\left(\frac{b_{os,r}}{g}\right)^2}\right] \approx \frac{\left(2\frac{b_{os,r}}{g}\right)^2}{5+2\frac{b_{os,r}}{g}} \tag{5.4}$$

for $b_{os,r}/g \gg 1$. In general, $b_{os,r} \approx (3-8)g$. Where $b_{os,r}$ is the stator(rotor) slot opening.

With a good approximation, the total Carter coefficient for double slotting is

$$K_C = K_{C1} \cdot K_{C2} \tag{5.5}$$

The distribution of airgap flux density for single-sided slotting is shown in Figure 5.2. Again, the iron permeability is considered to be infinite. As the magnetic circuit becomes heavily saturated, some of the flux lines touch the slot bottom (Figure 5.3) and the Carter coefficient formula has to be changed [2].

In such cases, however, we think that using FEM is the best solution.

If we introduce the relation

$$B_{\sim} = B_{g\,max} - B_{g\,min} = 2\beta B_{g\,max} \tag{5.6}$$

the flux drop (Figure 5.2) due to slotting, $\Delta\Phi$, is

$$\Delta\Phi_{s,r} = \sigma \frac{b_{os,r}}{2} B_{\sim s,r} \tag{5.7}$$

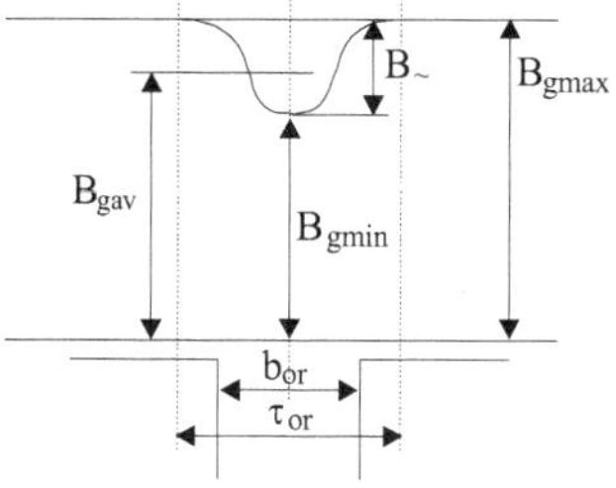

FIGURE 5.2 Airgap flux density for single slotting.

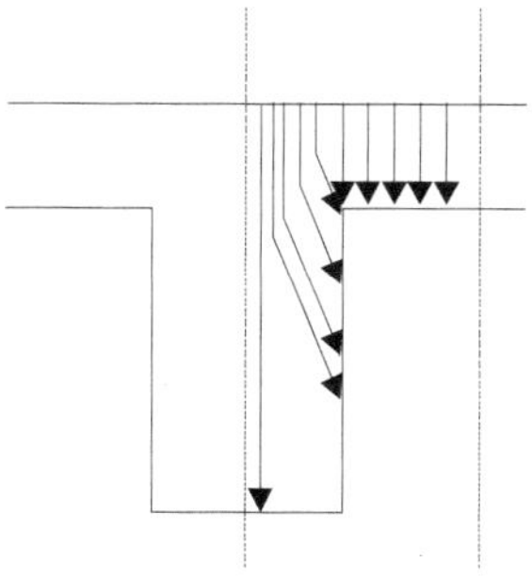

FIGURE 5.3 Flux lines in a saturated magnetic circuit.

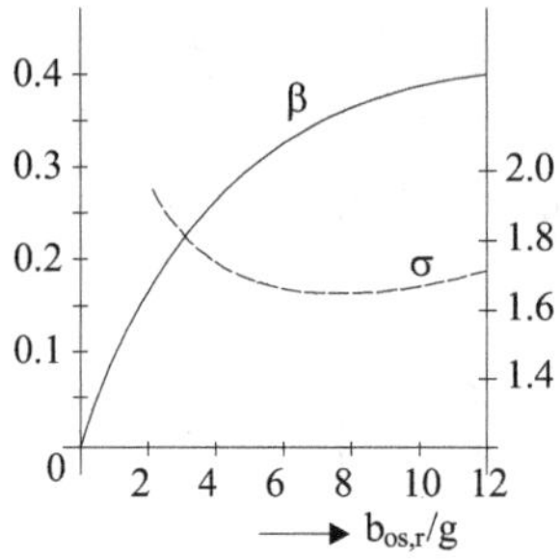

FIGURE 5.4 The factor β and σ as function of $b_{os,r}/(g/2)$.

From [3],

$$\beta\sigma b_{os,r} = \frac{g}{2}\gamma_{1,2} \tag{5.8}$$

The two factors β and σ are shown in Figure 5.4 as obtained through conformal transformation [3].

When single slotting is present, g/2 should be replaced by g.

Another slot-like situation occurs in long stacks when radial channels are placed for cooling purposes. This problem is approached next.

5.3 EFFECTIVE STACK LENGTH

Actual stator and rotor stacks are not equal in length to avoid notable axial forces, should any axial displacement of rotor occurred. In general, the rotor stack is longer than the stator stack by a few airgaps (Figure 5.5).

$$l_r = l_s + (4-6)g \tag{5.9}$$

Flux fringing occurs at stator stack ends. This effect may be accounted for by apparently increasing the stator stack by (2–3)g,

$$l_{se} = l_s + (2 \div 3)g \tag{5.10}$$

The average stack length, l_{av}, is thus

$$l_{av} \approx \frac{l_s + l_r}{2} \approx l_{se} \tag{5.11}$$

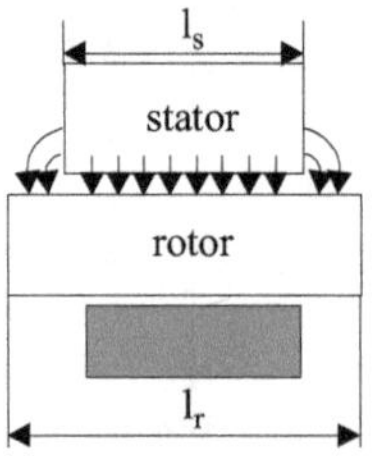

FIGURE 5.5 Single stack of stator and rotor.

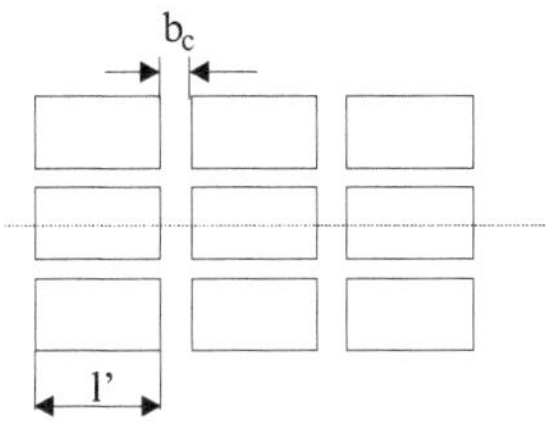

FIGURE 5.6 Multistack arrangement for radial cooling channels.

As the stacks are made of radial laminations insulated axially from each other through an enamel, the magnetic length of the stack L_e is

$$L_e = l_{av} \cdot K_{Fe} \tag{5.12}$$

The stacking factor K_{Fe} ($K_{Fe} = 0.9 - 0.95$ for $(0.35 - 0.5)$ mm thick laminations) takes into account the presence of nonmagnetic insulation between laminations.

When radial cooling channels (ducts) are used by dividing the stator into n elementary ones, the equivalent stator stack length L_e is (Figure 5.6)

$$L_e \approx (n+1)l' \cdot K_{Fe} + 2ng + 2g \tag{5.12a}$$

with

$$b_c = 5 - 10 \text{ mm}; \, l' = 100 - 250 \text{ mm} \tag{5.13}$$

It should be noted that recently, with axial cooling, longer single stacks up to 500 mm and more have been successfully built. Still, for induction motors in the MW power range, radial channels with radial cooling are in favour.

5.4 THE BASIC MAGNETISATION CURVE

The dependence of airgap flux density fundamental B_{g1} on stator mmf fundamental amplitude F_{1m} for zero rotor currents is called the magnetisation curve.

For mild levels of magnetic saturation, usual in general purpose induction motors, the stator mmf fundamental produces a sinusoidal distribution of the flux density in the airgap (slotting is neglected). As shown later in this chapter, by balancing the magnetic saturation of teeth and back cores, rather sinusoidal airgap flux density is maintained, even for heavy saturation levels.

The basic magnetisation curve ($F_{1m}(B_{g1})$ or $I_0(B_{g1})$ or I_0/I_n versus B_{g1}) is very important when designing an induction motor and influences notably the power factor and the core loss. Notice that I_0 and I_n are no-load and full-load stator phase currents (RMS) and F_{1m0} is

$$F_{1m0} = \frac{3\sqrt{2}W_1K_{w1}I_0}{\pi p_1} \tag{5.14}$$

The no-load (zero rotor current) design airgap flux density is $B_{g1} = 0.6 - 0.8$ T for 50 (60) Hz induction motors and goes down to 0.4–0.6 T for (400–1000) Hz high-speed induction motors to keep core loss within limits.

On the other hand, for 50 (60) Hz motors, I_0/I_n (no-load current/rated current) decreases with motor power from 0.5 to 0.8 (in subkW power range) to 0.2 to 0.3 in the high power range, but it increases with the number of pole pairs.

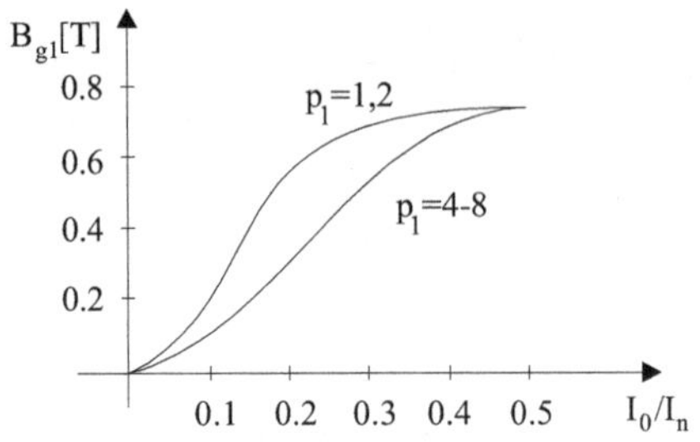

FIGURE 5.7 Typical magnetisation curves.

For low airgap flux densities, the no-load current tends to be smaller. A typical magnetisation curve is shown in Figure 5.7 for motors in the kW power range at 50 (60) Hz.

Now that we do have a general impression on the magnetising (mag.) curve, let us present a few analytical methods to calculate it.

5.4.1 The Magnetisation Curve via the Basic Magnetic Circuit

We shall examine first the flux lines corresponding to maximum flux density in the airgap and assume a sinusoidal variation of the latter along the pole pitch (Figure 5.8a and b).

$$B_{g1}(\theta_e, t) = B_{g1m}\cos(p_1\theta - \omega_1 t); \quad \theta_e = p_1\theta \tag{5.15}$$

For t = 0,

$$B_g(\theta, 0) = B_{g1m}\cos p_1\theta \tag{5.16}$$

The stator (rotor) back iron flux density $B_{cs,r}$ is

$$B_{cs,r} = \frac{1}{2h_{cs,r}}\int_0^\theta B_{g1}(\theta, t)d\theta \cdot \frac{D}{2} \tag{5.17}$$

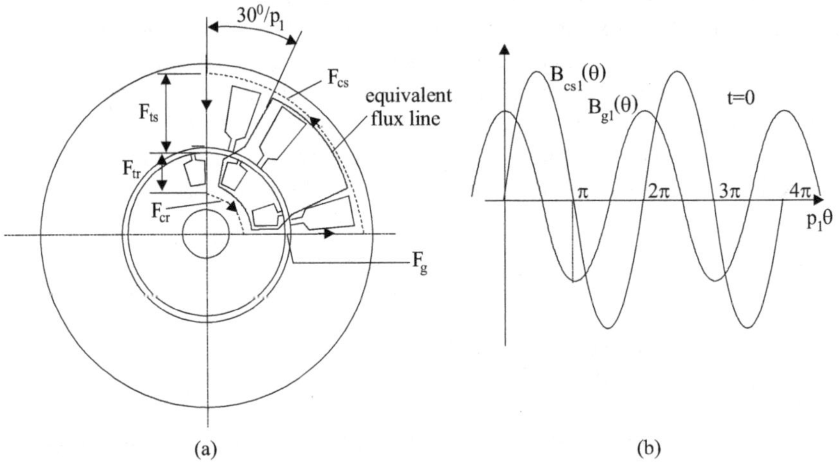

FIGURE 5.8 Flux path (a) and flux density types (b): ideal distribution in the airgap and stator core.

where $h_{cs,r}$ is the back core height in the stator (rotor). For the flux line in Figure 5.8a ($\theta = 0$ to π/p_1),

$$B_{cs1}(\theta, t) = \frac{1}{2}\frac{2}{\pi}\frac{\tau}{h_{cs}} \cdot B_{g1m} \sin(p_1\theta - \omega_1 t); \quad \tau = \frac{\pi D}{2p_1} \tag{5.18}$$

Due to mmf and airgap flux density sinusoidal distribution along the motor periphery, it suffices to analyse the mmf iron and airgap components F_{ts}, F_{tr} in teeth, F_g in the airgap, and F_{cs}, F_{cr} in the back cores. The total mmf is represented by F_{1m} (peak values).

$$2F_{1m} = 2F_g + 2F_{ts} + 2F_{tr} + F_{cs} + F_{cr} \tag{5.19}$$

Equation (5.19) reflects the application of the magnetic circuit (Ampere's) law along the flux line in Figure 5.8a.

In industry, to account for the flattening of the airgap flux density due to teeth saturation, B_{g1m} is replaced by the actual (designed) maximum flattened flux density B_{gm}, at an angle $\theta = 30°/p_1$, which makes the length of the flux lines in the back core 2/3 of their maximum length.

Then finally, the calculated I_{1m} is multiplied by $2/\sqrt{3}$ (1/cos30°) to find the maximum mmf fundamental.

At $\theta_{er} = p_1\theta = 30°$, it is supposed that the flattened and sinusoidal flux densities are equal to each other (Figure 5.9).

We have to write again Ampere's law for this case (interior flux line in Figure 5.8a).

$$2(F_1)_{30°} = 2F_g(B_{gm}) + 2F_{ts} + 2F_{tr} + F_{cs} + F_{cr} \tag{5.20}$$

and finally,

$$2F_{1m} = \frac{2(F_1)_{30°}}{\cos 30°} \tag{5.21}$$

For the sake of generality, we will use (5.20) and (5.21), remembering that the length of average flux line in the back cores is 2/3 of its maximum.

Let us proceed directly with a numerical example by considering an induction motor with the geometry in Figure 5.10.

$$\begin{gathered} 2p_1 = 4;\ D = 0.1\text{ m};\ D_e = 0.176\text{ m};\ h_s = 0.025\text{ m};\ N_s = 24; \\ N_r = 18;\ b_{r1} = 1.2b_{tr1};\ b_{s1} = 1.4b_{ts1};\ g = 0.5\cdot 10^{-3}\text{ m}; \\ h_r = 0.018\text{ m};\ D_{shaft} = 0.035\text{ m};\ B_{gm} = 0.7\text{ T} \end{gathered} \tag{5.22}$$

The B/H curve of the rotor and stator laminations is given in Table 5.1.

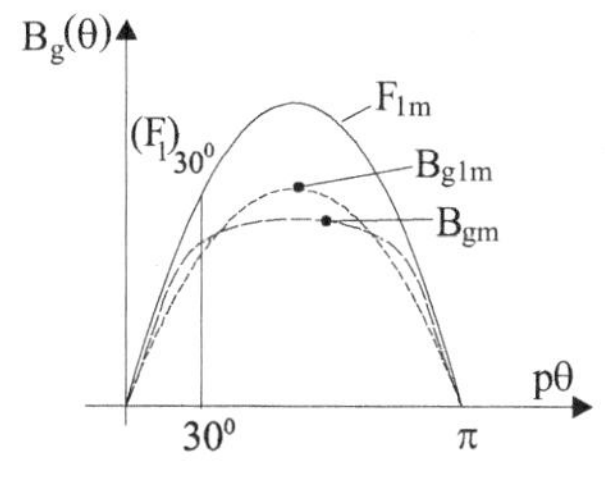

FIGURE 5.9 Sinusoidal and flat airgap flux density.

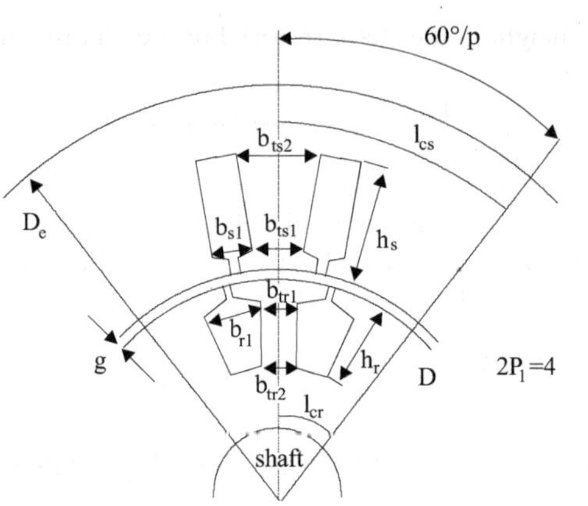

FIGURE 5.10 IM geometry for magnetisation curve calculation.

TABLE 5.1
B/H Curve for a Typical IM Lamination

B[T]	0.0	0.05	0.1	0.15	0.2	0.25	0.3	0.35	0.4	0.45	0.5	0.55	0.6
H[A/m]	0	22.8	35	45	49	57	65	70	76	83	90	98	206
B[T]	0.65	0.7	0.75	0.8	0.85	0.9	0.95	1	1.05	1.1	1.15	1.2	1.25
H[A/m]	115	124	135	148	177	198	198	220	237	273	310	356	417
B[T]	1.3	1.35	1.4	1.45	1.5	1.55	1.6	1.65	1.7	1.75	1.8	1.85	1.9
H[A/m]	482	585	760	1050	1340	1760	2460	3460	4800	6160	8270	11,170	15,220
B[T]	1.95	2.0											
H[A/m]	22,000	34,000											

Based on (5.20) and (5.21), Gauss law, and B/H curve in Table 5.1, let us calculate the value of F_{1m}.

To solve the problem in a rather simple way, we still assume a sinusoidal flux distribution in the back cores, based on the fundamental of the airgap flux density B_{g1m}.

$$B_{g1m} = \frac{B_{gm}}{\cos 30^\circ} = 0.7\frac{2}{\sqrt{3}} = 0.809\ \text{T} \tag{5.23}$$

The maximum stator and rotor back core flux densities are obtained from (5.18)

$$B_{csm} = \frac{1}{\pi}\cdot\frac{\pi D}{2p_1}\frac{1}{h_{cs}}B_{g1m} \tag{5.24}$$

$$B_{crm} = \frac{1}{\pi}\cdot\frac{\pi D}{2p_1}\frac{1}{h_{cr}}B_{g1m} \tag{5.25}$$

with

$$h_{cs} = \frac{(D_e - D - 2h_s)}{2} = \frac{0.174 - 0.100 - 2\cdot 0.025}{2} = 0.013\ \text{m} \tag{5.26}$$

$$h_{cr} = \frac{(D - 2g - D_{shaft} - 2h_r)}{2} = \frac{0.100 - 0.001 - 2\cdot 0.018 - 0.036}{2} = 0.0135\ \text{m} \tag{5.27}$$

Now from (5.24) and (5.25),

$$B_{csm} = \frac{0.1 \cdot 0.809}{4 \cdot 0.013} = 1.555\ \text{T} \tag{5.28}$$

$$B_{crm} = \frac{0.1 \cdot 0.809}{4 \cdot 0.0135} = 1.498\ \text{T} \tag{5.29}$$

As the core flux density varies from the maximum value cosinusoidally, we may calculate an average value of three points, say B_{csm}, $B_{csm}\cos 60°$, and $B_{csm}\cos 30°$

$$B_{csav} = B_{csm}\left(\frac{1 + 4\cos 60° + \cos 30°}{6}\right) = 1.555 \cdot 0.8266 = 1.285\ \text{T} \tag{5.30}$$

$$B_{crav} = B_{crm}\left(1 + \frac{4\cos 60° + \cos 30°}{6}\right) = 1.498 \cdot 0.8266 = 1.238\ \text{T} \tag{5.31}$$

From Table 5.1, we obtain the magnetic fields corresponding to the above flux densities. Finally, H_{csav} (1.285) = 460 A/m and H_{crav} (1.238) = 400 A/m.

Now the average length of flux lines in the two back irons are

$$l_{csav} \approx \frac{2}{3} \cdot \frac{\pi(D_e - h_{cs})}{2P_1} = \frac{2}{3}\pi\frac{(0.176 - 0.013)}{4} = 0.0853\ \text{m} \tag{5.32}$$

$$l_{crav} \approx \frac{2}{3} \cdot \frac{\pi(D_{shaft} + h_{cr})}{2P_1} = \frac{2}{3}\pi\frac{(0.035 + 0.0135)}{4} = 0.02593\ \text{m} \tag{5.33}$$

Consequently, the back core mmfs are

$$F_{cs} = l_{csav} \cdot H_{csav} = 0.0853 \cdot 460 = 39.238\ \text{Aturns} \tag{5.34}$$

$$F_{cr} = l_{crav} \cdot H_{crav} = 0.0259 \cdot 400 = 10.362\ \text{Aturns} \tag{5.35}$$

The airgap mmf F_g is straightforward.

$$2F_g = 2g \cdot \frac{B_{gm}}{\mu_0} = 2 \cdot 0.5 \cdot 10^{-3} \cdot \frac{0.7}{1.256 \cdot 10^{-6}} = 557.324\ \text{Aturns} \tag{5.36}$$

Assuming that all the airgap flux per slot pitch traverses the stator and rotor teeth, we have

$$B_{gm} \cdot \frac{\pi D}{N_{s,r}} = (B_{ts,r})_{av} \cdot (b_{ts,r})_{av}; \quad b_{ts,r} = \frac{\pi(D \pm h_{s,r})}{N_{s,r}(1 + b_{s,r}/b_{ts,r})} \tag{5.37}$$

Considering that the teeth flux is conserved (it is purely radial), we may calculate the flux density at the tooth bottom and top as we know the average tooth flux density for the average tooth width $(b_{ts,r})_{av}$. An average can be applied here again. For our case, let us consider $(B_{ts,r})_{av}$ all over the teeth height to obtain

$$(b_{ts})_{av} = \frac{\pi(0.1 + 0.025)}{(1 + 1.4) \cdot 24} = 6.81 \cdot 10^{-3}\ \text{m};$$

$$(b_{tr})_{av} = \frac{\pi(0.1 - 0.018)}{(1 + 1.2) \cdot 18} = 6.00 \cdot 10^{-3}\ \text{m} \tag{5.38}$$

$$
\begin{aligned}
(B_{ts})_{av} &= \frac{0.7\cdot\pi\cdot100}{24\cdot7.43\cdot10^{-3}} = 1.344\ \text{T} \\
(B_{tr})_{av} &= \frac{0.7\cdot\pi\cdot100}{18\cdot6.56\cdot10^{-3}} = 1.878\ \text{T}
\end{aligned} \tag{5.39}
$$

From Table 5.1, the corresponding values of $H_{tsav}(B_{tsav})$ and $H_{trav}(B_{trav})$ are found to be H_{tsav} = 520 A/m and H_{trav} = 13,600 A/m. Now the teeth mmfs are

$$2F_{ts} \approx H_{tsav}(2h_s) = 520\cdot(0.025\cdot2) = 26\ \text{Aturns} \tag{5.40}$$

$$2F_{tr} \approx H_{trav}(2h_r) = 13{,}600\cdot(0.018\cdot2) = 489.6\ \text{Aturns} \tag{5.41}$$

The total stator mmf $(F_1)_{30^\circ}$ is calculated from (5.20)

$$2(F_1)_{30^\circ} = 557.324 + 26 + 489.6 + 39.238 + 10.362 = 1122.56\ \text{Aturns} \tag{5.42}$$

The mmf amplitude F_{m10} (from 5.21) is

$$2F_{1m0} = \frac{2(F_1)_{30^\circ}}{\cos 30^\circ} = 1122.56\cdot\frac{2}{\sqrt{3}} = 1297.764\ \text{Aturns} \tag{5.43}$$

Based on (5.14), the no-load current may be calculated with the number of turns/phase W_1 and the stator winding factor K_{w1} already known.

Varying the value of B_{gm} as desired, the magnetisation curve – $B_{gm}(F_{1m})$ – is obtained.

Before leaving this subject, let us remember the numerous approximations we operated with and define two partial and one equivalent saturation factor as K_{st}, K_{sc}, and K_s.

$$K_{st} = 1 + \frac{2(F_{ts}+F_{tr})}{2F_g};\quad K_{sc} = 1 + \frac{(F_{cs}+F_{cr})}{2F_g} \tag{5.44}$$

$$K_s = \frac{(F_1)_{30^\circ}}{2F_g} = K_{st} + K_{sc} - 1 \tag{5.45}$$

The total saturation factor K_s accounts for all iron mmfs as divided by the airgap mmf. Consequently, we may consider the presence of iron as an increased airgap g_{es}.

$$g_{es} = gK_cK_s;\quad K_s = K_c\left(\frac{I_0}{I_n}\right) \tag{5.46}$$

Let us notice that in our case,

$$
\begin{aligned}
&K_{st} = 1 + \frac{26+489.6}{557.324} = 1.925;\quad K_{ct} = 1 + \frac{39.23+10.362}{557.324} = 1.089 \\
&K_s = 1.925 + 1.089 - 1 = 2.103
\end{aligned} \tag{5.47}
$$

A few remarks are in order.

- The teeth saturation factor K_{st} is notable while the core saturation factor is low; so the tooth are much more saturated (especially in the rotor, in our case); as shown later in this chapter, this is consistent with the flattened airgap flux density.

- In a rather proper design, the teeth and core saturation factors K_{st} and K_{sc} are close to each other: $K_{st} \approx K_{sc}$; in this case, both the airgap and core flux densities remain rather sinusoidal even if rather high levels of saturation are encountered.
- In two-pole machines, however, K_{sc} tends to be higher than K_{st} as the back core height tends to be large (large pole pitch) and its reduction in size is required to reduce motor weight.
- In mildly saturated IMs, the total saturation factor is smaller than in our case $K_s = 1.3 - 1.6$.

Based on the above theory, iterative methods, to obtain the airgap flux density distribution and its departure from a sinusoid (for a sinusoidal core flux density), have been recently introduced [2,4]. However, the radial flux density components in the back cores are still neglected.

5.4.2 Teeth Defluxing by Slots

So far, we did assume that all the flux per slot pitch goes radially through the teeth. Especially with heavily saturated teeth, a good part of magnetic path passes through the slot itself. Thus, the tooth is slightly "discharged" of flux.

We may consider that the following are approximates

$$B_t = B_{ti} - c_1 B_g b_{s,r}/b_{ts,r}\,;\; c_1 \ll 1.0 \tag{5.48}$$

$$B_{ti} = B_g \frac{b_{ts,r} + b_{s,r}}{b_{ts,r}} \tag{5.49}$$

The coefficient c_1 is, in general, adopted from experience, but it is strongly dependent on the flux density in the teeth, B_{ti}, and on the slotting geometry (including slot depth [2]).

5.4.3 Third Harmonic Flux Modulation due to Saturation

As only inferred above, heavy saturation in stator (rotor) teeth and/or back cores tends to flatten or peak, respectively, the airgap flux distribution.

This proposition can be demonstrated by noting that the back core flux density $B_{cs,r}$ is related to airgap (implicitly teeth) flux density by the equation

$$B_{cs,r} = C_{s,r} \int_0^{\theta_{er}} B_{ts,r}(\theta_{er})\, d\theta_{er}; \quad \theta_{er} = p_1\theta \tag{5.50}$$

Magnetic saturation in the teeth means flattening $B_{ts,r}(\theta)$ curve.

$$B_{ts,r}(\theta) = B_{ts,r1}\cos(\theta_{er} - \omega_1 t) + B_{ts,r3}\cos(3\theta_{er} - \omega_1 t) \tag{5.51}$$

Consequently, $B_{ts,r3} > 0$ means unsaturated teeth (peaked flux density, Figure 5.11a). With (5.51), Equation (5.50) becomes

$$B_{cs,r} = C_{s,r}\left[B_{ts,r1}\sin(\theta_{er} - \omega_1 t) + \frac{B_{ts,r3}}{3}\sin(3\theta_{er} - \omega_1 t)\right] \tag{5.52}$$

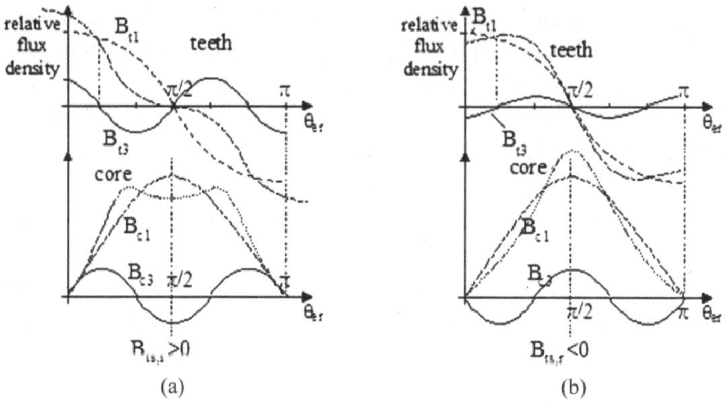

FIGURE 5.11 Tooth and core flux density distribution: (a) saturated back core ($B_{ts,r3} > 0$) and (b) saturated teeth ($B_{ts,r3} < 0$).

Analysing Figure 5.11, based on (5.51) and (5.52), leads to remarks such as

- Oversaturation of a domain (teeth or core) means flattened flux density in that domain (Figure 5.11b).
- In Section 5.4.1, we have considered flattened airgap flux density – that is also flattened tooth flux density – and thus oversaturated teeth is the case treated.
- The flattened flux density in the teeth (Figure 5.11b) leads to only a slightly peaked core flux density as the denominator 3 occurs in the second term of (5.52).
- On the contrary, a peaked teeth flux density (Figure 5.11a) leads to a flat core density. The back core is now oversaturated.
- We should also mention that the phase connection is important in third harmonic flux modulation. For sinusoidal voltage supply and delta connection, the third harmonic of flux (and its induced voltage) cannot exist, while it can for star connection. This phenomenon will also have consequences in the phase current waveforms for the two connections. Finally, the saturation produced third and other harmonics influence, notably the core loss in the machine. This aspect will be discussed in Chapter 11 dedicated to losses.

After describing some aspects of saturation-caused distribution modulation, let us present a more complete analytical non-linear field model, which allows for the calculation of actual spatial flux density distribution in the airgap, though with smoothed airgap.

5.4.4 The Analytical Iterative Model (AIM)

Let us remind here that essentially only FEM [5] or extended magnetic circuit methods (EMCMs) [6] are able to produce a rather fully realistic field distribution in the induction machine. However, they do so with large computation efforts and may be used for design refinements rather than for preliminary or direct optimisation design algorithms.

A fast analytical iterative (non-linear) model (AIM) [7] is introduced here for preliminary or optimisation design uses.

The following assumptions are made: only the fundamental of mmf distribution is considered; the stator and rotor currents are symmetric; the IM cross section is divided into five circular domains (Figure 5.12) with unique (but adjustable) magnetic permeabilities, essentially distinct along radial (r) and tangential (θ) directions: μ_r and μ_θ.

The magnetic vector potential A lays along the shaft direction, and thus, the model is two-dimensional; furthermore, the separation of variables is performed.

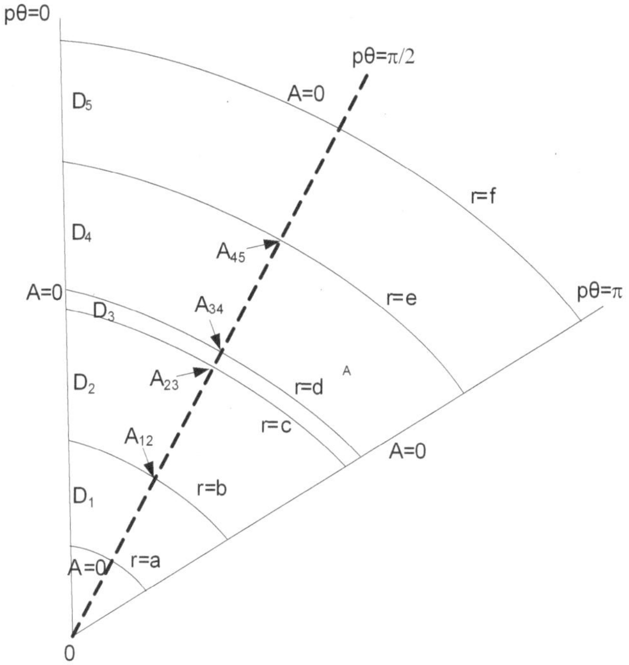

FIGURE 5.12 The IM cross section divided into five domains.

The Poisson equation in polar coordinates for magnetic vector potential A writes

$$\frac{1}{\mu_\theta}\left(\frac{1}{r}\frac{\partial A}{\partial r}+\frac{\partial^2 A}{\partial r^2}\right)+\frac{1}{\mu_r}\frac{1}{r^2}\frac{\partial^2 A}{\partial \theta^2}=-J \tag{5.53}$$

Separating the variables, we obtain:

$$A(r,\theta)=R(r)\cdot T(\theta) \tag{5.54}$$

Now, for the domains with zero current (D_1, D_3, D_5) – J = O, Equation (5.53) with (5.54) yields:

$$\frac{1}{R(r)}\left(r\frac{dR(r)}{dr}+r^2\frac{d^2R(r)}{dr^2}\right)=\lambda^2;\ \frac{\alpha^2}{T(\theta)}\frac{d^2T(\theta)}{d\theta^2}=-\lambda^2 \tag{5.55}$$

with $\alpha^2 = \mu_\theta/\mu_r$ and λ a constant.

5.4.4.1 Magnetic Potential, A, Solution

Also a harmonic distribution along θ direction was assumed. From (5.55),

$$r^2\frac{d^2R(r)}{dr^2}+r\frac{dR(r)}{dr}-\lambda^2R=0 \tag{5.56}$$

and

$$\frac{d^2T(\theta)}{d\theta^2}+\frac{\lambda^2}{\alpha^2}T=0 \tag{5.57}$$

The solutions of (5.56) and (5.57) are of the form

$$R(r) = C_1 r^{\lambda} + C_2 r^{-\lambda} \tag{5.58}$$

$$T(\theta) = C_3 \cos\left(\frac{\lambda}{\alpha}\theta\right) + C_4 \sin\left(\frac{\lambda}{\alpha}\theta\right) \tag{5.59}$$

as long as $r \neq 0$.

Assuming further symmetric windings and currents, the magnetic potential is a periodic function and thus,

$$A(r,0) = 0 \tag{5.60}$$

$$A\left(r, \frac{\pi}{P_1}\right) = 0 \tag{5.61}$$

Consequently, from (5.59), (5.57), and (5.53), $A(r,\theta)$ is

$$A(r,\theta) = \left(g \cdot r^{P_1\alpha} + h \cdot r^{-P_1\alpha}\right)\sin(P_1\theta) \tag{5.62}$$

Now, if the domain contains a homogenous current density J,

$$J = J_m \sin(P_1\theta) \tag{5.63}$$

the particular solution $A_p(r,\theta)$ of (5.54) is

$$A_p(r,\theta) = K_r^{\,2} \sin(P_1\theta) \tag{5.64}$$

with

$$K = -\frac{\mu_r \mu_\theta}{4\mu_r - P_1^2 \mu_\theta} J_m \tag{5.65}$$

Finally, the general solution of A (5.53) is

$$A(r,\theta) = \left(g \cdot r^{p_1\alpha} + h \cdot r^{-p_1\alpha} + Kr^2\right)\sin(p_1\theta) \tag{5.66}$$

As (5.66) is valid for homogenous media, we have to homogenise the slotting domains D_2 and D_4, as the rotor and stator yokes (D_1, D_5) and the airgap (D_3) are homogenous.

5.4.4.1.1 Homogenising the Slotting Domains

The main practical slot geometries (Figure 5.13) are defined by equivalent centre angles θ_s and θ_t for an equivalent (defined) radius r_{m4} (for the stator) and r_{m2} (for the rotor). Assuming that the radial magnetic field H is constant along the circles r_{m2} and r_{m4}, the flux linkage equivalence between the homogenised and slotting areas yields

$$\mu_t H\theta_t r_x L_1 + \mu_0 H\theta_s r_x L_1 = \mu_r H(\theta_t + \theta_s) r_x L_1 \tag{5.67}$$

Consequently, the equivalent radial permeability μ_r is

$$\mu_r = \frac{\mu_t\theta_t + \mu_0\theta_s}{\theta_t + \theta_s} \tag{5.68}$$

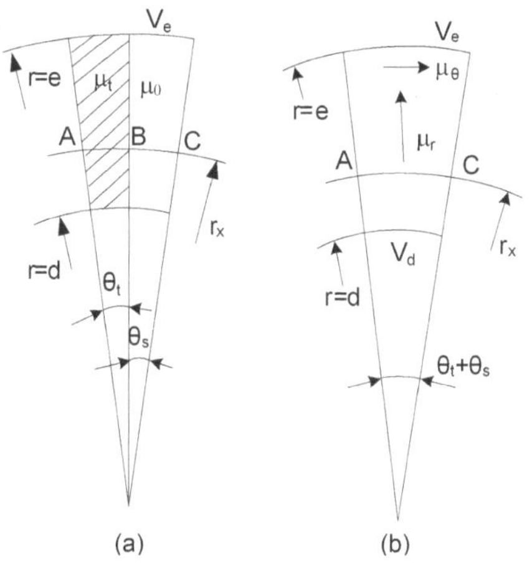

FIGURE 5.13 Stator (a) and rotor (b) slotting.

For the tangential field, the magnetic voltage relationship (along A, B, C trajectory on Figure 5.13) is

$$V_{mAC} = V_{mAB} + V_{mBC} \tag{5.69}$$

With B_θ the same, we obtain

$$\frac{B_\theta}{\mu_\theta}(\theta_t + \theta_s) r_x = \frac{B_\theta}{\mu_t}\theta_t r_x + \frac{B_\theta}{\mu_0}\theta_s r_x \tag{5.70}$$

Consequently,

$$\mu_\theta = \frac{\mu_t \mu_0 (\theta_t + \theta_s)}{\mu_0 \theta_t + \mu_t \theta_s} \tag{5.71}$$

Thus, the slotting domains are homogenised to be characterised by distinct permeabilities μ_r and μ_θ along the radial and tangential directions, respectively.

We may now summarise the magnetic potential expressions for the five domains:

$$\begin{aligned}
A_1(r,\theta) &= \left(g_1 r^{p_1} + h_1 r^{-p_1}\right)\sin(p_1\theta); \quad a' = \frac{a}{2} < r < b \\
A_3(r,\theta) &= \left(g_3 r^{p_1} + h_3 r^{-p_1}\right)\sin(p_1\theta); \quad c < r < d \\
A_5(r,\theta) &= \left(g_5 r^{p_1} + h_5 r^{-p_1}\right)\sin(p_1\theta); \quad e < r < f \\
A_2(r,\theta) &= \left(g_2 r^{p_1\alpha_2} + h_2 r^{-p_1\alpha_2} + k_2 r^2\right)\sin(p_1\theta); \quad b < r < c \\
A_4(r,\theta) &= \left(g_4 r^{p_1\alpha_4} + h_4 r^{-p_1\alpha_4} + k_4 r^2\right)\sin(p_1\theta); \quad d < r < e
\end{aligned} \tag{5.72}$$

with

$$\begin{aligned}
K_4 &= -\frac{\mu_{r4}\mu_{\theta 4}}{4\mu_{r4} - P_1^2 \mu_{\theta 4}} J_{m4} \\
K_2 &= +\frac{\mu_{r2}\mu_{\theta 2}}{4\mu_{r2} - P_1^2 \mu_{\theta 2}} J_{m2}
\end{aligned} \tag{5.73}$$

J_{m2} represents the equivalent demagnetising rotor equivalent current density which justifies the ⊕ sign in the second equation of (5.73). From geometrical considerations, J_{m2} and J_{m4} are related to the reactive stator and rotor phase currents I_{1s} and I'_{2r}, by the expressions (for the three-phase motor),

$$J_{m2} = \frac{6\sqrt{2}}{\pi}\frac{1}{c^2 - b^2}W_1K_{w1}I'_{2r}$$
$$J_{m4} = \frac{6\sqrt{2}}{\pi}\frac{1}{e^2 - d^2}W_1K_{w1}I_{1r} \tag{5.74}$$

The main pole-flux linkage Ψ_{m1} is obtained through the line integral of A_3 around a pole contour Γ (L_1, the stack length):

$$\psi_{m1} = \oint_{\Gamma} \overline{A}_3 d\overline{l} = 2L_1A_3\left(d, \frac{\pi}{2P_1}\right) \tag{5.75}$$

Notice that $A_3(d, \pi/2p_1,) = -A_3(d, -\pi/2p_1,)$ because of symmetry. With A_3 from (5.72), ψ_{m1} becomes

$$\psi_{m1} = 2L_1\left(g_3d^{p_1} + h_3d^{-p_1}\right) \tag{5.76}$$

Finally, the emf, E_1, (RMS value) is

$$E_1 = \pi\sqrt{2}f_1W_1K_{w1}\psi_{m1} \tag{5.77}$$

Now from boundary conditions, the integration constants, g_i and h_i, are calculated as shown in the appendix of Ref. [7].

5.4.4.2 The Computer Program

To prepare the computer program, we have to specify a few very important details. First, instead of a (the shaft radius), the first domain starts at a′ = a/2 to account for the shaft field for the case when the rotor laminations are placed directly on the shaft. Further, each domain is characterised by an equivalent (but adjustable) magnetic permeability. Here we define it. For the rotor and stator domains D_1 and D_5, the equivalent permeability would correspond to $r_{m1} = (a + b)/2$ and $r_{m5} = (e + f)/2$ and tangential flux density (and $\theta_0 = \pi/4$).

$$B_{1\theta}(r_{m1}, \theta_0) = -P_1\left[g_1\left(\frac{a+b}{2}\right)^{P_1-1} - h_1\left(\frac{a+b}{2}\right)^{-P_1-1}\right]\sin\frac{\pi}{4}$$
$$B_{5\theta}(r_{m5}, \theta_0) = -P_1\left[g_5\left(\frac{e+f}{2}\right)^{P_1-1} - h_5\left(\frac{e+f}{2}\right)^{-P_1-1}\right]\sin\frac{\pi}{4} \tag{5.78}$$

For the slotting domains D_2 and D_4, the equivalent magnetic permeabilities correspond to the radiuses r_{m2} and r_{m4} (Figure 5.13) and the radial flux densities B_{2r} and B_{4r}.

$$B_{2r}(r_{m2}, \theta_0) = p_1\left(g_2{r_{m2}}^{p_1\alpha_2-1} + h_2{r_{m2}}^{-p_1\alpha_2-1} + K_2r_{m2}\right)\cos p_1\theta_0$$
$$B_{4r}(r_{m4}, \theta_0) = p_1\left(g_4{r_{m4}}^{p_1\alpha_4-1} + h_4{r_{m4}}^{-p_1\alpha_4-1} + K_4r_{m4}\right)\cos p_1\theta_0 \tag{5.79}$$

with $\cos P_1\theta_0 = 0.9 \ldots 0.95$. Now to keep track of the actual saturation level, the actual tooth flux densities B_{4t} and B_{2t}, corresponding to B_{2r}, $H_{2r,}$ and B_{4r}, H_{4r} are

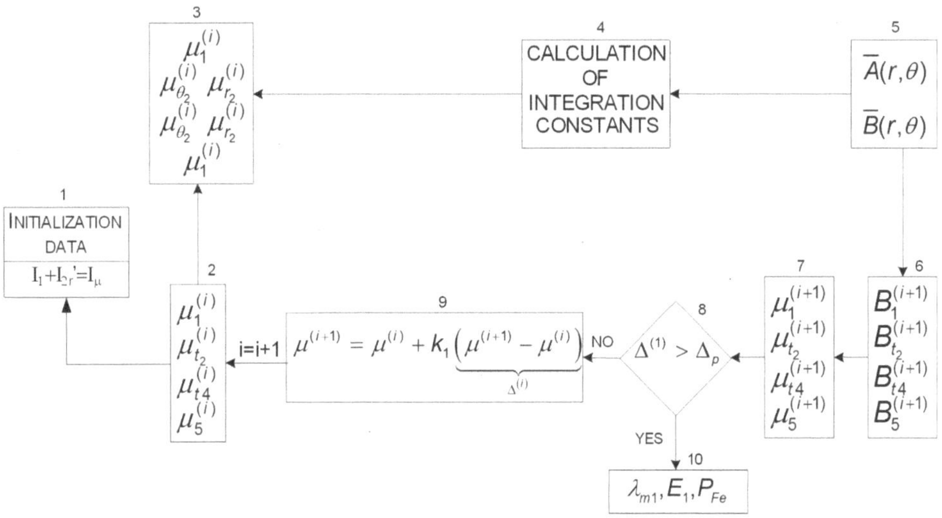

FIGURE 5.14 The computation algorithm.

$$B_{2t} = \frac{\theta_{t1} + \theta_{s2}}{\theta_{t2}} B_{2r} - c_0 \frac{\theta_{s2}}{\theta_{t2}} \mu_0 H_{2r}$$

$$B_{4t} = \frac{\theta_{t4} + \theta_{s4}}{\theta_{t4}} B_{4r} - c_0 \frac{\theta_{s4}}{\theta_{t4}} \mu_0 H_{4r} \quad (5.80)$$

The empirical coefficient c_0 takes into account the tooth magnetic unloading due to the slot flux density contribution.

Finally, the computing algorithm is shown in Figure 5.14 and starts with initial equivalent permeabilities.

For the next cycle of computation, each permeability is changed according to

$$\mu^{(i+1)} = \mu^{(i)} + c_1\left(\mu^{(i+1)} - \mu^{(i)}\right) \quad (5.81)$$

It has been proved that $c_1 = 0.3$ is an adequate value, for a wide power range.

5.4.4.3 Model Validation on No Load

The AIM has been applied to 12 different three-phase IMs from 0.75 to 15 kW (two-pole and four-pole motors) to calculate both the magnetisation curve $I_{10} = f(E_1)$ and the core losses on no load for various voltage levels.

The magnetising current is $I_\mu = I_{1r}$ and the no-load active current I_{0A} is

$$I_{0A} = \frac{P_{Fe} + P_{mv} + P_{copper}}{3V_0} \quad (5.82)$$

The no-load current I_{10} is thus

$$I_{10} = \sqrt{I_\mu^{\ 2} + I_{0A}^{\ 2}} \quad (5.83)$$

and

$$E_1 \approx V_0 - \omega_1 L_{1\sigma} I_{10} \quad (5.84)$$

where $L_{1\sigma}$ is the stator leakage inductance (known). Complete expressions of leakage inductances are introduced in Chapter 6.

Figure 5.15 exhibits computation results obtained with the conventional non-linear model, the proposed model, and experimental data, for up to 130% of rated voltage. It seems clear that AIM produces very good agreement with experiments even for high saturation levels.

A few remarks on AIM seem in order

- The slotting is only globally accounted for by defining different tangential and radial permeabilities in the stator and rotor teeth.
- A single, but variable, permeability characterises each of the machine domains (teeth and back cores).
- AIM is a bidirectional field approach, and thus, both radial and tangential flux density components are calculated.
- Provided the rotor equivalent current I_r (RMS value and phase shift with respect to stator current) is given, AIM allows to calculate the distribution of main flux in the machine on load.
- Heavy saturation levels ($V_0/V_n > 1$) are handled satisfactorily by AIM.
- By skewing the stack axially, the effect of skewing on main flux distribution can be handled.
- The computation effort is minimal (a few seconds per run on a contemporary PC).

So far, AIM was used considering that the spatial field distribution is sinusoidal along stator bore. In reality, it may depart from this situation as shown in the previous paragraph. We may use repeatedly AIM to produce the actual spatial flux distribution or the airgap flux harmonics.

AIM may be used to calculate saturation-caused harmonics. The total mmf F_{1m} is still considered sinusoidal (Figure 5.16).

The maximum airgap flux density B_{g1m} (sinusoidal in nature) is considered known by using AIM for given stator (and eventually also rotor) current RMS values and phase shifts.

By repeatedly using the Ampere's law on contours such as those in Figure 5.17 at different position θ, we may find the actual distribution of airgap flux density by admitting that the tangential flux density in the back core retains the sinusoidal distribution along θ. This assumption is not, in general, far from reality as was shown in Section 5.4.3.

Ampere's law on contour Γ (Figure 5.16) is

$$\oint_{\Gamma} \overline{H}\overline{dl} = \int_{-P_1\theta}^{P_1\theta} F_{1m} \sin p_1\theta d(P_1\theta) = 2F_{1m} \cos p_1\theta \tag{5.85}$$

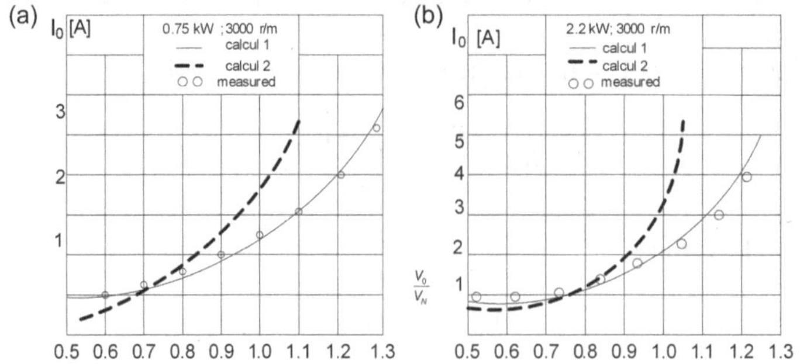

FIGURE 5.15 Magnetisation characteristic validation on no load: (a) AIM; (b) conventional non-linear model (Section 5.4.1).

(*Continued*)

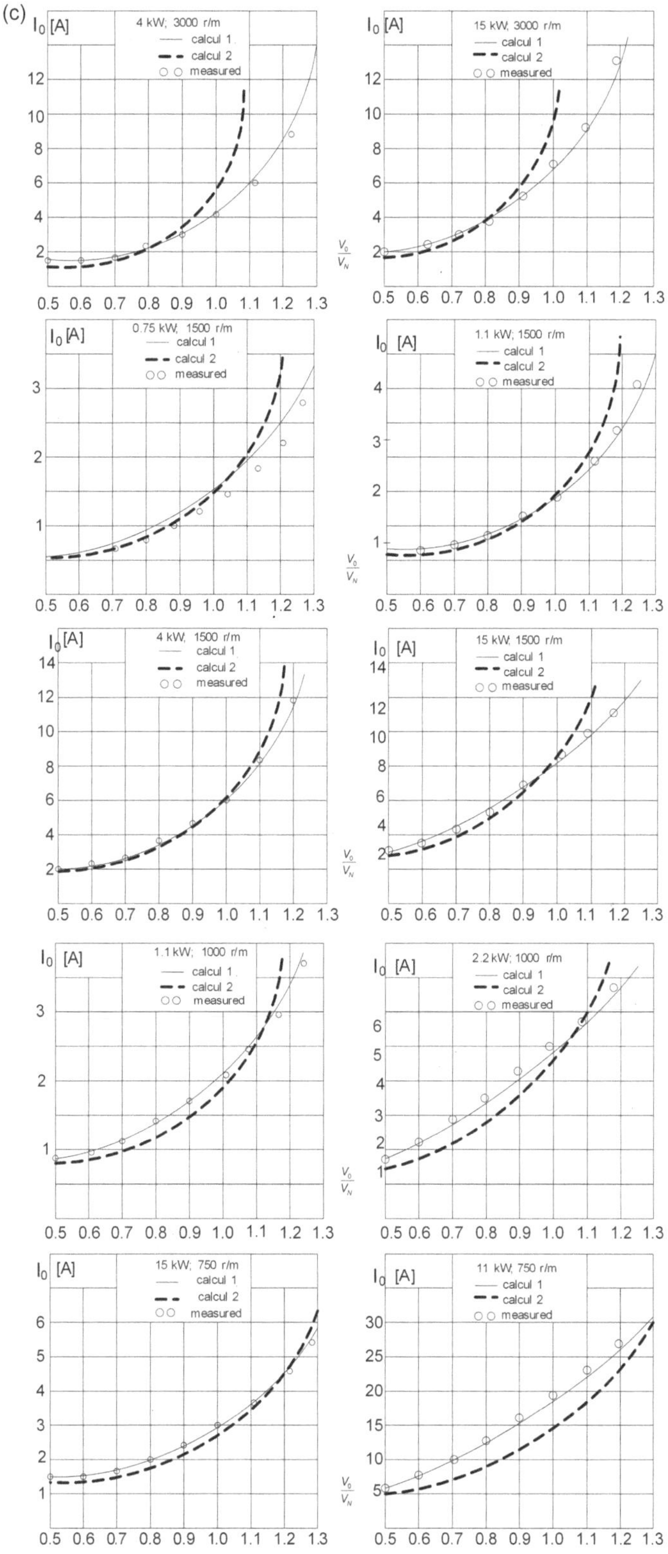

FIGURE 5.15 (CONTINUED) (c) experiments.

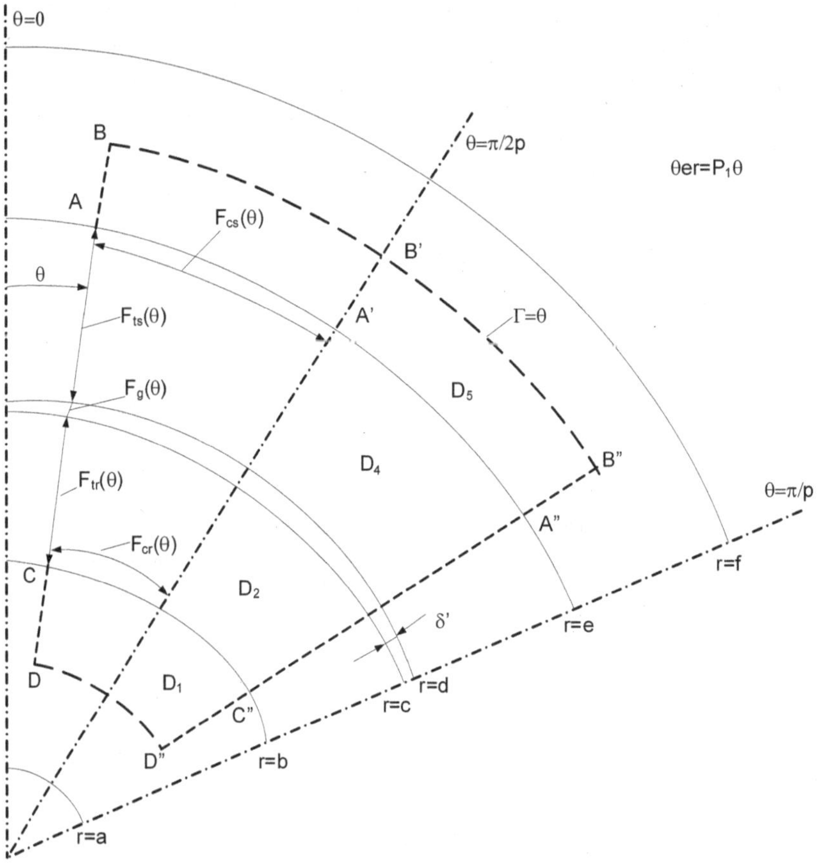

FIGURE 5.16 Ampere's law contours.

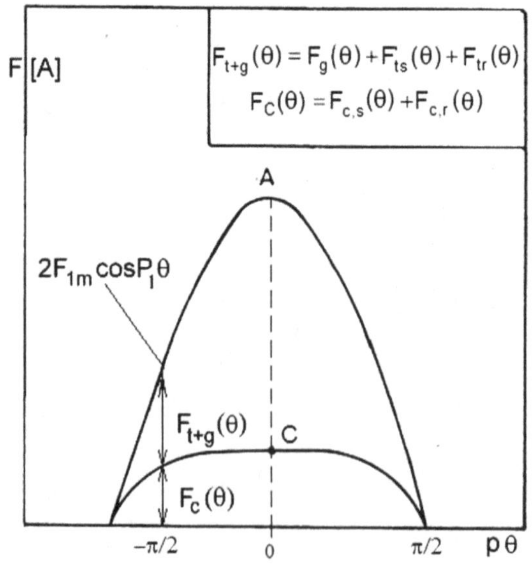

FIGURE 5.17 mmf – total and back core component F.

The left side of (5.85) may be broken into various parts (Figure 5.16). It may easily be shown that, due to the absence of any mmf within contours ABB′B″A″A′A and DCC″D″,

$$\int_{ABB'A''B''A''} \overline{H}\overline{dl} = \int_{AA''} \overline{H}\overline{dl}; \quad \int_{CDD''C''} \overline{H}\overline{dl} = \int_{CC''} \overline{H}\overline{dl} \tag{5.86}$$

We may now divide the mmf components into two categories: those which depend directly on the airgap flux density $B_g(\theta)$ and those which do not (Figure 5.17).

$$2F_{t+g} = 2F_g(\theta) + 2F_{ts}(\theta) + 2F_{tr}(\theta) \tag{5.87}$$

$$F_c(\theta) = F_{cs}(\theta) + F_{cr}(\theta) \tag{5.88}$$

Also, the airgap mmf $F_g(\theta)$ is

$$F_g(\theta) = \frac{gK_c}{\mu_0} B_g(\theta) \tag{5.89}$$

Suppose we know the maximum value of the stator core flux density, for given F_{1m}, as obtained from AIM, B_{csm},

$$B_{cs}(\theta) = B_{csm} \sin p\theta \tag{5.90}$$

We may now calculate $F_{cs}(\theta)$ as

$$F_{cs}(\theta) = 2\int_{p_1\theta}^{\pi/2} \left[H_{cs}(B_{csm} \sin\alpha)\right] e\, d\alpha \tag{5.91}$$

The rotor core maximum flux density B_{crm} is also known from AIM for same mmf F_{1m}. The rotor core mmf $F_{cr}(\theta)$ is thus

$$F_{cr}(\theta) = 2\int_{p_1\theta}^{\pi/2} \left[H_{cr}(B_{crm} \sin\alpha)\right] b\, d\alpha \tag{5.92}$$

From (5.88), $F_c(\theta)$ is obtained. Consequently,

$$F_{t+g} = 2F_{1m} \cos p\theta - F_{cr}(\theta) - F_{cs}(\theta) \tag{5.93}$$

Now in (5.87), we have only to express $F_{ts}(\theta)$ and $F_{tr}(\theta)$ – tooth mmfs – as functions of airgap flux density $B_g(\theta)$. This is straightforward.

$$B_{ts}(\theta) = \frac{\beta_s}{\xi_s} B_g(\theta); \quad \xi_s = \frac{\tau_s}{b_{ts}} \tag{5.94}$$

τ_s – stator slot pitch, b_{ts} – stator tooth, $\beta_s < 1$ – ratio of real and apparent tooth flux densities (due to the flux deviation through slot).

Similarly, for the rotor,

$$B_{tr}(\theta) = \frac{\beta_r}{\xi_r} B_g(\theta); \quad \xi_r = \frac{\tau_r}{b_{tr}} \tag{5.95}$$

Now,

$$F_{t+g} = 2B_{g0}(\theta)\frac{g}{\mu_0} + 2h_s H_{ts}\left[\frac{\beta_s}{\xi_s}B_g(\theta)\right] + 2h_r H_{tr}\left[\frac{\beta_r}{\xi_r}B_g(\theta)\right] \tag{5.96}$$

In (5.96), $F_{t+g}(\theta)$ is known for given θ from (5.93). $B_{g0}(\theta)$ is assigned an initial value in (5.96) and then, based on the lamination B/H curve, H_{ts} and H_{tr} are calculated. Based on this, F_{t+g} is recalculated and compared to the value from (5.93). The calculation cycle is reiterated until sufficient convergence is reached.

The same operation is done for 20–50 values of θ per half-pole to obtain smooth $B_g(\theta)$, $B_{ts}(\theta)$, and $B_{tr}(\theta)$ curves per given stator (and rotor) mmfs (currents). Fourier analysis is used then to calculate the harmonics contents. Sample results concerning the airgap flux density distribution for three designs of same motor are given in Figures 5.18–5.21 [8].

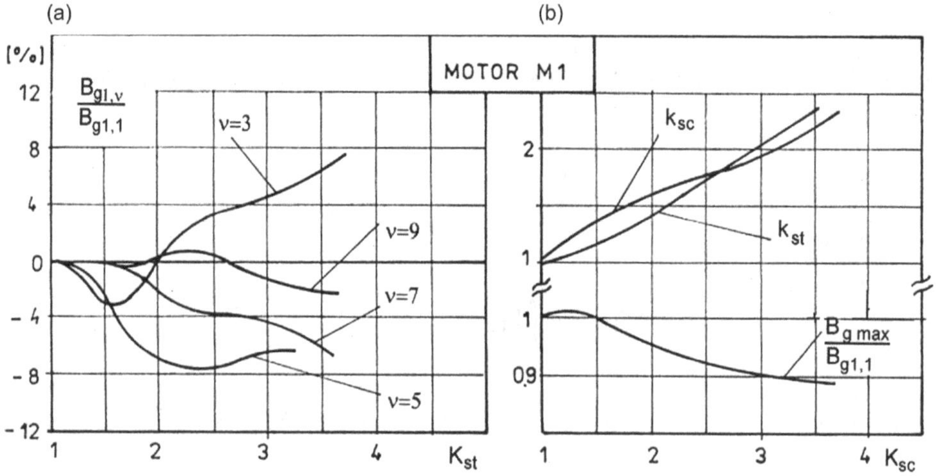

FIGURE 5.18 IM with equally saturated teeth (a) and back cores (b).

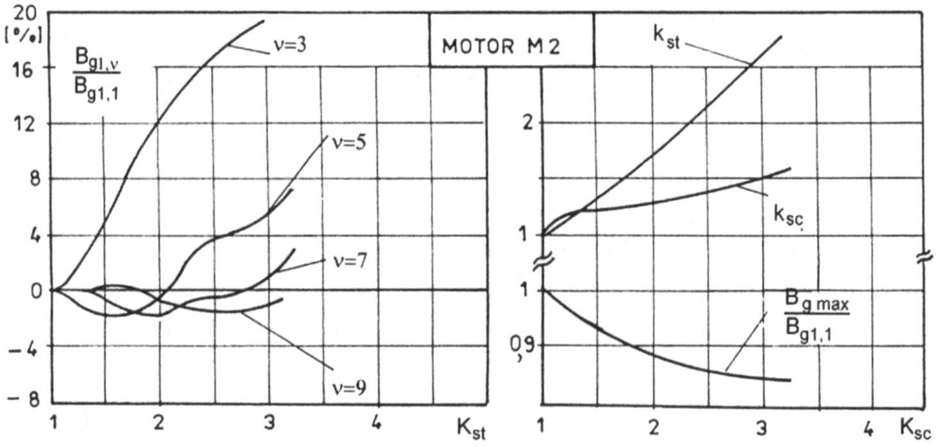

FIGURE 5.19 IM with saturated teeth ($K_{st} > K_{sc}$).

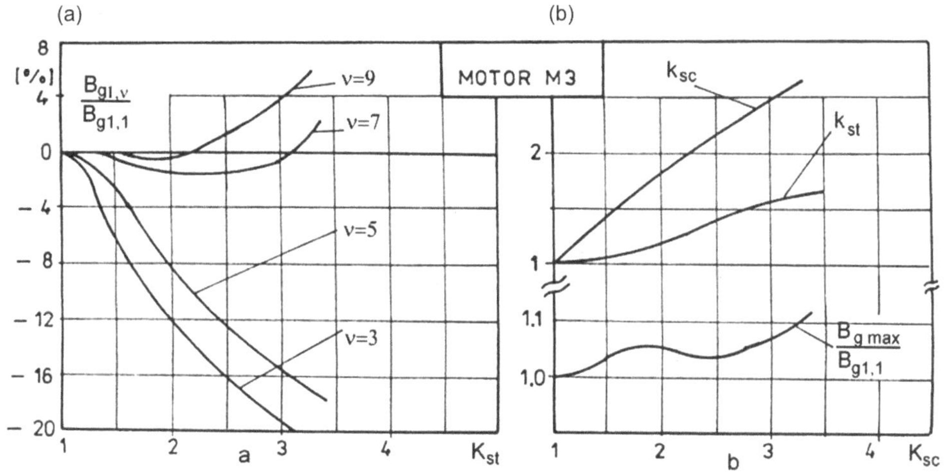

FIGURE 5.20 IM with saturated back cores (a) airgap flux density and (b) saturation factors K_{st} and K_{sc}, versus total saturation factor K_s.

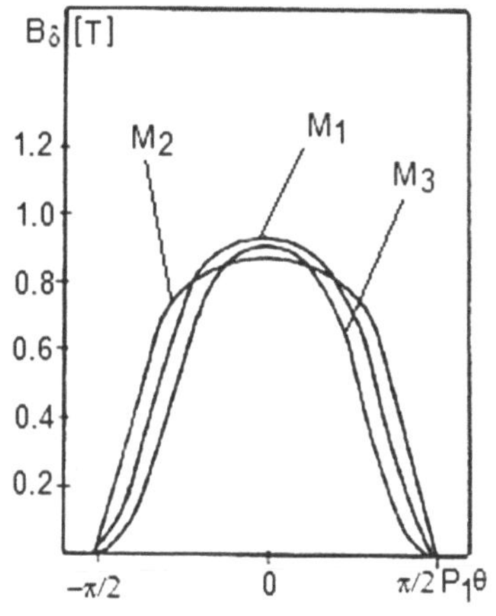

FIGURE 5.21 Airgap flux density distribution for three IM designs: M1, M2, and M3.

The geometrical data are

M1 – optimised motor = equal saturation of teeth and cores: a = 0.015 m, b = 0.036 m, c = 0.0505 m, d = 0.051 m, e = 0.066 m, f = 0.091 m, $\xi_s = 0.54$, and $\xi_r = 0.6$.
M2 – oversaturated stator teeth ($K_{st} > K_{sc}$): b = 0.04 m, e = 0.06 m, $\xi_s = 0.42$, and $\xi_r = 0.48$.
M3 – oversaturated back cores ($K_{sc} > K_{st}$): b = 0.033 m, e = 0.069 m, $\xi_s = 0.6$, and $\xi_r = 0.65$.

The results on Figures 5.18–5.21 may be interpreted as follows:

- If the saturation level (factor) in the teeth and back cores is about the same (motor M1 on Figure 5.21), then even for large degrees of saturation ($K_s = 2.6$ on Figure 5.18b), the airgap flux harmonics are small (Figure 5.18a). This is why a design may be called optimised.
- If the teeth and back core saturation levels are notably different (oversaturated teeth is common), then the harmonics content of airgap flux density is rich. Flattened airgap flux distribution is obtained for oversaturated back cores (Figure 5.20).

- The above remarks seem to suggest that equal saturation level in teeth and back iron is to be provided to reduce airgap flux harmonics content. However, if core loss minimisation is aimed at, this might not be the case, as discussed in Chapter 11 on core losses.

5.5 THE EMF IN AN A.C. WINDING

As already shown, the placement of coils in slots, slot openings, and saturation cause airgap flux density harmonics (Figure 5.22).

The voltage (emf) induced in the stator and rotor phases (bars) is of paramount importance for the IM behaviour. We will derive its general expression based on the flux per pole (Φ_ν) for the ν harmonic (ν = 1 for the fundamental here; it is an "electrical" harmonic).

$$\Phi_\nu = \frac{2}{\pi}\tau_\nu L_e B_{g\nu} \tag{5.97}$$

For low- and medium-power IMs, the rotor slots are skewed by a distance c (in metres here). Consequently, the phase angle of the airgap flux density along the inclined rotor (stator) slot varies continuously (Figure 5.23).

Consequently, the skewing factor $K_{skew\nu}$ is (Figure 5.23b)

$$K_{skew\,\nu} = \frac{\overline{AB}}{\overset{\frown}{AOB}} = \frac{\sin\frac{c}{\tau}\nu\frac{\pi}{2}}{\frac{c}{\tau}\nu\frac{\pi}{2}} \le 1 \tag{5.98}$$

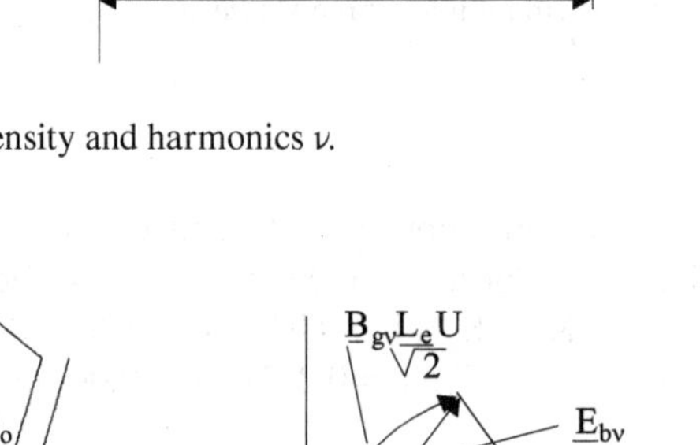

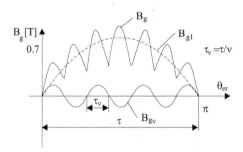

FIGURE 5.22 Airgap flux density and harmonics ν.

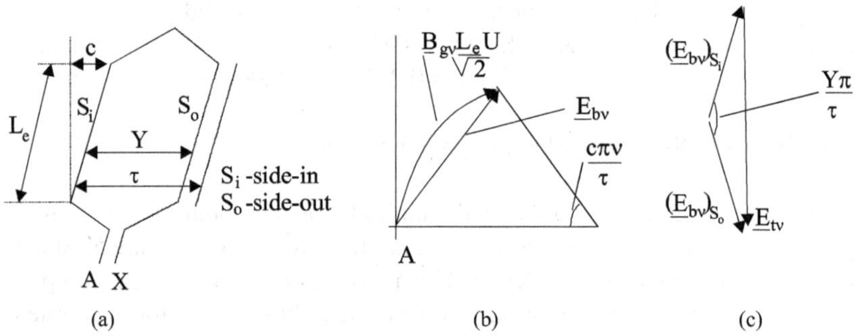

FIGURE 5.23 Chorded coil in skewed slots (a) coil; (b) emf per conductor; and (c) emf per turn.

The emf induced in a bar (conductor) placed in a skewed slot E_{bv} (RMS) is

$$E_{bv} = \frac{1}{2\sqrt{2}}\omega_1 K_{skew\,v}\Phi_v \tag{5.99}$$

with ω_1 – the primary frequency (time harmonics are not considered here).

As a turn may have chorded throw (Figure 5.23a), the emfs in the two turn sides S_i and S_o (in Figure 5.23c) are dephased by less than 180°. A vector composition is required, and thus, E_{tv} is

$$E_{tv} = 2K_{yv}E_{bv}; \quad K_{yv} = \sin v\frac{y}{2}\frac{\pi}{2} \le 1 \tag{5.100}$$

K_{yv} is the chording factor.

There are n_c turns per coil, so the coil emf E_{cv} is

$$E_{cv} = n_c E_{tv} \tag{5.101}$$

In all, there are q neighbouring coils per pole per phase. Let us consider them in series and with their emfs phase shifted by $\alpha_{ecv} = 2\pi p_1 v/N_s$ (Figure 5.24). The resultant emf for the q coils (in series in our case) E_{cv} is

$$E_{cv} = qE_{cv}K_{qv} \tag{5.102}$$

K_{qv} is called the distribution (or zone) factor. For an integer q (Figure 5.24), K_{qv} is

$$K_{qv} = \frac{\sin v\pi/6}{q\sin\left(v\pi/6q\right)} \le 1 \tag{5.103}$$

The total number of q coil groups is p_1 (pole pairs) in single layer and $2p_1$ in double-layer windings. Now we may introduce $a \le p_1$ current paths in parallel.

The total number of turns/phase is $W_1 = p_1 q n_c$ for single layer and $W_1 = 2p_1 q n_c$ for double-layer windings, respectively. Per current path, we do have W_a turns:

$$W_a = \frac{W_1}{a} \tag{5.104}$$

Consequently, the emf per current path E_a is

$$E_{av} = \pi\sqrt{2}f_1 W_a K_{qv} K_{yv} K_{cv} \Phi_v \tag{5.105}$$

For the stator phase emf, when the stator slots are straight, $K_{cv} = 1$.

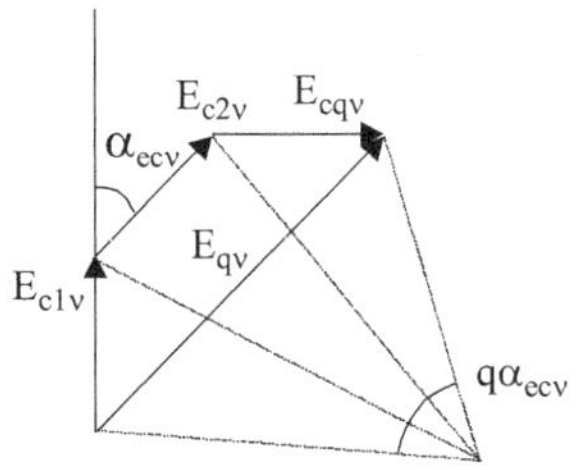

FIGURE 5.24 Distribution factor K_{qv}.

For the emf induced in the rotor phase (bar) with inclined rotor slots, $K_{cv} \neq 1$ and is calculated with (5.98).

$$(E_{av})_{stator} = \pi\sqrt{2}f_1 W_a K^s_{qv} K^s_{yv} \Phi_v \quad (5.106)$$

$$(E_{av})_{rotor} = \frac{\pi}{\sqrt{2}} f_1 K_{skew\,v} \Phi_v \quad (5.107)$$

$$(E_{av})_{rotor} = \pi\sqrt{2}f_1 W_a K^r_{qv} K^r_{yv} \Phi_v \quad (5.108)$$

The distribution K_{qv} and chording (K_{yv}) factors may differ in the wound rotor from stator.

Let us note that the distribution, chording, and skewing factors are identical to those derived for mmf harmonics. This is so because the source of airgap field distribution is the mmf distribution. The slot openings or magnetic saturation influence on field harmonics is lumped into Φ_v.

The harmonic attenuation or cancellation through chording or skewing is also evident in (5.102)–(5.105). Harmonics that may be reduced or cancelled by chording or skewing are of the order

$$\nu = Km \pm 1 \quad (5.109)$$

For one of them,

$$\nu\frac{y}{\tau}\frac{\pi}{2} = K'\pi \quad \text{or} \quad \nu\frac{c}{\tau}\frac{\pi}{2} = K''\pi \quad (5.110)$$

These are called mmf step (belt) harmonics (5, 7, 11, 13, 17, 19). Third mmf harmonics do not exist for star connection. However, as shown earlier, magnetic saturation may cause third-order harmonics, even 5th, 7th, and 11th order harmonics, which may increase or decrease the corresponding mmf step harmonics.

Even order harmonics may occur in two- or three-phase windings with fractionary q. Finally, the slot harmonics orders ν_c are

$$\nu_c = 2K_c qm \pm 1 \quad (5.111)$$

For fractionary $q = a + b/c_1$,

$$\nu_c = 2(ac_1 + b)\left(\frac{K}{c_1}\right)m \pm 1 \quad (5.112)$$

We should notice that for all slot harmonics,

$$K_{qv} = \frac{\sin(2Kqm \pm 1)\pi/6}{q\sin\left[(2Kqm \pm 1)\pi/6q\right]} = K_{q1} \quad (5.113)$$

Slot harmonics are related to slot openings' presence or the corresponding airgap permeance modulation. All slot harmonics have the same distribution factor with the fundamental, so they may not be destroyed or reduced. However, as their amplitude (in airgap flux density) decreases with increasing ν_c, the only thing we may do is to increase their lowest order. Here the fractionary windings come into play (5.112). With q increased from 2 to 5/2, the lowest slot integer harmonic is increased (see (5.111)–(5.112)) from $\nu_{cmin} = 11$ to $\nu_{cmin} = 29$! ($K/c_1 = 2/2$). As K is increased, the order ν of step (phase belt) harmonics (5.111) may become equal to ν_c.

Especially, the first slot harmonics ($K = 1$, $K/c_1 = 1$) may thus interact and amplify (or attenuate) the effect of mmf step harmonics of the same order. Intricate aspects like this will be dealt with in Chapter 10.

5.6 THE MAGNETISATION INDUCTANCE

Let us consider first a sinusoidal airgap flux density in the airgap, at no load (zero rotor currents). Making use of Ampere's law, we get, for airgap flux amplitude,

$$B_{g1} \approx \frac{\mu_0 (F_1)_{phase}}{gK_cK_s}; \quad K_s > 1 \tag{5.114}$$

where F_1 is the phase mmf/pole amplitude,

$$(F_1)_{phase} = \frac{2W_1 I_0 \sqrt{2} K_{q1} K_{y1}}{\pi P_1} \tag{5.115}$$

For a three-phase machine on no load,

$$F_{1m} = \frac{3W_1 I_0 \sqrt{2} K_{q1} K_{y1}}{\pi p_1} \tag{5.116}$$

Now the flux per pole, for the fundamental Φ_1, is

$$\Phi_1 = \frac{2}{\pi} B_{g1} \tau L_e \tag{5.117}$$

So the airgap flux linkage per phase (a = 1) Ψ_{11h} is

$$\psi_{11h} = W_1 K_{q1} K_{y1} \Phi_1 \tag{5.118}$$

The magnetisation inductance of a phase, when all other phases on stator and rotor are open, L_{11m} is

$$L_{11m} = \frac{\psi_{11h}}{I_0 \sqrt{2}} = \frac{4\mu_0 \left(W_1 K_{q1} K_{y1}\right)^2}{\pi^2} \frac{L_e \tau}{P_1 g K_c K_s} \tag{5.119}$$

K_s – total saturation factor.

In a similar manner, for a harmonic ν, we obtain

$$L_{11m\nu} = \frac{4\mu_0 \left(W_1 K_{q\nu} K_{y\nu}\right)^2}{\pi^2 * \nu^2} \frac{L_e \tau}{p_1 g K_c K_{s\nu}}; \quad K_{s\nu} \approx (1 - 1.2) \tag{5.120}$$

The saturation coefficient $K_{s\nu}$ tends to be smaller than K_s as the length of harmonics flux line in iron is shorter in general than for the fundamental.

Now, for three-phase supply, we use the same rationale, but, in the (5.114), F_{1m} will replace $(F_1)_{phase}$.

$$L_{1m\nu} = \frac{6\mu_0 \left(W_1 K_{q\nu} K_{y\nu}\right)^2}{\pi^2 \nu^2} \frac{L_e \tau}{p_1 g K_c K_{s\nu}} \tag{5.121}$$

When the no-load current (I_0/I_n) increases, the saturation factor increases, so the magnetisation inductance L_{1m} decreases (Figure 5.25).

$$L_{1m} = \frac{6\mu_0\left(W_1K_{q1}K_{y1}\right)^2}{\pi^2}\frac{L_e\tau}{P_1gK_cK_s\left(I_0/I_n\right)} \tag{5.122}$$

In general, a normal inductance L_n, is defined as

$$L_n = \frac{V_n}{I_n\omega_{1n}} : l_{1m} = \frac{L_{1m}}{L_n} \tag{5.123}$$

V_n, I_n, ω_{1n} rated phase voltage, current, and frequency of the respective machine.

Well-designed medium-power induction motors have l_{1m} (Figure 5.25) in the interval 2.5–4; in general, it increases with power level and decreases with large number of poles.

It should be noticed that Figure 5.25 shows yet another way (scale) of representing the magnetisation curve.

The magnetisation inductance of the rotor L_{22m} has expressions similar to (5.119) and (5.120) but with corresponding rotor number of turns phase, distribution, and chording factors.

However, for the mutual magnetising inductance between a stator and rotor phase L_{12m}, we obtain

$$L_{12mv} = \frac{4\mu_0}{\pi^2}\left(W_1K_{qv}{}^sK_{yv}{}^sW_2K_{qv}{}^rK_{yv}{}^rK_{cv}{}^r\right)\frac{L_e\tau}{p_1gK_{sv}v^2} \tag{5.124}$$

This time, the skewing factor is present (skewed rotor slots).

From (5.124) and (5.120),

$$\frac{L_{12mv}}{L_{11mv}} = \frac{W_2K_{qv}{}^rK_{yv}{}^rK_{cv}{}^r}{W_1K_{qv}{}^sK_{yv}{}^s} \tag{5.125}$$

It is now evident that for a rotor cage, $K_{qv}{}^r = K_{yv}{}^r = 1$ as each bar (slot) may be considered as a separate phase with $W_2 = ½$ turns.

The inductance L_{11m} or L_{22m} is also called main (magnetisation) self-inductances, while L_{1m} is called the cyclic (multiphase supply) magnetising inductance. In general, for m phases,

$$L_{1m} = \frac{m}{2}L_{11m}; \quad m = 3 \text{ for three phases} \tag{5.126}$$

Typical FEM versus analytical [4] airgap flux density, Figure 5.26, shows clearly the slot opening effects.

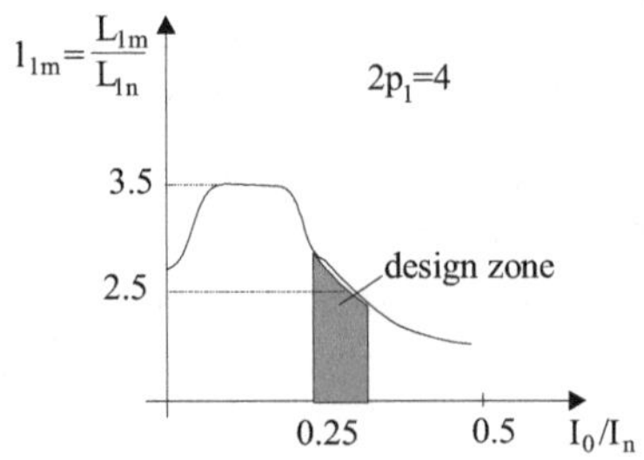

FIGURE 5.25 Magnetisation inductance versus no-load current I_0/I_n.

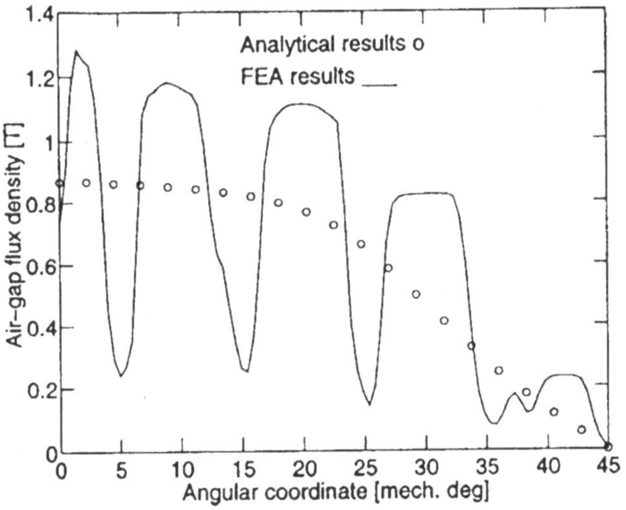

FIGURE 5.26 No-load airgap flux density [4].

Example 5.1 The Magnetisation Inductance L_{1m} Calculation

For the motor M1 geometry given in the previous paragraph, with double-layer winding and $N_s = 36$ slots, $N_r = 30$ slots, $2p_1 = 4$, $y/\tau = 8/9$, let us calculate L_{1m} and no-load phase current I_0 for a saturation factor $K_s = 2.6$ and $B_{g1} = 0.9$ T, $K_c \approx 1.3$ (Figures 5.20 and 5.21).

Stator (rotor) slot openings are: $b_{os} = 6g$, $b_{or} = 3g$, $W_1 = 120$ turns/phase, $L_e = 0.12$ m, and stator bore diameter $D_i/2 = 0.051$ m.

Solution

- The Carter coefficient K_c ((5.4), (5.3)) is

$$K_{c1,2} = \frac{\tau_{s,r}}{\tau_{s,r} - \gamma_{1,2} \cdot g/2} = \begin{cases} 1.280 \\ 1.0837 \end{cases} \tag{5.127}$$

$$\tau_s = \frac{\pi D_1}{N_s} = \frac{\pi \cdot 2 \cdot 0.051}{36} = 9.59 \cdot 10^{-3} \text{ m}$$
$$\tau_r = \frac{\pi(D_i - 2g)}{N_r} = \frac{\pi(2 \cdot 0.051 - 2 \cdot 0.0005)}{30} = 10.57 \cdot 10^{-3} \text{ m} \tag{5.128}$$

$$\gamma_1 = \frac{(2b_{os}/g)^2}{5 + 2b_{os}/g} = \frac{12^2}{5+12} = 8.47$$
$$\gamma_2 = \frac{(2b_{or}/g)^2}{5 + 2b_{or}/g} = \frac{6^2}{5+6} = 3.27 \tag{5.129}$$

$$K_c = K_{c1} \cdot K_{c2} = 1.280 \cdot 1.0837 = 1.327 \tag{5.130}$$

- The distribution and chording factors K_{q1}, K_{y1}, (5.100), (5.103) are

$$K_{q1} = \frac{\sin \pi/6}{3 \sin \pi/6 \cdot 3} = 0.9598; \quad K_{y1} = \sin\frac{8}{9}\frac{\pi}{2} = 0.9848 \tag{5.131}$$

From (5.122),

$$L_{1m} = 6 \cdot \frac{1.25 \cdot 10^{-6}}{\pi^2} \cdot (300 \cdot 0.9598 \cdot 0.9848)^2 \cdot \frac{9 \cdot 9.59 \cdot 10^{-3} \cdot 0.12}{2 \cdot 0.5 \cdot 10^{-3} \cdot 1.327 \cdot 2.6} = 0.1711\,\text{H} \quad (5.132)$$

- The no-load current I_0 from (5.114) is

$$B_{g1} = \frac{\mu_0 F_{1m}}{gK_cK_s} = \frac{1.256 \cdot 10^{-3} \cdot 3\sqrt{2} \cdot 300 \cdot I_0 \cdot 0.9598 \cdot 0.9848}{3.14 \cdot 2 \cdot 0.5 \cdot 10^{-3} \cdot 1.327 \cdot 2.6} = 0.9\,\text{T} \quad (5.133)$$

So the no-load phase current I_0 is

$$I_0 = 3.12\,\text{A} \quad (5.134)$$

Let us consider rated current $I_n = 2.5I_0$.

- The emf per phase for $f_1 = 60\,\text{Hz}$ is

$$E_1 = 2\pi f_1 L_{1m} I_0 = 2\pi 60 \cdot 0.1711 \cdot 3.12 = 201.17\,\text{V} \quad (5.135)$$

With a rated voltage/phase $V_1 = 210\,\text{V}$, the relative value of L_{1m} is

$$l_{1m} = \frac{L_{1m}}{V_1/\omega_1 I_n} = \frac{0.1711}{210/(2\pi 60 \cdot 2.5 \cdot 3.12)} = 2.39\,\text{p.u.} \quad (5.136)$$

It is obviously a small relative value, mainly due to the large value of saturation factor $K_s = 2.6$ (in the absence of saturation: $K_s = 1.0$).

In general, $E_1/V_1 \approx 0.9–0.98$; higher values correspond to larger power machines.

5.7 SATURATED MAGNETISATION INDUCTANCE BY CURVE FITTING

Though in the computation of magnetisation induction we calculated the influence of magnetic cores saturation on the magnetisation inductance, in the above example, we have neglected it for the sake of simplicity.

However, in reality, the magnetisation inductance l_m (p.u.) depends on quite a few factors:

- Magnetic core saturation
- Ratio of pole pitch/airgap (that is: machine size and the number of poles)
- Stator and rotor slot openings W_{0s} and W_{0r} to airgap and slot pitches ratio (by Carter coefficient, approximately).

A smaller airgap (g) leads to a higher p.u. magnetisation inductance (smaller no-load current), but it also results in higher core surface and cage additional losses through stator and rotor mmf harmonics-produced airgap flux density harmonics.

A larger airgap leads to the reversed effects:

Higher no-load current, that is lower power factor and efficiency, but lower additional core and rotor cage losses, is shown in subsequent dedicated chapters.

Magnetic saturation reduces $l_{m_{unsat}}$ by $K_s>1$ to $l_{m_{unsat}}/K_s$; K_s – saturation factor – is allowed in designing a two-pole machine, in general, values of up to 1.5 while in four and more pole machines (lower τ/g ratio), it is kept below 1.5–1.7, to keep the machine power factor and efficiency practically large enough.

Numerous curve fitting methods to approximate $L_m(i_m)$ function have been proposed [9]:

$$\begin{aligned}
L_m(i_m) &= L_{m\,unsat} \quad \text{for } i_m/i_{m\,sat} < 1 \\
L_m(i_m) &= \frac{2}{\pi}[\alpha + 0.5 \cdot \sin 2\alpha] \cdot L_{m\,unsat} \quad \text{for } i_m/i_{m\,sat} \geq 1
\end{aligned} \tag{5.137}$$

which, however, is not differentiable at $i_{m\,sat}$, and the $\Psi_m(i_m)$ curve has a zero ramp under saturated conditions.

To circumvent these inconveniences, Ref. [10] proposes another approximation [10]:

$$L_m(i_m) = \frac{L_{m\,unsat} - L_{m\,sat}}{\left(1 + \left(i_m/i_{m\,sat}\right)^p\right)^{1/p}} + L_{m\,sat} \tag{5.138}$$

As already mentioned, FEM could be used directly to calculate the magnetic field distribution and the magnetisation inductance, at standstill with zero rotor currents. To reduce the large FEM computation time, the non-linear magnetic equivalent circuit method has been introduced [6,11–15]. Local magnetic saturation and slot openings may be considered in some but in less detail than with FEM, at one order of magnitude less computation time, though.

5.8 SUMMARY

- In the absence of rotor currents, the stator winding currents produce an airgap flux density which contains a strong fundamental and spatial harmonics due to the placement of coils in slots (mmf harmonics), magnetic saturation, and slot opening presence. Essentially, FEM could produce a fully realistic solution to this problem.
- To simplify the study, a simplified analytical approach is conventionally used; only the mmf fundamental is considered.
- The effect of slotting is "removed" by increasing the actual airgap g to gK_c ($K_c > 1$); K_c – the Carter coefficient.
- The presence of eventual ventilating radial channels (ducts) is considered through a correction coefficient applied to the geometrical stack axial length.
- The dependence of peak airgap flux density on stator mmf amplitude (or stator current) for zero rotor currents is called the magnetisation curve.
- When the teeth are designed as the heavily saturated part, the airgap flux "flows" partially through the slots, thus "defluxing" the teeth to some extent.
- To account for heavily saturated teeth designs, the standard practice is to calculate the mmf $(F_1)_{30°}$ required to produce the maximum (flat) airgap flux density and then increase its value to $F_{1m} = (F_1)_{30°}/\cos 30°$.
- A more elaborate two-dimensional AIM, valid both for zero and nonzero rotor currents, is introduced to refine the results of the standard method. The slotting is considered indirectly through given but different radial and axial permeabilities in the slot zones. The results are validated on numerous low-power motors and sinusoidal airgap and core flux densities.
- Further on, AIM is used to calculate the actual airgap flux density distribution accounting for heavy magnetic saturation. For more saturated teeth, flat airgap flux density is obtained, while for heavier back core saturation, a peaked airgap and teeth flux density distribution is obtained.

- The presence of saturation harmonics is bound to influence the total core losses (as investigated in Chapter 11). It seems that if teeth and back iron cores are heavily, but equally, saturated, the flux density is still rather sinusoidal all over along the stator bore. The stator connection is also to be considered as for star connection, the stator no-load current is sinusoidal, and third harmonics of flux may occur, while for the delta connection, the opposite is true.
- The expression of emf in A.C. windings exhibits the distribution, chording, and skewing factors already derived for mmfs in Chapter 4.
- The emf harmonics include mmf space (step) harmonics, saturation-caused space harmonics, and slot opening (airgap permeance variation) harmonics ν_c.
- Slot flux (emf) harmonics ν_c show a distribution factor K_{qv} equal to that of the fundamental ($K_{qv} = K_{q1}$), so they cannot be destroyed. They may be attenuated by increasing the order of first slot harmonics $\nu_{cmin} = 2qm - 1$ with larger q slots per pole per phase or/and by increased airgap or by fractional q.
- The magnetisation inductance L_{1m} valid for the fundamental is the ratio of phase emf to angular stator frequency to stator current. L_{1m} is decreasing with airgap, number of pole pairs, and saturation level (factor), but it is proportional to pole pitch and stack length and equivalent number of turns per phase squared.
- In relative values (p.u.), the magnetisation inductance, l_{1m}, increases with motor power and decreases with larger number of poles, in general.
- Advanced saturation in $L_m(i_m)$ is required especially in IM generators for stand-alone applications, for self-excitation by capacitors [9].

REFERENCES

1. F. W. Carter, Airgap induction, *Electrical World and Engineering*, Vol. 38, No. 22, 1901, pp. 884–888.
2. T. A. Lipo, *Introduction to A.C. Machine Design*, Vol. 1, WEMPEC – University of Wisconsin, Madison, WI, 1996, p. 84.
3. B. Heller, V. Hamata, *Harmonic Field Effects in Induction Machines*, Elsevier Scientific Publishing Company, Amsterdam, 1977.
4. D. M. Ionel, M. V. Cistelecan, T. J. E. Miller, M. I. McGilp, A new analytical method for the computation of airgap reactances in three-phase induction machines, *Record of IEEE-IAS, Annual Meeting*, 1998, Vol. 1, pp. 65–72.
5. S. L. Ho, W. N. Fu, Review and future application of finite element methods in induction motors, *EMPS*, Vol. 26, 1998, pp. 111–125.
6. V. Ostovic, *Dynamics of Saturated Electric Machines*, Springer Verlag, New York, 1989.
7. G. Madescu, I. Boldea, T. J. E. Miller, An Analytical Iterative Model (AIM) for induction motor design, *Record of IEEE-IAS, Annual Meeting*, 1996, Vol. 1, pp. 566–573.
8. G. Madescu, Contributions to the modelling and design of induction motors, *PhD Thesis*, University Politehnica, Timisoara, Romania, 1995.
9. N. K. S. Naidu, B. Singh, Experimental implementation of doubly fed induction generator-based stand-alone wind energy conversion system, *IEEE Transactions on Industry Applications*, Vol. 52, No. 4, 2016, pp. 3332–3339.
10. R. L. J. Sprangers, J. J. H. Paulides, K. O. Boynov, J. Waarma, E. A. Lomonova, Comparison of two anisotropic layers models applied to induction motors, *Record of IEEE-IEMDC*, 2013, pp. 701–708.
11. P. Kundur, *Power System Stability and Control*, McGraw-Hill, New York, 1994.
12. L. Monjo, F. Corcoles, J. Pedra, Saturation effects on torque - and current - slip curves of squirrel – Cage induction motors, *IEEE Transactions on Energy Conversion*, Vol. 28, No. 1, 2013, pp. 243–254.
13. M. Amrhein, P. T. Krein, 3-D magnetic equivalent circuit framework for modeling electromechanical devices, *IEEE Transactions on Energy Conversion*, Vol. 24, No. 2, 2009, pp. 397–405.
14. N. R. Tavana, V. Dinavahi, Real – time nonlinear magnetic equivalent circuit model of induction machine on FPGA for hardware-in-the-loop simulation, *IEEE Transactions on Energy Conversion*, Vol. 31, No. 2, 2016, pp. 520–530.
15. J-N. Bae, Y-E. Kim, Ch-H. Yoo, Torque control topologies of a self-excited induction generator as an electric retarder, *IEEE Transactions on Energy Conversion*, Vol. 31, No. 2, 2016, pp. 557–565.

6 Leakage Inductances and Resistances

6.1 LEAKAGE FIELDS

Any magnetic field (H_i, B_i) zone within the IM is characterised by its stored magnetic energy (or coenergy) W_m (W_{mco}).

$$W_{mi} = \frac{1}{2}\iiint_V \left(\overline{B}\cdot\overline{H}\right)dV = L_i\frac{I_i^2}{2} \tag{6.1}$$

Equation (6.1) is valid as it is when, in that region, the magnetic field is produced by a single current source, so the inductance "translates" the field effects into circuit elements.

Besides the magnetic energy related to the magnetisation field (investigated in Chapter 5), there are flux lines that encircle only the stator or the rotor coils (Figure 6.1). They are characterised by some equivalent inductances called leakage inductances L_{sl} and L_{rl}.

There are leakage flux lines which cross the stator and, respectively, the rotor slots, end-turn flux lines, zig-zag flux lines, and airgap flux lines (Figure 6.1). In many cases, the differential leakage is included in the zig-zag leakage. Finally, the airgap flux space harmonics produce a stator emf as shown in Chapter 5, at power source frequency, so it should also be considered in the leakage category. Its torques occur at low speeds (high slips) and, thus, are not there at no-load operation.

6.2 DIFFERENTIAL LEAKAGE INDUCTANCES

As both the stator and rotor currents may produce space flux density harmonics in the airgap (only step mmf harmonics are considered here), there will be both a stator and a rotor differential inductance. For the stator, it is sufficient to add all $L_{1m\nu}$ harmonics, but the fundamental (5.122) to get L_{ds}

$$L_{ds} = \frac{6\mu_0\tau L_e W_1^2}{\pi^2 p_1 g K_c}\sum_{\nu\neq 1}\frac{K_{w\nu}^2}{\nu^2 K_{s\nu}}; \quad \nu = Km_1 \pm 1 \tag{6.2}$$

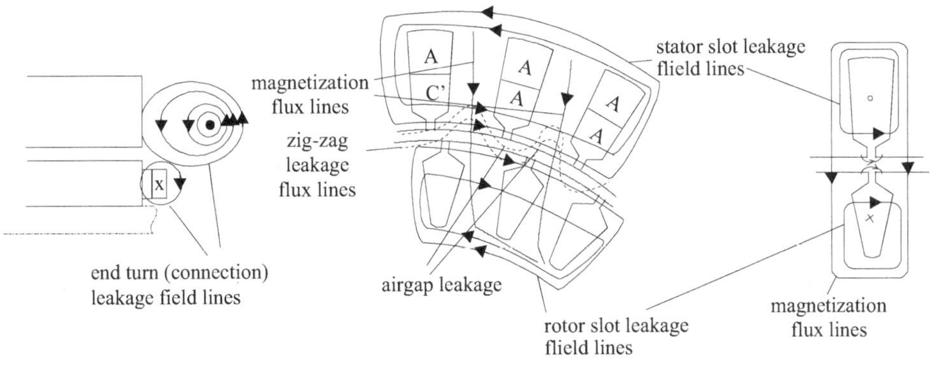

FIGURE 6.1 Leakage flux lines and components.

The ratio σ_d of L_{ds} to the magnetisation inductance L_{1m} is

$$\sigma_{dS0} = \frac{L_{ds}}{L_{1m}} = \sum_{\nu \neq 1} \left(\frac{K_{wS\nu}^{\ 2}}{\nu^2 K_{wS1}^{\ 2}} \right) \cdot \frac{K_s}{K_{s\nu}} \tag{6.3}$$

where $K_{Ws\nu}$ and K_{Ws1} are the stator winding factors for the harmonic ν and the fundamental, respectively.

K_s and $K_{s\nu}$ are the saturation factors for the fundamental and harmonics, respectively.

As the pole pitch of the harmonics is τ/ν, their fields do not reach the back cores, and thus, their saturation factor $K_{s\nu}$ is smaller than K_s. The higher the ν, the closer $K_{s\nu}$ is to unity. In a first approximation,

$$K_{s\nu} \approx K_{st} < K_s \tag{6.4}$$

That is, the harmonics field is retained within the slot zones so the teeth saturation factor K_{st} may be used (K_s and K_{st} have been calculated in Chapter 5). A similar formula for the differential leakage factor can be defined for the rotor winding.

$$\sigma_{dr0} = \sum_{\mu \neq 1} \left(\frac{K_{wr\mu}^{\ 2}}{\mu^2 K_{wr1}^2} \right) \cdot \frac{K_s}{K_{s\mu}} \tag{6.5}$$

As for the stator, the order μ of rotor harmonics is

$$\mu = K_2 m_2 \pm 1 \tag{6.6}$$

m_2 – number of rotor phases; for a cage rotor $m_2 = Z_2/p_1$, also $K_{wr1} = K_{wr\mu} = 1$.

The infinite sums in (6.3) and (6.5) are not easy to handle. To avoid this, the airgap magnetic energy for these harmonics fields can be calculated. Using (6.1),

$$B_{g\nu} = \frac{\mu_0 F_{m\nu}}{g K_c K_{s\nu}} \tag{6.7}$$

We consider the stepwise distribution of mmf for maximum phase A current, (Figure 6.2), and thus

$$\sigma_{ds0} = \frac{\int_0^{2\pi} F^2(\theta)\, d\theta}{F_{1m}^{\ 2}\, 2\pi} = \frac{1}{N_s} \frac{\sum_1^{N_s} F_j^{\ 2}(\theta)}{F_{1m}^{\ 2}} \tag{6.8}$$

The final result for the case in Figure 6.2 is $\sigma_{ds0} = 0.0285$.

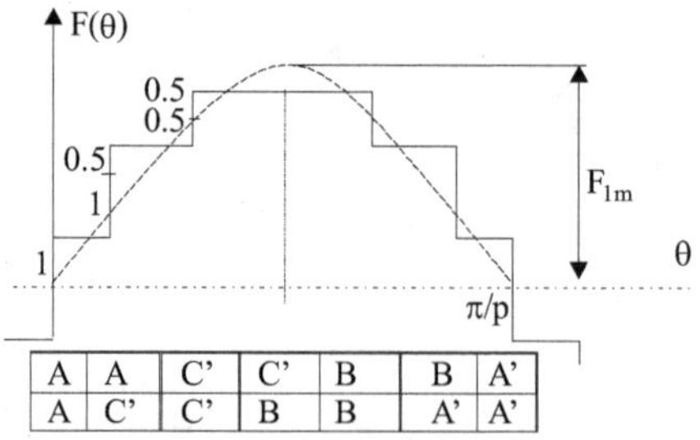

FIGURE 6.2 Stepwise mmf waveform ($q = 2$, $y/\tau = 5/6$).

This method may be used for any kind of winding once we know the number of turns per coil and its current in every slot.

For full-pitch coil three-phase windings [1],

$$\sigma_{ds} \approx \frac{5q^2+1}{12q^2} \cdot \frac{2\pi^2}{m_1{}^2 K_{w1}{}^2} - 1 \tag{6.9}$$

Also, for standard two-layer windings with chorded coils with chording length ε, σ_{ds} is [1]

$$\varepsilon = \tau - y = (1 - y/\tau)3q$$

$$\sigma_{ds} = \frac{2\pi^2}{m_1{}^2 K_{w1}{}^2} \cdot \frac{5q^2+1-\frac{3}{4}\left(1-\frac{y}{\tau}\right)\left[9q^2\left(1-\left(\frac{y}{\tau}\right)^2\right)+1\right]}{12q^2} - 1 \tag{6.10}$$

In a similar way, for the cage rotor with skewed slots,

$$\sigma_{r0} = \frac{1}{K'_{skew}{}^2 \eta_{r1}{}^2} - 1 \tag{6.11}$$

with

$$K'_{skew} = \frac{\sin\left(\alpha_{er} \cdot \frac{c}{\tau_r}\right)}{\alpha_{er} \cdot \frac{c}{\tau_r}}; \quad \eta_{r1} = \frac{\sin(\alpha_{er}/2)}{\alpha_{er}/2}; \quad \alpha_{er} = 2\pi \frac{p_1}{N_r} \tag{6.12}$$

The above expressions are valid for three-phase windings. For a single-phase winding, there are two distinct situations. At standstill, the A.C. field produced by one phase is decomposed in two equal travelling waves. They both produce a differential inductance and, thus, the total differential leakage inductance $(L_{ds1})_{s=1} = 2L_{ds}$.

On the other hand, at S = 0 (synchronism), basically the inverse (backward) field wave is almost zero and, thus, $(L_{ds1})_{S=0} \approx L_{ds}$.

The values of differential leakage factor σ_{ds} (for three- and two-phase machines) and σ_{dr}, as calculated from (6.9) and (6.10), are shown on Figures 6.3 and 6.4 [1].

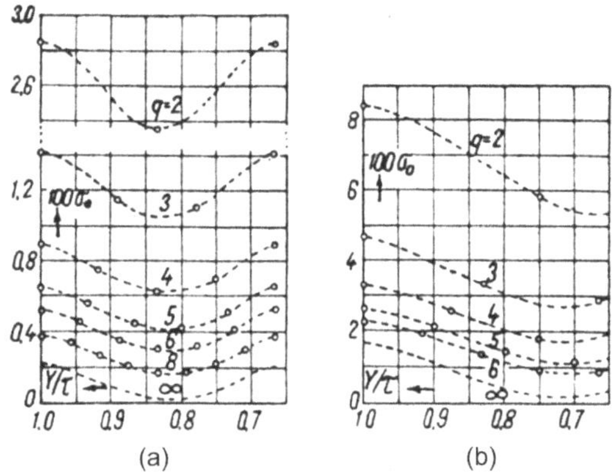

FIGURE 6.3 Stator differential leakage coefficient σ_{ds} for three phases (a) and two phases (b) for various q_s.

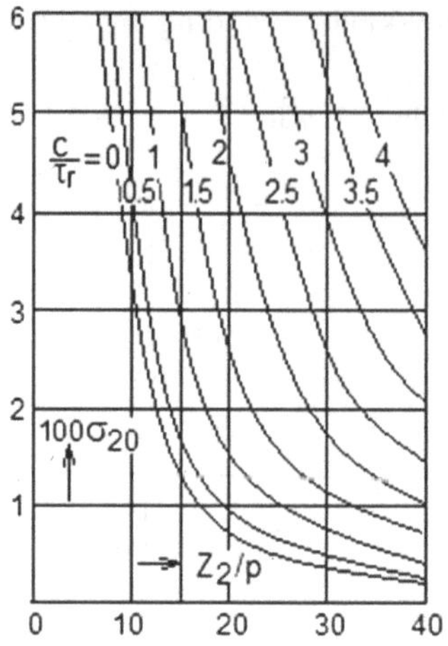

FIGURE 6.4 Rotor cage differential leakage coefficient σ_{dr} for various q_s and straight and single slot pitch skewing rotor slots ($c/\tau_r = 0,1,\ldots$).

A few remarks are in order.

- For q = 1, the differential leakage coefficient σ_{ds0} is about 10%, which means it is too large to be practical.
- The minimum value of σ_{ds0} is obtained for chorded coils with $y/\tau \approx 0.8$ for all q_s (slots/pole/phase).
- For same q, the differential leakage coefficient for two-phase windings is larger than for three-phase windings.
- Increasing the number of rotor slots is beneficial as it reduces σ_{dr0} (Figure 6.4).

Figures 6.3 do not contain the influence of magnetic saturation. In heavily saturated teeth induction machines (IMs) as evident in (6.3), $K_s/K_{st} > 1$, the value of σ_{ds} increases further.

The stator differential leakage flux (inductance) is attenuated by the reaction of the rotor cage currents.

Coefficient Δ_d for the stator differential leakage is [1]

$$\Delta_d \approx 1 - \frac{1}{\sigma_{ds0}} \sum_{\nu \neq 1}^{2mq_1+1} \left(K'_{skew\,\nu} \eta_{r\nu} \frac{K_{w\nu}}{\nu K_{w1}} \right)^2; \quad \sigma_{ds} = \sigma_{ds0}\Delta_d \tag{6.13}$$

As the stator winding-induced harmonic currents do not attenuate the rotor differential leakage, $\sigma_{dr} = \sigma_{dr0}$.

A rather complete study of various factors influencing the differential leakage may be found in [2].

Example 6.1 Leakage Inductance Calculation

For the IM in Example 5.1, with q = 3, $N_s = 36$, $2p_1 = 4$, $y/\tau = 8/9$, $K_{w1} = 0.965$, $K_s = 2.6$, $K_{st} = 1.8$, $N_r = 30$, stack length $L_e = 0.12$m, $L_{1m} = 0.1711$H, and $W_1 = 300$ turns/phase, let us calculate the stator differential leakage inductance L_{ds} including the saturation and attenuation coefficient Δ_d of rotor cage currents.

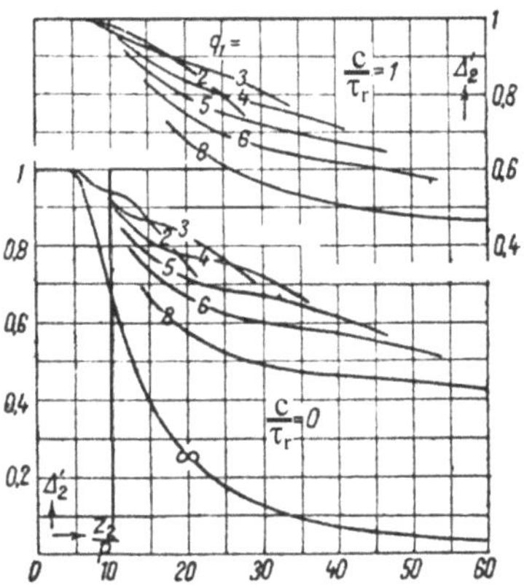

FIGURE 6.5 Differential leakage attenuation coefficient Δ_d for cage rotors with straight ($c/\tau_r = 0$) and skewed slots ($c/\tau_s = 1$).

Solution

We will first find from Figure 6.3 (for $q = 3$, $y/\tau = 0.88$) that $\sigma_{ds0} = 1.16{\cdot}10^{-2}$. Also from Figure 6.5 for $c = 1\tau_s$ (skewing), $Z_2/p_1 = 30/2 = 15$, $\Delta_d = 0.92$. Accounting for both saturation and attenuation coefficient Δ_d, the differential leakage stator coefficient K_{ds} is

$$\sigma_{ds} = \sigma_{ds0} \cdot \frac{\sigma_s}{K_{st}} \Delta_d = 1.16 \cdot 10^{-2} \cdot \frac{2.6}{1.8} \cdot 0.92 = 1.5415 \cdot 10^{-2} \tag{6.14}$$

Now the differential leakage inductance L_{ds} is

$$L_{ds} = \sigma_{ds} L_{1m} = 1.5415 \cdot 10^{-2} \cdot 0.1711 = 0.2637 \cdot 10^{-3}\ \text{H} \tag{6.15}$$

As seen from (6.13), due to the rather large q, the value of K_{ds} is rather small, but, as the number of rotor slots/pole pair is small, the attenuation factor Δ_d is large.

Values of $q = 1,2$ lead to large differential leakage inductances. The rotor cage differential leakage inductance (as reduced to the stator) L_{dr} is

$$L_{dr} = \sigma_{dr0} \frac{K_s}{K_{ts}} L_{1m} = 2.8 \cdot 10^{-2} \cdot \frac{2.6}{1.8} L_{1m} = 4.04 \cdot 10^{-2} L_{1m} \tag{6.16}$$

σ_{dr0} is taken from Figure 6.4 for $Z_2/p_1 = 15$, $c/\tau_r = 1$: $\sigma_{dr0} = 2.8{\cdot}10^{-2}$.

It is now evident that the rotor (reduced to stator) differential leakage inductance is, for this case, notable and greater than that of the stator.

6.3 RECTANGULAR SLOT LEAKAGE INDUCTANCE/SINGLE LAYER

The slot leakage flux distribution depends notably on slot geometry and less on teeth and back core saturation. It also depends on the current density distribution in the slot which may become non-uniform due to eddy currents (skin effect) induced in the conductors in slot by their A.C. leakage flux.

Let us consider the case of a rectangular stator slot where both saturation and skin effect are neglected (Figure 6.6).

Ampere's law on the contours in Figure 6.6 yields

$$\begin{aligned} H(x)b_s &= \frac{n_s \cdot i \cdot x}{h_s}; \quad 0 \le x \le h_s \\ H(x)b_s &= n_s \cdot i; \quad h_s \le x \le h_s + h_{0s} \end{aligned} \tag{6.17}$$

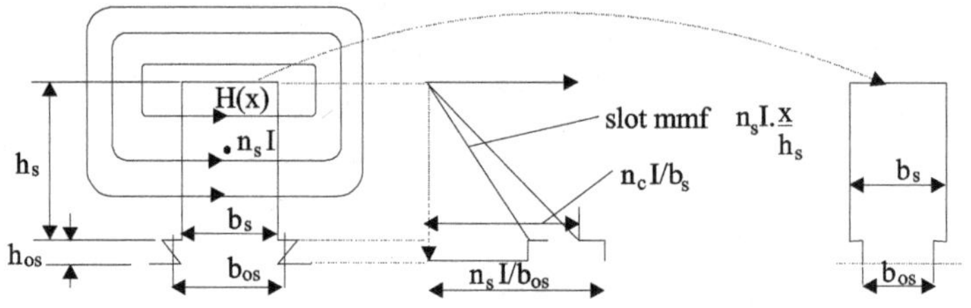

FIGURE 6.6 Rectangular slot leakage.

The leakage inductance per slot, L_{sls}, is obtained from the magnetic energy formula per slot volume.

$$L_{sls} = \frac{2}{i^2} W_{ms} = \frac{2}{i^2} \cdot \frac{1}{2} \int_0^{h_s+h_{os}} \mu_0 H(x)^2 \, dx \cdot L_e b_s = \mu_0 n_s{}^2 L_e \left[\frac{h_s}{3b_s} + \frac{h_{os}}{b_{os}} \right] \tag{6.18}$$

The term in square parenthesis is called the geometrical specific slot permeance.

$$\lambda_s = \frac{h_s}{3b_s} + \frac{h_{os}}{b_{os}} \approx 0.5 \div 2.5; \quad h_{os} = (1 \div 3)10^{-3} \text{ m} \tag{6.19}$$

It depends solely on the aspect of the slot. In general, the ratio $h_s/b_s < (5–6)$ to limit the slot leakage inductance to reasonable values.

The machine has N_s stator slots, and N_s/m_1 of them belong to one phase. So the slot leakage inductance per phase L_{sl} is

$$L_{sl} = \frac{N_s}{m_1} L_{sls} = \frac{2p_1 q m_1}{m_1} L_{sls} = 2\mu_0 W_1{}^2 L_e \frac{\lambda_s}{p_1 q} \tag{6.20}$$

The wedge location has been replaced by a rectangular equivalent area on Figure 6.6. A more exact approach is also possible.

The ratio of slot leakage inductance L_{sl} to magnetising inductance L_{1m} is (same number of turns/phase)

$$\frac{L_{sl}}{L_{1m}} = \frac{\pi^2}{3} \frac{g K_c K_s}{(K W_1)^2} \frac{1}{q\tau} \lambda_s \tag{6.21}$$

Suppose we keep a constant stator bore diameter D_i and increase the number of poles two times.

The pole pitch is thus reduced two times as $\tau = \pi D/2p_1$. If we keep the number of slots and their height constant, q will be reduced twice, and if the airgap and the winding factor are the same, the saturation stays low as already for the low number of poles. Consequently, L_{sl}/L_{1m} increases two times (as λ_s is doubled for same slot height).

Increasing q (and the number of slots/pole) is bound to reduce the slot leakage inductance (6.20) to the extent that λ_s does not increase by the same ratio. Our case here refers to a single-layer winding and rectangular slot.

Two-layer windings with chorded coils may be investigated the same way.

6.4 RECTANGULAR SLOT LEAKAGE INDUCTANCE/TWO LAYERS

We consider the coils are chorded (Figure 6.7)

Let us consider that both layers contribute a field in the slot and add the effects. The total magnetic energy in the slot volume is used to calculate the leakage inductance L_{sls}.

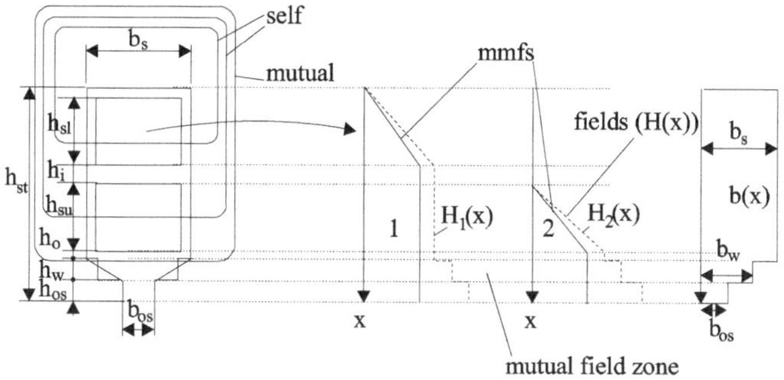

FIGURE 6.7 Two-layer rectangular semiclosed slots: leakage field.

$$L_{sls} = \frac{2L_e}{i^2}\int_0^{h_{st}} \mu_0\left[H_1(x)+H_2(x)\right]^2 dx \cdot b(x) \tag{6.22}$$

$$H_1(x)+H_2(x) = \begin{cases} \dfrac{n_{cl}I}{b_s}\cdot\dfrac{x}{h_{sl}}; & \text{for } 0 < x < h_{sl} \\ \dfrac{n_{cl}I}{b_s} & \text{for } h_{sl} < x < h_{sl}+h_i \\ \dfrac{n_{cl}I}{b_s}+\dfrac{n_{cu}I\cos\gamma_K}{b_s}\dfrac{(x-h_{sl}-h_i)}{h_{su}}; & \text{for } h_{sl}+h_i < x < h_{sl}+h_i+h_{su} \\ \dfrac{n_{cl}I}{b_i}+\dfrac{n_{cu}I\cos\gamma_K}{b_i}; & \text{for } x > h_{sl}+h_i+h_{su} \end{cases} \tag{6.23}$$

with $b_i = b_w$ or b_{os}

The phase shift between currents in lower- and upper-layer coils of slot K is γ_K, and n_{cl} and n_{cu} are the number of turns of the two coils. Adding up the effect of all slots per phase (1/3 of total number of slots), the average slot leakage inductance per phase L_{sl} is obtained.

While (6.23) is valid for general windings with different number of turns/coil and different phases in same slots, we may obtain simplified solutions for identical coils in slots: $n_{cl} = n_{cu} = n_c$.

$$\lambda_{sk} = \frac{L_{sl}}{\mu_0(2n_c)^2 L_e} = \frac{1}{4}\left[\frac{\left(h_{sl}+h_{su}\cos^2\gamma_K\right)}{3b_s}+\frac{h_{su}}{b_s}+\frac{h_{su}\cos\gamma_K}{b_s}\right.$$
$$\left. +\frac{h_i}{b_s}+(1+\cos\gamma)^2\left(\frac{h_o}{b_s}+\frac{h_w}{b_w}+\frac{h_{os}}{b_{os}}\right)\right] \tag{6.24}$$

Although (6.24) is quite general – for two-layer windings with equal coils in slots – the eventual different number of turns per coil can be lumped into cosγ as Kcosγ with $K = n_{cu}/n_{cl}$. In this latter case, the factor 4 will be replaced by $(1 + K)^2$.

In integer and fractionary slot windings with random coil throws, (6.24) should prove expeditious. All phase slot contributions are added up.

Other realistic rectangular slot shapes for large-power IMs (Figure 6.8) may also be handled via (6.24) with minor adaptations.

For full-pitch coils ($\cos\gamma_K = 1.0$), symmetric winding ($h_{su} = h_{sl} = h_s'$) (6.24) becomes

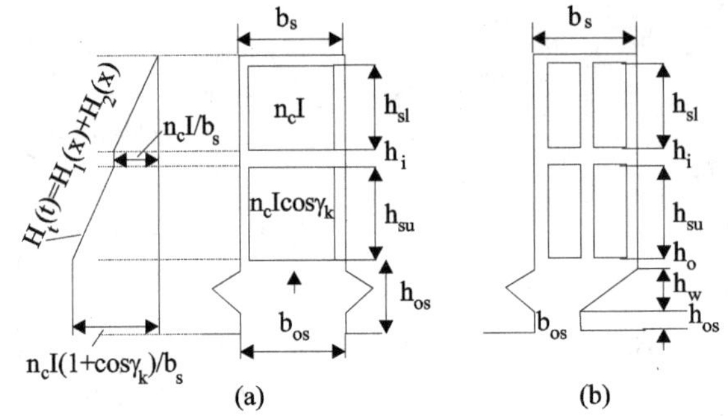

FIGURE 6.8 Typical high-power IM stator slots.

$$\left(\lambda_{sk}\right)_{\substack{\gamma_k=0\\ h_{su}=h_{sl}=h'_s}} = \frac{2h'_s}{3b_s}+\frac{h_i}{4b_s}+\frac{h_0}{b_s}+\frac{h_w}{b_w}+\frac{h_{os}}{b_{os}} \tag{6.25}$$

Further on with $h_i = h_o = h_w = 0$ and $2h'_s = h_s$, we reobtain (6.19), as expected.

6.5 ROUNDED SHAPE SLOT LEAKAGE INDUCTANCE/TWO LAYERS

Although the integral in (6.2) does not have exact analytical solutions for slots with rounded corners, or purely circular slots (Figure 6.9), so typical to low-power IMs, some approximate solutions have become standard for design purposes:

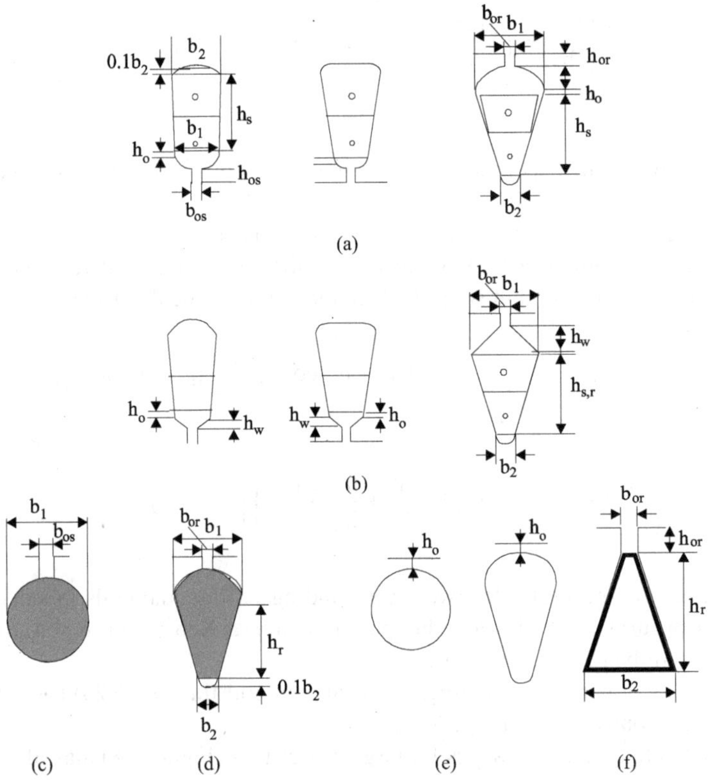

FIGURE 6.9 Rounded slots: oval, trapezoidal, and round.

- For slot (a) in Figure 6.9,

$$\lambda_{s,r} \approx \frac{2h_{s,r}K_1}{3(b_1+b_2)} + \left(\frac{h_{os,r}}{b_{os,r}} + \frac{h_o}{b_1} - \frac{b_{os,r}}{2b_1} + 0.785\right)K_2 \tag{6.26}$$

with

$$\begin{aligned} K_2 &\approx \frac{1+3\beta_y}{4}; \quad \text{for } \frac{2}{3} \le \beta_y = \frac{y}{\tau} \le 1 \\ K_2 &\approx \frac{6\beta_y - 1}{4}; \quad \text{for } \frac{1}{3} \le \beta_y \le \frac{2}{3} \\ K_2 &\approx \frac{3(2-\beta_y)+1}{4}; \quad \text{for } 1 \le \beta_y \le 2 \\ K_1 &\approx \frac{1}{4} + \frac{3}{4}K_2 \end{aligned} \tag{6.27}$$

- For slot (b),

$$\lambda_{s,r} \approx \frac{2h_{s,r}K_1}{3(b_1+b_2)} + \left(\frac{h_{os,r}}{b_{os,r}} + \frac{h_o}{b_1} + \frac{3h_w}{b_1 + 2b_{os,r}}\right)K_2 \tag{6.28}$$

- For slot (c),

$$\lambda_r = 0.785 - \frac{b_{or}}{2b_1} + \frac{h_{or}}{b_{or}} \approx 0.66 + \frac{h_{or}}{b_{or}} \tag{6.29}$$

- For slot (d),

$$\lambda_r = \frac{h_r}{3b_1}\left(1 - \frac{\pi b_1^2}{8A_b}\right)^2 + 0.66 - \frac{h_{or}}{2b_1} + \frac{h_{or}}{b_{or}} \tag{6.30}$$

where A_b is the bar cross section.

If the slots in Figure 6.9c and d are closed ($h_o \neq 0$) (Figure 6.9e), the terms h_{or}/b_{or} in Equations (6.29 and 6.30) may be replaced by a term dependent on the bar current which saturates the iron bridge.

$$\frac{h_{or}}{b_{or}} \rightarrow \approx 0.3 + 1.12h_{or}\frac{10^3}{I_b^{\,2}}; \quad I_b > 5b_1 10^3; \quad b_1 \text{ in [m]} \tag{6.31}$$

This is only an empirical approximation for saturation effects in closed rotor slots, potentially useful for preliminary design purposes.

For the trapezoidal slot (Figure 6.9f), typical for deep rotor bars in high-power IMs, by conformal transformation, the slot permeance is approximately [3]

$$\lambda_r = \frac{1}{\pi}\left[\ln\frac{\left(\frac{b_2}{b_{or}}\right)^2 - 1}{4\frac{b_2}{b_{or}}} + \frac{\left(\frac{b_2}{b_{or}}\right)^2 + 1}{\frac{b_2}{b_{or}}} \cdot \ln\frac{\frac{b_2}{b_{or}} - 1}{\frac{b_2}{b_{or}} + 1}\right] + \frac{h_{or}}{b_{or}} \tag{6.32}$$

The term in square brackets may be used to calculate the geometrical permeance of any trapezoidal slot section (wedge section, for example).

Finally, for stator (and rotor) with radial ventilation ducts (channels), additional slot leakage terms have to be added [4–6].

For more complicated rotor cage slots used in high skin effect (low starting current, high starting torque) applications, where the skin effect is to be considered, pure analytical solutions are hardly feasible, although many are still in industrial use. Realistic computer-aided methods are given in Chapter 8.

6.6 ZIG-ZAG AIRGAP LEAKAGE INDUCTANCES

The airgap flux does not reach the other slotted structure (Figure 6.10a) while the zig-zag flux "snakes out" through the teeth around slot openings.

In general, they may be treated together either by conformal transformation or by FEM. From conformal transformation, the following approximation is given for the geometric permeance $\lambda_{zs,r}$ [3].

$$\lambda_{zs,r} \approx \frac{5gK_c/b_{os,r}}{5+4gK_c/b_{os,r}} \cdot \frac{3\beta_y+1}{4} < 1.0; \quad \beta_y = 1 \text{ for cage rotors} \tag{6.33}$$

The airgap zig-zag leakage inductance per phase in stator–rotor is

$$L_{zls,r} \approx 2\mu_0 \frac{W_1 L_e}{p_1 q} \lambda_{zs,r} \tag{6.34}$$

In [7], different formulas are given

$$L_{zls} = L_{1m} \cdot \frac{\pi^2 p_1^2}{12 N_s^2}\left[1 - \frac{a(1+a)(1-K')}{2K'}\right] \tag{6.35}$$

for the stator, and

$$L_{zlr} = L_{1m} \cdot \frac{\pi^2 p^2}{12 N_s^2}\left[\frac{N_s^2}{N_r^2} - \frac{a(1+a)(1-K')}{2K'}\right] \tag{6.36}$$

for the rotor with $K' = 1/K_c$, $a = b_{ts,r}/\tau_{s,r}$, $\tau_{s,r}$ = stator (rotor) slot pitch, and $b_{ts,r}$ – stator (rotor) tooth top width.

It should be noticed that while expression (6.33) is dependent only on the airgap/slot opening, in (6.35) and (6.36), the airgap directly enters the denominator of L_{1m} (magnetisation inductance) and, in general, (6.35) and (6.36) include the number of slots of stator and rotor, N_s and N_r.

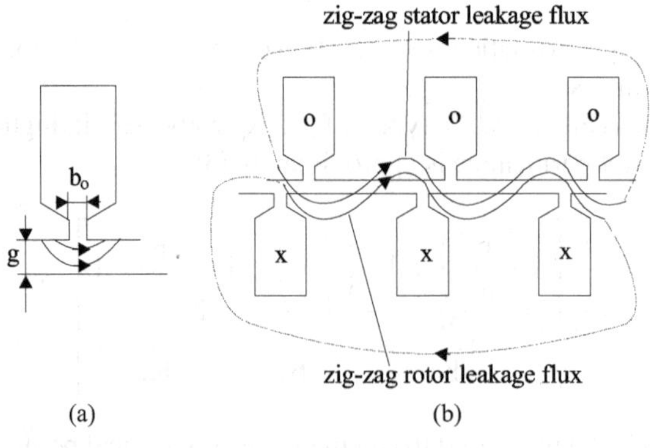

FIGURE 6.10 Airgap (a) and zig-zag (b) leakage fields.

As the term in parenthesis is a very small number, an error here will "contaminate" notably the results. On the other hand, iron saturation will influence the zig-zag flux path but to a much lower extent than the magnetisation flux as the airgap is crossed many times (Figure 6.10b). Finally, the influence of chorded coils is not included in (6.35) and (6.36).

We suggest to use an average of the two expressions (6.34) and (6.35 or 6.36).

In Chapter 7, we revisit this subject for heavy currents (at standstill) including the actual saturation in the tooth tops.

Example 6.2 Zig-Zag Leakage Inductance

For the machine in Example 6.1, with $g = 0.5\cdot10^{-3}$m, $b_{os} = 6g$, $b_{or} = 3g$, $K_c = 1.32$, $L_{1m} = 0.1711$H, $p_1 = 2$, $N_s = 36$ stator slots, $N_r = 30$ rotor slots, stator bore $D_i = 0.102$m, $B_y = y/\tau = 8/9$ (chorded coils), and $W_1 = 300$ turns/phase, let us calculate the zig-zag leakage inductance both from (6.32, 6.34) and (6.35, 6.36).

Solution

Let us first prepare the values of $K' = 1/K_c = 1/1.32 = 0.7575$.

$$a_{s,r} = \frac{\tau_{s,r} - b_{os,r}}{\tau_{s,r}} = \begin{cases} 1 - \dfrac{b_{os}N_s}{\pi D_i} = 1 - \dfrac{6\cdot0.5\cdot10^{-3}\cdot36}{\pi\cdot0.102} = 0.6628 \\ 1 - \dfrac{b_{or}N_r}{\pi(D_i - 2g)} = 1 - \dfrac{3\cdot0.5\cdot10^{-3}\cdot30}{\pi\cdot0.101} = 0.858 \end{cases} \quad (6.37)$$

From (6.32),

$$\lambda_{Zs} = \frac{\dfrac{5\cdot0.5\cdot10^{-3}\cdot1.32}{6\cdot0.1\cdot10^{-3}}}{5 + \dfrac{4\cdot0.1\cdot1.32}{6\cdot0.1}} = 0.187 \quad (6.38)$$

$$\lambda_{Zr} = \frac{\dfrac{5\cdot0.1\cdot10^{-3}\cdot1.32}{3\cdot0.1\cdot10^{-3}}}{5 + \dfrac{4\cdot0.1\cdot1.32}{3\cdot0.1}} = 0.101 \quad (6.39)$$

The zig-zag inductances per phase $L_{zls,r}$ are calculated from (6.34).

$$L_{zls,r} = \begin{cases} \dfrac{2\cdot1.256\cdot10^{-6}\cdot300^2\cdot0.12}{2\cdot3}\cdot0.187 = 8.455\cdot10^{-4}\ \mathrm{H} \\ \dfrac{2\cdot1.256\cdot10^{-6}\cdot300^2\cdot0.12}{2\cdot3}\cdot0.101 = 4.566\cdot10^{-4}\ \mathrm{H} \end{cases} \quad (6.40)$$

Now from (6.35) and (6.36),

$$L_{zls} = 0.1711\cdot\frac{\pi^2 2^2}{12\cdot36^2}\left[1 - \frac{0.6628(1+0.6628)(1-0.7575)}{2\cdot0.7575}\right] = 3.573\cdot10^{-4}\ \mathrm{H} \quad (6.41)$$

$$L_{zlr} = 0.1711\cdot\frac{\pi^2 2^2}{12\cdot36^2}\left[\left(\frac{36}{30}\right)^2 - \frac{0.858(1+0.858)(1-0.7575)}{2\cdot0.7575}\right] = 3.723\cdot10^{-4}\ \mathrm{H} \quad (6.42)$$

All values are small in comparison to $L_{1m} = 1711 \cdot 10^{-4}$ H, but there are notable differences between the two methods. In addition, it may be inferred that the zig-zag flux leakage also includes the differential leakage flux.

6.7 END-CONNECTION LEAKAGE INDUCTANCE

As seen in Figure 6.11, the three-dimensional character of end-connection field makes the computation of its magnetic energy and its leakage inductance per phase a formidable task.

Analytical field solutions need bold simplification [8]. Biot–Savart inductance formula [9] and 3D FEM have all been also tried for particular cases.

Some widely used expressions for the end-connection geometrical permeances are as follows:

- Single-layer windings (with end turns in two "stores").

$$\lambda_{es,r} = 0.67 \cdot \frac{g_{s,r}}{L_e}\left(l_{es,r} - 0.64\tau\right) \tag{6.43}$$

- Single-layer windings (with end connections in three "stores")

$$\lambda_{es,r} = 0.47 \cdot \frac{g_{s,r}}{L_e}\left(l_{es,r} - 0.64\tau\right) \tag{6.44}$$

- Double-layer (or single-layer) chain windings.

$$\lambda_{es,r} = 0.34 \cdot \frac{g_{s,r}}{L_e}\left(l_{es,r} - 0.64y\right) \tag{6.45}$$

with $q_{s,r}$ – slots/pole/phase in the stator/rotor, L_e – stack length, y – coil throw, and $l_{es,r}$ – end-connection length per motor side.

For cage rotors,

- With end rings attached to the rotor stack,

$$\lambda_{ei} \approx \frac{2 \cdot 3 \cdot D_{ir}}{4N_r L_e \sin^2\left(\frac{\pi p}{N_r}\right)} * l_{0g} * \left(4.7\frac{D_{ir}}{a + 2b}\right) \tag{6.46}$$

- With end rings distanced from the rotor stack,

$$\lambda_{ei} \approx \frac{2 \cdot 3 \cdot D_i}{4N_r L_e \sin^2\left(\frac{\pi p}{N_r}\right)} * l_{0g} * \left(4.7\frac{D_{ir}}{2(a + b)}\right) \tag{6.47}$$

with a and b the ring axial and radial dimensions and D_{ir} the average end ring diameter.

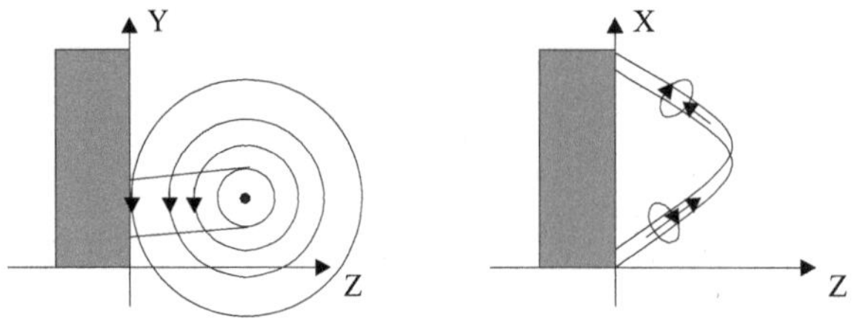

(a) (b)

FIGURE 6.11 (a and b) Three-dimensional end-connection field.

6.8 SKEWING LEAKAGE INDUCTANCE

In Chapter 4 on windings, we did introduce the concept of skewing (uncompensated) rotor mmf, variable along axial length, which acts along the main flux path and produces a flux which may be considered of leakage character. Its magnetic energy in the airgap may be used to calculate the equivalent inductance.

This skewing inductance depends on the local level of saturation of rotor and stator teeth and cores and acts simultaneously with the magnetisation flux, from which is phase shifted with an angle dependent on axial position and slip. We will revisit this complex problem in Chapter 8. In a first approximation, we can make use of the skewing factor K_{skew} and define $L_{skew,r}$ as

$$L_{skew,r} = \left(1 - K^2_{skew}\right) L_{1m} \tag{6.48}$$

$$K_{skew} = \frac{\sin \alpha_{skew}}{\alpha_{skew}}; \quad \alpha_{skew} = \frac{c}{\tau}\frac{\pi}{2} \tag{6.49}$$

As this inductance "does not act" when the rotor current is zero, we feel it should all be added to the rotor.

Finally, the total leakage inductance of stator (rotor) is

$$L_{sl} = L_{dls} + L_{zls} + L_{sls} + L_{els} = 2\mu_0 \frac{(W_1)^2 L_e}{pq} \sum \lambda_{si} \tag{6.50}$$

$$L_{rl} = L_{dlr} + L_{zlr} + L_{slr} + L_{elr} + L_{skew,r} \tag{6.51}$$

Although we discussed about rotor leakage inductance as if basically reduced to the stator, this operation is still due later in this chapter.

6.9 ROTOR BAR AND END RING EQUIVALENT LEAKAGE INDUCTANCE

The differential leakage L_{dlr}, zig-zag leakage $L_{zl,r}$ and $L_{skew,r}$ in (6.51) are already considered in stator terms (reduced to the stator). However, the terms L_{slr} and L_{elr} related to rotor slots (bar) leakage and end-connection (end ring) leakage are not clarified enough.

To do so, we use the equivalent bar (slot) leakage inductance expression L_{be},

$$L_{be} = L_b + 2L_i \tag{6.52}$$

where L_b is the slot (bar) leakage inductance and L_i is the end ring (end connection) segment leakage inductance (Figure 6.13).

Using (6.18) for one slot with $n_c = 1$ conductors yields:

$$L_b = \mu_0 L_b \lambda_b \tag{6.53}$$

$$L_{ei} = \mu_0 L_i \lambda_{er} \tag{6.54}$$

where λ_b is the slot geometrical permeance for a single layer in slot as calculated in Sections 6.4 and 6.5 for various slot shapes and λ_{er} is the geometrical permeance of end ring segment as calculated in (6.46) and (6.47).

Now all it remains to obtain L_{slr} and L_{elr} in (6.51) is to reduce L_{be} in (6.52) to the stator. This will be done in Section 6.13.

6.10 BASIC PHASE RESISTANCE

The stator resistance R_s is plainly

$$R_s = \rho_{Co} l_c \frac{W_a}{a} \frac{1}{A_{cos}} K_R \tag{6.55}$$

where ρ_{Co} is copper resistivity ($\rho_{Co} = 1.8 \cdot 10^{-8}\ \Omega$m at 25°C) and l_c the turn length:

$$l_c = 2L_e + 2b + 2l_{ec} \tag{6.56}$$

b – axial length of coil outside the core per coil side; l_{ec} – end-connection length per stack side, L_e – stack length; A_{cos} – actual conductor area, a – number of current paths, and W_a – number of turns per path. With a = 1, $W_a = W_1$ turns/phase. K_R is the ratio between the A.C. and D.C. resistance of the phase resistance.

$$K_R = \frac{R_{sac}}{R_{sdc}} \tag{6.57}$$

For 50 Hz in low- and medium-power motors, the conductor size is small with respect to field penetration depth δ_{Co} in it.

$$\delta_{Co} = \sqrt{\frac{\rho_{Co}}{\mu_0 2\pi f_1}} > d_{Co} \tag{6.58}$$

In other words, the skin effect is negligible. However, in high-frequency (speed) special IMs, this effect may be considerable unless many thin conductors are transposed (as in Litz wire). On the other hand, in large-power motors, there are large cross sections (even above 60 mm²) where a few elementary conductors are connected in parallel, even transposed in the end-connection zone, to reduce the skin effect (Figure 6.12). For K_R expressions, check Chapter 9. For wound rotors, (6.55) is valid.

6.11 THE CAGE ROTOR RESISTANCE

The rotor cage geometry is shown in Figure 6.13.

Let us denote by R_b the bar resistance and by R_i the ring segment resistance:

$$R_b = \rho_b \frac{l_b}{A_b}; \quad R_i = \rho_i \frac{l_i}{A_i}; \quad A_i = a \cdot b \tag{6.59}$$

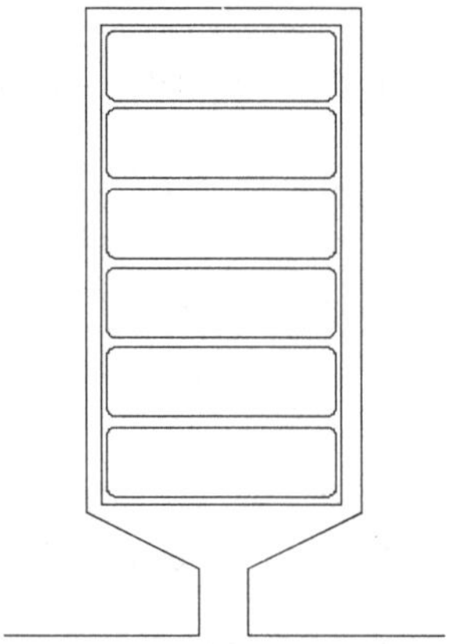

FIGURE 6.12 Multi-conductor single-turn coils for high-power IMs.

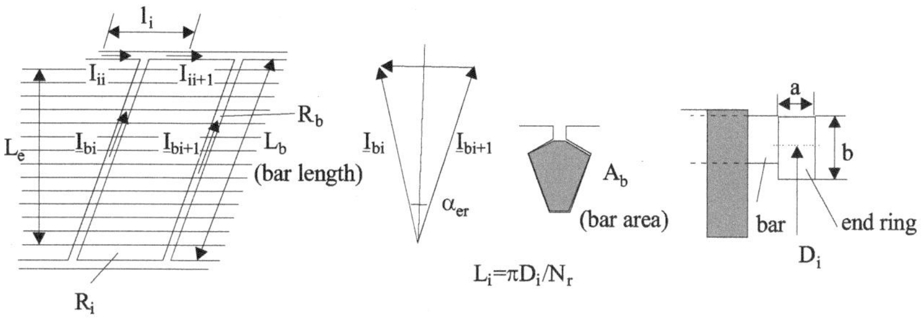

FIGURE 6.13 Rotor cage geometry.

As the emf in rotor bars is basically sinusoidal, the phase shift α_{er} between neighbouring bars emf is

$$\alpha_{er} = \frac{2\pi P_1}{N_r} \tag{6.60}$$

The current in a bar, I_{bi}, is the difference between currents in neighbouring ring segments I_{ii} and I_{ii+1} (Figure 6.13).

$$I_{bi} = I_{ii+1} - I_{ii} = 2I_i \sin\left(\alpha_{er}/2\right) \tag{6.61}$$

The bar and ring segments may be lumped into an equivalent bar with a resistance R_{be}.

$$R_{be} I_b^2 = R_b I_b^2 + 2R_i I_i^2 \tag{6.62}$$

With (6.61), Equation (6.62), leads to

$$R_{be} = R_b + \frac{R_i}{2\sin^2\left(\frac{\pi p}{N_r}\right)} \tag{6.63}$$

When the number of rotor slots/pole pair is small or fractionary, (6.49) becomes less reliable.

Example 6.3 Bar and Ring Resistance

For $N_r = 30$ slots per rotor $2p_1 = 4$, a bar current $I_b = 1000$ A with a current density $j_{cob} = 6$ A/mm^2 in the bar and $j_{coi} = 5$ A/mm^2 in the end ring, the average ring diameter $D_{ir} = 0.15$m, and $l_b = 0.14$m, let us calculate the bar, ring cross section, the end ring current, and bar and equivalent bar resistance.

Solution

The bar cross section A_b is

$$A_b = \frac{I_b}{j_{cob}} = 1000/6 = 166 \text{ mm}^2 \tag{6.64}$$

The current in the end ring I_i (from (6.61)) is

$$I_i = \frac{I_b}{2\sin\alpha_{er}/2} = \frac{1000}{2\sin\frac{2\pi}{30}} = 2404.86 \text{ A} \tag{6.65}$$

In general, the ring current is greater than the bar current. The end ring cross section A_i is

$$A_i = \frac{I_i}{j_{coi}} = \frac{2404.86}{5} = 480.97\ \text{mm}^2 \tag{6.66}$$

The end ring segment length is

$$L_i = \frac{\pi D_{ir}}{N_r} = \frac{\pi \cdot 0.150}{30} = 0.0157\ \text{m} \tag{6.67}$$

The end ring segment resistance R_i is

$$R_i = \rho_{Al}\frac{L_i}{A_i} = \frac{3\cdot 10^{-8}\cdot 0.0157}{480.97\cdot 10^{-6}} = 0.9\cdot 10^{-6}\ \Omega \tag{6.68}$$

The rotor bar resistance R_b is

$$R_b = \rho_{Al}\frac{L_b}{A_b} = \frac{3\cdot 10^{-8}\cdot 0.15}{166\cdot 10^{-6}} = 2.7\cdot 10^{-5}\ \Omega \tag{6.69}$$

Finally, (from (6.63)), the equivalent bar resistance R_{be} is

$$R_{be} = 2.7\cdot 10^{-5} + \frac{0.9\cdot 10^{-6}}{2\sin\frac{2\pi}{30}} = 3.804\cdot 10^{-5}\ \Omega \tag{6.70}$$

As we can see from (6.70), the contribution of the end ring to the equivalent bar resistance is around 30%. This is more than a rather typical proportion. So far, we did consider that the distribution of the currents in the bars and end rings is uniform. However, there are A.C. currents in the rotor bars and end rings. Consequently, the distribution of the current in the bar is not uniform and depends essentially on the rotor (slip) frequency $f_2 = Sf_1$, (S-slip).

Globally, the skin effect translates into resistance and slot leakage correction coefficients $K_R > 1$ and $K_x < 1$ (see Chapter 9).

In general, the skin effect in the end rings is neglected. The bar resistance R_b in (6.69) and the slot permeance λ_b in (6.53) are modified to

$$R_b = \rho_{Al}\frac{L_b}{A_b}K_R \tag{6.71}$$

$$L_b = \mu_0 L_b \lambda_b K_x \tag{6.72}$$

6.12 SIMPLIFIED LEAKAGE SATURATION CORRECTIONS

Further on, for large values of currents (large slips), the stator (rotor) tooth tops tend to be saturated by the slot leakage flux (Figure 6.14).

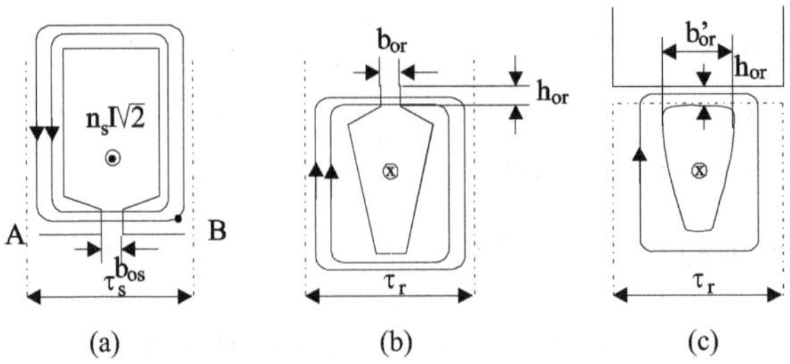

FIGURE 6.14 Slot neck leakage flux (a) stator semiclosed slot, (b) rotor semiclosed slot, and (c) rotor closed slot.

Neglecting the magnetic saturation along the tooth height, we may consider that only the tooth top saturates due to the slot neck leakage flux produced by the entire slot mmf.

A simple way to account for this leakage saturation that is used widely in industry consists of increasing the slot opening $b_{os,r}$ by the tooth top length t_t divided by the relative iron permeability μ_{rel} within it.

$$\begin{aligned} t_t &= \tau_{s,r} - b_{os,r} \\ b'_{os,r} &= b_{os,r} + \frac{t_t}{\mu_{rel}} \end{aligned} \tag{6.73}$$

For the closed slot (Figure 6.14a), we may directly use (6.31). To calculate μ_{rel}, we apply the Ampere's and flux laws.

$$\begin{aligned} B_t &= \mu_0 H_0 \\ n_s I\sqrt{2} &= t_t H_t + H_0 b_{os,r} \end{aligned} \tag{6.74}$$

$$\mu_0 n_s I\sqrt{2} = \mu_0 t_t H_t + B_t b_{os,r}; \quad B_0 = B_t \tag{6.75}$$

We have to add the lamination magnetisation curve, $B_t(H_t)$, to (6.75) and intersect them (Figure 6.15) to find B_t and H_t, and finally,

$$\mu_{rel} = \frac{B_t}{\mu_0 H_t} \tag{6.76}$$

H_0 – magnetic field in the slot neck.

Given the slot mmf $n_s I\sqrt{2}$ (the current, in fact), we may calculate iteratively μ_{rel} (6.76) and, from (6.73), the corrected $b_{os,r}$ ($b'_{os,r}$) then to be used in geometrical slot permeance λ_b calculations.

As expected, for open slots, the slot leakage saturation is negligible. Simple as it may seem, a second iteration process is required to find the current, as, in general, a voltage type supply is used to feed the IM.

It may be inferred that both differential and skewing leakage are influenced by saturation, in the sense of reducing them. For the former, we already introduced the partial teeth saturation factor $K_{st} = K_{sd}$ to account for it (6.14). For the latter, in (6.48), we have assumed that the level of saturation is implicitly accounted for L_{1m} (that is, it is produced by the magnetisation current in the machine). In reality, for large rotor currents, the skewing rotor mmf field, dependent on rotor current and axial position along stack, is quite different from that of the main flux path, at standstill.

However, to keep the formulas simple, Equation (6.73) has become rather standard for design purposes.

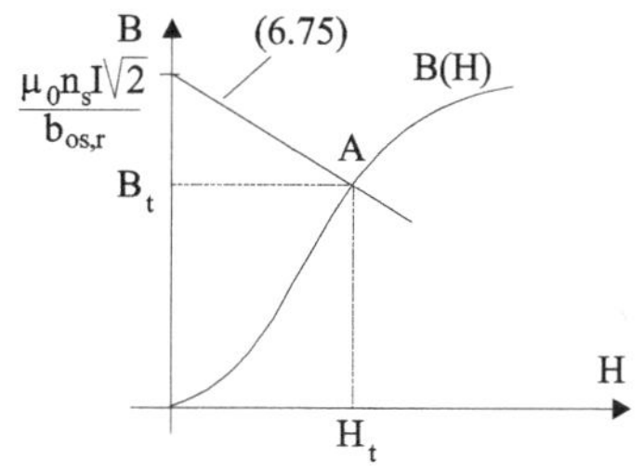

FIGURE 6.15 Tooth top flux density B_t.

Zig-zag flux path also tends to be saturated at high currents. Notice that the zig-zag flux path occupies the teeth tops (Figure 6.10b). This aspect is conventionally neglected as the zig-zag leakage inductance tends to be a small part of total leakage inductance.

So far, we considered the leakage saturation through simple approximate correction factors and treated skin effects only for rectangular coils. In Chapter 9, we will present comprehensive methods to treat these phenomena for slots of general shapes. Such slot shapes may be the result of design optimisation methods based on various cost (objective) functions such as high starting torque, low starting current, and large peak torque.

6.13 REDUCING THE ROTOR TO STATOR

The main flux paths embrace both the stator and rotor slots passing through airgap. So the two emfs per phase $\underline{E}_1$ (Chapter 5.10) and $\underline{E}_2$ are

$$\underline{E}_1 = -j\pi\sqrt{2}W_1K_{q1}K_{y1}f_1\underline{\Phi}_1; \quad \Phi_1 = \frac{2}{\pi}\underline{B}_{g1}\tau L_e \tag{6.77}$$

$$\underline{E}_2 = -j\pi\sqrt{2}W_2K_{q2}K_{y2}K_{c2}f_1\underline{\Phi}_1 \tag{6.78}$$

We have used phasors in (6.77) and (6.78) as E_1 and E_2 are sinusoidal in time (only the fundamental is accounted for).

Also in (6.78), E_2 is already calculated as "seen from the stator side", in terms of frequency because only in this case, Φ_1 is the same in both stator and rotor phases. In reality, the flux in the rotor varies at Sf_1 frequency, while in the stator at f_1 frequency. The amplitude is the same and, with respect to each other, they are at stall.

Now we may proceed as for transformers by dividing the two equations to obtain

$$\frac{E_1}{E_2} = \frac{W_1K_{w1}}{W_2K_{w2}} = K_e = \frac{V_r'}{V_r} \tag{6.79}$$

K_e is the voltage reduction factor to stator.

Eventually, the actual and the stator reduced rotor mmfs should be identical:

$$\frac{m_1\sqrt{2}K_{w1}I_r'W_1}{\pi p_1} = \frac{m_2\sqrt{2}K_{w2}I_rW_2}{\pi p_1} \tag{6.80}$$

From (6.80), we find the current reduction factor K_i.

$$K_i = \frac{I_r'}{I_r} = \frac{m_2K_{w2}W_2}{m_1K_{w1}W_1} \tag{6.81}$$

For the rotor resistance and reactance equivalence, we have to conserve the conductor losses and leakage field energy.

$$m_1R_r'I_r'^2 = m_2R_rI_r^2 \tag{6.82}$$

$$m_1\frac{L_{rl}'}{2}I_r'^2 = m_2\frac{L_{rl}}{2}I_r^2 \tag{6.83}$$

So,

$$R_r' = R_r\frac{m_2}{m_1}\frac{1}{K_i^2}; \quad L_{rl}' = L_{rl}\frac{m_2}{m_1}\frac{1}{K_i^2} \tag{6.84}$$

For cage rotors, we may still use (6.81) and (6.84) but with $m_2 = N_r$, $W_2 = 1/2$, and $K_{w2} = K_{skew}$ (skewing factor); that is, each bar (slot) represents a phase.

$$R'_r = R_{be}\frac{12K_{w1}^2W_1^2}{N_rK_{skew}^2} \tag{6.85}$$

$$L'_{be} = L_{be}\frac{12K_{w1}^2W_1^2}{N_rK_{skew}^2} \tag{6.86}$$

Notice that only the slot and ring leakage inductances in (6.86) have to be reduced to the stator in L_{rl} (6.51).

$$L'_{rl} = \left(L_{dlr} + L_{zlr} + L_{skew,r}\right) + L'_{be} \tag{6.87}$$

Example 6.4 Bar and Ring Resistance and Leakage Inductance Calculation

Let us consider an IM with $q = 3$, $p_1 = 2$, ($N_s = 36$ slots), $W_1 = 300$ turns/phase, $K_{w1} = 0.965$, $N_r = 30$ slots, one stator slot pitch rotor skewing ($c/\tau = 1/3q$) ($K_{skew} = 0.9954$), and the rotor bar (slot) and end ring cross sections are rectangular ($h_r/b_1 = 2/1$) and $h_{or} = 1.5{\cdot}10^{-3}$m and $b_{or} = 1.5{\cdot}10^{-3}$m, $L_e = 0.12$ m, and $D_{ir} = 0.15$m as in Example 6.3. Let us find the bar ring resistance and leakage inductance reduced to the stator.

Solution

From Example 6.3, we took $D_{ir} = 0.15$ m and $Ai = a \cdot b = 481$ mm² for the end ring. We may assume $a/b = 1/2$ and, thus, $a = 15.51$ mm and $b = 31$ mm.

The bar cross section in example 6.3 is $A_b = 166\,\text{mm}^2$.

$$b_1 = \sqrt{\frac{A_b}{\left(h_r/b_1\right)}} = \sqrt{\frac{166}{2}} \approx 9.1\ \text{mm};\quad h_r = 18.2\ \text{mm} \tag{6.88}$$

From Example 6.3, we already know the equivalent bar resistance (6.70) $R_{be} = 3.804{\cdot}10^{-5}\ \Omega$.

Also from (6.19), the rounded geometrical slot permeance λ_r is

$$\lambda_r = \lambda_{bar} \approx \frac{h_r}{3b_1} + \frac{h_{or}}{b_{or}} = \frac{18.2}{3\cdot 9.1} + \frac{1.5}{1.5} = 1.66 \tag{6.89}$$

From (6.46), the end ring segment geometrical permeance λ_{ei} may be found.

$$\lambda_{ei} = \frac{2.3D_{ir}}{4N_rL_e\sin^2\dfrac{\pi P_1}{N_r}}\lg\frac{4.7D_i}{a+2b}$$

$$= \frac{2.3\cdot 0.15}{4\cdot 30\cdot 0.12\cdot \sin^2\dfrac{\pi 2}{18}}\lg\frac{4.7\cdot 0.15}{(1.5+2\cdot 31)\cdot 10^{-3}} = 0.1964 \tag{6.90}$$

with $L_i = \pi\, D_i/N_r = \pi{\cdot}150/30 = 15.7{\cdot}10^{-3}$m.

Now from (6.40) and (6.41), the bar and end ring leakage inductances are

$$\begin{aligned} L_b &= \mu_0 l_b \lambda_{bar} = 1.256\cdot 10^{-6}\cdot 0.14\cdot 1.66 = 0.292\cdot 10^{-6}\ \text{H} \\ L_{li} &= \mu_0 l_i \lambda_{ei} = 1.256\cdot 10^{-6}\cdot 0.0157\cdot 1.964 = 0.3872\cdot 10^{-8}\ \text{H} \end{aligned} \tag{6.91}$$

The equivalent bar leakage inductance L_{be} (6.52) is written as

$$L_{be} = L_b + 2L_{ei} = 0.292 \cdot 10^{-6} + 2 \cdot 0.3872 \cdot 10^{-8} = 0.2997 \cdot 10^{-6}\ H \tag{6.92}$$

From (6.85) and (6.86), we may now obtain the rotor slot (bar) and end ring equivalent resistance and leakage inductance reduced to the stator

$$R_r' = R_{be}\frac{12K_{w1}^2W_1^2}{N_rK_{skew}^2} = \frac{3.804 \cdot 10^{-5} \cdot 12 \cdot 0.965^2 \cdot 300^2}{30 \cdot 0.99^2} = 1.28\ \Omega \tag{6.93}$$

$$L_{be}' = L_{be}\frac{12K_{w1}^2W_1^2}{N_rK_{skew}^2} = 0.2997 \cdot 10^{-6} \cdot 0.36 \cdot 10^{-5} = 1.008 \cdot 10^{-2}\ H \tag{6.94}$$

As a bonus, knowing from the Example 6.3 that the magnetisation inductance $L_{1m} = 0.1711H$, we may calculate the rotor skewing leakage inductance (from 6.48), which is already reduced to the stator because it is a fraction of L_{1m}

$$L_{skew,r} = \left(1 - K_{skew}^2\right) \cdot L_{1m} = \left(1 - 0.9954^2\right) \cdot 0.1711 = 1.57 \cdot 10^{-3}\ H \tag{6.95}$$

In this particular case, the skewing leakage is more than 6 times smaller than the slot (bar) and end ring leakage inductance. Notice also that $L_{be}'/L_{1m} = 0.059$, a rather practical value.

6.14 THE BRUSHLESS DOUBLY FED INDUCTION MACHINE (BDFIM)

For limited variable speed range (motoring and generating), the dual stator winding (with p_1 and p_2 pole pairs) in the stator (Figure 6.16a) is complemented by a $(p_1 + p_2)$ pole nested cage with internal loops (Figure 6.16b) or a nested-loop rotor (Figure 6.16b) [10].

In general, the main stator winding (with p_1 pole pairs) is connected directly to the grid (or load), while the control stator winding (with p_2 pole pairs) is interfaced to the grid (or load) by a dual PWM converter sized at a fraction of the main winding power commensurate with variable speed range. The relationship between the two stator frequencies ω_1 and ω_2 and the number of rotor poles $(p_1 + p_2)$ for synchronous operation speed ω_r is

$$\omega_r = \frac{\omega_1 + \omega_2}{p_1 + p_2};\ \text{in general}\ |\omega_2| < \frac{1}{3}\omega_1 \tag{6.96}$$

The two slips of rotor S_1 and S_2 with respect to the rotor are

$$S_1 = \frac{\omega_1 - p_1\omega_r}{\omega_1};\ S_2 = \frac{\omega_2 - p_2\omega_r}{\omega_2} \tag{6.97}$$

The rotor leakage inductance and resistance, reduced to the main stator winding, together with stator windings main (magnetisation) and leakage inductances per phase are crucial in calculating the performance of such a special IM without brushes, with potential application primarily (but not only) in wind energy conversion.

Minimum rotor leakage inductance and resistance is crucial for best magnetic coupling between the main and control stator winding.

This is how the cage types in the rotor (Figure 6.16b and c) have been introduced. Besides, the rotor turns ratio n_r is crucial to maximum power condition [10].

$$n_r = \left(\frac{p_1}{p_2}\right)^{1/2} \tag{6.98}$$

The self, coupling and conventional equivalent leakage inductance and resistance per phase (reduced to the main stator winding) computation is more involved than for standard IMs [10]. However, the

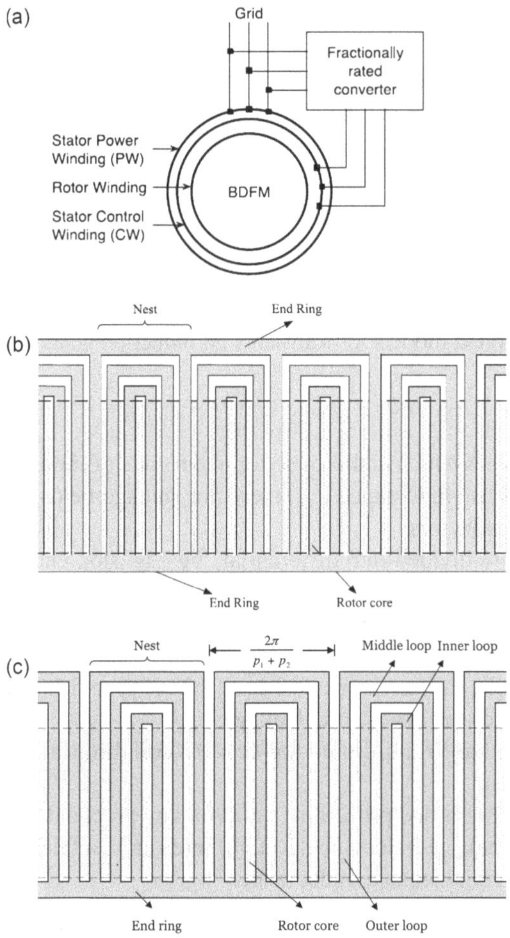

FIGURE 6.16 Brushless doubly fed induction machine (BDFIM): (a) basic topology, (b) $p_1 + p_2$ bar cage rotor, and (c) nested-loop rotor (after [10]).

conventional leakage inductance of rotor refers to slot leakage, overhang, and zig-zag leakage, such that for each nest of a nested-loop rotor (Figure 6.16c) they are

$$\left(L_{r\,leak}\right)_{nest} = \begin{bmatrix} L_{loop} & 0 & 0 \\ 0 & L_{loop} & 0 \\ 0 & 0 & L_{loop} \end{bmatrix} \tag{6.99}$$

The rotor resistance of each nest is similarly

$$\left[R_r\right] = \begin{bmatrix} R_{loop} & 0 & 0 \\ 0 & R_{loop} & 0 \\ 0 & 0 & R_{loop} \end{bmatrix} \tag{6.100}$$

for a three loop nest.

For $p_1 = 2$ and $p_2 = 4$ pole pairs, $p_1 + p_2 = 6$ rotor poles, and a rotor with three loops per rotor pole, the order of total $L_{r\,leak}$ and R_r diagonal matrixes will be $3 \times 6 = 18$.

While calculating (6.99 and 6.100) terms is straightforward, reducing them to the main stator phases has to be done.

The effective turns ratio is required, and it is the square root of fundamental space harmonics leakage terms for the rotor and the stator. It has to be noted that the rotor leakage inductance contains besides the conventional component illustrated above, the space harmonics manifested as magnetisation and mutual inductances (as a kind of differential leakage component).

The combination of pole pairs p_1 and p_2 plays a crucial role in reducing R_r' and L_{rl}' (reduced to the stator).

The paragraph here serves only as a starting enquiry in the study of special topology IMs rotor leakage inductance and resistance. The BDFIM will be investigated in the chapters dedicated to equivalent steady state circuits to IM models for transients on induction generators and IM optimal design.

6.15 SUMMARY

- Besides main path lines which embrace stator and rotor slots and cross the airgap to define the magnetisation phase inductance L_{1m}, there are leakage fields which encircle either the stator or the rotor conductors.
- The leakage fields are divided into differential, zig-zag, slot, end-turn, and skewing leakages. Their corresponding inductances are calculated from their stored magnetic energy.
- Step mmf harmonic fields through airgap induce emfs in the stator, while the space rotor mmf harmonics do the same. These space harmonics produced emfs have the supply frequency and this is why they are considered of leakage type. Magnetic saturation of stator and rotor teeth reduces the differential leakage.
- Stator differential leakage is minimum for $y/\tau = 0.8$ coil throw and decreases with increasing q (slots/pole/phase).
- A quite general graphic-based procedure, valid for any practical winding, is used to calculate the differential leakage inductance. Ref. [6] introduces a fairly general approach to the calculation of differential leakage inductance.
- The slot leakage inductance is based on the definition of a geometrical permeance λ_s dependent on the aspect ratio. In general, $\lambda_s \in (0.5$ to $2.5)$ to keep the slot leakage inductance within reasonable limits. For a rectangular slot, some rather simple analytical expressions are obtained even for double-layer windings with chorded, unequal coils.
- The saturation of teeth tops due to high currents at large slips reduces λ_s, and it is accounted for by a pertinent increase in slot opening which is dependent on stator (rotor) current (mmf).
- The zig-zag leakage flux lines of stator and rotor mmf snake through airgap around slot openings and close out through the two back cores. At high values of currents (large slip values), the zig-zag flux path, mainly the teeth tops, tends to saturate because of the combined action of slot neck saturation and zig-zag mmf contribution.
- The rotor slot skewing leads to the existence of a skewing (uncompensated) rotor mmf which produces a leakage flux along the main paths, but its maximum is phase shifted with respect to the magnetisation mmf maximum and is dependent on slip and position along the stack length. As an approximation only, a simple analytical expression for an additional rotor-only skewing leakage inductance $L_{skew,r}$ is given.
- Leakage path saturation reduces the leakage inductance.
- The A.C. stator resistance is higher than the D.C. because of skin effect, accounted for by a correction coefficient K_R, calculated in Chapter 9. In most IMs, even at higher power but at 50 (60) Hz, the skin effect in the stator is negligible for the fundamental. It is not so for current harmonics present in converter-fed IMs or high-speed (high-frequency) IMs.
- The rotor bar resistance in squirrel cage motors is, in general, increased notably by skin effect, for rotor (slip) frequencies $f_2 = Sf_1 > (4\text{–}5)$Hz; $K_R > 1$.

- The skin effect also reduces the slot geometrical permeance ($K_x < 1$) and, finally, the leakage inductance of the rotor.
- The rotor cage (or winding) has to be reduced to the stator to prepare the rotor resistance and leakage inductance for utilisation in the equivalent circuit of IMs. The equivalent circuit is widely used for IM performance computation.
- Accounting for leakage saturation and skin effect in a comprehensive way for general shape slots (with deep bars or double rotor cage) is a subject revisited in Chapter 9.
- Now with Chapters 5 and 6 in place, we do have all basic parameters – magnetisation inductance L_{1m}, leakage inductances L_{sl} and L_{rl}', and phase resistances R_s and R_r'.
- With rotor parameters reduced to the stator, we are ready to approach the basic equivalent circuit as a "vehicle" for performance computation.
- Yet new analytical approaches to more precisely calculate the IM slot leakage inductances are proposed [11,12].

REFERENCES

1. R. Richter, *Electric Machines* – Vol.4. – Induction Machines, Verlag Birkhauser, Bassel/Stuttgart, 1954, (in German).
2. B. Heller and V. Hamata, *Harmonic Field Effects in Induction Machines*, Elsevier Scientific Publishing Company, Amsterdam, 1977.
3. I. B. Danilevici, V. V. Dombrovski, E. I. Kazovski, A.C. *Machines Parameters*, Science Publishers, St. Petersburg, 1965 (in Russian).
4. R. H. Engelmann and W. H. Middendorf (editors), *Handbook of Electric Motors*, Marcel Dekker Inc., New York, 1995, Chapter 4.
5. A. Demenko and K. Oberretl, Calculation of additional slot leakage permeance in electrical machines due to radial ventilation ducts, *Compel*, Vol. 11. No. 1, James & James Science Publishers Ltd, 1991, pp. 93–96(c).
6. H. Xueliang, D. Qiang, H. Minqiang, A novel exact and universal approach for calculating the differential leakage related to harmonics waves in A.C. electric motors, *IEEE Transactions on Energy Conversion*, Vol. 19, No. 1, 2004, pp. 1–6.
7. P. L. Alger, *The Nature of Induction Machines*, 2nd edition, Gordon & Breach, New York, 1970.
8. I. A. Voldek, The computation of end-connection leakage inductances of A.C. machine, *Electriceskvo*, No. 1, 1963 (in Russian).
9. S. Williamson and M. A. Mueller, Induction motor end – Winding leakage reactance calculation using the Biot – Sawart method, taking rotor currents into account, *Record of ICEM*, 1990, MIT, Vol. 2., pp. 480–484.
10. A. Oraee, E. Abdi, S. Abdi, R. McMahon, P.J. Tavner, Effects of rotor winding structure on the BDFM equivalent circuit parameters, *IEEE Transactions on Energy Conversion*, Vol. 30, No. 4, 2015, pp. 1660–1669.
11. A. Tessarolo, Analytical determination of slot leakage field and inductances of electric machines with double-layer windings and semiclosed slots, *IEEE Transactions on Energy Conversion*, Vol. 30, No. 4, 2015, pp. 1528–1536.
12. A. Cavagnino, Accuracy – enhanced algorithms for the slot leakage inductance computation of double – layer windings, *IEEE Transactions on Industry Applications*, Vol. 53, No. 3, 2017, pp. 4422–4430.

- The skin effect also reduces the slot geometrical permeance (Eq. 6.1) and, finally, the leakage inductance of the rotor.
- The rotor cage (or winding) has to be reduced to the stator to prepare the rotor resistance and leakage inductance for the equivalent circuit of IMs. The equivalent circuit is [illegible] used for IM performance computation.
- [illegible]
- [illegible]

REFERENCES

[illegible]

7 Steady-State Equivalent Circuit and Performance

7.1 BASIC STEADY-STATE EQUIVALENT CIRCUIT

When the induction machine (IM) is fed in the stator from a three-phase balanced power source, the three-phase currents produce a travelling field in the airgap. This field produces emfs both in the stator and the rotor windings: E_1 and E_{2s}. A symmetrical cage in the rotor may be reduced to an equivalent three-phase winding.

The frequency of E_{2s} is f_2:

$$f_2 = f_1 - np_1 = \left(\frac{f_1}{p_1} - n\right)p_1 = \frac{n_1 - n}{n_1} n_1 p_1 = Sf_1 \tag{7.1}$$

This is so because the stator mmf travels around the airgap with a speed $n_1 = f_1/p_1$, while the rotor travels with a speed of n. Consequently, the relative speed of the mmf wave with respect to rotor conductors is $(n_1 - n)$, and thus, the rotor frequency f_2 in (7.1) is obtained.

Now the emf in the short-circuited rotor "acts upon" the rotor resistance R_r and leakage inductance L_{rl}

$$\underline{E}_{2s} = S\underline{E}_2 = (R_r + jS\omega_1 L_{rl})\underline{I}_r \tag{7.2}$$

$$\frac{\underline{E}_{2s}}{S} = \underline{E}_2 = \left(\frac{R_r}{S} + j\omega_1 L_{rl}\right)\underline{I}_r \tag{7.3}$$

If Equation (7.2) includes variables at rotor frequency $S\omega_1$, with the rotor in motion, Equation (7.3) refers to a circuit at stator frequency ω_1, that is, with a "fictitious" rotor at standstill.

Now after reducing $\underline{E}_2$, $\underline{I}_r$, R_r, and L_{rl} to the stator by the procedure shown in Chapter 6, Equation (7.3) yields

$$\underline{E}_2' = \underline{E}_1 = \left[R_r' + R_r'\left(\frac{1}{S} - 1\right) + j\omega_1 L_{rl}'\right]\underline{I}_r' \tag{7.4}$$

$$\frac{E_2}{E_1} = \frac{K_{w2}W_2}{K_{w1}W_1} = K_E; \quad K_{w2} = 1, \quad W_2 = 1/2 \text{ for cage rotors}$$

$$\frac{\underline{I}_r}{\underline{I}_r'} = \frac{m_1 K_{w1} W_1}{m_2 K_{w2} W_2} = K_I; \quad W_2 = 1/2, \quad m_2 = N_r \text{ for cage rotors} \tag{7.5}$$

$$\frac{R_r}{R_r'} = \frac{L_{rl}}{L_{rl}'} = \frac{K_E}{K_I} \tag{7.6}$$

$$K_{w1} = K_{q1} \cdot K_{y1}; \quad K_{w2} = K_{q2} \cdot K_{y2} \cdot K_{skew} \tag{7.7}$$

W_1 and W_2 are turns per phase (or per current path).

K_{w1} and K_{w2} are winding factors for the fundamental mmf waves.

m_1 and m_2 are the numbers of stator and rotor phases and N_r is the number of rotor slots.

The stator phase equation is easily written as

$$-\underline{E}_1 = \underline{V}_s - \underline{I}_s\left(R_s + j\omega_1 L_{sl}\right) \tag{7.8}$$

because in addition to the emf, there is only the stator resistance and leakage inductance voltage drop.

Finally, as there is current (mmf) in the rotor, the emf E_1 is produced concurrently by the two mmfs ($\underline{I}_s$ and $\underline{I}_r'$).

$$\underline{E}_1 = -\frac{d\psi_{1m}}{dt} = -j\omega_1 L_{1m}\left(\underline{I}_s + \underline{I}_r'\right) \tag{7.9}$$

If the rotor is not short-circuited, Equation (7.4) becomes

$$\underline{E}_1 - \frac{\underline{V}_r'}{S} = \left[R_r' + R_r'\left(\frac{1}{S} - 1\right) + j\omega_1 \underline{L}_{rl}'\right]\underline{I}_r' \tag{7.10}$$

The division of V_r (rotor applied voltage) by slip (S) comes into place as the derivation of (7.10) starts in (7.2) where

$$S\underline{E}_2 - \underline{V}_r = \left(R_r + jS\omega_1 L_{rl}\right)\underline{I}_r \tag{7.11}$$

The rotor circuit is considered as a source, while the stator circuit is a sink. Now Equations (7.8)–(7.11) constitute the IM equations per phase reduced to the stator for the rotor circuit.

Notice that in these equations there is only one frequency, the stator frequency ω_1, which means that they refer to an equivalent rotor at standstill, but with an additional "virtual" rotor resistance per phase, $R_r(1/S - 1)$ dependent on slip (speed).

It is now evident that the active power in this additional resistance is in fact the electromechanical power of the actual motor

$$P_m = T_e \cdot 2\pi n = 3R_r'\left(\frac{1}{S} - 1\right)\left(\underline{I}_r'\right)^2 \tag{7.12}$$

with

$$n = \frac{f_1}{p_1}(1 - S) \tag{7.13}$$

$$\frac{\omega_1}{p_1}T_e = \frac{3R_r'\left(\underline{I}_r'\right)^2}{S} = P_{elm} \tag{7.14}$$

P_{elm} is called the electromagnetic power, the active power which crosses the airgap, from stator to rotor for motoring and vice versa for generating.

Equation (7.14) provides an alternative definition of slip which is very useful for design purposes:

$$S = \frac{3R_r'\underline{I}_r'}{P_{elm}} = \frac{P_{Cor}}{P_{elm}} \tag{7.15}$$

Equation (7.15) signifies that, for a given electromagnetic power P_{elm} (or torque, for given frequency), the slip is proportional to rotor winding losses.

Equations (7.8)–(7.11) lead progressively to the ideal equivalent circuit in Figure 7.1.

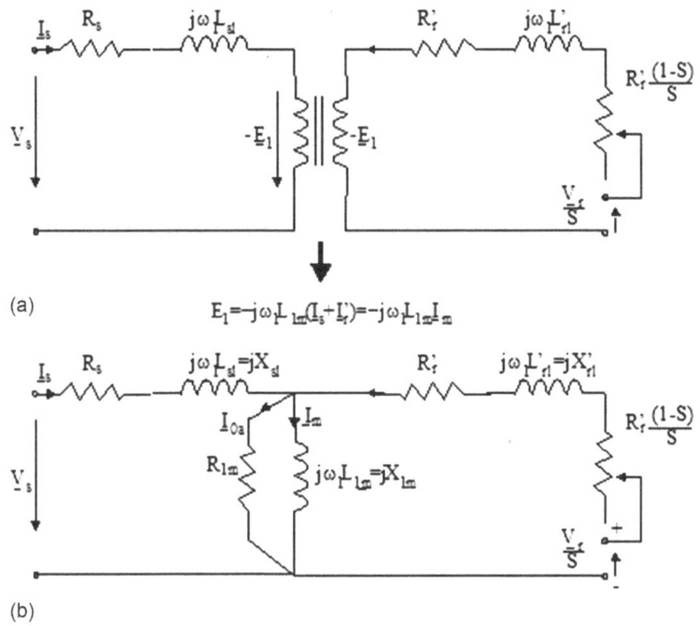

FIGURE 7.1 (a and b) The equivalent circuit.

Still missing in Figure 7.1a are the parameters to account for core losses, additional losses (in the cores and windings due to harmonics), and the mechanical losses.

The additional losses P_{ad} will be left out and considered separately in Chapter 11 as they amount, in general, to up to 3% of rated power in well-designed IM.

The mechanical and fundamental core losses may be combined in a resistance R_{1m} in parallel with X_{1m} in Figure 7.1b, as at least core losses are produced by the main path flux (and magnetisation current $\underline{I}_m$). R_{1m} may also be combined as a resistance in series with X_{1m}, for convenience in constant frequency IMs. For variable frequency IMs, however, the parallel resistance R_{1m} varies only slightly with frequency as the power in it (mainly core losses) is proportional to $E_1^2 = \omega_1^2 K_{w1}^2 \Phi_1^2$, which is consistent to eddy current core loss variation with both frequency and flux squared.

R_{1m} may be calculated in the design stage or may be found through standard measurements.

$$R_{1m} = \frac{3E_1^2}{P_{iron}} = \frac{3X_{1m}^2 I_m^2}{P_{iron}}; \quad I_{oa} \ll I_m \tag{7.16}$$

7.2 CLASSIFICATION OF OPERATION MODES

The electromagnetic (active) power crossing the airgap P_{elm} (7.14) is positive for $S > 0$ and negative for $S < 0$.

That is, for $S < 0$, the electromagnetic power flows from the rotor to the stator. After covering the stator losses, the rest of it is sent back to the power source. For $\omega_1 > 0$ (7.14), $S < 0$ means negative torque T_e. Also, $S < 0$ means $n > n_1 = f_1/p_1$. For $S > 1$ from the slip definition, $S = (n_1 - n)/n_1$, it means that either $n < 0$ and $n_1(f_1) > 0$ or $n > 0$ and $n_1(f_1) < 0$.

In both cases, as $S > 1$ ($S > 0$), the electromagnetic power $P_{elm} > 0$ and thus flows from the power source into the machine.

On the other hand, with $n > 0$, $n_1(\omega_1) < 0$, the torque T_e is negative; it is opposite to motion direction. That is braking. The same is true for $n < 0$ and $n_1(\omega_1) > 0$. In this case, the machine absorbs electric power through the stator and mechanical power from the shaft and transforms them both into heat in the rotor circuit total resistances.

TABLE 7.1
Operation Modes ($f_1/p_1 > 0$)

S	– – – –	0	+ + + +	1	+ + + +
N	+ + + +	f_1/p_1	+ + + +	0	– – – –
T_e	– – – –	0	+ + + +	+ + + +	+ + + +
P_{elm}	– – – –	0	+ + + +	+ + + +	+ + + +
Operation mode	Generator	|	Motor	|	Braking
		|		|	

Now for $0 < S < 1$, $T_e > 0$, $0 < n < n_1$, and $\omega_1 > 0$, the IM is motoring as the torque acts along the direction of motion.

The above reasoning is summarised in Table 7.1.

Positive $\omega_1(f_1)$ means positive sequence-forward mmf travelling wave. For negative $\omega_1(f_1)$, a companion table for reverse motion may be obtained.

7.3 IDEAL NO-LOAD OPERATION

The ideal no-load operation mode corresponds to zero rotor current. From (7.11), for $I_{r0} = 0$, we obtain

$$S_0\underline{E}_2 - \underline{V}_R = 0; \quad S_0 = \frac{\underline{V}_R}{\underline{E}_2} \tag{7.17}$$

The slip S_0 for ideal no load depends on the value and phase of the rotor applied voltage $\underline{V}_R$. For $\underline{V}_R$ in phase with E_2, $S_0 > 0$ and, with them in opposite phase, $S_0 < 0$.

The conventional ideal no-load synchronism for the short-circuited rotor ($V_R = 0$) corresponds to $S_0 = 0$ and $n_0 = f_1/p_1$. If the rotor windings (in a wound rotor) are supplied with a forward voltage sequence of adequate frequency $f_2 = Sf_1$ ($f_1 > 0$, $f_2 > 0$), subsynchronous operation (motoring and generating) may be obtained. If the rotor voltage sequence is changed, $f_2 = Sf_1 < 0$ ($f_1 > 0$), supersynchronous operation may be obtained. This is the case of doubly fed (wound rotor) induction machine. For the time being, we will deal, however, with the conventional ideal no load (conventional synchronism) for which $S_0 = 0$.

The equivalent circuit degenerates into the one in Figure 7.2a (rotor circuit is open).

Building the phasor diagram (Figure 7.2b) starts with $\underline{I}_m$, continues with $jX_{1m}\underline{I}_m$, then $\underline{I}_{0a}$

$$\underline{I}_{oa} = \frac{jX_{1m}\underline{I}_m}{R_{1m}} \tag{7.18}$$

and

$$\underline{I}_{s0} = \underline{I}_{oa} + \underline{I}_m \tag{7.19}$$

Finally, the stator phase voltage $\underline{V}_s$ (Figure 7.2b) is

$$\underline{V}_s = jX_{1m}\underline{I}_m + R_s\underline{I}_{s0} + jX_{sl}\underline{I}_{s0} \tag{7.20}$$

The input (active) power P_{s0} is dissipated into electromagnetic loss, fundamental and harmonics stator core losses, and stator windings and space harmonic-caused rotor core and cage losses. The driving motor covers the mechanical losses and whatever losses would occur in the rotor core and squirrel cage due to space harmonics fields and hysteresis.

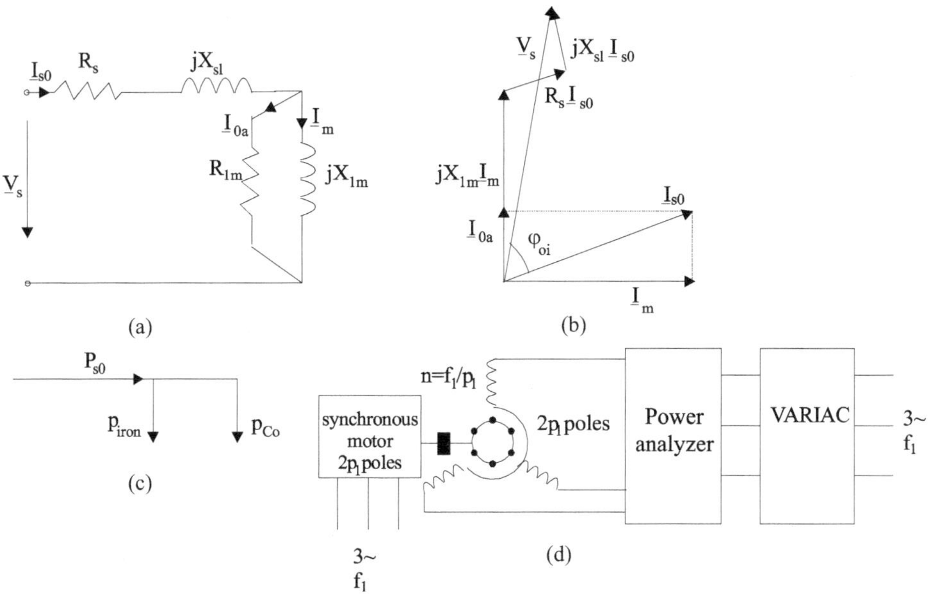

FIGURE 7.2 Ideal no-load operation ($V_R = 0$): (a) equivalent circuit, (b) phasor diagram, (c) power balance and (d) test rig.

For the time being, when doing the measurements, we will consider only stator core and winding losses.

$$P_{s0} \approx 3R_{1m}I_{0a}^2 + 3R_sI_{s0}^2 = \left(3R_{1m}\frac{X_{1m}^2}{X_{1m}^2 + R_{1m}^2} + 3R_s\right)I_{s0}^2$$

$$= p_{iron} + 3R_sI_{s0}^2 \tag{7.21}$$

From D.C. measurements, we may determine the stator resistance R_s. Then, from (7.21), with P_{s0}, I_{s0} measured with a power analyser, we may calculate the iron losses p_{iron} for given stator voltage V_s and frequency f_1.

We may repeat the testing for variable f_1 and V_1 (or variable V_1/f_1) to determine the core loss dependence on frequency and voltage.

The input reactive power is

$$Q_{s0} = \left(3X_{1m}\frac{R_{1m}^2}{X_{1m}^2 + R_{1m}^2} + 3X_{sl}\right)I_{s0}^2 \tag{7.22}$$

From (7.21) and (7.22), with R_s known, Q_{s0}, I_{s0}, and P_{s0} measured, we may calculate only two out of the three unknowns (parameters): X_{1m}, R_{1m}, and X_{sl}.

We know that $R_{1m} \gg X_{1m} \gg X_{sl}$. However, X_{sl} may be taken by the design value or the value measured with the rotor out or half the stall rotor ($S = 1$) reactance X_{sc}, as shown later in this chapter.

Consequently, X_{1m} and R_{1m} may be found with good precision from the ideal no-load test (Figure 7.2d). Unfortunately, a synchronous motor with the same number of poles is needed to provide driving at synchronism. This is why the no-load motoring operation mode has become standard for industrial use, while the ideal no-load test is mainly used for prototyping.

Example 7.1 Ideal No-Load Parameters

An induction motor driven at synchronism ($n = n_1 = 1800\,rpm$, $f_1 = 60\,Hz$, $p_1 = 2$) is fed at rated voltage $V_1 = 240\,V$ (phase RMS) and draws a phase current $I_{s0} = 3A$, the power analyser measures $P_{s0} = 36W$, $Q_{s0} = 700\ VAR$, the stator resistance $R_s = 0.1\ \Omega$, and $X_{sl} = 0.3\ \Omega$. Let us calculate the core loss p_{iron}, X_{1m}, and R_{1m}.

Solution

From (7.21), the core loss p_{iron} is

$$p_{iron} = P_{s0} - 3R_sI_{s0}^2 = 36 - 3\cdot 0.1\cdot 3^2 = 33.3\ W \tag{7.23}$$

Now, from (7.21) and (7.22), we get

$$\frac{R_{1m}X_{1m}^2}{X_{1m}^2 + R_{1m}^2} = \frac{P_{s0} - 3R_sI_{s0}^2}{3I_{s0}^2} = \frac{36 - 3\cdot 0.1\cdot 3^2}{3\cdot 3^2} = \frac{33.3}{27} = 1.233\ \Omega \tag{7.24}$$

$$\frac{R_{1m}^2X_{1m}}{X_{1m}^2 + R_{1m}^2} = \frac{Q_{s0} - 3X_{sl}I_{s0}^2}{3I_{s0}^2} = \frac{700 - 3\cdot 0.3\cdot 3^2}{3\cdot 3^2} = 25.626\ \Omega \tag{7.25}$$

Dividing (7.25) by (7.26) we get

$$\frac{R_{1m}}{X_{1m}} = \frac{25.626}{1.233} = 20.78 \tag{7.26}$$

From (7.25),

$$\frac{X_{1m}}{\left(\frac{X_{1m}}{R_{1m}}\right)^2 + 1} = 25.626;\quad \frac{X_{1m}}{(20.78)^{-2} + 1} = 25.626 \Rightarrow X_{1m} = 25.685\ \Omega \tag{7.27}$$

R_{1m} is calculated from (7.26),

$$R_{1m} = X_{1m}\cdot 20.78 = 25.626\cdot 20.78 = 533.74\ \Omega \tag{7.28}$$

By doing experiments for various frequencies f_1 (and V_s/f_1 ratios), the slight dependence of R_{1m} on frequency f_1 and on the magnetisation current I_m may be proved.

As a bonus, the power factor $\cos\varphi_{0i}$ may be obtained as

$$\cos\varphi_{0i} = \cos\left(\tan^{-1}\left(\frac{Q_{s0}}{P_{s0}}\right)\right) = 0.05136 \tag{7.29}$$

The lower the power factor at ideal no load, the lower the core loss in the machine (the winding losses are low in this case).

In general, when the machine is driven under load, the value of emf ($E_1 = X_{1m}I_m$) does not vary notably up to rated load and thus the core loss found from ideal no-load testing may be used for assessing performance during loading, through the loss segregation procedure. Note, however, that, on load, additional losses, produced by space field harmonics, occur. For a precise efficiency computation, these "stray load losses" have to be added to the core loss measured under ideal no load or even for no-load motoring operation.

7.4 SHORT-CIRCUIT (ZERO SPEED) OPERATION

At start, the IM speed is zero (S = 1), but the electromagnetic torque is positive (Table 7.1), so when three-phase fed, the IM tends to start (rotate); to prevent this, the rotor has to be stalled.

First, we just adapt the equivalent circuit by letting S = 1 and replacing R'_r and X_{sl}, X'_{rl} by their values as affected by skin effect and magnetic saturation (mainly leakage saturation as shown in Chapter 6): $X_{slstart}$, R'_{rstart}, $X'_{rlstart}$ (Figure 7.3).

For standard frequencies of 50 (60) Hz, and above, $X_{1m} >> R'_{rstart}$. Also, $X_{1m} >> X'_{rlstart}$, so it is acceptable to neglect it in the classical short-circuit equivalent circuit (Figure 7.3b).

For low frequencies, however, this is not so; that is, $X_{1m} <> R'_{rstart}$, so the complete equivalent circuit (Figure 7.3a) is mandatory.

The power balance and the phasor diagram (for the simplified circuit of Figure 7.3b) are shown in Figure 7.3c and d. The test rigs for three-phase and single-phase supply testing are presented in Figure 7.3e and f.

It is evident that for single-phase supply, there is no starting torque as we do have a non-travelling stator mmf aligned to phase a. Consequently, no rotor stalling is required.

The equivalent impedance is now $(3/2)\underline{Z}_{sc}$ because phase a is in series with phases b and c in parallel. The simplified equivalent circuit (Figure 7.3b) may be used consistently for supply frequencies

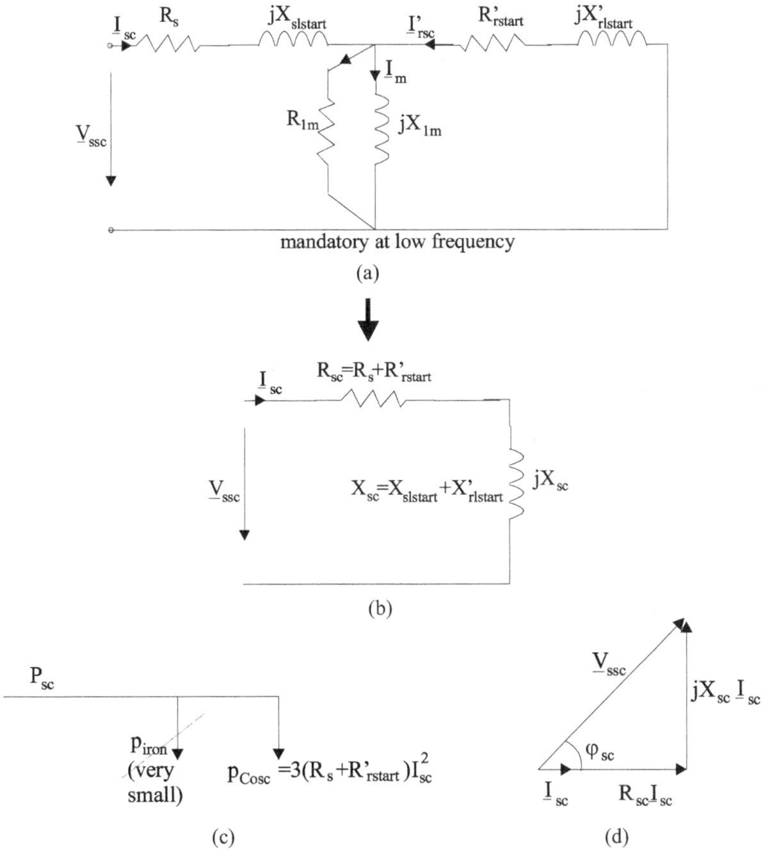

FIGURE 7.3 Short-circuit (zero speed) operation: (a) complete equivalent circuit at S = 1, (b) simplified equivalent circuit S = 1, (c) power balance, (d) phasor diagram.

(Continued)

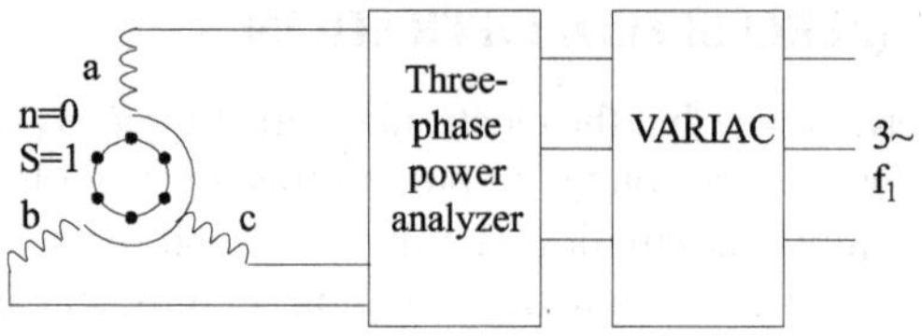

$\underline{Z}_{sc}$ – impedance per phase

(e)

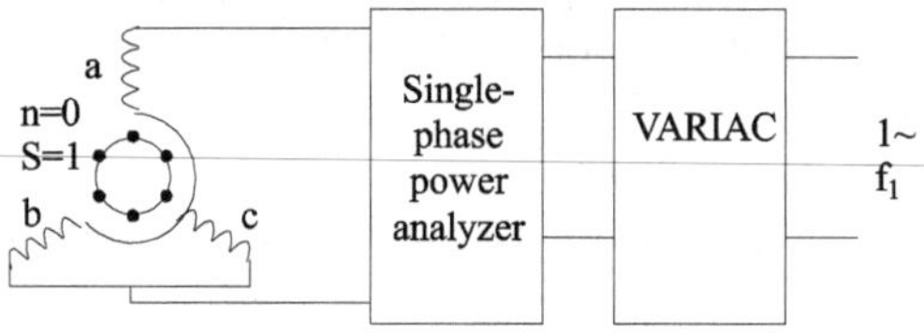

$\frac{3}{2}\underline{Z}_{sc}$ – impedance per phase

(f)

FIGURE 7.3 (CONTINUED) (e) three-phase zero speed testing, and (f) single-phase supply zero speed testing.

above 50 (60) Hz. For lower frequencies, the complete equivalent circuit has to be known. Still, the core loss may be neglected ($R_{1m} \approx \infty$), but from ideal no-load test at various voltages, we have to have $L_{1m}(I_m)$ function. A rather cumbersome iterative (optimisation) procedure is then required to find R'_{rstart}, $X'_{rlstart}$, and $X_{slstart}$ with only two equations from measurements of R_{sc} and V_{sc}/I_{sc}.

$$P_{sc} = 3R_s I_{sc}^2 + 3R'_{rstart} I'^2_{rsc} \tag{7.30}$$

$$\underline{Z}_{sc} = R_s + jX_{slstart} + \frac{jX_{1m}\left(R'_{rlstart} + jX'_{rlstart}\right)}{R'_{rlstart} + j\left(X_{1m} + X'_{rlstart}\right)} \tag{7.31}$$

This particular case, so typical with variable frequency IM supplies, will not be pursued further. For frequencies above 50 (60) Hz, the short-circuit impedance is simply

$$\underline{Z}_{sc} \approx R_{sc} + jX_{sc}; \quad R_{sc} = R_s + R'_{rlstart}; \quad X_{sc} = X_{slstart} + X'_{rlstart} \tag{7.32}$$

And with P_{sc}, V_{ssc}, I_{sc} measured, we may calculate

$$R_{sc} = \frac{P_{sc}}{3I_{sc}^2}; \quad X_{sc} = \sqrt{\left(\frac{V_{ssc}}{I_{sc}}\right)^2 - R_{sc}^2}, \tag{7.33}$$

for three-phase zero speed testing and

$$R_{sc} = \frac{2}{3}\frac{P_{sc\sim}}{I_{sc\sim}^2}; \quad X_{sc} = \frac{2}{3}\sqrt{\left(\frac{V_{ssc\sim}}{I_{sc\sim}}\right)^2 - \left(\frac{3}{2}R_{sc}\right)^2} \tag{7.34}$$

for single-phase zero speed testing.

If the test is done at rated voltage, the starting current I_{start} $(I_{sc})_{Vsn}$ is much larger than the rated current,

$$\frac{I_{start}}{I_n} \approx 4.5 \div 8.0 \tag{7.35}$$

for cage rotors, larger for high efficiency motors, and

$$\frac{I_{start}}{I_n} \approx 10 \div 12 \tag{7.36}$$

for short-circuited rotor windings.

The starting torque T_{es} is

$$T_{es} \approx \frac{3R'_{rstart} I_{start}^{2}}{\omega_1} p_1 \tag{7.37}$$

with

$$T_{es} = (0.7 \div 2.3) T_{en} \tag{7.38}$$

for cage rotors and

$$T_{es} = (0.1 \div 0.3) T_{en} \tag{7.39}$$

for short-circuited wound rotors.

Thorough testing of IM at zero speed generally is completed up to rated current. Consequently, lower voltages are required, thus avoiding machine overheating. A quick test at rated voltage is done on prototypes or random IMs from the production line to measure the peak starting current and the starting torque. This is a must for heavy starting applications (nuclear power plant cooling pump motors, for example) as both skin effect and leakage saturation notably modify the motor starting parameters: $X_{slstart}$, $X'_{rlstart}$, and R'_{rstart}.

Also, closed-slot rotors have their slot leakage flux path saturated at rotor currents notably less than rated current, so a low voltage test at zero speed should produce results as in Figure 7.4.

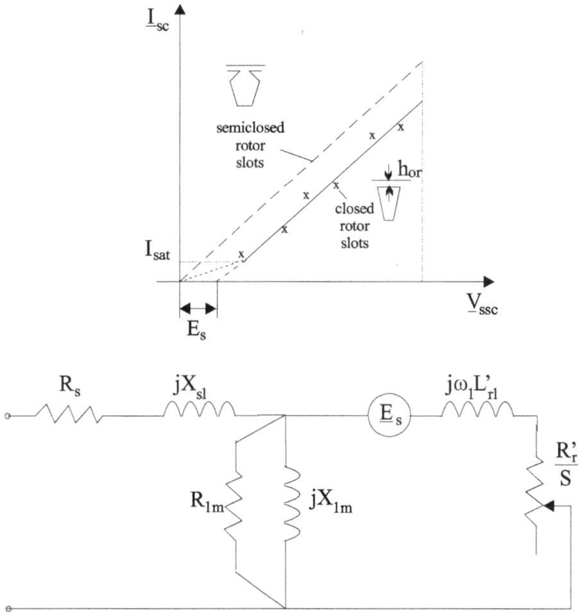

FIGURE 7.4 Stator voltage versus short-circuit current.

Intercepting the I_{sc}/V_{sc} curve with abscissa, we obtain, for the closed slot rotor, a nonzero emf E_s [1]. E_s is in the order of 6 to 12 V for 220 V phase RMS, 50 (60) Hz motors. This additional emf, E_s, is sometimes introduced in the equivalent circuit together with a constant rotor leakage inductance to account for rotor slot–bridge saturation. E_s is 90° ahead of rotor current I_r' and is equal to

$$E_s \approx \frac{4}{\pi}\pi\sqrt{2}f_1(2W_1)\Phi_{bridge}K_{w1}; \quad \Phi_{bridge} = B_{sbridge} \cdot h_{or} \cdot L_e \tag{7.40}$$

$B_{sbridge}$ is the saturation flux density in the rotor slot bridges ($B_{sbridge} = (2-2.2)T$). The bridge height is $h_{or} = 0.3–1$ mm depending on rotor peripheral speed. The smaller, the better.

A more complete investigation of combined skin and saturation effects on leakage inductances is to be found in Chapter 9 for both semiclosed and closed rotor slots.

Example 7.2 Parameters from Zero Speed Testing

An induction motor with a cage semiclosed slot rotor has been tested at zero speed for $V_{ssc} = 30$ V (phase RMS, 60 Hz). The input power and phase current are: $P_{sc} = 810$ W and $I_{sc} = 30$ A. The A.C. stator resistance $R_s = 0.1\ \Omega$. The rotor resistance, without skin effect, good for running conditions, is $R_r = 0.1\ \Omega$. Let us determine the short-circuit (and rotor) resistance and leakage reactance at zero speed and the start-to-load rotor resistance ratio due to skin effect.

Solution

From (7.33), we may directly determine the values of short-circuit resistance and reactance, R_{sc} and X_{sc},

$$R_{sc} = \frac{P_{sc}}{3I_{sc}^2} = \frac{810}{3 \cdot 30^2} = 0.3\ \Omega;$$
$$X_{sc} = \sqrt{\left(\frac{V_{ssc}}{I_{sc}}\right)^2 - R_{sc}^2} = \sqrt{\left(\frac{30}{30}\right)^2 - 0.3^2} = 0.954\ \Omega \tag{7.41}$$

The rotor resistance at start R'_{rstart} is

$$R'_{rstart} = R_{sc} - R_s = 0.3 - 0.1 = 0.2\ \Omega \tag{7.42}$$

So, the rotor resistance at start is two times that of full-load conditions.

$$\frac{R'_{rstart}}{R'_r} = \frac{0.2}{0.1} = 2.0 \tag{7.43}$$

The skin effect is responsible for this increase.

Separating the two leakage reactances $X_{slstart}$ and $X'_{rlstart}$ from X_{sc} is hardly possible. In general, $X_{slstart}$ is affected at start by leakage saturation, if the stator slots are not open, while $X'_{rlstart}$ is affected by both leakage saturation and skin effect. However, at low voltage (at about rated current), the stator and rotor leakage reactances are not affected by leakage saturation; only skin effect affects both R'_{rstart} and $X'_{rlstart}$. In this case, it is common practice to consider

$$\frac{1}{2}X_{sc} \approx X'_{rlstart} \tag{7.44}$$

7.5 NO-LOAD MOTOR OPERATION

When no mechanical load is applied to the shaft, the IM works on no load. In terms of energy conversion, the IM input power has to cover the core, winding, and mechanical losses. The IM has to develop some torque to cover the mechanical losses. So there are some currents in the rotor. However, they tend to be small, and consequently, they are usually neglected.

The equivalent circuit for this case is similar to the case of ideal no load, but now the core loss resistance R_{1m} may be paralleled by a fictitious resistance R_{mec} which includes the effect of mechanical losses.

The measured values are P_0, I_0, and V_s. Voltage is varied from, in general, $1/3V_{sn}$ to $1.2V_{sn}$ through a Variac.

As the core loss (eddy current loss, in fact) varies with $(\omega_1\Psi_1)^2$, that is approximately with V_s^2, we may end up with a straight line if we represent the function

$$P_0 - 3R_sI_0^2 = p_{iron} + p_{mec} = f\left(V_s^2\right) \tag{7.45}$$

The intersection of this line with the vertical axis represents the mechanical losses p_{mec} which are independent of voltage (Figure 7.5b). But for all voltages, the rotor speed has to remain constant and very close to synchronous speed.

Subsequently, the core losses p_{iron} are found. We may compare the core losses from the ideal no load and the no-load motoring operation modes. Now that we have p_{mec} and the rotor circuit is resistive ($R_r'/S_{0n} \gg X_{rl}'$), we may calculate approximately the actual rotor current I_{r0}.

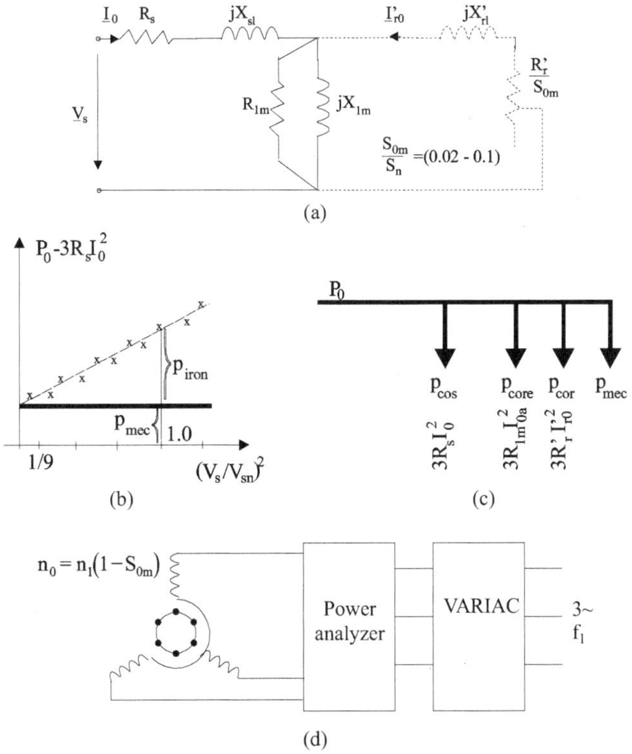

FIGURE 7.5 No-load motor operation: (a) equivalent circuit, (b) no-load loss segregation, (c) power balance, and (d) test arrangement.

$$\frac{R'_r}{S_{0n}} I'_{r0} \approx V_{sn}$$
$$p_{mec} \approx \frac{3R'_r I'^2_{r0}}{S_{0n}}(1-S_{0n}) \approx \frac{3R'_r I'^2_{r0}}{S_{0n}} \approx 3V_{sn} I'_{r0} \tag{7.46}$$

Now with I'_{r0} from (7.46) and R'_r known, S_{0n} may be determined. After a few such calculation cycles, convergence towards more precise values of I'_{r0} and S_{0n} is obtained.

Example 7.3 No-Load Motoring

An induction motor has been tested at no load at two voltage levels: $V_{sn} = 220\,V$, $P_0 = 300$ W, and $I_0 = 5$ A and, respectively, $V'_s = 65\,V$, $P'_0 = 100$ W, and $I'_0 = 4$ A. With $R_s = 0.1\ \Omega$, let us calculate the core and mechanical losses at rated voltage p_{iron} and p_{mec}. It is assumed that the core losses are proportional to voltage squared.

Solution

The power balance for the two cases (7.45) yields

$$P_0 - 3R_s I_0^2 = (p_{iron})_n + p_{mec}$$
$$P'_0 - 3R_s I'^2_0 = (p_{iron})_n \left(\frac{V'_s}{V_{sn}}\right)^2 + p_{mec} \tag{7.47}$$

$$300 - 3\cdot 0.1\cdot 5^2 = (p_{iron})_n + p_{mec} = 292.5$$
$$100 - 3\cdot 0.1\cdot 4^2 = (p_{iron})_n \left(\frac{65}{220}\right)^2 + p_{mec} = 95.2$$

From (7.47),

$$p_{iron} = \frac{292.5 - 95.25}{1 - 0.0873} = 216.17\ \text{W}$$
$$p_{mec} = 292.5 - 216.17 = 76.328\ \text{W} \tag{7.48}$$

Now, as a bonus, from (7.46), we may calculate approximately the no-load current I'_{r0}.

$$I'_{r0} \approx \frac{p_{mec}}{3V_{sn}} = \frac{76.328}{3\cdot 220} = 0.1156\ \text{A} \tag{7.49}$$

It should be noted that the rotor current is much smaller than the no-load stator current $I_0 = 5$A! During the no-load motoring tests, especially for rated voltage and above, due to teeth or/and back core saturation, there is a notable third flux and emf harmonic for the star connection. However, in this case, the third harmonic in current does not occur. The 5th and 7th saturation harmonics occur in the current for the star connection.

For the delta connection, the emf (and flux) third saturation harmonic does not occur. It occurs only in the phase currents (Figure 7.6).

As expected, the no-load current includes various harmonics (due to mmf and slot openings).

They are, in general, smaller for larger q slot/pole/phase chorded coils and skewing of rotor slots. More details are given in Chapter 10. In general, the current harmonics content decreases with increasing load.

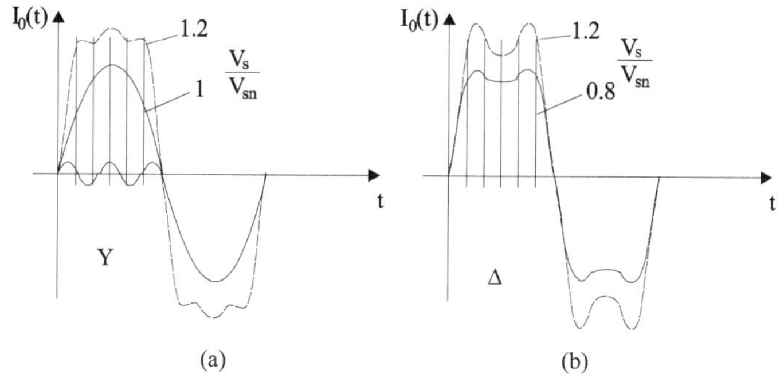

FIGURE 7.6 No-load currents: (a) star connection and (b) delta connection.

7.6 THE MOTOR MODE OF OPERATION

The motor mode of operation occurs when the motor drives a mechanical load (a pump, compressor, drive train, machine tool, electrical generator, etc.). For motoring, $T_e > 0$ for $0 < n < f_1/p_1$ and $T_e < 0$ for $0 > n > -f_1/p_1$. So the electromagnetic torque acts along the direction of motion. In general, the slip is $0 < S < 1$ (see Section 7.2). This time the complete equivalent circuit is used (Figure 7.1).

The power balance for motoring is shown in Figure 7.7.

The motor efficiency η is

$$\eta = \frac{\text{shaft power}}{\text{input electric power}} = \frac{P_m}{P_{1e}} = \frac{P_m}{P_m + p_{Cos} + p_{iron} + p_{Cor} + p_s + p_{mec}} \tag{7.50}$$

The rated power speed n is

$$n = \frac{f_1}{P_1}(1 - S) \tag{7.51}$$

The rated slip is $S_n = (0.08 - 0.006)$, larger for lower power motors.

The stray load losses p_s refer to additional core and winding losses due to space (and eventual time) field and voltage time harmonics. They occur both in the stator and rotor. In general, due to difficulties in computation, the stray load losses are still often assigned a constant value in some standards (0.5% or 1% of rated power). More on stray losses is presented in Chapter 11.

The slip definition (7.15) is a bit confusing as P_{elm} is defined as active power crossing the airgap. As the stray load losses occur both in the stator and rotor, part of them should be counted in the stator. Consequently, a more realistic definition of slip S (from 7.15) is

$$S = \frac{p_{cor}}{P_{elm} - p_s} = \frac{p_{cor}}{P_{1e} - p_{Cos} - p_{iron} - p_s} \tag{7.52}$$

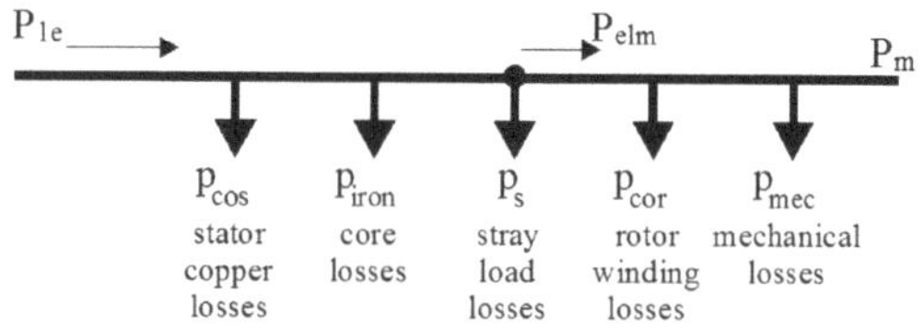

FIGURE 7.7 Power balance for motoring operation.

As slip frequency (rotor current fundamental frequency) $f_{2n} = Sf_{1n}$, it means that in general for $f_1 = 60$ (50) Hz, $f_{2n} = 4.8(4)$ to 0.36(0.3) Hz.

For high-speed (frequency) IMs, the value of f_2 is much higher. For example, for $f_{1n} = 300\,Hz$ (18,000 rpm, $2p_1 = 2$), $f_{2n} = 4$–8 Hz, while for $f_{1n} = 1200\,Hz$, it may reach values in the interval of 16–32 Hz. So, for high-frequency (speed) IMs, some skin effect is present even at rated speed (slip). Not so, in general, in 60 (50) Hz low-power motors.

7.7 GENERATING TO POWER GRID

As shown in Section 7.2, with $S < 0$, the electromagnetic power "travels" from rotor to stator ($P_{eml} < 0$), and thus, after covering the stator losses, the rest of active power is sent back to the power grid. As expected, the machine has to be driven at the shaft at a speed $n > f_1/p_1$ as the electromagnetic torque T_e (and P_{elm}) is negative (Figure 7.8).

The driving motor could be a wind turbine, a diesel motor, a hydraulic turbine, etc. or an electric motor (in laboratory tests).

The power grid is considered in general stiff (constant voltage and frequency), but, sometimes, in remote areas, it may also be rather weak.

To calculate the performance, we may again use the complete equivalent circuit (Figure 7.1) with $S < 0$. It may be easily proved that the equivalent resistance R_e and reactance X_e, as seen from the power grid, vary with slip as shown in Figure 7.9.

It should be noted that the equivalent reactance remains positive at all slips. Consequently, the IM draws reactive power in any conditions. This is necessary for a short-circuited rotor. If the IM is doubly fed as shown in Chapter 19, the situation changes as reactive power may be infused in the machine through the rotor slip rings by proper rotor voltage phasing, at $f_2 = Sf_1$.

Between S_{0g1} and S_{0g2} (both negative), Figure 7.9, the equivalent resistance R_e of IM is negative. This means that it delivers active power to the power grid.

The power balance is, in a way, opposite to that for motoring (Figure 7.10).

The efficiency η_g is now

$$\eta = \frac{\text{electric power output}}{\text{shaft power input}} = \frac{P_{2electric}}{P_{1shaft}} = \frac{P_{2electric}}{P_{2electric} + p_{Cus} + p_{iron} + p_{Cor} + p_s + p_{mec}} \quad (7.53)$$

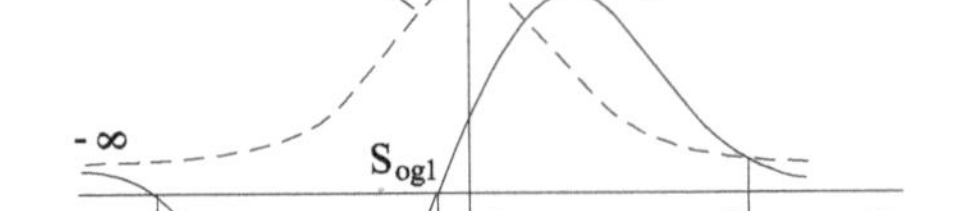

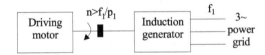

FIGURE 7.8 IG at power grid.

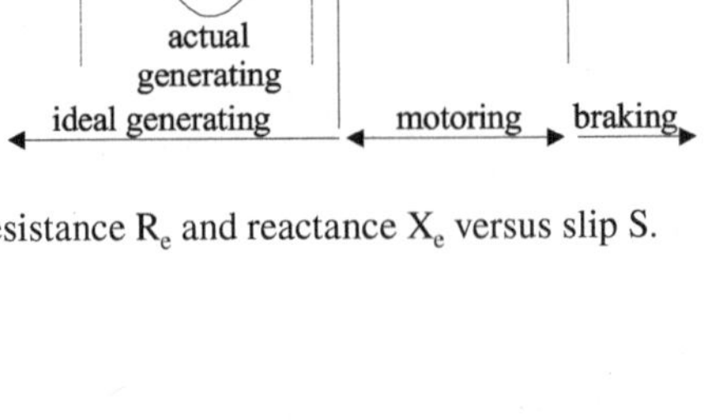

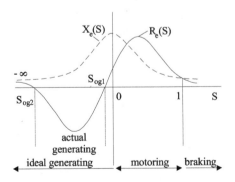

FIGURE 7.9 Equivalent IM resistance R_e and reactance X_e versus slip S.

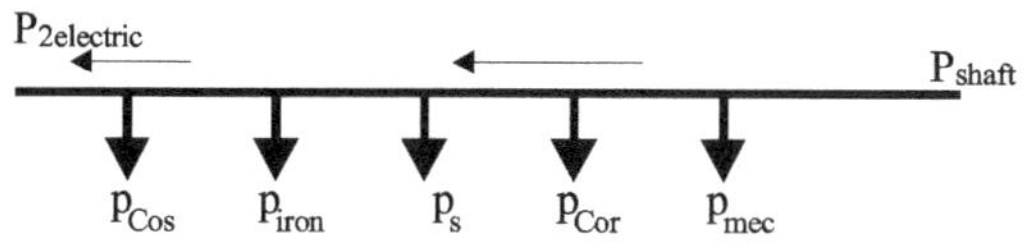

FIGURE 7.10 Power balance for generating.

Above the speed,

$$n_{max} = \frac{f_1}{P_1}(1 - S_{0g2}); \quad S_{0g2} < 0, \tag{7.54}$$

as evident in Figure 7.9, the IM remains in the generator mode but all the electric power produced is dissipated as loss in the machine itself.

Induction generators (IGs) are used more and more for industrial generation to produce part of the plant energy at convenient times and costs. However, as it still draws reactive power, "sources" of reactive power (synchronous generators or synchronous capacitors or capacitors) are also required to regulate the voltage in the power grid.

Example 7.4 Generator at Power Grid

A squirrel-cage IM with the parameters $R_s = R_r' = 0.6\ \Omega$, $X_{sl} = X_{rl}' = 2\ \Omega$, $X_{1ms} = 60\ \Omega$, and $R_{1ms} = 3\ \Omega$ (the equivalent series resistance to cover the core losses, instead of a parallel one) – Figure 7.11 – works as a generator at the power grid. Let us find the two slip values S_{0g1} and S_{0g2} between which it delivers power to the grid.

Solution

The switch from parallel to series connection in the magnetisation branch, used here for convenience, is widely used.

The condition to find the values of S_{0g1} and S_{0g2} for which, in fact, (Figure 7.9) the delivered power is zero is

$$R_e(S_{0g}) = 0.0 \tag{7.55}$$

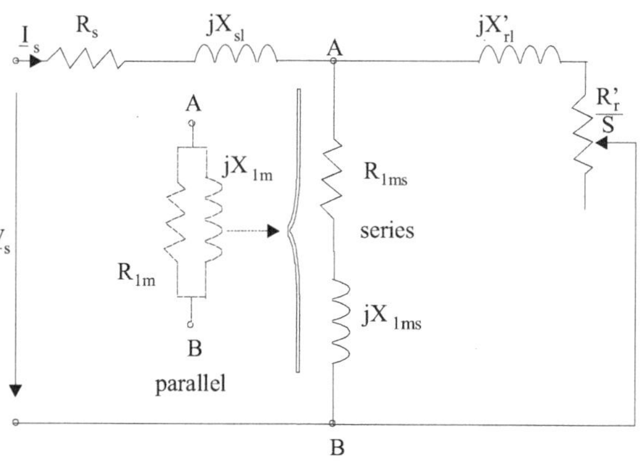

FIGURE 7.11 Equivalent circuit with series magnetisation branch.

Equation (7.55) translates into

$$\left(R_{1ms}+R_s\right)\left(\frac{R_r'}{S_{0g}}\right)^2+\frac{R_r'}{S_{0g}}\left(R_{1ms}{}^2+X_{1ms}{}^2+2R_{1ms}R_s\right)$$
$$+R_{1ms}X_{rl}'{}^2+R_{1ms}{}^2R_s+R_s\left(X_{1ms}+X_{rl}'\right)^2=0 \tag{7.56}$$

In numbers,

$$3.6\cdot 0.6^2\left(\frac{1}{S_{0g}}\right)^2+0.6\left(3^2+60^2+2\cdot 3\cdot 0.6\right)\frac{1}{S_{0g}}$$
$$+3\cdot 2^2+3^2\cdot 0.6+0.6(60+2)^2=0 \tag{7.57}$$

with the solutions $S_{og1} = -0.33 \cdot 10^{-3}$ and $S_{0g2} = -0.3877$.

Now with $f_1 = 60\,\text{Hz}$ and with $2p_1 = 4$ poles, the corresponding speeds (in rpm) are

$$n_{og1}=\frac{f_1}{P_1}\left(1-S_{0g1}\right)\cdot 60=\frac{60}{2}\left(1-\left(-0.33\cdot 10^{-3}\right)\right)\cdot 60=1800.594\text{ rpm}$$
$$n_{og2}=\frac{f_1}{P_1}\left(1-S_{0g2}\right)\cdot 60=\frac{60}{2}\left(1-(-0.3877)\right)\cdot 60=2497.86\text{ rpm} \tag{7.58}$$

7.8 AUTONOMOUS INDUCTION GENERATOR MODE

As shown in Section 7.7, to become a generator, the IM needs to be driven above no-load ideal speed n_1 ($n_1 = f_1/p_1$ with short-circuited rotor) and to be provided with reactive power to produce and maintain the magnetic field in the machine.

As known, this reactive power may be "produced" with synchronous condensers (or capacitors) – Figure 7.12.

The capacitors are Δ connected to reduce their capacitance as they are supplied by line voltage. Now the voltage V_s and frequency f_1 of the IG on no load and on load depend essentially on machine parameters, capacitors C_Δ, and speed n. Still $n > f_1/p_1$.

Let us explore the principle of IG capacitor excitation on no load. The machine is driven at a speed n.

The D.C. remanent magnetisation in the rotor, if any (if none, D.C. magnetisation may be provided as a few D.C. current surges through the stator with one phase in series and the other two in parallel), produces an A.C. emf in the stator phases. Then the three-phase emfs of frequency

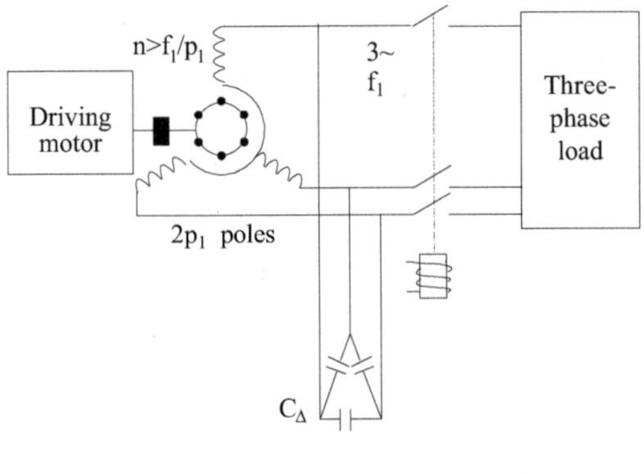

FIGURE 7.12 Autonomous IG with capacitor magnetisation.

$f_1 = p_1 \cdot n$ cause currents to flow in the stator phases and capacitors. Their phase angle is such that they are producing an airgap field that always increases the remanent field; then again, this field produces a higher stator emf and so on until the machine settles at a certain voltage V_s and frequency $f_1 \approx p_1 n$. Changing the speed will change both the frequency f_1 and the no-load voltage V_{s0}. The same effect is produced when the capacitors C_Δ are changed.

A quasi-quantitative analysis of this self-excitation process may be produced by neglecting the resistances and leakage reactances in the machine. The equivalent circuit per phase degenerates into that in Figure 7.13.

The presence of rotor remanent flux density (from prior action) is depicted by the resultant small emf E_{rem} ($E_{rem} = 2$–$4\,V$) whose frequency is $f_1 = np_1$. The frequency f_1 is imposed by speed.

The machine equation becomes simply

$$V_{s0} = jX_{1m}\underline{I}_m + \underline{E}_{rem} = -j\frac{1}{\omega_1 C_Y}\underline{I}_m = V_{s0}(I_m) \tag{7.59}$$

As we know, the magnetisation characteristic (curve) – $V_{s0}(I_m)$ – is, in general, non-linear due to magnetic saturation (Chapter 5) and may be obtained through the ideal no-load test at $f_1 = p_1 n$. On the other hand, the capacitor voltage depends linearly on capacitor current. The capacitor current on no load is, however, equal to the motor stator current. Graphically, Equation (7.59) is depicted in Figure 7.14.

Point A represents the no-load voltage V_{s0} for given speed, n, and capacitor C_Y. If the self-excitation process is performed at a lower speed n′ (n′ < n), a lower no-load voltage (point A′), V'_{s0}, at a lower frequency $f'_1 \approx p_1 n'$ is obtained.

Changing (reducing) the capacitor C_Y produces similar effects. The self-excitation process requires, as seen in Figure 7.14, the presence of remanent magnetisation ($E_{rem} \neq 0$) and magnetic saturation to be successful, that is, to produce a clear intersection of the two curves.

When the magnetisation curve $V_{1m}(I_m)$ is available, we may use the complete equivalent circuit (Figure 7.1) with a parallel capacitor C_Y over the terminals to explore the load characteristics. For a given capacitor bank and speed n, the output voltage V_s versus load current I_s depends on the load power factor (Figure 7.15).

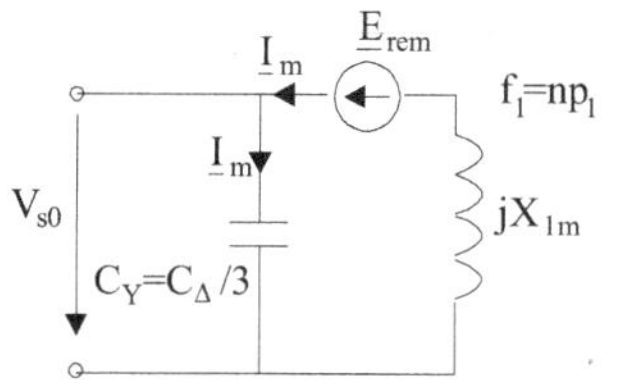

FIGURE 7.13 Ideal no-load IG per phase equivalent circuit with capacitor excitation.

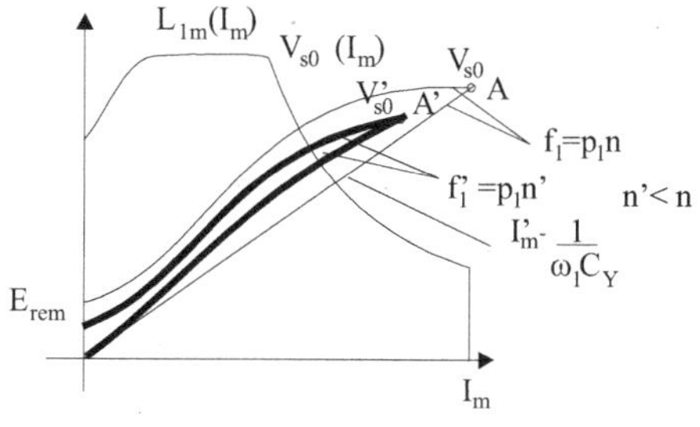

FIGURE 7.14 Capacitor self-excitation of IG on no load.

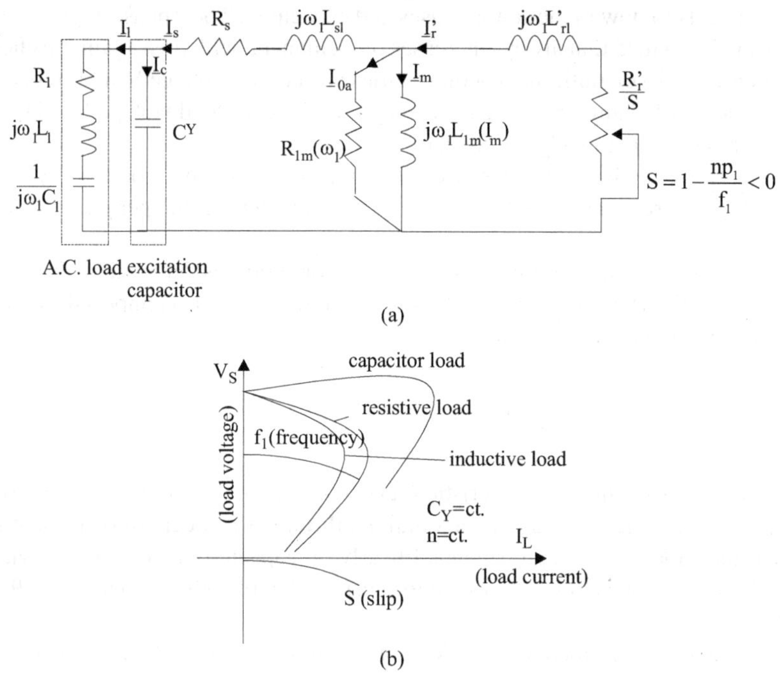

FIGURE 7.15 Autonomous IG on load: (a) equivalent circuit and (b) load curves.

The load curves in Figure 7.15 may be obtained directly by solving the equivalent circuit in Figure 7.15a for load current, for given load impedance, speed n, and slip S. However, the necessary non-linearity of magnetisation curve $L_{1m}(I_m)$, Figure 7.14, imposes an iterative procedure to solve the equivalent circuit. This is now at hand with existing application software such as MATLAB®.

Above a certain load, the machine voltage drops gradually to zero as there will be a deficit of capacitor energy to produce machine magnetisation. Point A on the magnetisation curve will drop gradually to zero.

As the load increases, the slip (negative) increases and, for given speed n, the frequency decreases. So the IG can produce power above a certain level of magnetic saturation and above a certain speed for given capacitors. The voltage and frequency decrease notably with load.

A variable capacitor would keep the voltage constant with load. Still, the frequency, by principle, at constant speed, will decrease with load.

Only simultaneous capacitor and speed control may produce constant voltage and frequency for variable load. More on autonomous IGs is presented in Chapters 10–11, Volume 2.

7.9 THE ELECTROMAGNETIC TORQUE

By electromagnetic torque, T_e, we mean, the torque produced by the fundamental airgap flux density in interaction with the fundamental rotor current.

In Section 7.1, we have already derived the expression of T_e (7.14) for the singly fed IM. By singly fed IM, we understand the short-circuited or wound rotors with a passive impedance at its terminals. For the general case, an $R_lL_lC_l$ impedance could be connected to the slip rings of a wound rotor (Figure 7.16).

Even for this case, the electromagnetic torque T_e may be calculated (7.14) where instead of rotor resistance R_r', the total series rotor resistance $R_r' + R_l'$ is introduced.

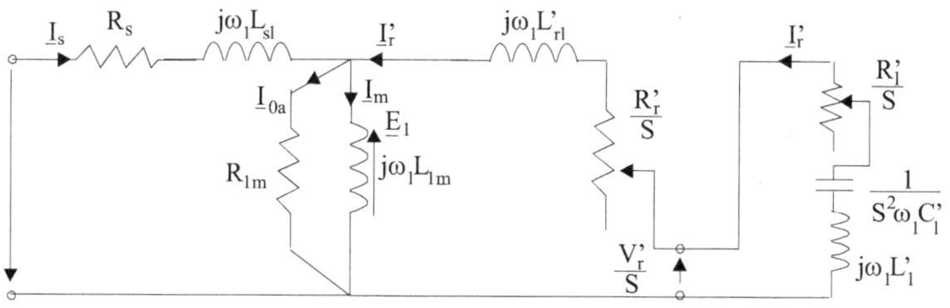

FIGURE 7.16 Singly fed IM with additional rotor impedance.

Notice that the rotor circuit and thus the R_l, C_l, L_l have been reduced to primary, becoming R_l', C_l', L_l'.

Both rotor and stator circuit blocks in Figure 7.16 are characterised by the stator frequency ω_1.

Again, Figure 7.16 refers to a fictitious IM at standstill which behaves as the real machine but delivers active power in a resistance $(R_r'+R_l')(1-S)/S$ dependent on slip, instead of producing mechanical (shaft) power.

$$T_e = \frac{3R_{rl}'}{S} I_r'^2 \frac{P_1}{\omega_1}; \quad R_{rl}' = R_r' + R_l' \tag{7.60}$$

In industry, the wound rotor slip rings may be connected through brushes to a three-phase variable resistance or a diode rectifier, or a D.C. – D.C. static converter and a single constant resistance, or a semi- (or fully) controlled rectifier and a constant single resistance for starting and (or) limited range speed control as shown later in this section. In essence, however, such devices have models that can be reduced to the situation shown in Figure 7.16. To further explore the torque, we will consider the rotor with a total resistance R_{re}' but without any additional inductance and capacitance connected at the brushes (for the wound rotor).

From Figure 7.16, with $V_r' = 0$ and R_r' replaced by R_{re}', we can easily calculate the rotor and stator currents $\underline{I}_r'$ and $\underline{I}_s$ as

$$\underline{I}_r' = -\frac{\underline{I}_s \cdot \underline{Z}_{1m}}{\frac{R_{re}'}{S} + jX_{rl}' + \underline{Z}_{1m}} \tag{7.61}$$

$$\underline{I}_s = \frac{\underline{V}_s}{R_s + jX_{sl} + \frac{\underline{Z}_{1m}\left(\frac{R_{re}'}{S} + jX_{rl}'\right)}{\frac{R_{re}'}{S} + jX_{rl}' + \underline{Z}_{1m}}} \tag{7.62}$$

$$\underline{I}_{0a} + \underline{I}_m = \underline{I}_s + \underline{I}_r' = \underline{I}_{0s} \tag{7.63}$$

$$\underline{Z}_{1m} = \frac{R_{1m} jX_{1m}}{R_{1m} + jX_{1m}} = R_{1ml} + jX_{1ml} \tag{7.64}$$

With constant parameters, we may approximate

$$\frac{\underline{Z}_{1m} + jX_{sl}}{\underline{Z}_{1m}} \approx \frac{X_{1ml} + X_{sl}}{X_{1ml}} = C_1 \approx (1.02 - 1.08) \tag{7.65}$$

Consequently,

$$\underline{I}_s \approx \frac{\underline{V}_s}{R_s + \frac{C_1 R'_{re}}{S} + j(X_{sl} + C_1 X'_{rl})} \tag{7.66}$$

This approximation leads to acceptable errors in current amplitude but to unacceptable errors in the power factor!.

Substituting I_s from (7.66) into (7.61) and then I'_r from (7.61) into (7.60), T_e becomes

$$T_e = \frac{3V_s^2 p_1}{\omega_1} \frac{\frac{R'_{rl}}{S}}{\left(R_s + \frac{C_1 R'_{rl}}{S}\right)^2 + (X_{sl} + C_1 X'_{rl})^2} \tag{7.67}$$

As we can see, T_e is a function of slip S. Its peak values are obtained for

$$\frac{\partial T_e}{\partial S} = 0 \rightarrow S_k = \frac{\pm C_1 R'_{rl}}{\sqrt{R_s^2 + (X_{sl} + C_1 X'_{rl})^2}} \tag{7.68}$$

Finally, the peak (breakdown) torque T_{ek} is obtained from the critical slip S_k in (7.67)

$$T_{ek} = (T_e)_{sk} = \frac{3p_1}{\omega_1} \frac{V_s^2}{2C_1\left[R_s \pm \sqrt{R_s^2 + (X_{sl} + C_1 X'_{rl})}\right]} \tag{7.69}$$

The whole torque/slip (or speed) curve is shown in Figure 7.17.

Concerning the T_e versus S curve, a few remarks are in order

- The peak (breakdown) torque T_{ek} value is independent of rotor equivalent (cumulated) resistance, but the critical slip S_k at which it occurs is proportional to it.
- When the stator resistance R_s is notable (low power, sub-kW motors), the generator breakdown torque is slightly larger than the motor breakdown torque. With R_s neglected, the two are equal to each other.

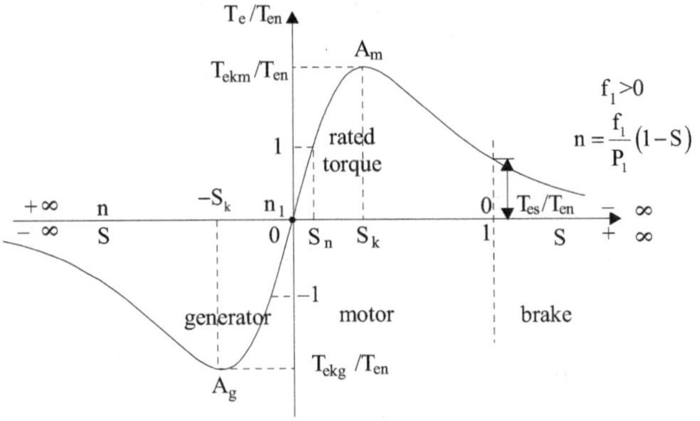

FIGURE 7.17 The electromagnetic torque T_e (in relative units) versus slip (speed).

- With R_s neglected (power above 10 kW), S_k from (7.68) and T_{ek} from (7.69) become

$$S_k \approx \frac{\pm C_1 R'_{re}}{(X_{sl} + C_1 X'_{rl})} \approx \frac{\pm C_1 R'_{re}}{\omega_1 (L_{sl} + C_1 L'_{rl})} \approx \frac{\pm R'_{re}}{\omega_1 L_{sc}} \tag{7.70}$$

$$T_{ek} \approx \pm 3p_1 \left(\frac{V_s}{\omega_1}\right)^2 \frac{1}{2C_1 (L_{sl} + C_1 L'_{rl})} \approx 3p_1 \left(\frac{V_s}{\omega_1}\right)^2 \frac{1}{2L_{sc}} \tag{7.71}$$

- In general, as long as $I_s R_s / V_s < 0.05$, we may safely approximate the breakdown torque to Equation (7.71).
- The critical slip speed in (7.70) $S_k \omega_1 = \pm R'_{re}/L_{sc}$ is dependent on rotor resistance (cumulated) and on the total leakage (short-circuit) inductance. Care must be exercised to calculate L_{sl} and L'_{rl} for the actual conditions at breakdown torque, where skin and leakage saturation effects are sometimes already notable.
- The breakdown torque in (7.71) is proportional to voltage per frequency squared and inversely proportional to equivalent leakage inductance L_{sc}.

  Notice that (7.70) and (7.71) are not valid for low values of voltage and frequency when $V_s \approx I_s R_s$.
- When designing an IM for high breakdown torque, a low short-circuit inductance configuration is needed.
- In (7.70) and (7.71), the final form, $C_1 = 1$, which, in fact, means that $L_{1m} \approx \infty$ or the iron is not saturated. For deeply saturated IMs, $C_1 \neq 1$ and the first form of T_{ek} in (7.71) is to be used.
- Stable operation under steady state is obtained when torque decreases with speed

$$\frac{\partial T_e}{\partial n} < 0 \text{ or } \frac{\partial T_e}{\partial S} > 0 \tag{7.72}$$

- With $R_s = 0$ and $C_1 = 1$, the relative torque T_e/T_{ek} is

$$\frac{T_e}{T_{ek}} \approx \frac{2}{\frac{S}{S_k} + \frac{S_k}{S}} \tag{7.73}$$

This is known as simplified Kloss formula.

- The margin of stability is often defined as the ratio between breakdown torque and rated torque.

$$T_{ek}/(T_e)_{S_n} > 1.6; \quad (T_e)_{S_n} - \text{rated torque} \tag{7.74}$$

- Values of $T_{ek}/(T_e)_{S_n}$ larger than 2.3–2.4 are not easy to obtain unless very low short-circuit inductance designs are applied. For loads with unexpected, frequent, large load bursts or load torque time pulsations (compressor loads), large relative breakdown torque values are recommended.
- On the other hand, the starting torque depends essentially on rotor resistance and short-circuit impedance Z_{sc} for given voltage and frequency

$$T_{es} = \frac{3p_1}{\omega_1} V_s^2 \frac{R'_{re}}{(R_s + C_1 R'_{re})^2 + (X_{sl} + C_1 X'_{rl})^2} \approx \frac{3p_1}{\omega_1} V_s^2 \frac{R'_{re}}{|\underline{Z}_{sc}|^2} \tag{7.75}$$

Cage rotor IMs have a relative starting torque $T_{es}/T_{en} = 0.7–1.1$ while deep bar or double cage rotors (as shown in the next chapter) have large starting torque but smaller breakdown torque and larger critical slip S_k because their short-circuit inductance (reactance) is larger.

Slip ring rotors (with the phases short-circuited over the slip rings) have much lower starting torque $T_{es}/T_{en} < 0.3$ in general, larger peak torque at lower critical slip S_k.

- The current/slip curve for a typical single cage and wound or short-circuited rotor is shown in Figure 7.18.

Due to negative slip, the stator current increases even more with absolute slip value in the generator than in the motor operation mode.

Also, while starting currents of $(5–7)I_n$ (as in squirrel cage rotors) are accepted for direct starting values in excess of 10, typical for wound short-circuited rotor are not acceptable. A wound rotor IM should be started either with a rotor resistance added or with a rotor power supply to absorb part of rotor losses and reduce rotor currents.

On the other hand, high-efficiency motors, favoured today for energy saving, are characterised by lower stator and rotor resistances under load but also at start. Thus, higher starting currents are expected: $I_s/I_n = 6.5–8$. To avoid large voltage sags at start, the local power grids have to be strengthened (by additional transformers in parallel) to allow for larger starting currents.

These additional hardware costs have to be added to high-efficiency motor costs and then compared to the better efficiency motor energy savings and maintenance cost reductions.

Also, high efficiency motors tend to have slightly lower starting torque (Figure 7.19).

More on these aspects are to be found in chapters on design methodologies (Chapters 4–9, Volume 2).

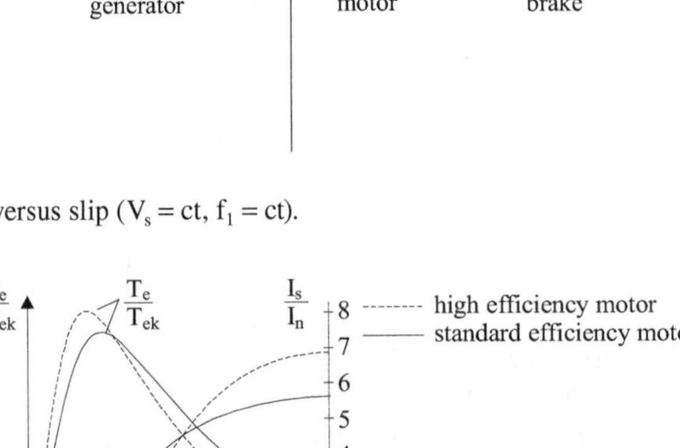

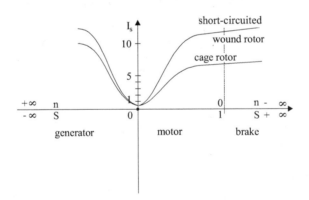

FIGURE 7.18 Current versus slip ($V_s = ct$, $f_1 = ct$).

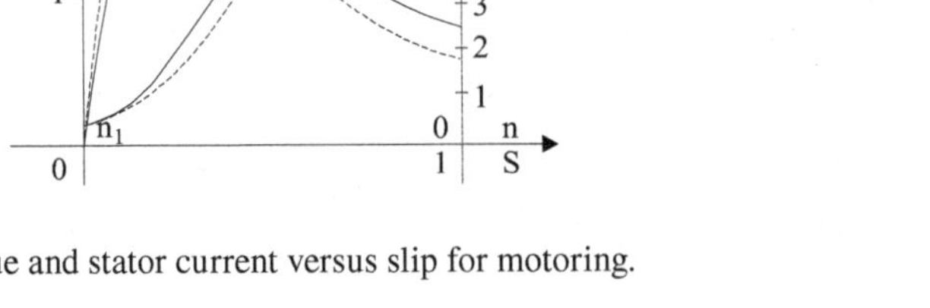

FIGURE 7.19 Torque and stator current versus slip for motoring.

7.10 EFFICIENCY AND POWER FACTOR

The efficiency η has already been defined, both for motoring and generating, earlier in this chapter. The power factor $\cos\varphi_1$ may be defined in relation to the equivalent circuit as

$$\cos\varphi_1 = \frac{|R_e|}{Z_e} \tag{7.76}$$

Figure 7.9 shows that R_e is negative for actual generating mode. This explains why in (7.76) the absolute value of R_e is used. The power factor, related to reactive power flow, has the same formula for both motoring and generating, while the efficiency needs dedicated formulas (7.50) and (7.53).

The dependence of efficiency and power factor on load (slip or speed) is essential when the machine is used for variable load (Figure 7.20).

Such curves may be calculated strictly based on the equivalent circuit (Figure 7.1) with known (constant or variable) parameters and given values of slip. Once the currents are known, all losses and input power may be calculated. The mechanical losses may be lumped into the core loss resistance R_{1m} or, if not, they have to be given from calculations or tests performed in advance. For applications with highly portable loads rated from 0.25 to 1.5, the design of the motor has to preserve a high efficiency.

A good power factor will reduce VAR compensation hardware ratings and costs, as industrial plant average power factor should not be lagging below 0.95 to avoid VAR penalty costs.

Example 7.5 Motor Performance

An induction motor with deep bars is characterised by: rated power $P_n = 20\,\text{kW}$, supply line voltage $V_{sl} = 380\,\text{V}$ (star connection), $f_1 = 50\,\text{Hz}$, $\eta_n = 0.92$, $p_{mec} = 0.005\ P_n$, $p_{iron} = 0.015\ P_n$, $p_s = 0.005\ P_n$, $p_{cosn} = 0.03\ P_n$, $\cos\varphi_n = 0.9$, $p_1 = 2$, starting current $I_{sc} = 5.2I_n$, power factor $\cos\varphi_{sc} = 0.4$, and no-load current $I_{on} = 0.3I_n$.

Let us calculate

a. Rotor cage losses p_{corn}, electromagnetic power P_{elm}, slip, S_n, speed, n_n, rated current I_n, and rotor resistance R_r' for rated load
b. Stator resistance R_s and rotor resistance at start R_{rs}'
c. Electromagnetic torque for rated power and at start
d. Breakdown torque

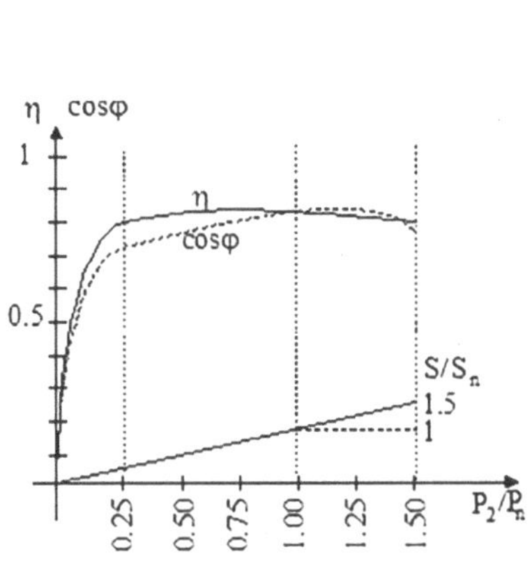

FIGURE 7.20 IM efficiency, η, power factor, $\cos\varphi$, and slip, S/S_n versus relative power load at the shaft: P_2/P_n.

Solution

a. Based on (7.50), we notice that, at rated load, from all losses, we lack the rotor cage losses, p_{com} losses.

$$p_{com} = \frac{P_n}{\eta_n} - \left(p_{cos\,n} + p_{Sn} + p_{iron} + p_{iron}\right) - P_n$$

$$= \frac{20{,}000}{0.92} - (0.03 + 0.005 + 0.015 + 0.015) \cdot 20{,}000 - 20{,}000 = 639.13\ \text{W} \tag{7.77}$$

On the other hand, the rated current I_n comes directly from the power input expression

$$I_n = \frac{P_n}{\eta_n \sqrt{3} V_{sl} \cos\varphi_n} = \frac{20{,}000}{0.92 \cdot \sqrt{3} \cdot 380 \cdot 0.9} = 36.742\ \text{A} \tag{7.78}$$

The slip expression (7.52) yields

$$S_n = \frac{p_{com}}{\frac{P_n}{\eta_n} - p_{cos\,n} - p_{Sn} - p_{iron}} = \frac{639.13}{\frac{20{,}000}{0.92} - (0.03 + 0.005 + 0.015)20{,}000} = 0.0308 \tag{7.79}$$

The rated speed n_n is

$$n_n = \frac{f_1}{P_1}(1 - S_n) = \frac{50}{2}(1 - 0.0308) = 24.23\ \text{rps} = 1453.8\ \text{rpm} \tag{7.80}$$

The electromagnetic power P_{elm} is

$$P_{elm} = \frac{p_{com}}{S_n} = \frac{639.13}{0.0308} = 20{,}739\ \text{W} \tag{7.81}$$

To easily find the rotor resistance, we need the rotor current. At rated slip, the rotor circuit is dominated by the resistance R_r'/S_n, and thus, the no-load (or magnetising current) I_{0n} is about 90° behind the rotor resistive current I_{rn}'. Thus, the rated current I_n is

$$I_n \approx \sqrt{I_{0n}^2 + I_{rn}'^2} \tag{7.82}$$

$$I_{rn}' = \sqrt{I_n^2 - I_{0n}^2} = \sqrt{36.742^2 - (0.3 \cdot 36.672)^2} = 35.05\ \text{A} \tag{7.83}$$

Now the rotor resistance at load R_r' is

$$R_r' = \frac{p_{com}}{3 I_{rn}'^2} = \frac{639.13}{3 \cdot 35.05^2} = 0.1734\ \Omega \tag{7.84}$$

b. The stator resistance R_s comes from

$$R_s = \frac{p_{cos\,n}}{3 \cdot I_n^2} = \frac{0.03 \cdot 20{,}000}{3 \cdot 36.742^2} = 0.148\ \Omega \tag{7.85}$$

The starting rotor resistance R'_{rs} is obtained based on starting data

$$R'_{rs} = \frac{V_{sl}}{\sqrt{3}}\frac{\cos\varphi_{sc}}{I_{sc}} - R_s = \frac{380}{\sqrt{3}}\frac{0.4}{5.2\cdot 36.742} - 0.148 = 0.31186\,\Omega \tag{7.86}$$

Notice that

$$K_r = \frac{R'_{rs}}{R'_r} = \frac{0.31186}{0.1734} = 1.758 \tag{7.87}$$

This is mainly due to skin effect.

c. The rated electromagnetic torque T_{en} is

$$T_{en} = \frac{P_{elm}}{\omega_1} p_1 = 20,739\cdot\frac{2}{2\pi 50} = 132.095\text{ Nm} \tag{7.88}$$

The starting torque T_{es} is

$$T_{es} \approx \frac{3R'_{rs}I_{sc}{}^2}{\omega_1} p_1 = \frac{3\cdot 0.31186\cdot(5.2\cdot 36.742)^2}{2\pi 50} = 108.76\text{ Nm} \tag{7.89}$$

$$T_{es}/T_{en} = 108.76/132.095 = 0.833$$

d. For the breakdown torque, even with approximate formula (7.71), the short-circuit reactance is needed. From starting data,

$$X_{sc} = \frac{V_{sl}}{\sqrt{3}}\frac{\sin\varphi_{sc}}{I_{sc}} = \frac{380}{\sqrt{3}}\frac{\sqrt{1-0.4^2}}{5.2\cdot 36.672} = 1.0537\,\Omega \tag{7.90}$$

The rotor leakage reactance $X_{rls} = X_{sc} - X_{sl} = 1.0537 - 0.65 = 0.4037\ \Omega$. Due to skin effect and leakage saturation, this reactance is smaller than the one "acting" at rated load and even at breakdown torque conditions. Knowing that $K_r = 1.758$ (7.87), resistance correction coefficient value due to skin effect, we could find $K_x \approx 0.9$ from Chapter 8, for a rectangular slot.

With leakage saturation effects neglected, the rotor leakage reactance at critical slip S_k (or lower) is

$$X'_{rl} = K_x{}^{-1}X'_{rls} = \frac{1}{0.9}\cdot 0.4 = 0.444\,\Omega \tag{7.91}$$

Now, with $C_1 \approx 1$ and $R_s \approx 0$, from (7.71), the breakdown torque T_{ek} is

$$T_{ek} \approx 3\left(\frac{V_{sl}}{\sqrt{3}}\right)^2 \frac{p_1}{\omega_1}\frac{1}{2(X_{sl}+X'_{rl})} = 3\left(\frac{380}{\sqrt{3}}\right)^2 \frac{2}{2\pi 50}\frac{1}{2(0.65+0.444)} = 422.68\text{ Nm} \tag{7.92}$$

The ratio T_{ek}/T_{en} is

$$T_{ek}/T_{en} = \frac{422.68}{132.095} = 3.20$$

This is an unusually large value, facilitated by a low starting current and a high power factor at start; that is, a high-resistance low leakage reactance X_{sc} was considered.

7.11 PHASOR DIAGRAMS: STANDARD AND NEW

The IM equations under steady state are, again,

$$\begin{cases} \underline{I}_s\left(R_s + jX_{sl}\right) - \underline{V}_s = \underline{E}_s \\ \underline{I}_r'\left(\dfrac{R_r'}{S} + jX_{rl}'\right) + \dfrac{\underline{V}_r'}{S} = \underline{E}_s \\ \underline{E}_s = -\underline{Z}_{1m}\left(\underline{I}_s + \underline{I}_r'\right) = -\underline{Z}_{1m}\underline{I}_{0s} \\ \underline{Z}_{1m} = \dfrac{R_{1m} \cdot jX_{1m}}{R_{1m} + jX_{1m}} = R_{1ms} + jX_{1ms} \end{cases}$$

$$R_{1m} \gg X_{1m}; \quad R_{1ms} \ll X_{1ms}; \quad R_{1ms} \ll R_{1m}; \quad X_{1ms} \approx X_{1m} \tag{7.93}$$

They can be and have been illustrated in phasor diagrams. Such diagrams for motoring ($S > 0$) and generating ($S < 0$) are shown in Figure 7.21a and b, for $\underline{V}_r' = 0$ (short-circuited rotor).

As expected, for the short-circuited rotor, the reactive power remains positive for generating ($Q_1 > 0$) as only the active stator current changes sign from motoring to generating.

$$\begin{aligned} P_1 &= 3\underline{V}_s\underline{I}_s\cos\varphi_1 <> 0 \\ Q_1 &= 3\underline{V}_s\underline{I}\sin\varphi_1 > 0 \end{aligned} \tag{7.94}$$

Today, the phasor diagrams are used mainly to explain IM action, while, before the occurrence of computers, they served as a graphical method to determine the steady-state performance without solving the equivalent circuit.

However, rather recently, the advancement in high-performance variable speed A.C. drives has led to new phasor (or vector) diagrams in orthogonal axis models of A.C. machine. Such vector diagrams

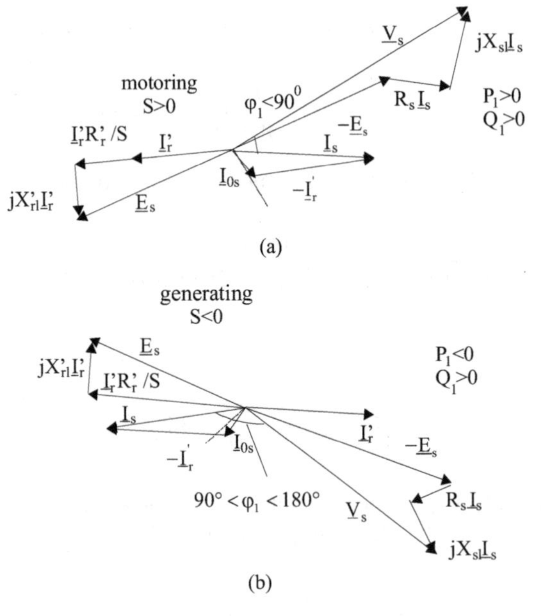

FIGURE 7.21 Standard phasor diagrams: (a) for motoring and (b) for generating.

may be accompanied by similar phasor diagrams valid for each phase of A.C. machines. To simplify the derivation, let us neglect the core loss ($R_{1m} \approx \infty$ or $R_{1ms} = 0$). We start by defining three flux linkages per phase: stator flux linkage $\underline{\Psi}_s$, rotor flux linkage $\underline{\Psi}_r'$, and airgap (magnetising) $\underline{\Psi}_m$.

$$\underline{\psi}_s = L_{sl}\underline{I}_s + \underline{\psi}_m$$

$$\underline{\psi}_r' = L_{rl}'\underline{I}_r' + \underline{\psi}_m \tag{7.95}$$

$$\underline{\psi}_m = L_{1m}\left(\underline{I}_s + \underline{I}_r'\right) = L_{1m}\underline{I}_m$$

$$\underline{E}_s = -j\omega_1 L_{1m}\underline{I}_m = -j\omega_1\underline{\psi}_m \tag{7.96}$$

The stator and rotor equations in (7.93) become

$$\underline{I}_s R_s - \underline{V}_s = -j\omega_1\underline{\psi}_s \tag{7.97}$$

$$\underline{I}_r'\frac{R_r'}{S} + \frac{\underline{V}_r'}{S} = -j\omega_1\underline{\psi}_r' \tag{7.98}$$

For zero rotor voltage (short-circuited rotor), Equation (7.98) becomes

$$\underline{I}_r'\frac{R_r'}{S} = -j\omega_1\underline{\psi}_r'; \quad \underline{V}_r' = 0 \tag{7.99}$$

Equation (7.99) shows that at steady state, for $\underline{V}_r' = 0$, the rotor current I_r' and rotor flux $\underline{\psi}_r'$ are phase shifted by 90°.

Consequently, the torque T_e expression (7.60) becomes

$$T_e = \frac{3R_r'I_r'^2}{S}\frac{p_1}{\omega_1} = 3p_1\psi_r'I_r' \tag{7.100}$$

Equation (7.100) may be considered the basis for modern (vector) IM control. For constant rotor flux per phase amplitude (RMS), the torque is proportional to rotor phase current amplitude (RMS). With these new variables, the phasor diagrams based on Equations (7.97) and (7.98) are shown in Figure 7.22.

Such phasor diagrams could be instrumental in computing IM performance when fed from a static power converter (variable voltage, variable frequency).

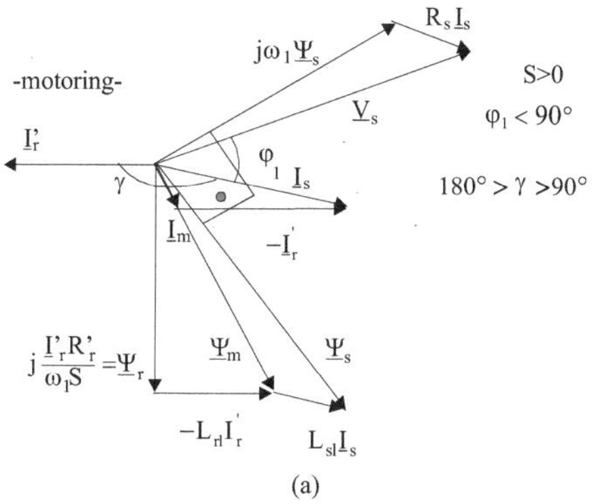

FIGURE 7.22 Phasor diagram with stator, rotor, and magnetisation flux linkages, $\underline{\Psi}_r'$, $\underline{\Psi}_m$, and $\underline{\Psi}_s$: (a) motoring and (b) generating.

(Continued)

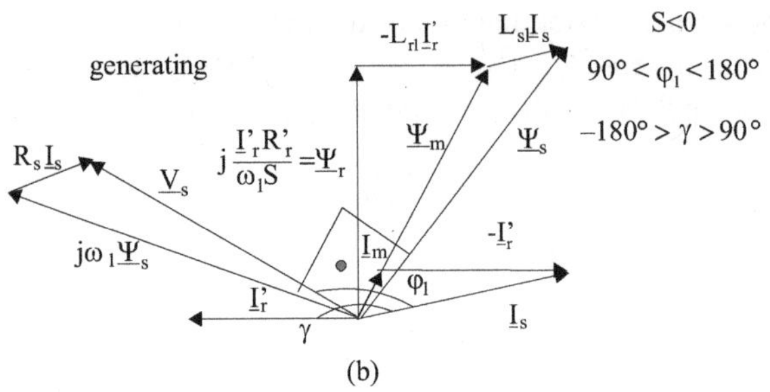

FIGURE 7.22 (CONTINUED) Phasor diagram with stator, rotor, and magnetisation flux linkages, $\underline{\psi}'_r$, $\underline{\Psi}_m$, and $\underline{\Psi}_s$: (a) motoring and (b) generating.

The torque expression may also be obtained from (7.97) by taking the real part of it after multiplying by I_s^*.

$$\text{Re}\left[3\underline{I}_s\underline{I}_s^* R_s + 3j\omega_1\underline{\psi}_s\underline{I}_s^*\right] = 3\,\text{Re}\left(\underline{V}_s\underline{I}_s^*\right) \tag{7.101}$$

The second term in (7.101), with core loss neglected, is, in fact, the electromagnetic power P_{elm}

$$P_{elm} = 3\omega_1\,\text{Imag}\left(\underline{\psi}_s\underline{I}_s^*\right) = T_e\left(\omega_1/p_1\right) \tag{7.102}$$

$$T_e = 3p_1\,\text{Imag}\left(\underline{\psi}_s\underline{I}_s^*\right) = 3p_1L_{1m}\,\text{Imag}\left(\underline{I}'_r\underline{I}_s^*\right) \tag{7.103}$$

or

$$T_e = 3p_1L_{1m}\underline{I}'_r\underline{I}_s\sin\gamma;\quad |\gamma| > 90° \tag{7.104}$$

In (7.104), γ is the angle between the rotor and stator phase current phasors with $\sin\gamma > 0$ for motoring and $\sin\gamma < 0$ for generating.

Now, from the standard equivalent circuit (Figure 7.1 with $R_{1m} \approx \infty$),

$$\underline{I}'_r = -\underline{I}_s\frac{j\omega_1L_{1m}}{\frac{R'_r}{S} + j\omega_1\left(L_{1m} + L'_{rl}\right)} \tag{7.105}$$

So the angle γ between $\underline{I}'_r$ and $\underline{I}_s^*$ depends on slip S and frequency ω_1 and motor parameters (Figure 7.23). It should be noted that the angle γ is close to ±180° for S = ±1 and close to ±90° towards |S| = 0.

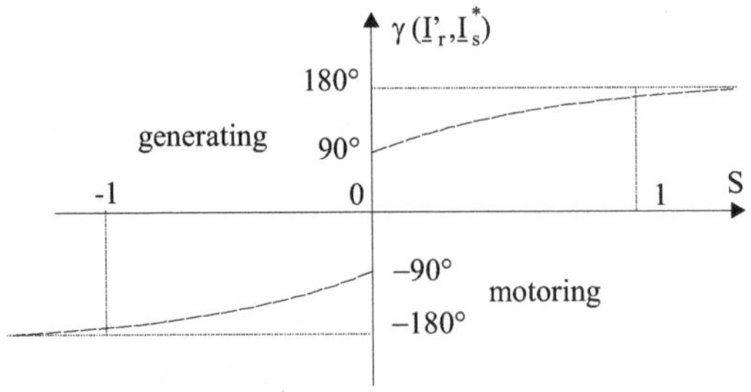

FIGURE 7.23 Angle γ ($\underline{I}'_r$, $\underline{I}_s^*$) versus slip S.

7.12 ALTERNATIVE EQUIVALENT CIRCUITS

Alternative equivalent circuits for IMs abound in the literature. Here we will deal with some that have become widely used in IM drives. In essence [2], a new rotor current is introduced:

$$\underline{I}'_{ra} = \frac{\underline{I}'_r}{a} \tag{7.106}$$

Using this new variable in (7.95)–(7.98), we easily obtain

$$\begin{aligned}
&\underline{I}_s R_s + j\omega_1 (L_s - aL_{1m}) \underline{I}_s - \underline{V}_s = \underline{E}_a \\
&\underline{E}_a = -j\omega_1 \underline{\psi}_{ma}; \quad \underline{\psi}_{ma} = aL_{1m}\left(\underline{I}_s + \underline{I}'_{ra}\right); \quad \underline{I}_{ma} = \underline{I}_s + \frac{\underline{I}'_r}{a} \\
&\underline{I}'_{ra} R'_r a^2 + j\omega_1 (aL'_r - L_{1m}) a\underline{I}'_{ra} + \frac{V'_r a}{S} = \underline{E}_a \\
&L_s = L_{sl} + L_{1m}; \quad L'_r = L'_{rl} + L_{1m}
\end{aligned} \tag{7.107}$$

For a = 1, we reobtain, as expected, Equations (7.95)–(7.98). L_s and L'_r represent the total stator and rotor inductances per phase when the IM is three-phase fed.

Now the general equivalent circuit of (7.107), similar to that of Figure 7.1, is shown in Figure 7.24. For a = 1, the standard equivalent circuit (Figure 7.1) is obtained. If $a = L_s/L_{1m} \approx 1.02 - 1.08$, the whole inductance (reactance) is occurring in the rotor side of equivalent circuit and $\underline{\Psi}_{ma} = \underline{\Psi}_s$ (Figure 7.25a). The equivalent circuit directly evidentiates the stator flux.

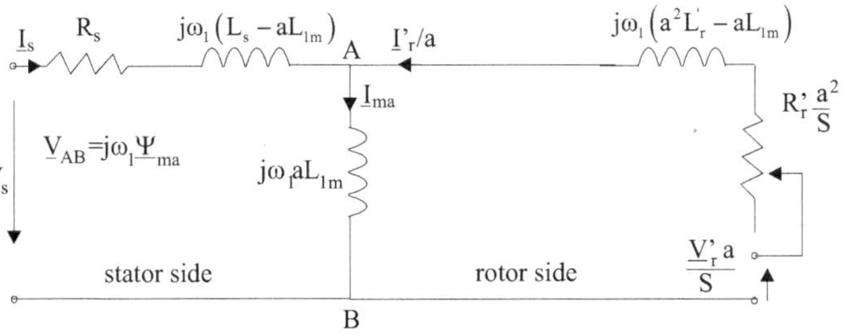

FIGURE 7.24 General equivalent circuit (core loss neglected).

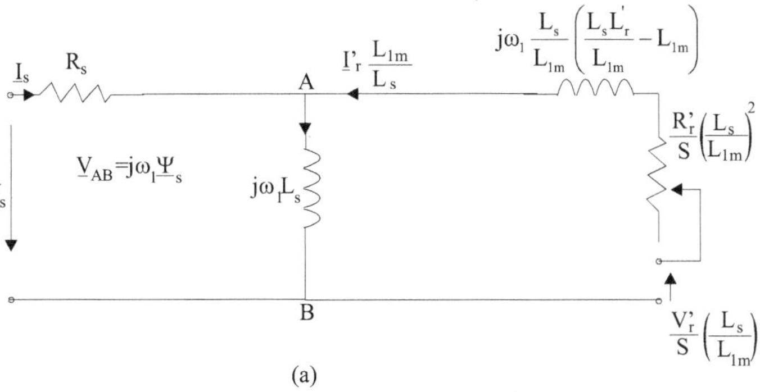

FIGURE 7.25 Equivalent circuit: (a) with stator and (b) rotor flux shown.

(Continued)

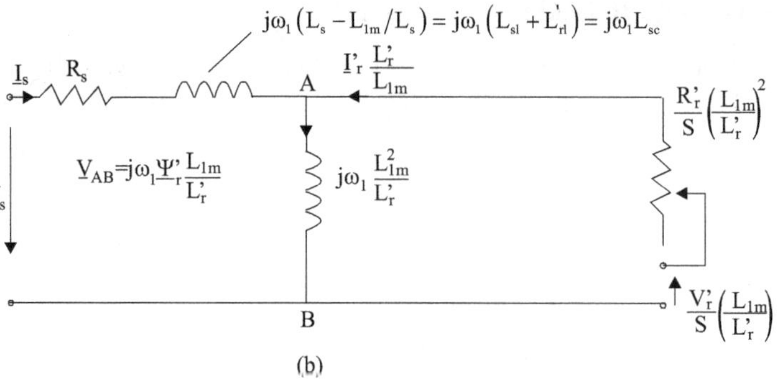

FIGURE 7.25 (CONTINUED) Equivalent circuit: (a) with stator and (b) rotor flux shown.

For $a = L_{1m}/L'_r \approx 0.93 - 0.97$, $\underline{\psi}_{ma} = \frac{L_{1m}}{L'_r}\psi'_r$, the equivalent circuit is said to evidentiate the rotor flux.

Advanced IM control is performed at constant stator, Ψ_s, rotor Ψ_r, or magnetisation flux, Ψ_m amplitudes. That is voltage and frequency, V_s and f_1, are changed to satisfy such conditions.

Consequently, circuits in Figure 7.25 should be instrumental in calculating IM steady-state performance under such flux linkage constraints.

Alternative equivalent circuits to include leakage and main saturation and skin effect are to be treated in Chapter 8.

Example 7.6 Flux Linkage Calculations

A cage-rotor induction motor has the following parameters: $R_s = 0.148\ \Omega$, $X_{sl} = X'_{rl} = 0.5\ \Omega$, $X_m = 20\ \Omega$, $2p_1 = 4$ poles, $V_s = 220$ V/phase (RMS) – star connection, $f_1 = 50$ Hz, rated current $I_{sn} = 36$ A, and $\cos\varphi_n = 0.9$. At start, the current $I_{start} = 5.8I_n$ and $\cos\varphi_{start} = 0.3$.

Let us calculate the stator, rotor, and magnetisation flux linkages Ψ_s, ψ'_r, and Ψ_m at rated speed and at standstill (core loss is neglected). With $S = 0.02$, calculate the rotor resistance at full load.

Solution

To calculate the RMS values of Ψ_s, ψ'_r, and Ψ_m, we have to determine, in fact, V_{AB} in Figure 7.24, for $a_s = L_s/L_{1m}$, $a_r = L_{1m}/L'_r$, and $a_m = 1$, respectively.

$$\psi_s\sqrt{2} = \frac{\sqrt{2}\left|\underline{V}_s - \underline{I}_s R_r\right|}{\omega_1} \tag{7.108}$$

$$\psi'_r\sqrt{2} = \sqrt{2}\frac{\left|\underline{V}_s - \underline{I}_s R_r - \underline{I}_s j\omega_1(L_{sl} + L'_{rl})\right|}{\omega_1\left(\frac{L_{1m}}{L'_r}\right)} \tag{7.109}$$

$$\psi_m\sqrt{2} = \frac{\sqrt{2}\left|\underline{V}_s - \underline{I}_s R_r - \underline{I}_s j\omega_1 L_{sl}\right|}{\omega_1} \tag{7.110}$$

$$\underline{I}_s = I_s(\cos\varphi - j\sin\varphi) \tag{7.111}$$

These operations have to be done twice, once for $I_s = I_n$, $\cos\varphi_n$ and once for $I_s = I_{start}$, $\cos\varphi_{start}$.

For $I_n = 36$A, $\cos\varphi_n = 0.9$,

$$\psi_s\sqrt{2} = \frac{\sqrt{2}\left|220 - 36\cdot(0.9 - j4359)0.148\right|}{2\pi 50} = 0.9654\ \text{Wb}$$

$$\psi_r'\sqrt{2} = \frac{\left|220 - 36\cdot(0.9 - j4359)(0.148 + j1)\right|}{2\pi 50\left(\frac{20}{20.5}\right)} = 0.9314\ \text{Wb} \tag{7.112}$$

$$\psi_m\sqrt{2} = \frac{\sqrt{2}\left|220 - 36\cdot(0.9 - j4359)(0.148 + j1)\right|}{2\pi 50} = 0.9369\ \text{Wb} \tag{7.113}$$

The same formulas are applied at zero speed (S = 1), with starting current and power factor,

$$\left(\psi_s\sqrt{2}\right)_{start} = \frac{\sqrt{2}\left|220 - 36\cdot 5.8\cdot(0.3 - j0.888)\cdot 0.148\right|}{2\pi 50} = 0.9523\ \text{Wb}$$

$$\left(\psi_r'\sqrt{2}\right)_{start} = \frac{\sqrt{2}\left|220 - 36\cdot 5.8\cdot(0.3 - j0.888)(0.148 + j1)\right|}{2\pi 50\left(\frac{20}{20.5}\right)} = 0.1998\ \text{Wb} \tag{7.114}$$

$$\left(\psi_m\sqrt{2}\right)_{start} = \frac{\sqrt{2}\left|220 - 36\cdot 5.8\cdot(0.3 - j0.888)(0.148 + j1)\right|}{2\pi 50} = 0.5306\ \text{Wb}$$

The above results show that while at full load the stator, magnetisation, and rotor flux linkage amplitudes do not differ much, they do so at standstill. In particular, the magnetisation flux linkage is reduced at start to 55%–65% of its value at full load, so the main magnetic circuit of IMs is not saturated at standstill.

7.13 UNBALANCED SUPPLY VOLTAGES

Steady-state performance with unbalanced supply voltages may be treated by the method of symmetrical components. The three-wire supply voltages $\underline{V}_a$, $\underline{V}_b$, and $\underline{V}_c$ are decomposed into forward and backward components

$$\underline{V}_{af} = \frac{1}{3}\left(\underline{V}_a + a\underline{V}_b + a^2\underline{V}_c\right);\quad a = e^{j\frac{2\pi}{3}}$$

$$\underline{V}_{ab} = \frac{1}{3}\left(\underline{V}_a + a^2\underline{V}_b + a\underline{V}_c\right); \tag{7.115}$$

$$\underline{V}_{bf} = a^2\underline{V}_{af};\quad \underline{V}_{cf} = a\underline{V}_{af};\quad \underline{V}_{bb} = a\underline{V}_{ab};\quad \underline{V}_{cb} = a^2\underline{V}_{ab}; \tag{7.116}$$

Also note that the slip for the forward component is $S_f = S$, while for the backward component, S_b is

$$S_b = \frac{-\left(\frac{f_1}{p_1}\right) - n}{-\left(\frac{f_1}{p_1}\right)} = 2 - S \tag{7.117}$$

So we obtain two equivalent circuits (Figure 7.26) as

$$\underline{V}_a = \underline{V}_{af} + \underline{V}_{ab} \tag{7.118}$$

$$\underline{I}_a = \underline{I}_{af} + \underline{I}_{ab} \tag{7.119}$$

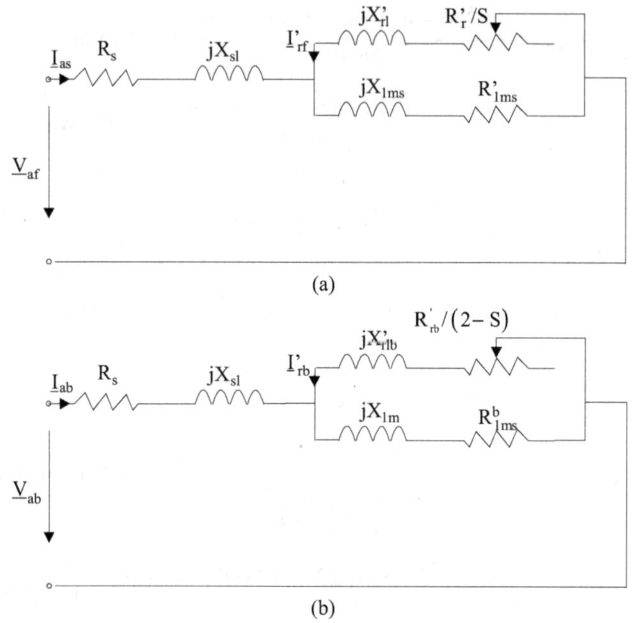

FIGURE 7.26 Forward (a) and backward (b) component equivalent circuits per phase.

For $S = 0.01 - 0.05$ and $f_1 = 60$ (50) Hz, the rotor frequency for the backward components is $f_2 = (2 - S)f_1 \approx 100$ (120) Hz. Consequently, the rotor parameters R_r' and L_{rl}' are notably influenced by the skin effect, so

$$R_{rb}' > R_r'; \quad X_{rlb}' < X_{rl}' \tag{7.120}$$

Also for the backward component, the core losses are notably less than those of forward component. The torque expression contains two components:

$$T_e = \frac{3R_r'(I_{rf}')^2}{S}\frac{p_1}{\omega_1} + \frac{3R_{rb}'(I_{rb}')^2}{(2-S)}\frac{p_1}{(-\omega_1)} \tag{7.121}$$

With phase voltages given as amplitudes and phase shifts, all steady-state performance may be calculated with the symmetrical component method.

The voltage unbalance index $V_{unbalance}$ (in %) may be defined as (NEMA MG1):

$$V_{unbalance} = \frac{\Delta V_{max}}{V_{ave}}100\%; \quad \Delta V_{max} = V_{max} - V_{min}; \quad V_{ave} = \frac{(V_a + V_b + V_c)}{3} \tag{7.122}$$

V_{max} = maximum phase voltage; V_{min} = minimum phase voltage.

An alternative definition (IEC) would be

$$V_{unb}[\%] = \frac{V_{ab}}{V_{af}}100 \tag{7.123}$$

Notice that the average line voltage in (7.123) or the positive sequence voltage $\underline{V}_{af}$ are not specified in the abovementioned standards yet.

Due to very different values of slip and rotor parameters for forward and backward components, a small voltage unbalance is likely to produce a rather large current unbalance. This is illustrated in Figure 7.27.

Also, for rated slip (power), the presence of backward (braking) torque and its losses leads to a lower efficiency η.

$$\eta = \frac{T_e \frac{\omega_1}{P_1}(1-S)}{3\,\mathrm{Re}\left[\underline{V}_{af}\underline{I}_{af}^{*} + \underline{V}_{ab}\underline{I}_{ab}^{*}\right]} \tag{7.124}$$

Apparently, cost-optimised motors – with larger rotor skin effect in general – are more sensitive to voltage unbalances than premium motors (Figure 7.28) [2].

As losses in the IM increase with voltage unbalance, machine derating is to be applied to maintain rated motor temperature. NEMA 14.35 Standard recommends an IM derating with voltage unbalance as shown in Figure 7.29 [3].

Voltage unbalance also produces, as expected, $2f_1$ frequency vibrations. As small voltage unbalance can easily occur in local power grids, care must be exercised in monitoring it and the motor currents and temperature.

As voltage unbalance derating also depends on the mean value of terminal voltage, derating in Figure 7.29 should be considered with caution. The influence of voltage angle and amplitude unbalance leads to intricate influences not only on efficiency but also on starting and pull-out torque and should be considered as such [4,5].

An extreme case of voltage unbalance is one phase open.

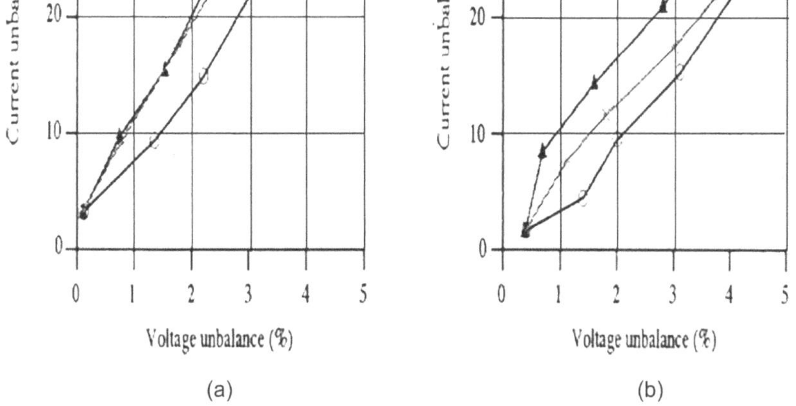

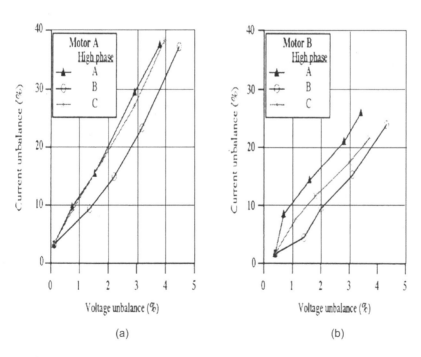

FIGURE 7.27 Current unbalance versus voltage unbalance [2]: (a) cost-optimised motor (motor A) and (b) premium motor (motor B).

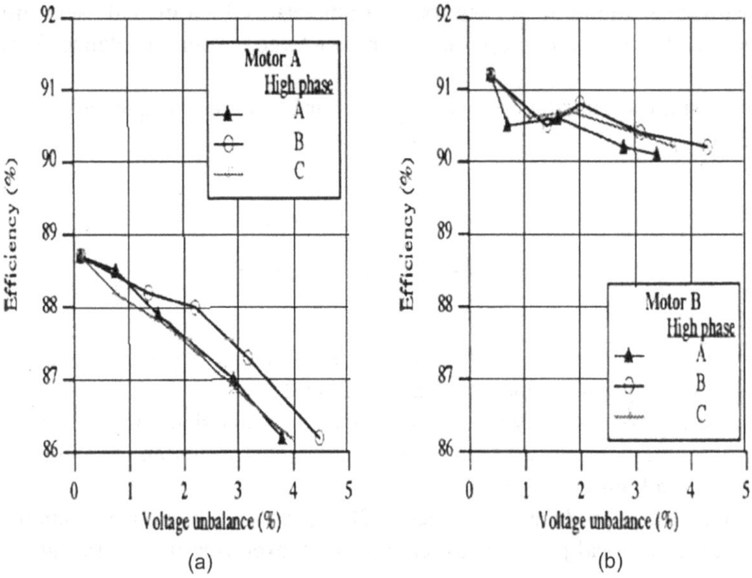

FIGURE 7.28 Efficiency versus voltage unbalance: (a) cost-optimised motor and (b) premium motor.

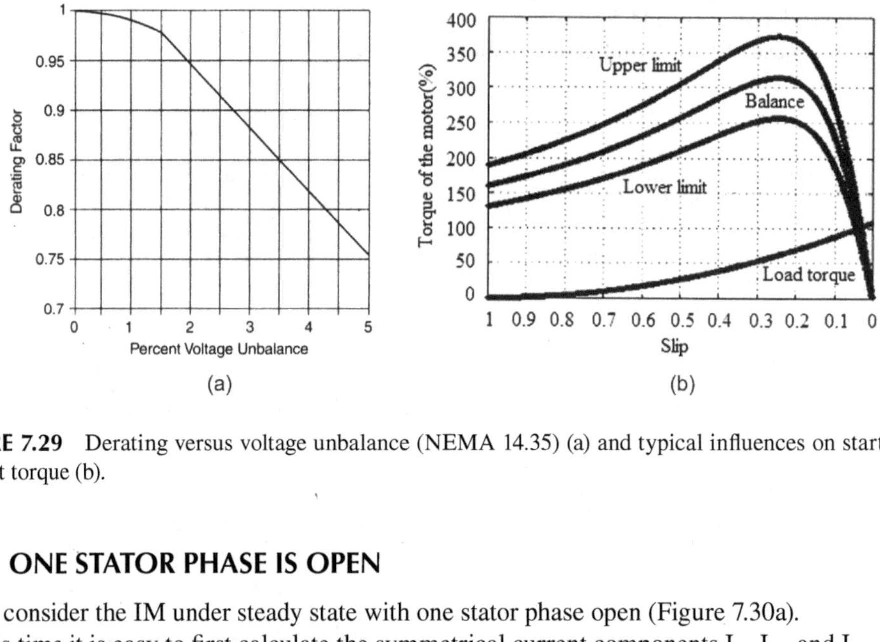

FIGURE 7.29 Derating versus voltage unbalance (NEMA 14.35) (a) and typical influences on starting and pull out torque (b).

7.14 ONE STATOR PHASE IS OPEN

Let us consider the IM under steady state with one stator phase open (Figure 7.30a).

This time it is easy to first calculate the symmetrical current components $\underline{I}_{af}$, $\underline{I}_{ab}$, and $\underline{I}_{ao}$.

$$\underline{I}_{af} = \frac{1}{3}\left(\underline{I}_a + a\underline{I}_b + a^2\underline{I}_c\right) = -\frac{a - a^2}{3}\underline{I}_b = \frac{j\underline{I}_b}{\sqrt{3}}$$

$$\underline{I}_{ab} = \frac{1}{3}\left(\underline{I}_a + a^2\underline{I}_b + a\underline{I}_c\right) = -\frac{a - a^2}{3}\underline{I}_b = -\underline{I}_{af} \tag{7.125}$$

Let us replace the equivalent circuits in Figure 7.26 by the forward and backward impedances Z_f and Z_b.

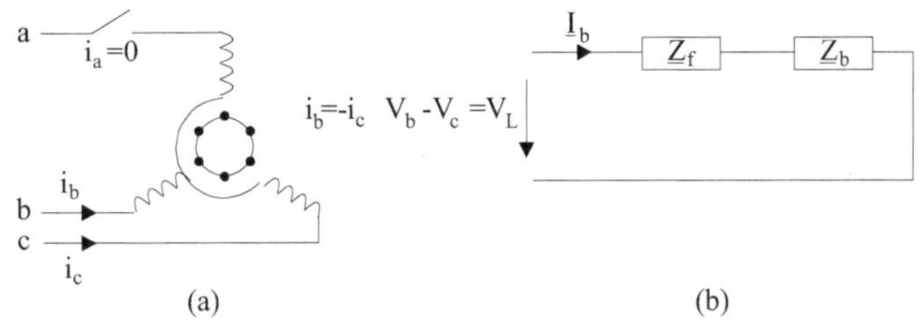

FIGURE 7.30 One stator phase is open: (a) phase *a* is open and (b) equivalent single-phase circuit.

$$\underline{V}_{af} = \underline{Z}_f \underline{I}_{af}; \quad \underline{V}_{ab} = \underline{Z}_b \underline{I}_{ab} \tag{7.126}$$

Similar relations are valid for $\underline{V}_{bf} = a^2\underline{V}_{af}$, $\underline{V}_{bb} = a\underline{V}_{ab}$, $\underline{V}_{cf} = a\underline{V}_{af}$, and $\underline{V}_{cb} = a^2\underline{V}_{ab}$

$$\begin{aligned}\underline{V}_b - \underline{V}_c = \underline{V}_{bf} + \underline{V}_{bb} - \underline{V}_{cf} - \underline{V}_{cb} &= a^2\underline{Z}_f\underline{I}_{af} + a\underline{Z}_b\underline{I}_{ab} - a\underline{Z}_f\underline{I}_{af} - a^2\underline{Z}_b\underline{I}_{ab} \\ &= \left(a^2 - a\right)\underline{I}_{af}\left(\underline{Z}_f + \underline{Z}_b\right)\end{aligned} \tag{7.127}$$

With (7.125),

$$\underline{V}_b - \underline{V}_c = \underline{I}_b\left(\underline{Z}_f + \underline{Z}_b\right) \tag{7.128}$$

The electromagnetic torque still retains Equation (7.121), but (7.128) allows a handy computation of current in phase b, and then from (7.125)–(7.126), I_{af} and I_{ab} are calculated.

At standstill (S = 1), $\underline{Z}_f = \underline{Z}_b = \underline{Z}_{sc}$ (Figure 7.26) and thus, the short-circuit current I_{sc1} is

$$\underline{I}_{sc1} = \frac{\underline{V}_L}{2\underline{Z}_{sc}} = \frac{\sqrt{3}}{2}\frac{\underline{V}_{phase}}{\underline{Z}_{sc}} = \frac{\sqrt{3}}{2}\underline{I}_{sc3} \tag{7.129}$$

The short-circuit current I_{sc1} with one phase open is thus $\sqrt{3}/2$ times smaller than for balanced supply. There is no danger from this point of view at start. However, as $\left(\underline{Z}_f\right)_{S=1} = \left(\underline{Z}_b\right)_{S=1}$, also $I'_{rf} = I'_{rb}$, and thus, the forward and backward torque components in (7.121) are cancelling each other. The IM will not start with one stator phase open.

The forward and backward torque/slip curves differ by the synchronous speed: ω_1/p_1 and, $-\omega_1/p_1$, respectively, and consequently, by the slips S and 2 – S, respectively. So they are antisymmetric (Figure 7.31).

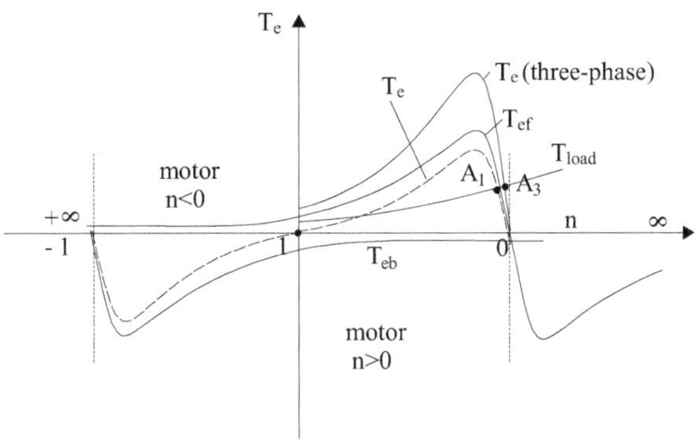

FIGURE 7.31 Torque components when one stator phase is open.

If the IM is supplied by three-phase balanced voltages and works in point A_3, and one phase is open, the steady-state operation point moves to A_1.

So, the slip increases (speed decreases). The torque slightly decreases. Now the current in phase b is expected to increase. Also, the power factor decreases because the backward equivalent circuit (for small S) is strongly inductive and the slip increases notably.

It is, therefore, not practical to let a fully loaded three-phase IM to operate a long time with one phase open.

Example 7.7 One Stator Phase Is Open

Let us consider a three-phase induction motor with the data $V_L = 220\,V$, $f_1 = 60\,Hz$, $2p_1 = 4$, star stator connection $R_s = R_r' = 1\ \Omega$, $X_{sl} = X_{rl}' = 2.5\ \Omega$, $X_{1m} = 75\ \Omega$, and $R_{1m} = \infty$ (no core losses). The motor is operating at $S = 0.03$ when one stator phase is opened. Calculate the stator current, torque, and power factor before one phase is opened and the slip, current, and power factor for same torque with one phase open.

Solution

We make use of Figure 7.26a.

With balanced voltages, only Z_f counts.

$$\underline{V}_a = \underline{Z}_f \underline{I}_a;\quad C_1 = \frac{X_{sl} + X_{1m}}{X_{1m}} = \frac{75 + 2.5}{75} = 1.033 \tag{7.130}$$

$$I_a = I_b = I_c \approx \frac{\frac{V_L}{\sqrt{3}}}{\sqrt{\left(R_s + C_1\frac{R_r'}{S}\right)^2 + (X_{sl} + C_1 X_{rl}')^2}}$$

$$= \frac{\frac{220}{\sqrt{3}}}{\sqrt{\left(1 + 1.033\frac{1}{0.03}\right)^2 + (2.5 + 1.033\cdot 2.5)^2}} = 3.55\text{ A};\quad \cos\varphi = 0.989! \tag{7.131}$$

The very high value of cosφ proves again that (7.131) does not approximate the phase of stator current while it calculates correctly the amplitude. Also,

$$\underline{I}_r' = \underline{I}_s\left|\frac{jX_{1m}}{\frac{R_r'}{S} + j(X_{1m} + X_{rl}')}\right| = 3.55\cdot\frac{75}{\sqrt{\left(\frac{1}{0.03}\right)^2 + (75 + 2.5)^2}} = 3.199\text{ A} \tag{7.132}$$

Now $\cos\varphi \approx \frac{I_r'}{I_a} = \frac{3.199}{3.55} = 0.901$, which is a reasonable value.

From (7.121),

$$T_{e3} = \frac{3R_r'}{S} I_r'^2 \frac{P_1}{\omega_1} = 3\cdot\frac{1}{0.03}\cdot 3.199^2 \frac{2}{2\pi 60} = 5.4338\text{ Nm} \tag{7.133}$$

It is not very simple to calculate the slip S for which the IM with one phase open will produce the torque as in (7.133).

To circumvent this difficulty, we might give the slip a few values higher than S = 0.03 and plot torque T_{e1} versus slip until $T_{e1} = T_{e3}$.

Here we take only one slip value, say, S = 0.05 and calculate first the current I_b from (7.128), then $I_{af} = -I_{ab}$ from (7.125), and finally, the torque T_{e1} from (7.121).

$$I_b = \frac{V_L}{|\underline{Z}_f + \underline{Z}_b|} \approx \frac{V_L}{\left| 2(R_s + jX_{sl}) + \frac{\frac{R_r'}{S} jX_{1m}}{\frac{R_r'}{S} + j(X_{1m} + X_{rl}')} + \frac{R_r'}{2-S} + jX_{rl}' \right|}$$

$$= \frac{220}{\left| 2(1 + j2.5) + \frac{\frac{1}{0.05} j75}{\frac{1}{0.05} + j(75 + 2.5)} + \frac{1}{2-S} + j2.5 \right|} = \frac{220}{23.709} = 9.2789\text{A}! \tag{7.134}$$

$$\cos\varphi_1 = \frac{20.34}{23.709} = 0.858$$

From (7.125),

$$I_{af} = -I_{ab} = \frac{I_b}{\sqrt{3}} = \frac{9.2789}{5.73} = 5.3635\text{A} \tag{7.135}$$

Figure (7.26) yields

$$\underline{I}_{rf}' = I_{af} \left| \frac{jX_{1m}}{\frac{R_r'}{S} + j(X_{1m} + X_{rl}')} \right| = \left| \frac{5.3635 \cdot j \cdot 7.5}{\frac{1}{0.05} + j(75 + 2.5)} \right| = 5.0258\text{A} \tag{7.136}$$

$$I_{rb}' \approx -I_{ab} = 5.3635\text{A}$$

Now, from (7.121), the torque T_{e1} is

$$T_{e1} = \frac{3 \cdot 2 \cdot 1}{2\pi 60} \left[\frac{5.0258^2}{0.05} - \frac{5.3635^2}{2 - 0.05} \right] = 7.809 \text{ Nm} \tag{7.137}$$

In this particular case, the influence of the backward component on torque has been small. A great deal of torque is obtained at S = 0.05 but at a notably large current and smaller power factor. The low values of leakage reactances and the large value of magnetisation reactance explain, in part, the impractically high power factor calculation. For even more correct results, the complete equivalent circuit is to be solved.

7.15 UNBALANCED ROTOR WINDINGS

Wound rotors may be provided with external three-phase resistances which may not be balanced. Also, broken bars in the rotor cage lead to unbalanced rotor windings. This latter case will be treated in Volume 2, Chapter 1, on transients.

However, the wound rotor unbalanced windings (Figure 7.32) may be treated here with the method of symmetrical components.

We may start decomposing the rotor phase currents $\underline{I}_{ar}$, $\underline{I}_{br}$, and $\underline{I}_{cr}$ into symmetrical components

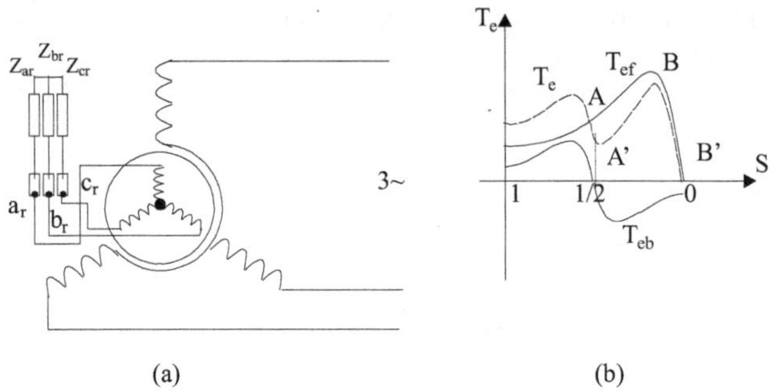

FIGURE 7.32 Induction motor with unbalanced wound rotor winding (a) and torque/speed curve (b).

$$\underline{I}_{ar}^{\ f} = \frac{1}{3}\left(\underline{I}_{ar} + a\underline{I}_{br} + a^2\underline{I}_{cr}\right)$$

$$\underline{I}_{ar}^{\ b} = \frac{1}{3}\left(\underline{I}_{ar} + a^2\underline{I}_{br} + a\underline{I}_{cr}\right) \quad (7.138)$$

$$\underline{I}_{ar}^{\ 0} = \frac{1}{3}\left(\underline{I}_{ar} + \underline{I}_{br} + \underline{I}_{cr}\right) = 0$$

Also, we do have

$$\underline{V}_{ar} = -\underline{Z}_{ar}\underline{I}_{ar}; \quad \underline{V}_{br} = -\underline{Z}_{br}\underline{I}_{br}; \quad \underline{V}_{cr} = -\underline{Z}_{cr}\underline{I}_{cr}; \quad (7.139)$$

In a first approximation, all rotor currents at steady state have the frequency $f_2 = Sf_1$. The forward mmf, produced by $\underline{I}_{ar}^{\ f}$, $\underline{I}_{br}^{\ f}$, and $\underline{I}_{cr}^{\ f}$, interacts as usual with the stator winding and its equations are

$$\underline{I}_r'^{\,f} R_r' - \underline{V}_r'^{\,f} = -jS\omega_1\underline{\psi}_r'^{\,f}; \quad \underline{\psi}_r'^{\,f} = L_r'\underline{I}_r'^{\,f} + L_{1m}\underline{I}_s^{\ f}$$

$$\underline{I}_s^{\ f} R_s - \underline{V}_s = -j\omega_1\underline{\psi}_s^{\ f}; \quad \underline{\psi}_s^{\ f} = L_s\underline{I}_s^{\ f} + L_{1m}\underline{I}_r'^{\,f} \quad (7.140)$$

The backward mmf component of rotor currents rotates with respect to the stator at the speed n_1'.

$$n_1' = n - S\frac{f_1}{P_1} = \frac{f_1}{P_1}(1-2S) \quad (7.141)$$

So it induces stator emfs of a frequency $f_1' = f_1(1-2S)$. The stator may be considered short-circuited for these emfs as the power grid internal impedance is very small in relative terms. The equations for the backward component are

$$\underline{I}_r'^{\,b} R_r' - \underline{V}_r'^{\,b} = -jS\omega_1\underline{\psi}_r'^{\,b}; \quad \underline{\psi}_r'^{\,b} = L_r'\underline{I}_r'^{\,b} + L_{1m}\underline{I}_s^{\ b}$$

$$\underline{I}_s^{\ b} R_s = -j(1-2S)\omega_1\underline{\psi}_s^{\ b}; \quad \underline{\psi}_s^{\ b} = L_s\underline{I}_s^{\ b} + L_{1m}\underline{I}_r'^{\,b} \quad (7.142)$$

For a given slip, rotor external impedances $\underline{Z}_{ar}$, $\underline{Z}_{br}$, $\underline{Z}_{cr}$, motor parameters, stator voltage, and frequency, Equation (7.139) and their counterparts for rotor voltages, (7.140)–(7.142) are used to determine $\underline{I}_r^{\ f}$, $\underline{I}_r^{\ b}$, $\underline{I}_s^{\ f}$, $\underline{I}_s^{\ b}$, $\underline{V}_r'^{\,f}$, and $\underline{V}_r'^{\,b}$. Note that (7.140) and (7.142) are per phase basis and are thus valid for all three phases in the rotor and stator.

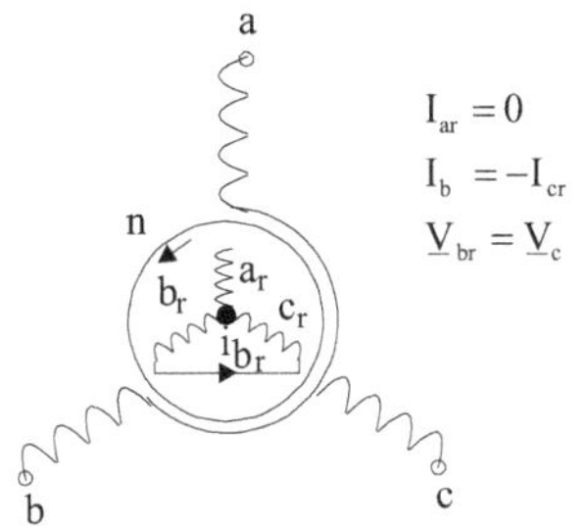

FIGURE 7.33 One rotor phase is open.

The torque expression is

$$T_e = 3p_1L_{1m}\left[\text{Imag}\left(\underline{I}_s{}^f\underline{I}_r'^{f*}\right) + \text{Imag}\left(\underline{I}_s{}^b\underline{I}_r'^{b*}\right)\right] = T_{ef} + T_{eb} \tag{7.143}$$

The backward component torque is positive (motoring) for $1–2S < 0$ or $S > ½$ and negative (braking) for $S < ½$. At start, the backward component torque is motoring. Also for $S = ½$, $\underline{I}_s{}^b = 0$, and thus, the backward torque is zero. This backward torque is also called monoaxial or Geörge's torque.

The torque/speed curve obtained is given in Figure 7.33b. A stable zone, AA′ around S = ½ occurs, and the motor may remain "hanged" around half ideal no-load speed.

The larger the stator resistance R_s, the larger the backward torque component. Consequently, the saddle in the torque–speed curve will be more visible in low power machines where the stator resistance is larger in relative values: $r_s = R_sI_n/V_n$. Moreover, the frequency f_1' of stator current is close to f_1, and thus, visible stator backward current pulsations occur. This may be used to determine the slip S as $f' - f_1 = 2Sf_1$. Low frequency and noise at $2Sf_1$ may be encountered in such cases.

7.16 ONE ROTOR PHASE IS OPEN

An extreme case of unbalanced rotor winding occurs when one rotor phase is open (Figure 7.33). Qualitatively, the phenomenon occurs as described for the general case in Section 7.13. After reducing the rotor variables to stator, we have

$$\underline{I}_{ar}'^{f} = -\underline{I}_{ar}'^{b} = -\frac{j}{\sqrt{3}}I_{br}' \tag{7.144}$$

$$\underline{V}_r'^f = \underline{V}_r'^b = \frac{1}{3}\left(\underline{V}_{ar}' - \underline{V}_{br}'\right) \tag{7.145}$$

Now, with (7.140) and (7.142), the unknowns are $\underline{I}_s{}^f$, $\underline{I}_s{}^b$, $\underline{I}_r'^f$, $\underline{I}_r'^b$, and $\underline{V}_r'^f$.

This time we may get (from (7.142) with (7.144 and 7.145))

$$\underline{I}_s{}^b = \frac{-j\omega_1(1-2S)L_{1m}\underline{I}_r'^b}{R_s + j\omega_1(1-2S)L_s} \tag{7.146}$$

It is again clear that with $R_s = 0$, the stator backward current is

$$\left(\underline{I}_s{}^b\right)_{R_s=0} = \frac{-L_{1m}}{L_s}\underline{I}_r'^b \tag{7.147}$$

Consequently, the backward torque component in (7.143) becomes zero.

7.17 WHEN VOLTAGE VARIES AROUND RATED VALUE

It is common practice that, in industry, the local power grid voltage varies around rated value as influenced by the connection (disconnection) of other loads and capacitors used for power factor correction. Higher supply voltages, even when balanced, notably influence the IM under various load levels.

A ±10% voltage variation around the rated value is, in general, accepted in many standards. However, the IMs have to be designed to incur such voltage variations without excessive temperature rise at rated power.

In essence, if the voltage increases above the rated value, the core losses p_{iron} in the machine increase notably as

$$p_{iron} \approx \left(C_h f_1 + C_e f_1^{\,2}\right) V^2 \tag{7.148}$$

At the same time, with the core notably more saturated, L_{1m} decreases accordingly.

The rated torque is to be obtained, thus, at lower slips. The power factor decreases so the stator current tends to increase, while, for lower slip, the rotor current does not change much. So the efficiency, not only the power factor, decreases for rated load.

The temperature tends to rise. In the design process, a temperature reserve is to be allowed for nonrated voltage safe operation.

On the other hand, if low-load operation at higher than rated voltage occurs, the high core losses mean excessive power loss. Motor disconnection instead of no-load operation is a practical solution in such cases.

If the voltage decreases below rated value, the rated torque is obtained at higher slip and, consequently, higher stator and rotor currents occur. The winding losses increase while the core losses decrease, because the voltage (and magnetising flux linkage) decreases.

The efficiency should decrease slowly, while the power factor might also decrease slightly or remain the same as with rated voltage.

A too big voltage reduction leads, however, to excessive losses and lower efficiency at rated load. With partial load, lower voltage might be beneficial in terms of both efficiency and power factor, as the core plus winding losses decrease towards a minimum.

When IMs are designed for two frequencies – 50 or 60Hz – care is exercised to meet the over temperature limitation for the most difficult situation (lower frequency).

7.18 WHEN STATOR VOLTAGES HAVE TIME HARMONICS

When the IM is PWM inverter fed for variable speed or grid connected but the grid is "infected" with time voltage harmonics from other PWM converter loads, the former exhibits: nonsinusoidal voltages, and as a consequence, nonsinusoidal currents.

The voltage time harmonics will produce additional copper losses in the stator and rotor windings and additional core losses as detailed in Chapter 11.

Finally, the IM over-temperature increases due to the presence of voltage time harmonics. Using active power filters in the local power grid is the way out of this problem.

However, when this is not done, the motor should be, in general, derated.

Treating the consequences of voltage time harmonics depends on the level of time harmonic frequency we are talking about: a few hundred or a few thousand hertz.

For the time being (the problem will be revised in Chapter 11), it suffices to draw the basic equivalent circuit for the speed ω_r

$$\omega_{r_t} = \nu^* \omega_1 \tag{7.149}$$

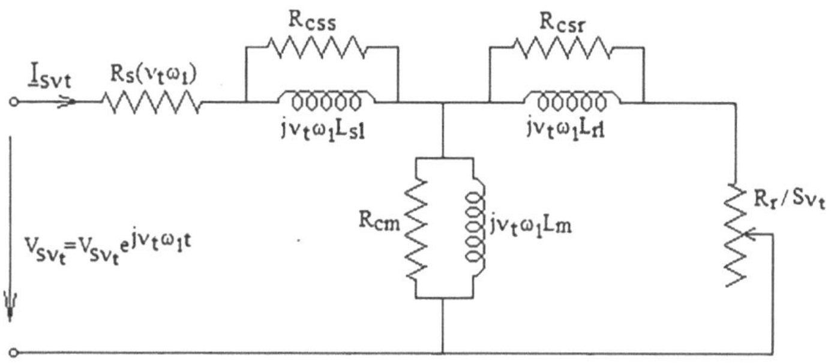

FIGURE 7.34 Equivalent circuit for voltage time harmonics.

Its slip is

$$S_{v_t} = \frac{v * \omega_1 - \omega_r}{v * \omega_1} = 1 - \frac{\omega_r}{v_t * \omega_1} = 1 - \frac{1-S}{v_t} \quad (7.150)$$

with S small or large $S_{v_t} \approx 1$

So, the frequency of rotor and stator currents for the time harmonics is the same and it is large, approximately equal to $v * \omega_1$.

The skin effects both in the stator and rotor laminations and bars will be notable. However, their torque will be negligible, but their losses will be important.

A simplified equivalent circuit to account for voltage time harmonics should duplicate the fundamental T circuit treated so far but with some modifications (Figure 7.34).

- $R_s(v_t\omega_1)$ and $R_r(v_t\omega_1)$ are stator and winding resistances as influenced by skin effects at $\approx v_t * \omega_1$.
- L_{sl}, L_{rl} are stator and rotor leakage inductances considered rather constant.
- R_{css}, R_{csr} are stator and rotor slot wall-induced core loss resistances by the $v_t * \omega_1$ frequency magnetic fields.
- R_{cm} is the stator and rotor core loss resistance due to airgap magnetic field.
- L_m is the magnetisation inductance.

It is not at all easy to identify the parameters in Figure 7.34, but most probably, a frequency response analysis and curve fitting would be appropriate. The increase in switching frequency of PWM converters has not only led to time harmonic losses reduction in IMs but, unfortunately, also the increase of switching losses in the PWM converters. To the point that six-pulse (rectangular 120° line voltage pulse) supply in the converter is considered worthy of consideration, when overall losses and costs are considered [6].

7.19 EQUIVALENT CIRCUITS FOR BRUSHLESS DOUBLY FED IMS

The most investigated brushless doubly fed induction machines (BDFIM) are the

- BDFIM with nested (loop) cage single rotor
- BDFIM with dual (cascaded) stator and rotor.

Both have two stator windings of p_1 and p_2 pole pairs but on a single core stator and rotor in brushless doubly fed induction generator (BDFIG) and on two cores and rigidly coupled rotors for cascaded BDFIM.

The speed formula for synchronous operation is the same for both

$$\omega_r = \frac{\omega_1 + \omega_2}{p_1 + p_2} \tag{7.151}$$

If for cascaded BDFIG (Figure 7.35b) the two three phase wound rotor windings (of p_1 and p_2 pole pairs) have the windings sequences reversed, then $\omega_1 - p_1{*}\,\omega_r = -(\omega_2 - p_2{*}\omega_r)$ to yield (7.151).

The equivalent circuits of the two special IMs in Figure 7.35 are given in Figure 7.36, as adapted from Refs. [8] and [9], respectively.

- Both schemes in Figure 7.36 are similar, with the two stators at the ends and the rotor circuit in the middle.
- S_1 and S_2 are the slips ($S_1 = 1 - p_1 \cdot \omega_r / \omega_1$, $S_2 = 1 - p_2 \cdot \omega_r / \omega_2$).
- In Figure 7.36b, stator (phase) coordinates are used, this is how the motion-induced voltages e_r and e_{s2}, proportional to $p_1 \cdot \omega_r$ and respectively to $(p_1+p_2) \cdot \omega_r$ occur.

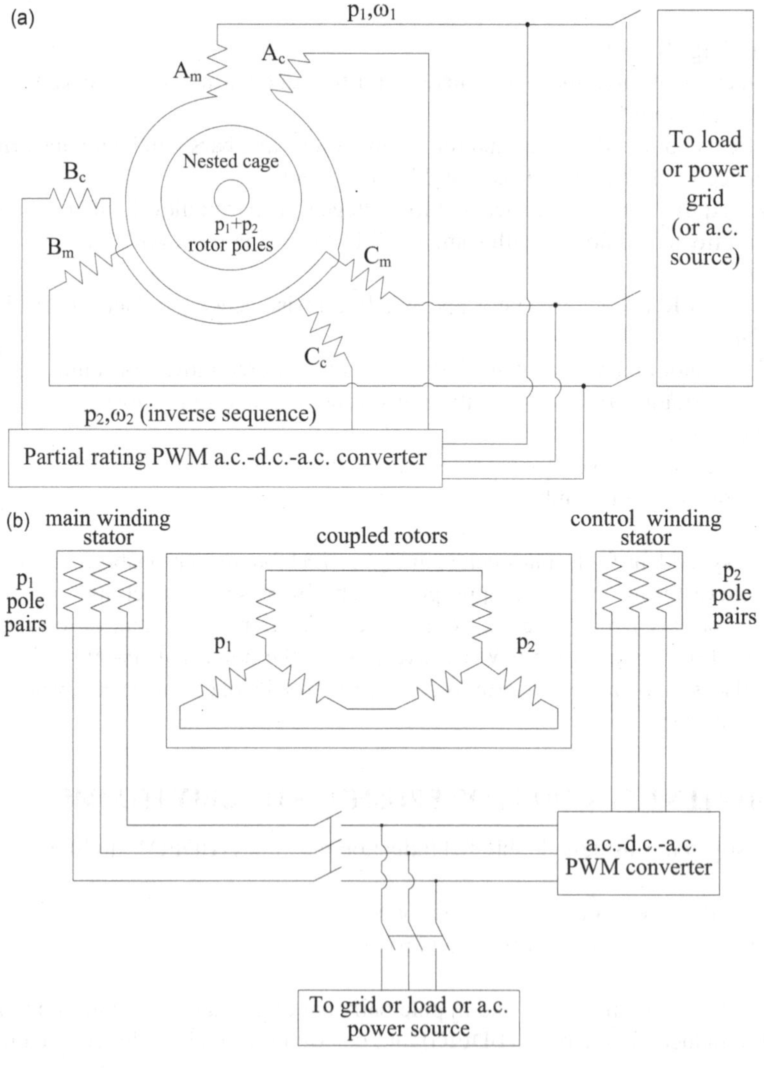

FIGURE 7.35 BDFIMs: (a) with single nested cage rotor and (b) cascaded.

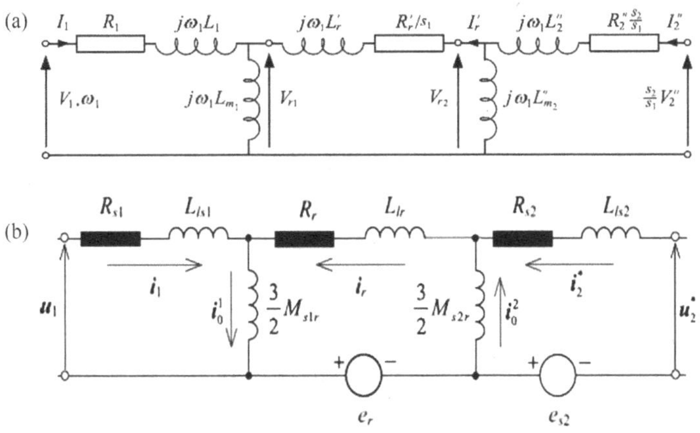

FIGURE 7.36 BDFIMs per-phase equivalent steady-state circuits: (a) BDFIM with nested cage (after [7]) and (b) cascaded BDFIM (after [8]).

As evident, the BDFIM equivalent circuits are more involved than for the standard-cage rotor IM, but some similarities are evident. Such machines are now actively proposed for wind or avionics generators or even for electric vehicles/hybrid electric vehicles (EVs/HEVs) due to their ruggedness and multifunctional characteristics at moderate power electronics costs and acceptable efficiency.

7.20 SUMMARY

- The relative difference between mmf speed $n_1 = f_1/p_1$ and the rotor speed n is called slip, $S = 1 - np_1/f_1$.
- The frequency of the emf induced by the stator mmf field in the rotor is $f_2 = Sf_1$.
- The rated slip S_n, corresponding to rated n_n, $S_n = 0.08 - 0.01$; larger values correspond to smaller power motors (under 1 kW).
- At zero slip (S = 0), for short-circuited rotor, the rotor current and the torque are zero; this is called the ideal no-load mode $n_0 = n_1 = f_1/p_1$.
- When the rotor windings are fed by balanced three-phase voltages V_r' of frequency $f_2 = Sf_1$, the zero rotor current and zero torque are obtained at a slip $S_0 = V_r'/E_1$, where E_1 is the stator phase emf. S_0 may be positive (subsynchronous operation) or negative (supersynchronous operation) depending on the phase angle between V_r' and $\underline{E}_1$.
- The active power travelling through an airgap (related to Poynting's vector) is called the electromagnetic power P_{elm}.
- The electromagnetic torque, for a short-circuited rotor (or in presence of a passive, additional, rotor impedance), is

$$T_e = P_{elm}\frac{p_1}{\omega_1}$$

- At ideal no-load speed (zero torque), the value of P_{elm} is zero, so the machine driven at that speed absorbs power in the stator to cover the winding and core loss.
- Motor no-load mode is when there is no-load torque at the shaft. The input active power now covers the core and stator winding losses and the mechanical losses.
- The induction motor operates as a motor when $(0 < S < 1)$.
- For generator mode $S < 0$ and for braking $S > 1$.

- In all operation modes, the singly fed IM motor "absorbs" reactive power for magnetisation.
- Autonomous generating mode may be obtained with capacitors at terminals to produce the reactive power for magnetisation.
- At zero speed, the torque is nonzero with stator balanced voltages. The starting torque $T_{es} = (0.5 - 2.2)T_{en}$. T_{en} – rated torque; starting current at rated voltage is $I_{start} = (5 - 7(8))I_n$; I_n – rated current. Higher values of starting current correspond to high efficiency motors.
- At high currents ($S \gg S_n$), the slot leakage flux path saturates and the short-circuit (zero speed) inductance decreases by 10%–40%. In the same time, the rotor frequency $f_2 = Sf_1 = f_1$ and the skin effect causes a reduction of rotor slot leakage inductance and a higher increase in rotor resistance.

  Lower starting current and larger starting torque are thus obtained.
- The closed-slot rotor leakage inductance saturates well below rated current. Still, it remains higher than with half-closed rotor slots.
- No-load and short-circuit (zero speed) tests may be used to determine the IM parameters – resistances and reactance with some classical approximations.
- It should be noticed that the basic T equivalent circuit of IM is not unique in the sense that any of the parameters R_r, L_r, L_m is a discretionary quantity [9].
- With a voltage source, the electromagnetic torque T_e versus slip curve has two breakdown points: one for motoring and one for generating. The breakdown torque is inversely proportional to short-circuit leakage reactance X_{sc} but independent of rotor resistance.
- Efficiency and power factor also have peak values both for motoring and generating at distinct slips. The rather flat maxima allow for a large plateau of good efficiency for loads from 25% to 150%.
- Adequate phasor diagrams evidentiating stator, rotor, and airgap (magnetisation) flux linkages (per phase) show that at steady state, the rotor flux linkage and rotor current per phase are time-phase shifted by 90°. It is +90° for motor and –90° for generating. If the rotor flux amplitude is maintained constant during transients, the 90° phase shift of rotor flux linkage and rotor currents would also stand for transients. Independent rotor flux and torque control may thus be obtained. This is the essence of vector control.
- Unbalanced stator voltages cause large unbalances in stator currents. Derating is required for sustained stator voltage unbalance.
- Higher than rated balanced voltages cause lower efficiency and power factor for rated power. In general, IMs are designed (thermally) to stand ±10% voltage variation around the rated value.
- Unbalanced rotor windings cause an additional stator current at the frequency $f_1(1 - 2S)$, besides the one at f_1. Also an additional rotor-initiated backward torque which is zero at $S = ½$ ($n = f_1/2p_1$), Geörge's torque is produced. A saddle in the torque versus slip occurs around $S = ½$ for IMs with relatively large stator resistance (low-power motors). The machine may be "hanged" (stuck) around half the ideal no-load speed.
- In this chapter, we dealt only with the single-cage rotor IM and the fundamental mmf and airgap field performance during steady state for constant voltage and frequency. The next chapter treats starting and speed control methods.

REFERENCES

1. P. T. Lagonotte, H. Al. Miah, N. Poloujadoff, Modelling and identification of parameters of saturated induction machine operating under motor and generator conditions, *EMPS*, Vol. 27, No. 2, 1999, pp. 107–121.
2. J. Kneck, D. A. Casada, P. J. Otday, A comparison of two energy efficient motors, *IEEE Transactions on Energy Conversion*, Vol. 13, No. 2., 1998, pp. 140–147.

3. A. H. Bonnett, G. C. Soukup, NEMA motor – Generator standards for three phase induction motors, *IEEE Industry Applications Magazine*, Vol. 5, No. 3, 1999, pp. 49–63.
4. J. Faiz, H. Ebrahimpour, P. Pillay, Influence of unbalanced voltage on the steady-state performance of a three-phase squirrel-cage induction motor, *IEEE Transactions on Energy Conversion*, Vol. 19, No. 4., 2004, pp. 657–662.
5. J. H. Dymond, N. Stranges, Operation on unbalanced voltage: One motor's experience and more, *IEEE Transactions on Industry Applications*, Vol. 43, No. 3, 2007, pp. 829–837.
6. D. G. Dorrell, C. Y. Leong, R. A. McMahon, Analysis and performance assessment of six-pulse inverter-fed three-phase and six-phase induction machines, *IEEE Transactions on Industry Applications*, Vol. 46, No. 6, 2006, pp. 1487–1495.
7. A. Oraee, E. Abdi, S. Abdi, R. McMahon, P.J. Tavner, Effects of rotor winding structure on the BDFIM equivalent circuit parameters, *IEEE Transactions on Energy Conversion*, Vol. 30, No. 4, 2015, pp. 1660–1669.
8. P. Han, M.Cheng, Z. Chen, Dual-electrical-port control of cascaded doubly fed induction machine for EV/HEV applications, *IEEE Transactions on Industry Applications*, Vol. 53, No. 2, 2017, pp.1390–1398.
9. K. R. Davey, The equivalent T circuit of the induction motor: Its nonuniqueness and use to the magnetic field analyst, *IEEE Transactions on Magnetics*, Vol. 43, No. 4, 2007, pp. 1745–1748.

8 Starting and Speed Control Methods

Starting refers to speed, current, and torque variations in an induction motor when fed directly or indirectly from a rather constant voltage and frequency local power grid.

A "stiff" local power grid would mean rather constant voltage even with large starting currents in the induction motors with direct full-voltage starting (5.8–6.5 times rated current is expected at zero speed at steady state). Full starting torque is produced in this case and starting over notable loads is possible.

A large design KVA (kilo volt ampere) in the local power grid, which means a large KVA power transformer, is required in this case. For starting under heavy loads, such a large design KVA power grid is mandatory.

On the other hand, for low-load starting, less stiff local power grids are acceptable. Decrease in voltage due to large starting currents will lead to a starting torque which decreases with voltage squared. As many local power grids are less stiff, for low starting loads, it means to reduce the starting currents, although in most situations even larger starting torque reduction is inherent for cage-rotor induction machines (IMs).

For wound-rotor IMs, additional hardware connected to the rotor brushes may provide even larger starting torque while starting currents are reduced. In what follows, various starting methods and their characteristics are presented. Speed control means speed variation with given constant or variable load torque. Speed control can be performed either by open loop (feedforward) or close loop (feedback) control. In this chapter, we will introduce the main methods for speed control and their corresponding steady-state characteristics.

Transients related to starting and speed control are treated in Volume 2, Chapter 1. Close loop speed control methods are beyond the scope of this book as they are fully covered by literature [1,2].

8.1 STARTING OF CAGE-ROTOR INDUCTION MOTORS

Starting of cage-rotor induction motors may be performed by

- Direct connection to power grid
- Low-voltage autotransformer
- Star-delta switch connection
- Additional resistance (reactance) in the stator
- Soft starting (through static variacs).

8.1.1 Direct Starting

Direct connection of cage-rotor induction motors to the power grid is used when the local power grid is stiff and rather large starting torque is required.

Typical variations of stator current and torque with slip (speed) under steady state are shown in Figure 8.1.

For single-cage induction motors, the rotor resistance and leakage inductance are widely influenced by skin effect and leakage saturation. At start, the current is reduced and the torque is increased due to skin effect and leakage saturation.

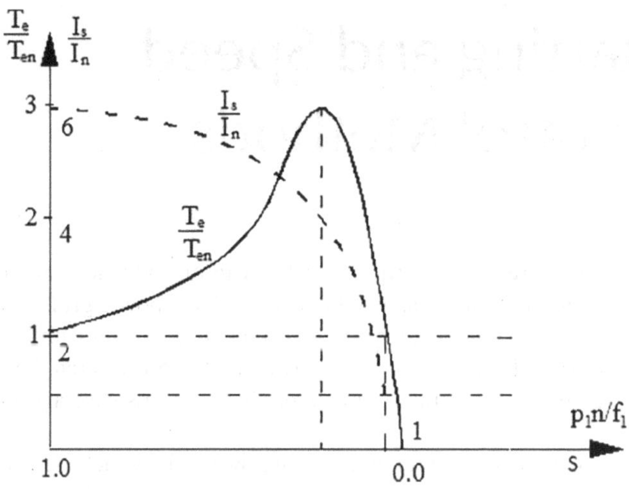

FIGURE 8.1 Current and torque versus slip (speed) in a single-cage induction motor.

In deep-bar or double-cage rotor induction motors, the skin effect is more influential, as shown in Chapter 9. When the load torque is notable from zero speed (>0.5 T_e) or the inertia is large ($J_{total} > 3J_{motor}$), the starting process is slow and the machine may be considered to advance from steady state to steady state until full-load speed is reached in a few seconds to minutes (in large motors).

If the induction motor remains at stall for a long time, the rotor and stator temperatures become too large, so there is a maximum stall time for each machine design.

On the other hand, for frequent start applications, it is important to count the rotor acceleration energy.

Let us consider applications with load torque proportional to squared speed (fans, ventilators). In this case, we may, for the time being, neglect the presence of load torque during acceleration. Also, a large inertia is considered, and thus, the steady-state torque/speed curve is used.

$$\frac{J}{p_1}\cdot\frac{d\omega_r}{dt} \approx T_e(\omega_r) \quad \omega_r = \omega_1(1-S) \tag{8.1}$$

The rotor winding loss p_{cor} is

$$p_{cor} = S\cdot P_{elm} = S\cdot T_e\left(\frac{\omega_1}{P}\right) \tag{8.2}$$

with T_e from (8.1), the rotor winding losses W_{cos} are

$$W_{cor} = \int_0^{t_s}\left(S\cdot T_e\cdot\frac{\omega_1}{p_1}\right)dt = \int_0^{t_s}\frac{J}{p_1}\cdot\frac{d\omega_r}{dt}\cdot\frac{\omega_1}{p_1}\cdot S\,dt \approx -\frac{J}{p_1^2}\int_1^0 \omega_1^2\cdot S\,dS$$

$$= +\frac{J}{2}\left(\frac{\omega_1}{p_1}\right)^2; \quad S_{initial} = 1.0; \quad S_{final} = 0.0 \tag{8.3}$$

On the other hand, the stator winding losses during motor acceleration under no-load W_{cos} are

$$W_{cos} = 3\int_0^{t_s} I_s^2(S)R_s\,dt \approx 3\int I_r'^2\cdot R_r'\cdot\frac{R_s}{R_r'}dt \approx W_{cor}\cdot\frac{R_s}{R_r'} \tag{8.4}$$

Consequently, the total winding energy losses W_{co} are

$$W_{co} = W_{cos} + W_{cor} = \frac{J}{2}\left(\frac{\omega_1}{p_1}\right)^2\left(1+\frac{R_s}{R_r'}\right) \tag{8.5}$$

A few remarks are in order.

- The rotor winding losses during rotor acceleration under no load are equal to the rotor kinetic energy at ideal no-load speed.
- Equation (8.5) tends to suggest that for given stator resistance R_s, a larger rotor resistance (due to skin effect) is beneficial.
- The temperature transients during such frequent starts are crucial for the motor design, with (8.5) as a basic evaluation equation for total winding losses.
- The larger the rotor-attached inertia J, the larger the total winding losses during no-load acceleration.

Returning to the starting current issue, it should be mentioned that even in a rather stiff local power grid, a voltage drop occurs during motor acceleration, as the current is rather large. A 10% voltage reduction in such cases is usual.

On the other hand, with an oversized reactive power compensation capacitor, the local power grid voltage may increase up to 20% during low-energy consumption hours.

Such a large voltage has an effect on voltage during motor starting in the sense of increasing both the starting current and torque.

Example 8.1 Voltage Reduction during Starting

The local grid power transformer in a small company has the data $S_n = 700$ KVA, secondary line voltage $V_{L2} = 440\,V$ (star connection), short-circuit voltage $V_{SC} = 4\%$, and $\cos\varphi_{SC} = 0.3$. An IM is planned to be installed for direct starting. The IM power $P_n = 100\,kW$, $V_L = 440\,V$ (star connection), rated efficiency $\eta_n = 95\%$, $\cos\varphi_n = 0.92$, starting current $\frac{I_{start}}{I_n} = \frac{6.5}{1}$, and $\cos\varphi_{start} = 0.3$.

Calculate the transformer short-circuit impedance, the motor starting current at rated voltage, and impedance at start. Finally, determine the voltage drop at start and the actual starting current in the motor.

Solution

First we have to calculate the rated transformer current in the secondary I_{2n}

$$I_{2n} = \frac{S_n}{\sqrt{3}\cdot V_{L2}} = \frac{700\times 10^3}{\sqrt{3}\cdot 440} = 919.6\ A \tag{8.6}$$

The short-circuit voltage V_{SC} corresponds to rated current

$$V_{SC2} = 0.04\cdot\frac{V_{L2}}{\sqrt{3}} = I_{2n}\left|\underline{Z}_{SC2}\right| \tag{8.7}$$

$$\left|\underline{Z}_{SC2}\right| = \frac{0.04\cdot 440}{\sqrt{3}\cdot 919.6} = 11.0628\cdot 10^{-3}\Omega \tag{8.8}$$

$$R_{SC} = \left|\underline{Z}_{SC}\right|\cos\varphi_{SC} = 11.0628\cdot 10^{-3}\cdot 0.3 = 3.3188\cdot 10^{-3}\ \Omega \tag{8.9}$$

$$X_{SC} = \left|\underline{Z}_{SC}\right|\sin\varphi_{SC} = 11.0628\cdot 10^{-3}\cdot\sqrt{1-0.3^2} = 10.5532\cdot 10^{-3}\ \Omega$$

For the rated voltage, the motor rated current I_{sn} is

$$I_{sn} = \frac{P_n}{\sqrt{3}\eta_n \cos\varphi_n V_L} = \frac{100 \cdot 10^3}{\sqrt{3} \cdot 0.95 \cdot 0.92 \cdot 440} = 150.3 \text{ A} \tag{8.10}$$

The starting current is

$$I_{start} = 6.5 \cdot 150.3 = 977\text{A} \tag{8.11}$$

Now the starting motor impedance $Z_{start} = R_{start} + jX_{start}$ is

$$R_{start} = \frac{V_L}{\sqrt{3}I_{start}} \cos\varphi_{start} = \frac{440}{\sqrt{3} \cdot 977} \cdot 0.3 = 78.097 \cdot 10^{-3}\,\Omega \tag{8.12}$$

$$X_{start} = \frac{V_L}{\sqrt{3}I_{start}} \sin\varphi_{start} = \frac{440}{\sqrt{3} \cdot 977} \cdot \sqrt{1-0.3^2} = 0.24833\,\Omega \tag{8.13}$$

The actual starting current in the motor/transformer I'_{start} is

$$\begin{aligned} I'_{start} &= \frac{V_L}{\sqrt{3}\left|R_{SC} + R_{start} + j(X_{SC} + X_{start})\right|} \\ &= \frac{440}{10^{-3}\sqrt{3}\left|3.3188 + 78.097 + j(10.5542 + 248.33)\right|} \\ &= 937[0.3 + j0.954] \end{aligned} \tag{8.14}$$

The voltage at motor terminal is

$$\frac{V'_L}{V_L} = \frac{I'_{start}}{I_{start}} = \frac{937}{977} = 0.959 \tag{8.15}$$

Consequently, the voltage reduction $\frac{\Delta V_L}{V_L}$ is

$$\frac{\Delta V_L}{V_L} = \frac{V_L - V'_L}{V_L} = 1.0 - 0.959 = 0.04094 \tag{8.16}$$

A 4.1% voltage reduction qualifies the local power grid as stiff for the starting of the newly purchased motor. The starting current is reduced from 977 A to 937 A, while the starting torque is reduced $\left(\frac{V'_L}{V_L}\right)^2$ times. That is $0.959^2 \approx 0.9197$ times. A smaller KVA transformer and/or a larger short-circuit transformer voltage would result in a larger voltage reduction during motor starting, which may jeopardize a safe start under heavy load.

Notice that high-efficiency motors have larger starting currents so the voltage reduction becomes more severe. Larger transformer KVA is required in such cases.

8.1.2 Autotransformer Starting

Although the induction motor power for direct starting has increased lately, for large induction motors (MW range) and not so stiff local power grids voltage, reductions below 10% are not acceptable for the rest of the loads and, thus, starting current reduction is mandatory. Unfortunately, in a cage rotor motor, this means a starting torque reduction, so reducing the stator voltage will reduce the stator current by K_i times and the torque by K_i^2

$$K_i = \frac{V_L}{V_L'} = \frac{I_L}{I_L'} = \sqrt{\frac{T_e}{T_e'}}; \quad I_s = \frac{\frac{V_L}{\sqrt{3}}}{|\underline{Z}_e(S)|}, \tag{8.17}$$

because the current is proportional to voltage and the torque with voltage squared.

Autotransformer voltage reduction is thus adequate only with light starting loads (fan, ventilator, and pump loads).

A typical arrangement with three-phase power switches is shown in Figure 8.2.

Before starting, C_4 and C_3 are closed. Finally, the general switch C_1 is closed and thus the induction motor is fed through the autotransformer at the voltage

$$\frac{V_L'}{V_L} \cong 0.5, 0.65, 0.8$$

To avoid large switching current transients when the transformer is bypassed and to connect the motor to the power grid directly, first C_4 is opened and C_2 is closed with C_3 still closed. Finally, C_3 is opened. The transition should occur after the motor accelerated to almost final speed or after a given time interval for start, based on experience at the commissioning place. Autotransformers are preferred due to their smaller size, especially with $\frac{V_L'}{V_L} = 0.5$ when the size is almost halved.

8.1.3 Wye-Delta Starting

In induction motors that are designed to operate with delta stator connection, it is possible, during starting, to reduce the phase voltage by switching to wye connection (Figure 8.3).

During wye connection, the phase voltage V_s becomes

$$V_S = \frac{V_L}{\sqrt{3}} \tag{8.18}$$

So the phase current, for same slip, I_{sY}, is reduced $\sqrt{3}$ times

$$I_{sY} = \frac{I_{s\Delta}}{\sqrt{3}} \tag{8.19}$$

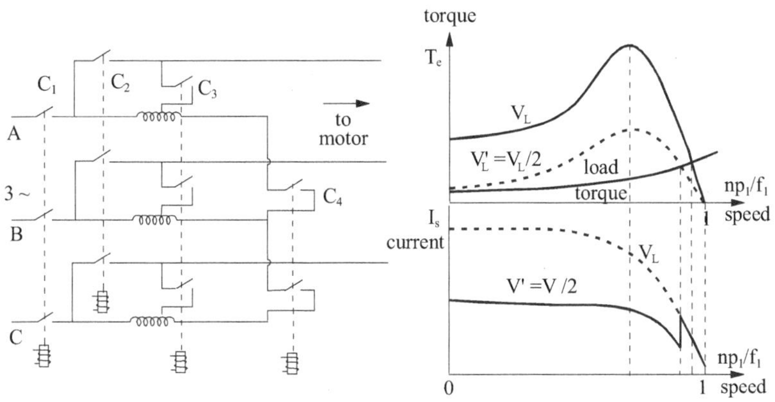

FIGURE 8.2 Autotransformer starting.

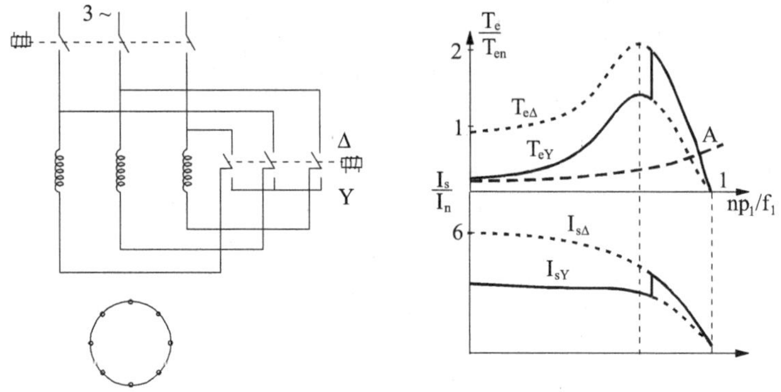

FIGURE 8.3 Wye-delta starting.

Now the line current in Δ connection $I_{l\Delta}$ is

$$I_{l\Delta} = \sqrt{3} I_{s\Delta} = 3 I_{sY} \tag{8.20}$$

so the line current is three times smaller for wye connection. The torque is proportional to phase voltage squared

$$\frac{T_{e\lambda}}{T_{e\Delta}} = \left(\frac{V_{sY}}{V_L}\right)^2 = \frac{1}{3} \tag{8.21}$$

Therefore, the wye-delta starting is equivalent to $\frac{\sqrt{3}}{1}$ reduction of phase voltage and a 3 to 1 reduction in torque. Only low-load torque (at low speeds) and mildly frequent start applications are adequate for this method.

A double-throw three-phase power switch is required and notable transients are expected to occur when switching from wye to delta connection takes place.

An educated guess in starting time is used to figure out when switching is to take place.

The series resistance and reactance starting methods behave in a similar way as voltage reduction in terms of current and torque. However, they are not easy to build, especially in high voltage (2.3, 4, and 6 kV) motors. At the end of starting process, they have to be short-circuited. With the advance of softstarters, such methods are used less and less frequently.

8.1.4 Softstarting

We use here the term softstarting for the method of A.C. voltage reduction through A.C. voltage controllers called softstarters.

In numerous applications such as fans, pumps, or conveyors, softstarters are now common practice when speed control is not required.

Two basic symmetric softstarter configurations are shown in Figure 8.4. They use thyristors and enjoy natural commutation (from the power grid). Consequently, their cost is reasonably low to be competitive.

In small (sub-kW) power motors, the antiparallel thyristor group may be replaced by a triac to cut costs.

Connection (a) in Figure 8.4 may also be used with delta connection, and this is why it became a standard in the marketplace.

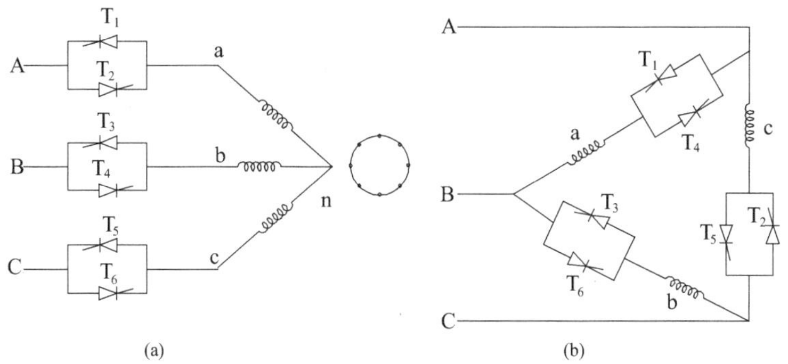

FIGURE 8.4 Softstarters for three-phase induction motors: (a) wye connection and (b) delta connection.

Connection (b) in Figure 8.4 reduces the current rating of the thyristor by $\sqrt{3}$ in comparison with connection (a). However, the voltage rating is basically the same as the line voltage and corresponds to a faulty condition when thyristors in one phase remain on, while all other phases would be off.

To apply connection (b), both phase ends should be available, which is not the case in many applications.

The six thyristors in Figure 8.4b are turned on in the order T_1, T_2, T_3, T_4, T_5, and T_6 every 60°. The firing angle α is measured with respect to zero crossing of V_{an} (Figure 8.5). The motor power factor angle is φ_{10}.

The stator current is continuous if $\alpha < \varphi_1$ and discontinuous (Figure 8.5) if $\alpha > \varphi_1$.

As the motor power factor angle varies with speed (Figure 8.5), care must be exercised to keep $\alpha > \varphi_1$ as a current (and voltage) reduction for starting is required.

So, besides voltage, current fundamentals, and torque reductions, the softstarters produce notable voltage and current low-order harmonics. Those harmonics pollute the power grid and produce additional losses in the motor. This is the main reason why softstarters do not produce notable energy savings when used to improve performance at low loads by voltage reduction [3].

However, for light-load starting, they are acceptable as the start currents are reduced. The acceleration into speed time is decreased, but it may be programmed (Figure 8.6).

During starting, either the stator current or the torque may be controlled. After the start ends, the softstarter will be isolated and a bypass power switch takes over. In some implementations, only a short-circuiting (bypass) power switch is closed after the start ends and the softstarter command circuits are disengaged (Figure 8.7). Dynamic braking is also performed with softstarters. The starting current may be reduced to twice rated current or more.

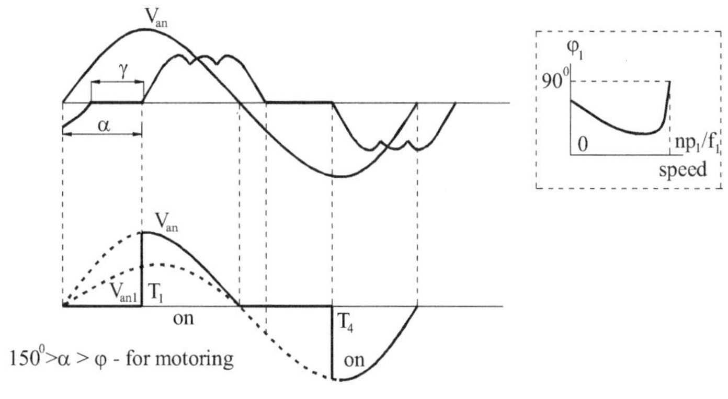

FIGURE 8.5 Softstarter phase voltage and current.

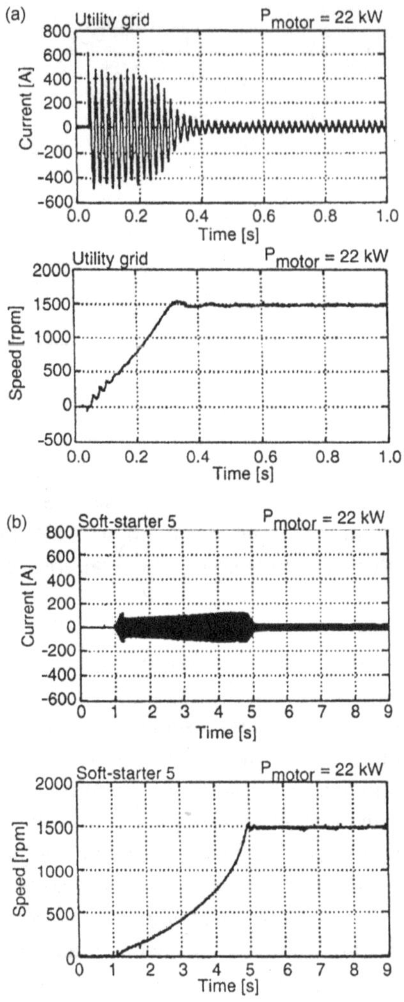

FIGURE 8.6 Start under no load of a 22 kW motor (a) direct starting and (b) softstarting.

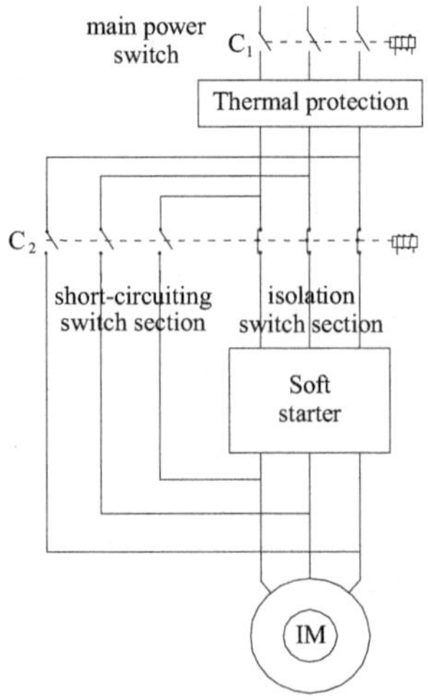

FIGURE 8.7 Softstarter with isolation and short-circuiting power switch C_2.

8.2 STARTING OF WOUND-ROTOR INDUCTION MOTORS

A wound-rotor induction motor is built for heavy-load frequent starts and (or) limited speed control motoring or generating at large power loads (above a few hundred kW).

Here we insist on the classical starting method used for such motors: variable resistance in the rotor circuit (Figure 8.8). As discussed in the previous chapter, the torque/speed curve changes with additional rotor resistance in terms of critical slip S_k, while the peak (breakdown) torque remains unchanged (Figure 8.8b).

$$S_K = \frac{C_1\left(R_r' + R_{ad}\right)}{\sqrt{R_s^2 + \left(X_{sl} + C_1 X_{rl}'\right)^2}}$$
$$T_{ek} = \frac{3p_1}{2C_1}\frac{V_s^2}{\omega_1}\cdot\frac{1}{R_1 \pm \sqrt{R_s^2 + \left(X_{sl} + C_1 X_{rl}'\right)^2}} \quad (8.22)$$

As expected, the stator current, for given slip, also decreases with increasing R'_{ad} (Figure 8.8c). This way, it is possible to start (S = 1) with peak torque by providing $S_K = 1.0$. While R'_{ad} increases, the power factor, especially at high slips, improves the torque/current ratio. The additional losses in R'_{ad} make this method poor in terms of energy conversion for starting or sustained low-speed operation.

However, the peak torque at start is an extraordinary feature for heavy starts, and this feature has made the method popular for driving elevators or overhead cranes.

There a few ways to implement the variable rotor resistance method as shown in Figure 8.9a–c.

The half-controlled rectifier and the diode-rectifier static-switch methods (Figure 8.8a and b) allow for continuous stator (or rotor) current close loop control during starting. Also, only a fixed resistance is needed.

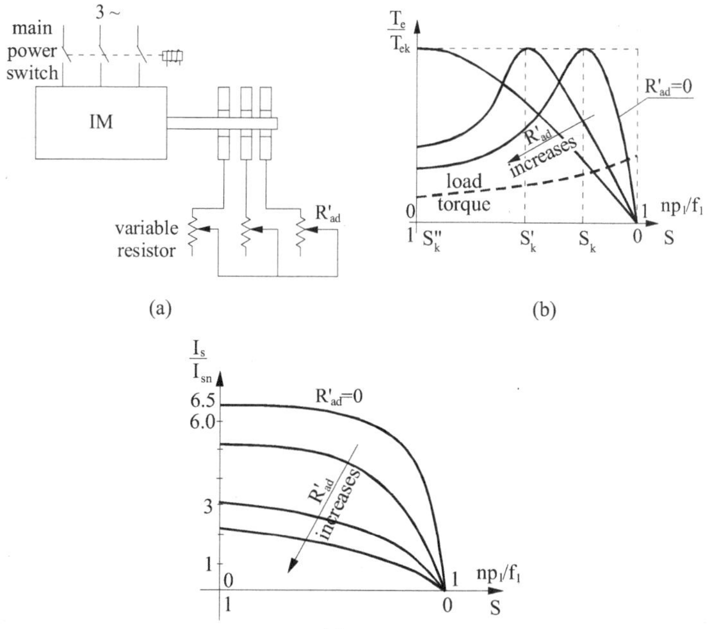

FIGURE 8.8 Starting through additional rotor resistance R'_{ad}: (a) general scheme, (b) torque/speed, and (c) current/speed.

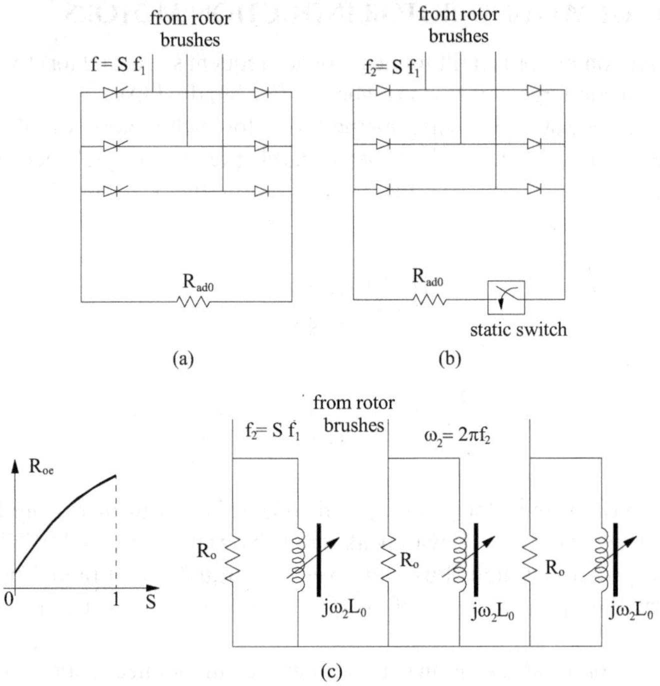

FIGURE 8.9 Practical implementations of additional rotor resistance: (a) with half-controlled rectifier, (b) with diode rectifier and static switch, and (c) with self-adjustable resistance.

The diode-rectifier static-switch method implies a better power factor and lower stator current harmonics but is slightly more expensive.

A low cost solution is shown in Figure 8.9c, where a three-phase pair of constant resistances and inductances leads to an equivalent resistance R_{oe} (Figure 8.9c), which increases when slip increases (or speed decreases).

The equivalent reactance of the circuit decreases with slip increases. In this case, there is no way to intervene in controlling the stator current unless the inductance L_0 is varied through a D.C. coil which controls the magnetic saturation in the coil-laminated magnetic core.

8.3 SPEED CONTROL METHODS FOR CAGE-ROTOR INDUCTION MOTORS

Speed control means speed variation under open loop (feedforward) or close loop conditions. We will discuss here only the principles and steady-state characteristics of speed control methods.

For cage-rotor induction motors, all speed control methods have to act on the stator windings as only they are available.

To start with, here is the speed/slip relationship.

$$n = \frac{f_1}{p_1}(1 - S) \tag{8.23}$$

Equation (8.23) suggests that the speed may be modified through

- Slip S variation: through voltage reduction
- Pole number $2p_1$ change: through pole-changing windings
- Frequency f_1 control: through frequency converters.

8.3.1 The Voltage Reduction Method

When reducing the stator-phase (line) voltage through an autotransformer or an A.C. voltage controller as inferred from (8.22), the critical slip s_K remains constant, but the peak (breakdown) torque varies with voltage V_s squared (Figure 8.10).

$$T_e = f(S) \cdot V_s^2 \tag{8.24}$$

The speed may be varied downwards until the critical slip speed n_K is reached.

$$n_K = \frac{f_1}{p_1}(1 - S_K) \tag{8.25}$$

In standard motors, this means an ideal speed control range of

$$(n_1 - n_k)/n_1 = S_K < 0.1 = 10\% \tag{8.26}$$

On the contrary, in high-resistance rotor motors, such as solid rotor motors where the critical slip is high, the speed control range may be as high as 100%.

However, in all cases, when increasing the slip, the rotor losses increase accordingly, so the wider the speed control range, the poorer the energy conversion ratio.

Finally, A.C. voltage controllers (softstarters) have been proposed to reduce voltage, when the load torque decreases, to reduce the flux level in the machine and thus reduce the core losses and increase the power factor and efficiency while stator current also decreases.

The slip remains around the rated value. Figure 8.11 shows a qualitative illustration of the above claims.

The improvement in performance at light loads, through voltage reduction, tends to be larger in low-power motors (less than 10 kW) and lower for larger power levels [3]. In fact, below 50% load, the overall efficiency decreases due to significant softstarter losses.

For motor designs dedicated to long light-load operation periods, the efficiency decreases only 3%–4% from 100% to 25% load and, thus, reduced voltage by softstarters does not produce significant performance improvements. In view of the above, voltage reduction has very limited potential for speed control.

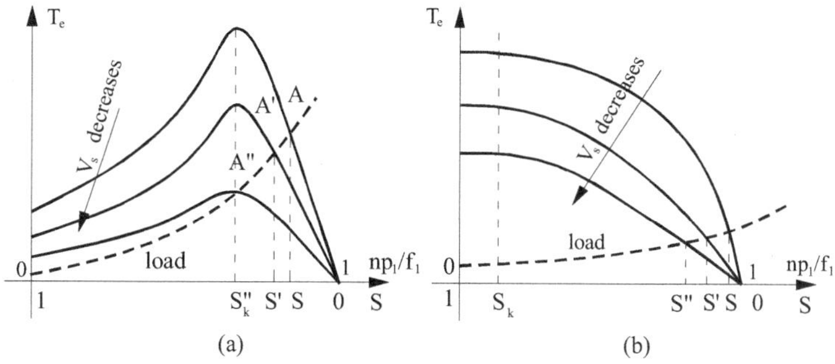

FIGURE 8.10 Torque versus speed for different voltage levels V_s (a) standard motor: $S_K = 0.04 - 0.10$ and (b) high-resistance (solid iron) rotor motor $S_K > 0.7 - 0.8$.

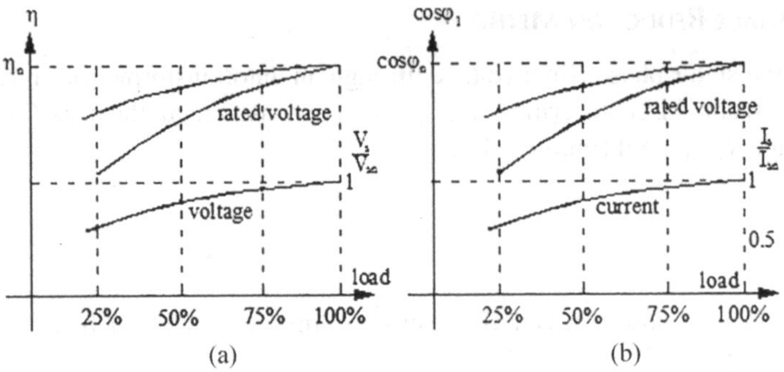

FIGURE 8.11 Performance versus load for reduced voltage (a) voltage V_s/V_{sn} and efficiency η and (b) cos φ_1 and stator current.

8.3.2 The Pole-Changing Method

Changing the number of poles, $2p_1$, changes the ideal no-load speed $n_1 = f_1/p_1$ accordingly. In Chapter 4, we discussed pole-changing windings and their connections of phases to produce constant power or constant torque for the two different pole numbers $2p_2$ and $2p_1$ (Figure 8.12).

The IM has to be sized carefully for the largest torque conditions and with careful checking for performance for both $2p_1$ and $2p_2$ poles.

Switching from $2p_2$ to $2p_1$ and back in standard pole-changing ($p_2/p_1 = 2$ Dahlander) windings implies complicated electromechanical power switches.

For better performance, new pole-changing windings (Chapter 4) that require only two single-throw power switches have been proposed recently.

For applications where the speed ratio is 3/2, 4/3, 6/4, etc. and the power drops dramatically for the lower speed (wind generators), dual windings may be used. The smaller power winding will

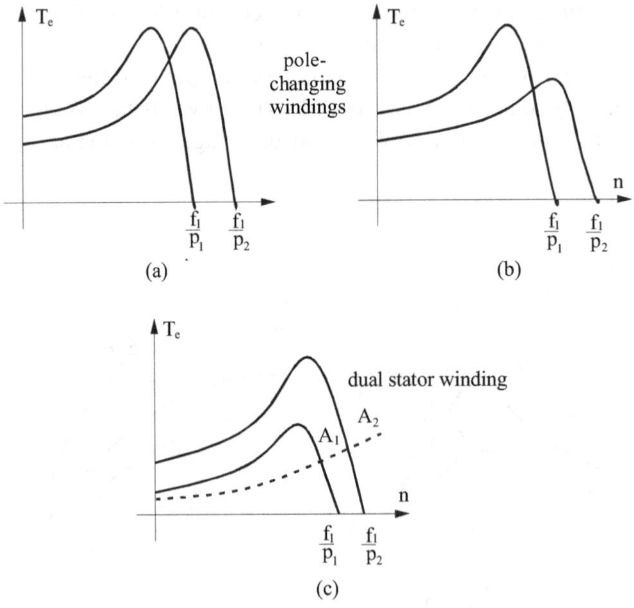

FIGURE 8.12 Pole-changing torque/speed curves (a) constant torque, (b) constant power, and (c) dual winding.

occupy only a small part of slot area. Again, only two power switches are required, the second one of notably smaller power rating.

Pole-changing windings are also useful for wide speed range $\omega_{max}/\omega_b > 3$ power induction motor drives (spindle drives or electric (or hybrid) automobile electric propulsion). This solution is a way to reduce motor size for $\omega_{max}/\omega_b > 3$.

8.4 VARIABLE FREQUENCY METHODS

When changing frequency f_1, the ideal no-load speed $n_1 = f_1/p_1$ changes and so does the motor speed for a given slip.

Frequency static converters are capable of producing variable voltage and frequency, V_s and f_1. A coordination of V_s with f_1 is required.

Such a coordination may be "driven" by an optimisation criterion or by flux linkage control in the machine to secure fast torque response.

The various voltage–frequency relationships may be classified into four main categories:

- V/f scalar control
- Rotor flux vector control
- Stator flux vector control
- Direct torque and flux control (DTFC).

Historically, the V/f scalar control was first introduced and is used widely today for open loop speed control in driving fans, pumps, etc., which have the load torque dependent on speed squared or more. The method is rather simple, but the torque response tends to be slow.

For high-torque response performance, separate flux and torque control, much like in a D.C. machine, is recommended. This is called vector control.

Either rotor or stator flux control is performed. In essence, the stator current is decomposed into two components. One is flux producing while the other one is torque producing. This time the current or voltage phase and amplitude and frequency are continuously controlled. DTFC [2] shows similar performance.

Any torque/speed curve could thus be obtained as long as voltage and current limitations are met. Also very quick torque response, as required in servo drives, is typical for vector control and DTFC.

All these technologies are now enjoying very dynamic markets worldwide.

8.4.1 V/f Scalar Control Characteristics

The frequency converter, which supplies the motor, produces sinusoidal symmetrical voltages whose frequency is ramped for starting. Their amplitude is related to frequency by a certain relationship of the form

$$V = V_0 + K_0(f_1)\cdot f_1 \tag{8.27}$$

V_0 is called the voltage boost destined to cover the stator resistance voltage at low frequency (speed).

Rather simple $K_0(f_1)$ functions are implemented into digitally controlled variable frequency converters (Figure 8.13).

As seen in Figure 8.13(a), a slip frequency compensator may be added to reduce speed drop with load (Figure 8.14).

Safe operation is provided above $f_{1min} = 3\,Hz$ as torque response is rather slow (above 20 ms).

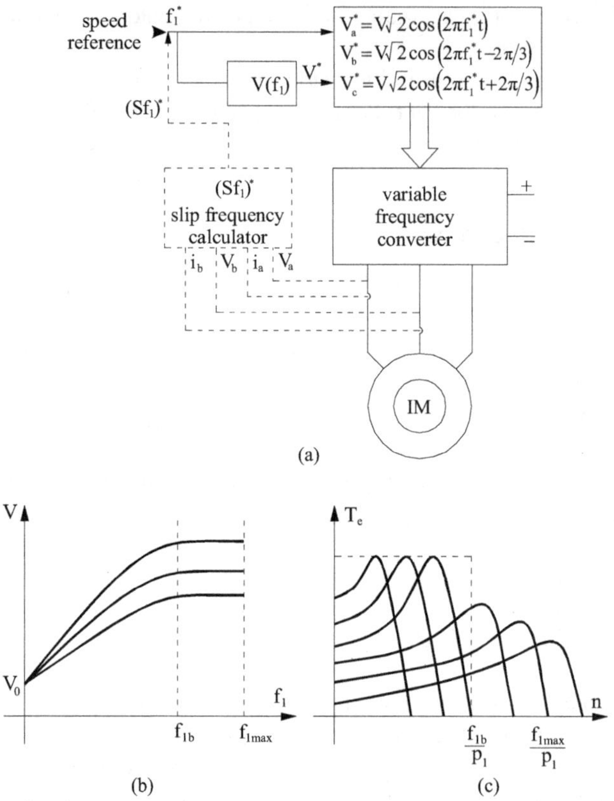

FIGURE 8.13 V/f speed control: (a) structural diagram, (b) V/f relationship, and (c) torque/speed curve.

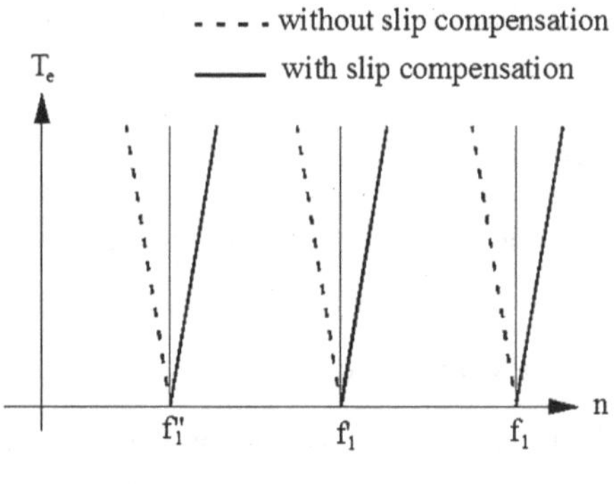

FIGURE 8.14 Torque/speed curves with slip compensation.

Example 8.2 V/f Speed Control

An induction motor has the following design data: $P_n = 10\,kW$, $V_{Ln} = 380\,V$ (Y), $f_{1b} = 50\,Hz$, $\eta_n = 0.92$, $\cos\varphi_n = 0.9$, $2p_1 = 4$, $I_{start}/I_n = 6/1$, $I_0/I_{sn} = 0.3$, $p_{mec} = 0.015P_n$, and $p_{add} = 0.015P_n$; core losses are neglected and $R_s = R'_r$ and $X_{sl} = X'_{rl}$. Such data are known from the manufacturer. Let us calculate rated current; motor parameters R_s, X_{sl}, and X_{1m}; critical slip S_K; and breakdown torque T_{eK} at f_{1b} and $f_{1max} = 100\,Hz$ for rated voltage; voltage for critical slip S_K; and minimum frequency $f_{1min} = 3\,Hz$ to provide rated breakdown torque.

Find the voltage boost V_0 for linear V/f dependence up to base speed.

Solution

$f_{1b} = 50\,Hz$

Based on the efficiency expression,

$$\eta_n = \frac{P_n}{\sqrt{3} V_L I_{sn} \cos\varphi_n} \tag{8.28}$$

The rated current

$$I_{sn} = \frac{10,000}{\sqrt{3}\cdot 0.92\cdot 380\cdot 0.9} = 18.37\ \text{A}$$

The rotor and stator winding losses $p_{cos} + p_{cor}$ are

$$p_{cos} + p_{cor} = \frac{P_n}{\eta_n} - P_n - p_{mec} - p_{add} \tag{8.29}$$

$$p_{cos} + p_{cor} = 10,000\left(\frac{1}{0.92} - 1 - 0.015 - 0.015\right) = 569.56\ \text{W}$$

Now,

$$p_{cos} + p_{cor} = 3R_s I_{sn}^2 + 3R_r I_{rn}'^2 \tag{8.30}$$

$$I_{rn}' \approx \sqrt{I_{sn}^2 - I_{0n}^2} = 18.37\cdot\sqrt{1 - 0.3^2} = 17.52\ \text{A} \tag{8.31}$$

From (8.30) and (8.31)

$$R_s = R_r' = \frac{569.56}{3\cdot\left(18.37^2 + 17.52^2\right)} = 0.4316\ \Omega$$

Neglecting the skin effect,

$$X_{sc} = X_{sl} + X_{rl}' \approx \sqrt{\left(\frac{V_{sn}}{I_{start}}\right)^2 - \left(R_s + R_r'\right)^2} = \sqrt{\left(\frac{380/\sqrt{3}}{6\cdot 18.37}\right)^2 - 0.8632^2} = 1.80\ \Omega \tag{8.32}$$

Therefore,

$$X_{sl} = X_{rl}' = 0.9\ \Omega$$

The critical slip S_K (8.22) and breakdown torque T_{eK} (8.22) are, for $f_1 = f_{1b} = 50\,Hz$,

$$(S_K)_{50\,Hz} = \frac{R_r'}{\sqrt{R_s^2 + X_{sc}^2}} = \frac{0.4316}{\sqrt{0.4316^2 + 1.8^2}} = 0.233 \tag{8.33}$$

$$(T_{eK})_{50\,Hz} = \frac{3p_1}{2}\frac{\left(V_{Ln}/\sqrt{3}\right)^2}{2\pi f_{1b}}\cdot\frac{1}{R_s + \sqrt{R_s^2 + X_{sc}^2}}$$

$$= 3\cdot\frac{2}{2}\cdot\frac{220^2}{2\pi 50}\cdot 0.4378 = 202.44\ \text{Nm} \tag{8.34}$$

for $f_{1max} = 100\,\text{Hz}$,

$$(s_K)_{100\,Hz} = \frac{R_r'}{\sqrt{R_s^2 + X_{sc}^2 \cdot \left(\frac{f_{1max}}{f_{1b}}\right)^2}} = \frac{0.4316}{\sqrt{0.4316^2 + 1.8^2 \cdot \left(\frac{100}{50}\right)^2}} = 0.119 \tag{8.35}$$

$$(T_{eK})_{100Hz} = \frac{3p_1}{2}\frac{\left(V_{Ln}/\sqrt{3}\right)^2}{2\pi f_{1max}} \cdot \frac{1}{R_s + \sqrt{R_s^2 + \left(X_{sc} \cdot \frac{f_{1max}}{f_{1b}}\right)^2}}$$

$$= 3 \cdot \frac{2}{2} \cdot \frac{220^2}{2\pi 100} \cdot \frac{1}{0.4316 + \sqrt{0.4316^2 + 1.8^2\left(\frac{100}{50}\right)^2}} = 113.97\text{ Nm} \tag{8.36}$$

$$(S_K)_{f1min} = \frac{R_r'}{\sqrt{R_s^2 + \left(X_{sc} \cdot \frac{f_{1min}}{f_{1b}}\right)^2}} = \frac{0.4316}{\sqrt{0.4316^2 + \left(1.8 \cdot \frac{3}{50}\right)^2}} = 0.97! \tag{8.37}$$

From (8.36), written for f_{1min} and $(T_{eK})_{50Hz}$, the stator voltage $(V_s)_{3Hz}$ is

$$(V_s)_{3Hz} = \sqrt{\frac{2(T_{eK})_{50\,Hz} \cdot 2\pi f_{1min}\left(\sqrt{R_s^2 + \left(X_{sc} \cdot \frac{f_{1min}}{f_{1b}}\right)^2} + R_s\right)}{3 \cdot p_1}}$$

$$= \sqrt{\frac{2 \cdot 202.44 \cdot 2\pi \cdot 3}{3 \cdot 2}\left(\sqrt{0.4316^2 + \left(1.8 \cdot \frac{3}{50}\right)^2} + 0.4316\right)} = 33.38\text{ V} \tag{8.38}$$

And, according to (8.27), the voltage increases linearly with frequency up to base (rated) frequency $f_{1b} = 50\,\text{Hz}$,

$$\begin{aligned} 33.38 &= V_0 + K_0 \cdot 3 \\ 220 &= V_0 + K_0 \cdot 50 \end{aligned} \tag{8.39}$$

$$K_0 = 3.97\text{ V/Hz};\ V_0 = 21.47\text{ V}$$

The results are synthesised in Figure 8.15. Towards minimum frequency $f_{1min} = 3\,\text{Hz}$, neglecting magnetising reactance branch in the critical slip calculation may produce notable errors.

A check of stator current and torque at $(S_K)_{3Hz} = 0.97$, $V_{sn} = 33.38\text{V}$, $f_{1min} = 3\,\text{Hz}$ is recommended. This suggests that assigning a value for voltage boost V_0 is a sensitive issue.

Any variation of parameters due to temperature (resistances) and magnetic saturation (inductances) may lead to serious stability problems. This is the main reason why V/f control method, though simple, is to be used only with light-load start applications or with stabilising loops.

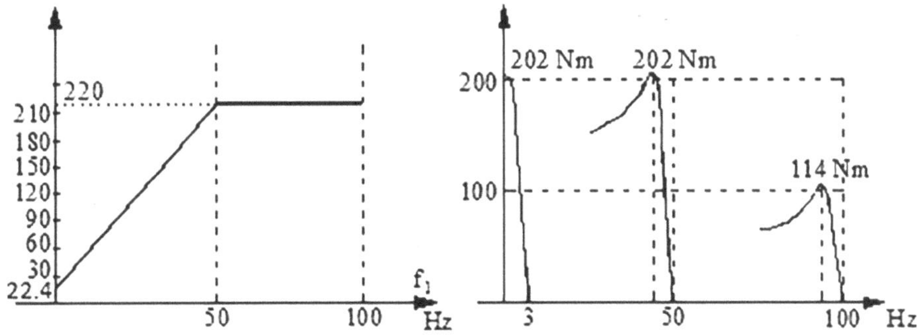

FIGURE 8.15 V/f and peak torque for V/f control.

8.4.2 Rotor Flux Vector Control

As already mentioned, when firm starting or stable low-speed performance is required, vector control is needed. To start with, let us reconsider the IM equations for steady state (Chapter 7, Section 7.10, Equations (7.97) and (7.98)).

$$\begin{aligned} &\underline{I}_s R_s - V_s = -j\omega_1 \underline{\Psi}_s \\ &\underline{I}_r' R_r' = -j\omega_1 S \underline{\Psi}_r \end{aligned} \quad \text{for } V_r' = 0 \text{ (cage rotor)} \tag{8.40}$$

Also from (7.95),

$$\underline{\Psi}_s = \frac{L_{1m}}{L_r'} \cdot \underline{\Psi}'_r + L_{sc}\underline{I}_s;\ \underline{I}_r' = \frac{\underline{\Psi}_r'}{L_r'} - \frac{L_{1m}}{L_r'} \cdot \underline{I}_s \tag{8.41}$$

The torque (7.100) is

$$T_e = 3p_1 \Psi_r' I_r' \tag{8.42}$$

It is evident in (8.40) that for steady state in a cage-rotor IM, the rotor flux and current per equivalent phase are phase shifted by $\pi/2$. This explains the torque formula (8.42) which is very similar to the case of a D.C. motor with separate excitation.

Separate (decoupled) rotor flux control represents the original vector control method [4].

Now from (8.41) and (8.40),

$$\underline{I}_s = \frac{\underline{\Psi}_r'}{L_{1m}} + jS\omega_1 \frac{L_r'}{R_r'} \cdot \frac{\underline{\Psi}_r'}{L_{1m}};\ T_r = L_r'/R_r' \tag{8.43}$$

or with Ψ_r' along real axis,

$$\underline{I}_s = I_M + jI_T;\quad I_M = \Psi_r/L_{1m};\quad I_T = jS\omega_1 T_r \cdot I_M \tag{8.44}$$

Equations (8.43) and (8.44) show that the stator current may be decomposed into two separate components, one, I_M, in phase with rotor flux $\underline{\Psi}_r$, called flux current, and the other, shifted ahead 90°, I_T, called torque current.

With (8.43) and (8.44), the torque equation (8.42) may be progressively written as

$$T_e = 3p_1 \frac{{L_{1m}}^2}{L_r'} \cdot I_M \cdot I_T = \frac{3p_1 \Psi_r'^2 S\omega_1}{R_r'} = 3p_1 (L_s - L_{sc}) \cdot I_M \cdot I_T \quad (8.45)$$

Consequently, for constant rotor flux, the torque/speed curve represents a straight line as for a separately excited D.C. motor (Figure 8.16).

As expected, keeping the rotor flux amplitude constant is feasible until the voltage ceiling in the frequency converter is reached. This happens above the base frequency f_{1b}. Above f_{1b}, Ψ_r has to be decreased, as the voltage is constant. Consequently, a kind of flux weakening occurs as in D.C. motors.

The IM torque–speed curve degenerates into V/f torque/speed curves above base speed.

As long as the rotor flux transients are kept at zero, even during machine transients, the torque expression (8.45) and rotor Equation (8.40) remain valid. This explains the fast torque response claims with rotor flux vector control. The bonus is that, for constant rotor flux, the mathematics to handle the control is the simplest. This explains its enormous commercial success.

A basic structural diagram for a rotor flux vector control is shown in Figure 8.17.

The rotor flux and torque reference values are used to calculate the flux and torque current components I_M and I_T as amplitudes. Then the slip frequency ($S\omega_1$) is calculated and added to the measured (or calculated on line) speed value ω_r to produce the primary reference frequency ω_1*. Its integral is the angle θ_1 of rotor flux position. With I_M, I_T, and θ_1, the three-phase reference currents i_a*, i_b*, and i_c* are calculated.

Then A.C. current controllers are used to produce a pulse-width modulation (PWM) strategy in the frequency converter to "copy" the reference currents. There are some delays in this "copying" process, but they are small, so fast response in torque is provided.

Three remarks are in order.

- To produce regenerative braking, it is sufficient to reduce the reference speed ω_r* below ω_r; I_T will become negative and so will be the torque.
- The calculation of slip frequency is heavily dependent on rotor resistance (T_r) variation with temperature.
- For low-speed good performance, the rotor resistance has to be corrected on line.

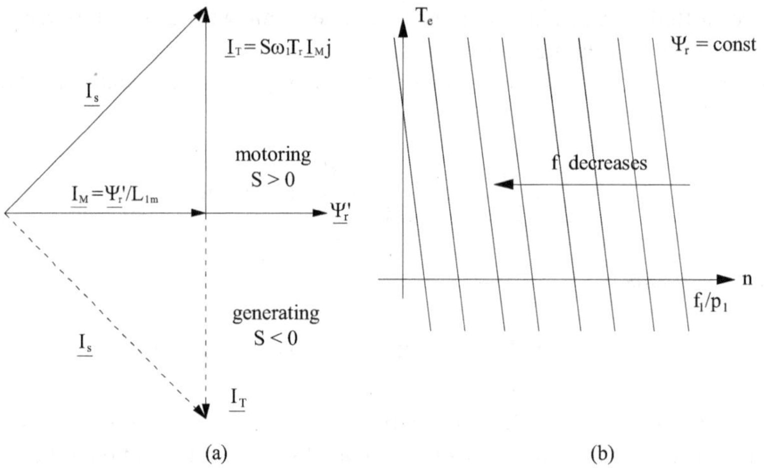

FIGURE 8.16 Rotor flux vector control: (a) stator current components and (b) torque versus speed curves for variable frequency f_1 at constant rotor flux.

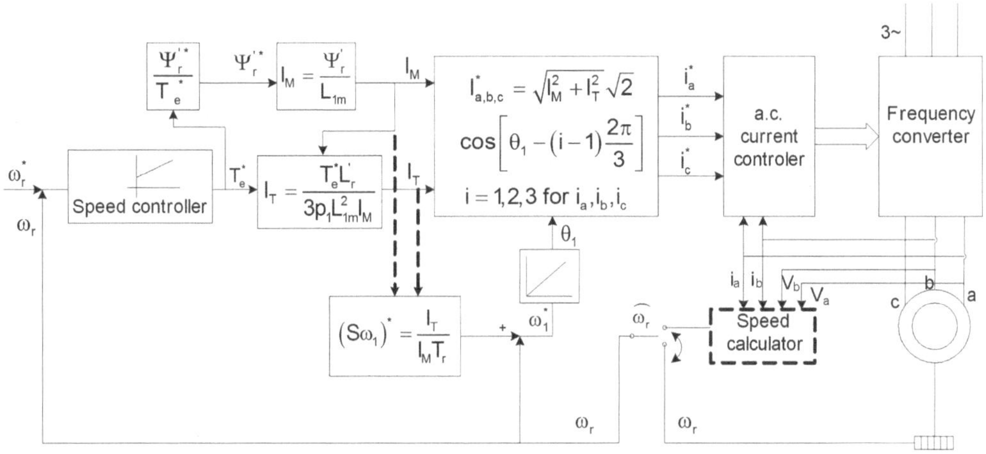

FIGURE 8.17 Basic rotor flux vector control system.

Example 8.3 Rotor Flux Vector Speed Control

For the induction motor in Example 8.2 with the data $2p_1 = 4$, $R_s = R_r = 0.4316\ \Omega$, $L_{sl} = L'_{rl} = \frac{X_{sc}}{2\cdot 2\pi\cdot f_{1b}} = \frac{1.8}{2\cdot 2\pi\cdot 50} = 2.866\cdot 10^{-3}$ H, and $L_{1m} \approx \frac{V_s}{I_{0m}\cdot 2\pi\cdot f_{1b}} - L_{sl} = \frac{220}{18.37\cdot 0.3\cdot 314} - 2.866\times 10^{-3} = 0.12427$ H, the rotor flux magnetising current $I_M = 6$ A and the torque current $I_T = 20$ A, for speed $n = 600$ rpm, calculate the torque, rotor flux Ψ_r, stator flux slip frequency $S\omega_1$, frequency ω_1, and voltage required.

Solution

The rotor flux Ψ'_r (8.44) is

$$\Psi'_r = L_{1m}\cdot I_M = 0.12427\cdot 6 = 0.74256 \text{ Wb}$$

Equation (8.45) yields

$$T_e = 3p_1\frac{L_{1m}^2}{L'_r} I_M\cdot I_T = 3\cdot 2\cdot\frac{0.12427^2}{2.866\times 10^{-3} + 0.12427}\cdot 6\times 20 = 87.1626 \text{ Nm}$$

The slip frequency is calculated from (8.44)

$$S\omega_1 = \frac{1}{T_r}\frac{I_T}{I_M} = \frac{20}{6}\cdot\frac{0.4316}{0.12417 + 2.866\times 10^{-3}} = 11.30 \text{ rad/s}$$

Now the frequency ω_1 is

$$\omega_1 = 2\pi n\cdot p_1 + S\omega_1 = 2\pi\cdot\frac{600}{60}\cdot 2 + 11.30 = 136.9 \text{ rad/s}$$

To calculate the voltage, we have to use Equations (8.40) and (8.41) progressively.

$$\underline{\Psi}_s = \frac{L_{1m}}{L'_r}\cdot L_{1m} I_M + L_{sc}\cdot(I_M + jI_T) = L_s I_M + jL_{sc} I_T \tag{8.46}$$

So the stator flux has two components, one produced by I_M through the no-load inductance $L_s = L_{sl} + L_{1m}$ and the other produced by the torque current through the short-circuit inductance L_{sc}.

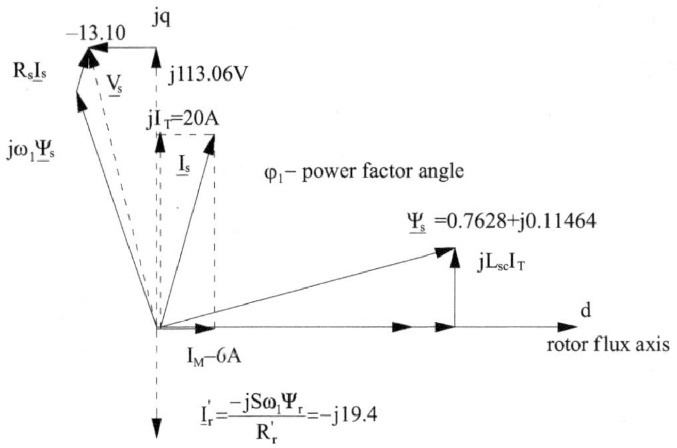

FIGURE 8.18 Phasor diagram with rotor and stator fluxes shown.

$$\underline{\Psi}_s = \left(2.866\times10^{-3} + 0.12427\right)\cdot 6 + j2\cdot 2.866\times10^{-3}\cdot 20 = 0.7628 + j0.11464$$

Now from (8.40):

$$\underline{V}_s = j\omega_1\underline{\Psi}_s + \underline{I}_s R_s = j136.9\cdot(0.7628 + j0.11464) + (6 + j20)\cdot 0.4316 = -13.10 + j113.06$$
$$V_s = \sqrt{13.10^2 + 113.06^2} = 113.81\text{ V} \tag{8.47}$$

The results in terms of fluxes and voltage are summarised in Figure 8.18.

For negative torque (regenerative braking), only I_T becomes negative and so does the slip frequency $S\omega_1$.

Let us further exploit the torque expression T_e which may be alternative to (8.42).

$$T_e = 3p_1 \,\text{Re}\left(j\underline{\Psi}_s \underline{I}_s^*\right) = 3p_1\left(L_s - L_{sc}\right) I_M \cdot I_T \tag{8.48}$$

Also the rotor current (8.38) is

$$I_r' = S\omega_1 \frac{L_{1m} I_M}{R_r'} \tag{8.49}$$

Torque equation (8.48) may be interpreted as pertaining to a reluctance synchronous motor with constant high magnetic saliency as $L_s/L_{sc} > 10$–20 in general.

This is true only for constant rotor flux.

The apparent magnetic saliency is created by the rotor current $\underline{I}_r'$ which is opposite to jI_T in the stator (Figure 8.18) to "kill" the flux in the rotor along axis q. The situation is similar with the short-circuited secondary effect on the transformer equivalent inductance.

Stator flux vector control may be treated in a similar way. For detailed information on advanced IM drives, see Refs. [2,5,6].

8.5 SPEED CONTROL METHODS FOR WOUND-ROTOR IMS

When the IMs are provided with a wound rotor, speed control is performed by

- Adding a variable resistor to the rotor circuit
- Connecting a power converter to the rotor to introduce or extract power from the rotor. Let us call this procedure as additional voltage to the rotor.

As the method of additional rotor resistance has been presented for starting, it will suffice to stress here once again that only for limited speed control range (say 10%) and medium power such a method is still used in some special applications. In the configuration of self-adjustable resistance, it may also be used for 10%–15% speed variation range in generators for wind power conversion systems.

In general, for this method, a unidirectional or a bidirectional power flow frequency converter is connected to the rotor brushes while the stator winding is fed directly, through a power switch, to the power grid (Figure 8.19).

Large power motors (2–3 MW to 15 MW) or very large power generators/motors, for pump-back hydropower plants (in the hundreds of MW/unit), are typical for this method.

A half-controlled rectifier-current source inverter has constituted for a long while one of the most convenient ways to extract energy from the rotor while the speed decreases to 75%–80% of its rated value. Such a drive is called a slip recovery drive, and it may work as a motor only below ideal no-load speed (S > 0).

8.5.1 Additional Voltage to the Rotor (the Doubly Fed Machine)

On the other hand, with bidirectional power flow, when the phase sequence of the voltages in the rotor may be changed, the machine may work both as a motor and as a generator, both below and above ideal no-load speed f_1/p_1.

With such indirect and direct frequency bidirectional converters, it is possible to keep the rotor slip frequency f_2 constant and adjust the rotor voltage phase angle δ with respect to stator voltage, after Park transformation to stator coordinates.

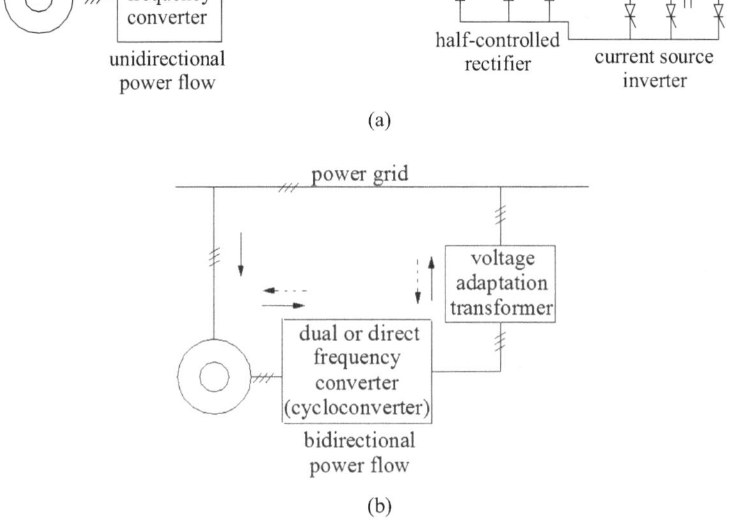

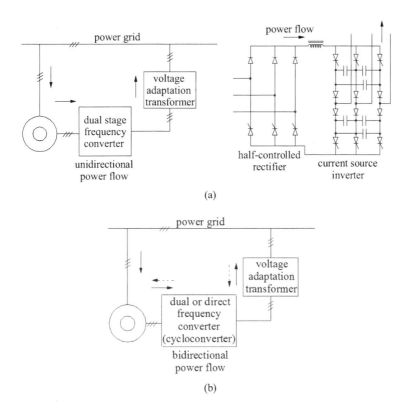

FIGURE 8.19 Additional voltage to the rotor: (a) with unidirectional power flow and (b) with bidirectional power flow.

The speed is constant under steady state and the machine behaves like a synchronous machine with the torque dependent on the power angle δ ($S = f_2/f_1 = ct.$)

On the other hand, the frequency f_2 may be varied with speed such that

$$f_2 = f_1 - np_1;\ S = f_2/f_1 \ - \text{variable} \tag{8.50}$$

In this case, the phase angle δ may be kept constant.

The rotor equation is

$$\underline{I}_r'R_r' + \underline{V}_r' = -jS\omega_1\underline{\Psi}_r' = S\underline{E}_r' \tag{8.51}$$

For $\underline{V}_r$ in phase (or in phase opposition, for $V_r' < 0$) with $\underline{E}_r'$

$$\underline{I}_r'R_r' = S\underline{E}_r' - \underline{V}_r' \tag{8.52}$$

For zero rotor current, the torque is zero. This corresponds to the ideal no-load speed (slip, S_0)

$$S_0E_r' - V_r' = 0;\ S_0 = \frac{V_r'}{E_r'} \tag{8.53}$$

S_0 may be positive if $V_r' > 0$, that is $\underline{V}_r'$ and $\underline{E}_r'$ are in phase while it is negative for $V_r' < 0$ or $\underline{V}_r'$ and $\underline{E}_r'$ in phase opposition.

For constant values of V_r' (+ or –), the torque–speed curves obtained by solving the equivalent circuit may be calculated as

$$T_e = \frac{P_{elm}p_1}{S\omega_1} = \frac{3p_1}{S\omega_1}\left[I_r'^2R_r' + V_r'I_r'\cos\theta_r'\right] \tag{8.54}$$

θ_r' is the angle between $\underline{V}_r'$ and I_r'. In our case, $\theta_r = 0$ and V_r' is positive or negative and given.

It is seen (Figure 8.20) that with such a converter, the machine works well as a motor above synchronous conventional speed f_1/p_1 and as a generator below f_1/p. This is true for voltage control. A different behaviour is obtained for constant rotor current control [7].

However, under subsynchronous motor and generator modes, all are feasible with adequate control [2].

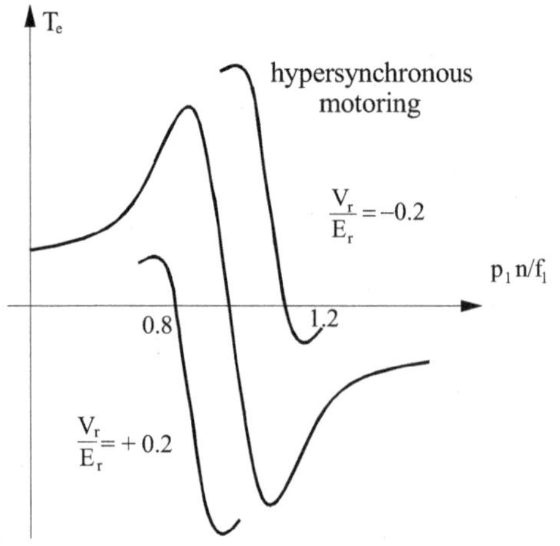

FIGURE 8.20 Torque–speed curves for $V_r/E_r = \pm 0.2$, $\theta_r' = 0$.

Example 8.4 Doubly Fed IM

An IM with wound rotor has the data: $V_{sn} = 220$ V/phase, $f_1 = 50$ Hz, $R_s = R'_s = 0.01\ \Omega$, $X_{sl} = X'_{rl} = 0.03\ \Omega$, $X_{1m} = 2.5\ \Omega$, $2p_1 = 4$, $V_{rn} = 300$ V/phase, and $S_n = 0.01$. Calculate

- The stator current and torque at $S_n = 0.01$ for the short-circuited rotor
- For the same stator current at n = 1200 rpm, calculate the rotor voltage V_r in phase with stator voltage V_{sn} (Figure 8.21) and the power extracted from the rotor by frequency converter.

Solution

The stator current (see Chapter 7) is

$$(I'_s)_{V'_r=0} \approx \frac{V_s}{\sqrt{\left(R_s + C_1\frac{R'_r}{S}\right)^2 + (X_{sl} + C_1X'_{rl})^2}}$$

$$= \frac{220}{\sqrt{\left(0.01 + 1.012\frac{0.01}{0.01}\right)^2 + (0.03 + 1.012\cdot 0.03)^2}} = 217.17\ \text{A}$$

with

$$C_1 = \frac{X_{ls}}{X_{1m}} + 1 = 1 + \frac{0.03}{2.5} = 1.012 \tag{8.55}$$

The actual rotor current I_r per phase is

$$I_r = I'_r \cdot \frac{V_s}{V_m} = 217.17 \cdot \frac{220}{300} \approx 152\ \text{A} \tag{8.56}$$

The torque T_e is

$$T_e = \frac{3R'_rI'^2_r}{S\omega_1}\cdot p_1 = \frac{3\cdot 0.01\cdot(217.172)^2}{0.01\cdot 2\pi\cdot 50}\cdot 2 = 901.2\ \text{Nm} \tag{8.57}$$

If V'_r is in phase with $\underline{V}_{sn}$, then

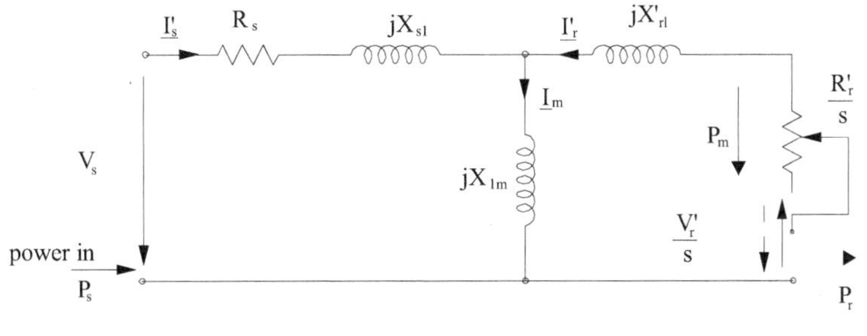

FIGURE 8.21 Equivalent circuit with additional rotor voltage, $\underline{V}'_r$.

$$(I'_r)_{S'} \approx I_s = \frac{V_s + V'_r/S'}{\sqrt{\left(R_s + C_1\frac{R'_r}{S'}\right)^2 + (X_{sl} + C_1 X'_{rl})^2}}$$

$$= \frac{220 + V'_r/0.2}{\sqrt{\left(0.01 + 1.012\frac{0.01}{0.2}\right)^2 + (0.03 + 1.012\cdot 0.03)^2}} = 217.17\text{ A} \quad (8.58)$$

$$S' = 1 - \frac{np_1}{f_1} = 1 - \frac{1200}{60}\cdot\frac{2}{50} = 0.2;\ f_2 = S'f_1 = 0.2\cdot 50 = 10\text{ Hz}$$

So $V'_r = -40.285$ V.

In this case, the angle between $\underline{V}'_r$ and $\underline{I}'_r$ corresponds to the angle $+\underline{V}'_s$ and $-\underline{I}'_s$ (from 8.58)

$$\theta'_r\left(\underline{V}'_r, \underline{I}'_r\right) \approx -45° + 180° = 135° \quad (8.59)$$

Now the torque (8.54) is

$$T_e = \frac{3\cdot 2}{0.2\cdot 2\pi\cdot 50}\left[0.01\cdot 217.17^2 + (-40.28)(217.17)(-0.707)\right] = 635.94\text{ Nm} \quad (8.60)$$

Note that the torque is larger for S = 0.01 with the short-circuited rotor.

The input active power P_s, electric power out of rotor P_r, and mechanical power P_m are

$$P_s = 3V_s I_s \cos\varphi_s = 3\cdot 220\times 217.17\times 0.707 = 101.335\times 10^3\text{ W}$$

$$P_r = 3V_r I_r \cos\theta_r = 3\cdot 40.38\times 217.17\cdot(-0.707) \approx -18.6\times 10^3\text{ W} \quad (8.61)$$

$$P_m = T_e\cdot 2\pi n = 635.94\cdot 2\pi\cdot 1200/60 = 79.874\times 10^3\text{ W}$$

The efficiency at 1200 rpm η is

$$\eta = \frac{P_m}{P_s + P_r} = \frac{79.874}{101.335 - 18.6} = 0.9654 \quad (8.62)$$

Note that core, mechanical, and stray load losses have been neglected.

$$-j\omega_1\underline{\Psi}_r = -\left(\frac{L'_r}{L_{1m}}\cdot\underline{\Psi}_s - L_{sc}\underline{I}_s\right)j\omega_1 = -\frac{L'_r}{L_{1m}}\left[\left(V_s - R_s\underline{I}_s\right) - jX_{sc}\underline{I}_s\right] = \underline{E}'_r \quad (8.63)$$

$$\underline{E}'_r = -\frac{(2.5 + 0.03)}{2.5}\left[220 - (217.17 - j217.17)0.707(0.01 + j0.06036)\right]$$

$$= -(209.49 - j7.74) \quad (8.64)$$

Now $\underline{I}'_r$ is recalculated from (8.51)

$$\underline{I}'_r = \frac{S\underline{E}'_r - \underline{V}'_r}{R'_r} = \frac{-0.2(211.767 - j7.74) - (-40.301)}{0.01}$$

$$= (-158.88 + j154.8) \approx -\underline{I}_s \quad (8.65)$$

As expected, both components of $\underline{I}'_r$ are close to those of (8.58).

When the rotor slip frequency f_2 is constant, the speed is constant so only the rotor voltage sequence, amplitude, and phase may be modified.

The phase sequence information is contained into S_2 sign:

$$S_2 = f_2/f_1 >< 0 \tag{8.66}$$

We may calculate the stator current $\underline{I}_s$ and the rotor current $\underline{I}_r'$ (reduced to the stator) from the equivalent circuit (Figure 8.21)

$$\underline{I}_s = \frac{\underline{V}_s\left(R_r'/S + jX_r'\right) - jX_{1m}\underline{V}_r'/S}{\left(R_s + jX_s\right)\left(R_r'/S + jX_r'\right) + X_{1m}^2}; \underline{X}_s = X_{sl} + X_{1m} \tag{8.67}$$

$$\underline{I}_r' = \frac{\underline{V}_r'\left(R_s + jX_s\right)/S - jX_{1m}\underline{V}_s}{\left(R_s + jX_s\right)\left(R_r'/S + jX_r'\right) + X_{1m}^2}; X_r' = X_{rl}' + X_{1m} \tag{8.68}$$

The stator (input) powers are

$$\begin{aligned} &P_s = 3\,\mathrm{Re}\left(\underline{V}_s\underline{I}_s^*\right); T_e \approx \left(P_s - 3R_sI_s^2\right)\frac{p_1}{\omega_1} \\ &Q_s = 3\,\mathrm{Imag}\left(\underline{V}_s\underline{I}_s^*\right) \end{aligned} \tag{8.69}$$

Similarly, the powers out of rotor are

$$\begin{aligned} &P_r = 3\,\mathrm{Re}\left(\underline{V}_r'\underline{I}_r'^*\right) \\ &Q_r = 3\,\mathrm{Imag}\left(\underline{V}_r'\underline{I}_r'^*\right) \end{aligned} \tag{8.70}$$

With S = constant, such a control method can work at constant rotor voltage V_r' but with variable phase γ (V_s, V_r'). Alternatively, it may operate at constant stator or rotor current [8–10]. Finally, vector control is also feasible [6]. Such schemes are currently proposed for powers above up to 300 MW in pump-storage power plants, where both for pumping and generating, a 10%–25% range in speed control improves the hydroturbine pump output.

The main advantage is the limited rating of the rotor-side frequency converter $S_r = S_{max} \cdot S_n$. As long as $S_{max} < 0.2$–0.15, the savings in converter costs are very good. Notice that resistive starting is required in such cases.

For medium- and lower-power limited speed control, dual stator winding stator nest cage rotor induction motors have also been proposed. The two windings have different pole numbers p_1 and p_2, and the rotor has a pole count $p_r = p_1 + p_2$. One winding is fed from the power grid at the frequency f_1 and the other at frequency f_2 from a limited power frequency converter. The machine speed n is

$$n = \frac{f_1 \pm f_2}{p_1 + p_2} \tag{8.71}$$

The smaller the f_2, the smaller the rating of the corresponding frequency converter. The behaviour is typical to that of a synchronous machine when f_2 is constant. Low-speed applications such as wind generators or some pump applications with low initial cost for limited speed control range (less than 20%) might be suitable for such configurations, with rather large rotor losses.

8.6 CONTROL BASICS OF DFIMS

A simplified diagram of grid-connected doubly fed induction machine (DFIM) with shunt-series connected grid side converter (GSC) is shown in Figure 8.22 [11].

Such a control system uses two back-to-back voltage source PWM converters (one rotor side connected and one GSC).

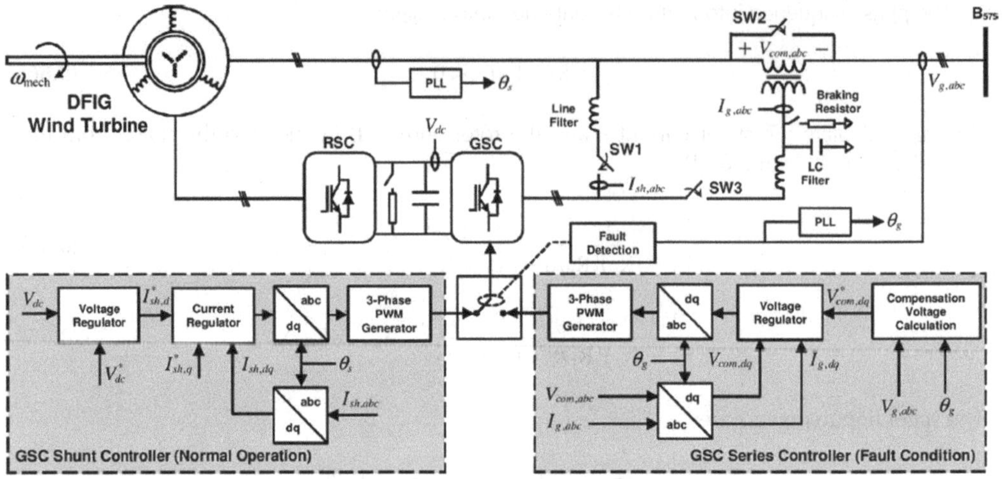

FIGURE 8.22 DFIG control system with shunt-series connected GSC, after [11].

In this implementation [11], the GSC may operate as usual (shunt connection) with SW3 open and SW1 and SW2 closed or in series connection (SW1 open, SW2 open, and SW3 closed) to compensate grid voltage sags when necessary.

Field-oriented control is performed on both converters (RSC and GSC). In Figure 8.22, only the GSC control is illustrated.

A typical RSC FOC is shown in Figure 8.23 [12].

The rotor active current reference is related to active stator power reference P_2^* and the rotor reactive current reference is dictated by the stator reactive power reference Q_s^* (which usually is set to zero). U_{rd}^* and U_{rq}^* are the d-q reference RSC voltages to be realised by open-loop PWM.

This chapter only referred to essentials of IM control; for details, books on Electric Drives are to be consulted [3, 5–6].

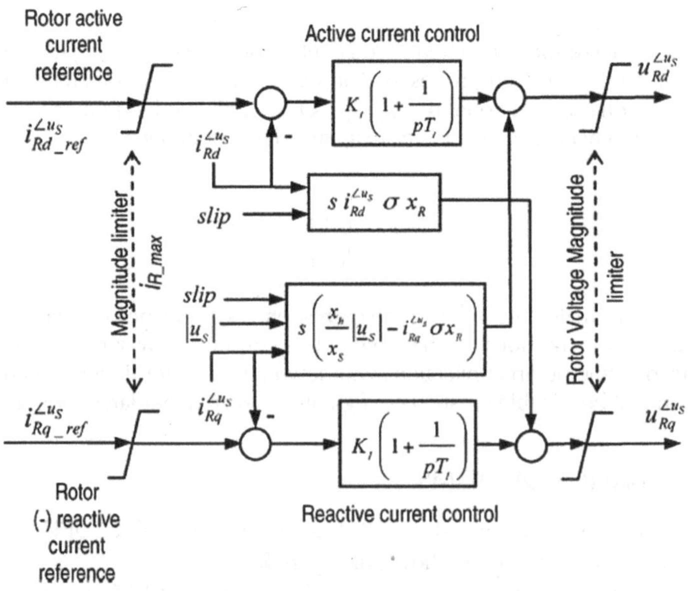

FIGURE 8.23 RSC FOC system (after [12]).

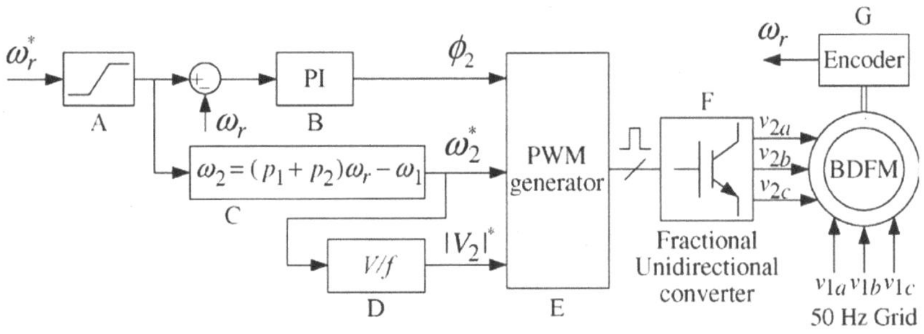

FIGURE 8.24 Scalar control of control-winding PWM converter in motoring DFIM (after [13]).

While the control systems in Figures 8.22 and 8.23 refer to the generator mode, motor mode control is similar and includes a self-starting sequence that observes the fractional kVA ratings of RSC and GSC.

Though the BDFIM refers mostly to generator mode, for the sake of variation, a simplified motoring scalar control system is illustrated in Figure 8.24 [13].

The machine may self-start as an induction motor with the control winding short-circuited and then self-synchronises at ω_r speed as in Equation (8.71).

The reference speed ω_r^* is controlled through a PI regulator that prescribes the control winding flux linkage ϕ_2 while ω_r^* is used, with ω_1 known, to calculate ω_2^* (the frequency in the control winding); finally, ω_2^* by integration gives the angle of the flux linkage and by V_2^*/f_2^* leads to control winding reference voltage V_2^* realised by the PWM generator.

8.7 SUMMARY

- Starting methods are related to IMs fed from the industrial power grid.
- With direct starting and stiff local power grids, the starting current is in the interval of 580%–650% (even higher for high-efficiency motors).
- For direct starting at no mechanical load, the rotor winding energy losses during machine acceleration equals the rotor-attached kinetic energy at no load.
- For direct frequent starting at no load, and same stator, rotors with higher rotor resistance lead to lower total energy input for acceleration.
- To avoid notable voltage sags during direct starting of a newly installed IM, the local transformer KVA has to be oversized accordingly.
- For light load, starting voltage reduction methods are used, as they reduce the line currents. However, the torque/current ratio is also reduced.
- Voltage reduction is performed through an autotransformer, wye/delta connection or through softstarters.
- Softstarters are now built to about 1 MW motors at reasonable costs as they are made with thyristors. Input current harmonics and motor additional losses are the main drawbacks of softstarters. However, they are continually being improved and are expected to be common practice in light-load starting applications (pumps, fans) where speed control is not required but soft (low current), slow but controlled, starting is required.
- Rotor resistive starting of wound-rotor IMs is traditional. The method produces up to peak torque at start, but at the expense of very large additional losses.
- Self-adjustable resistance–reactance paralleled pairs may also be used for the scope to cut the cost of controls.
- Speed control in cage-rotor IMs may be approached by voltage amplitude control, for a very limited range (up to 10%–15%, in general).

- Pole-changing windings in the stator can produce two-speed motors and are used in some applications, especially for low power levels.
- Dual stator windings and cage rotors can also produce two-speed operation efficiently if the power level for one speed is much smaller than for the other.
- Coordinated frequency/voltage speed control represents the modern solution to adjustable speed drives.
- V/f scalar control is characterised by an almost linear dependence of voltage amplitude on frequency; a voltage boost $V_0 = (V)_{f_1=0}$ is required to produce sufficient torque at lowest frequency $f_1 \approx 3\,\text{Hz}$. Slip feedforward compensation is added. Still, slow transient torque response is obtained. For pumps, fans, etc., such a method is, however, adequate. This explains its important market share.
- Rotor flux vector control keeps the rotor flux amplitude constant and requires the motor to produce two stator current components: a flux current I_M and a torque current I_T, 90° apart. These two components are decoupled to be controlled separately. Fast dynamics, quick torque availability, stable, high-performance drives are built and sold based on rotor flux vector control (FOC) or on other related forms of DTFC.
- Linear torque/speed curves, ideal for control, are obtained with rotor flux vector control.
- Flux/torque coordination through various optimisation criteria could be applied to cut energy losses or widen the torque–speed range.
- For constant rotor flux, the IM behaves like a high-saliency reluctance synchronous motor. However, the apparent saliency is produced by the rotor currents phase shifted by 90° with respect to rotor flux for cage rotors. As for the reluctance synchronous motor, a maximum ideal power factor for the lossless motor can be defined as

$$\cos\varphi_{max} = \frac{1 - L_{sc}/L_s}{1 + L_{sc}/L_s};\ L_s/L_{sc} > 15 \div 20 \tag{8.72}$$

  L_s – no-load inductance; L_{sc} – short-circuit inductance.
- Wound-rotor IM speed control is to be approached through frequency converters connected to the rotor brushes. The rotor converter rating is low. Considerable converter cost saving is obtained with limited speed control so characteristic to such drives. With adequate frequency converter, motoring and generating over or under conventional synchronous speed ($n_1 = f_1/p_1$) is possible. High and very high-power motor/generator systems are the main applications as separate active and reactive power control for limited speed variation at constant stator frequency is feasible. In fact, the largest electric motor/generator with no-load starting has been built for such purposes (for pump-storage power plants: 400 MW).
- Dual-stator windings, with $2p_1 \le 2p_1'$ nested-cage rotor IMs, with one winding grid fed and the other one inverter fed, to allow for limited speed range control, has also been proposed recently.
- Details on control of electric drives with IMs are to be found in the rich literature dedicated to this very dynamic field of engineering [2].

REFERENCES

1. W. Leonhard, *Control of Electric Drives*, 3rd edition, Springer Verlag, New York, 2001.
2. I. Boldea, S. A. Nasar, *Electric Drives*, 3rd edition, CRC Press, Taylor and Francis, New York, 2017.
3. F. Blaabjerg et al, Can softstarters save energy, *IEEE Industry Applications Magazine*, 1997, pp. 56–66.
4. F. Blaschke, The principle of field orientation as applied to the new transvector closed loop control system for rotating field machines, *Siemens Review*, Vol. 34, 1972, pp. 217–220 (in German).
5. D. W. Novotny, T. A. Lipo, *Vector Control and Dynamics of A.C. Drives*, Clarendon Press, Oxford, 1996.

6. I. Boldea, S. A. Nasar, *Vector Control of A.C. Drives*, CRC Press, Boca Raton, FL, 1992.
7. A. Masmoudi, M. Poloujadoff, Steady state operation of doubly fed synchronous machines under voltage and current control, *EMPS*, Vol. 27, 1999.
8. L. Schreir, M. Chomat, J. Bendl, Working regions of adjustable speeds unit with doubly fed machines, *Record of IEEE – IEMDC*, 1999, Seattle, USA, pp. 457–459.
9. L. Schreir, M. Chomat, J. Bendl, Analysis of working regions of doubly fed generators, *Record of ICEM*, 1998, Vol. 3, pp. 1892–1897.
10. O. Ojo, I. E. Davidson, PWM VSI inverter assisted standalone dual stator winding generator, *Record of IEEE – IAS, Annual Meeting*, 1996.
11. P. H. Huang, M. S. El Moursi, W. Xiao, J. L. Kirtley Jr, Novel fault ride-through configuration and transient management scheme for DFIG, *IEEE Transactions on Energy Conversion*, Vol. 28, No. 1, 2013, pp. 86–94.
12. S. Engelhardt, I. Erlich, C. Feltes, J. Kretschmann, F. Shewarega, Reactive power capability of wind turbines based on DFIGs, *IEEE Transactions on Energy Conversion*, Vol. 26, No. 1, 2011, pp. 364–372.
13. S. Shao, E. Abdi, R. McMahon, Low cost variable speed drive based on a brushless doubly-fed motor and a fractional unidirectional converter, *IEEE Transactions on Industrial Electronics*, Vol-59, No. 1, 2012, pp. 317–325.

9 Skin and On-Load Saturation Effects

9.1 INTRODUCTION

So far, we have considered that the resistances, leakage, and magnetisation inductances are invariable with load.

In reality, the magnetisation current I_m varies only slightly from no load to full load (from zero slip to rated slip $S_n \approx 0.01–0.06$), so the magnetisation inductance L_{1m} varies little in such conditions.

However, as the slip increases towards standstill, the stator current increases up to (5.8–6.5) times rated current at stall (S = 1).

In the same time, as the slip increases, even with constant resistances and leakage inductances, the magnetisation current I_m decreases.

So the magnetisation current decreases while the stator current increases when the slip increases (Figure 9.1).

When the rotor (stator) current increases with slip, the leakage magnetic field path in iron tends to saturate. With open slots on stator, this phenomenon is limited, but, with semiopen or semiclosed slots, the slot leakage magnetic flux path saturates the tooth tops both in the stator and rotor (Figure 9.2), above 2–3 times rated current.

Also, the differential leakage inductance which is related to main flux path is affected by the tooth top saturation caused by the circumferential flux produced by slot leakage flux lines (Figure 9.3). As the space harmonics flux paths are contained within τ_v/π from the airgap, only the teeth saturation affects them.

Further on, for large values of stator (and rotor) currents, the zig-zag flux becomes important and contributes notably to teeth top magnetic saturation in addition to slot leakage flux contribution.

Rotor slot skewing is also known to produce variable main flux path saturation along the stack length, together with the magnetisation current. However, the flux densities from the two contributions are phase shifted by an angle which varies and increases towards 90° at standstill. The skewing contribution to the main flux path saturation increases with slip and dominates the picture for $S > S_k$ as the magnetisation flux density, in fact, decreases with slip such that at standstill it is usually 55%–65% of its rated value.

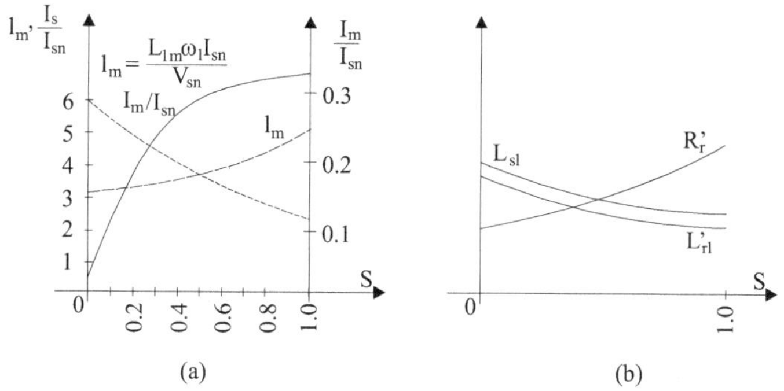

FIGURE 9.1 Stator I_s/I_{sn}, magnetisation I_m current, and magnetisation inductance (l_m) in p.u. (a) and leakage inductance and rotor resistance versus slip (b).

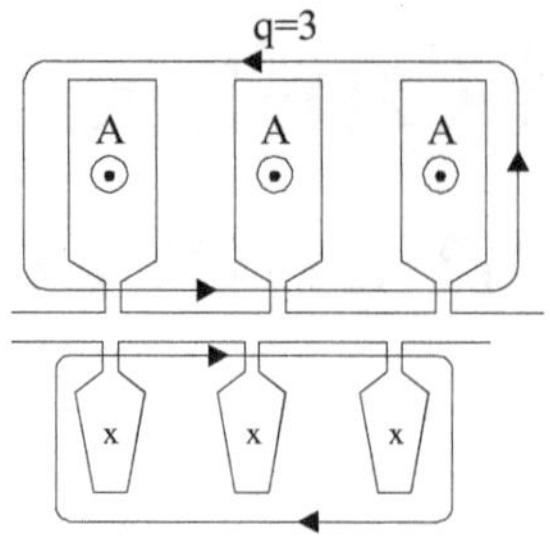

FIGURE 9.2 Slot leakage flux paths.

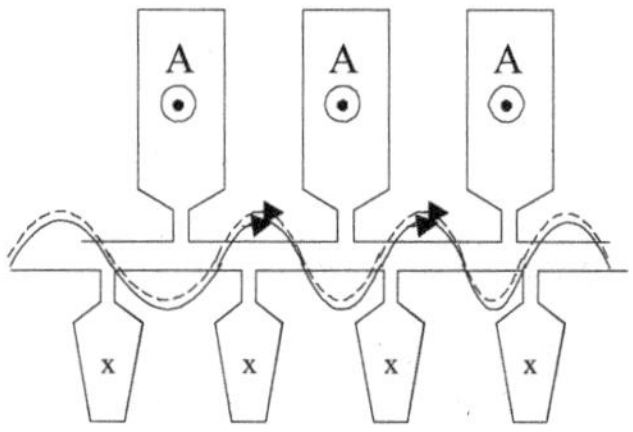

FIGURE 9.3 Zig-zag flux lines.

A few remarks are in order.

- The magnetisation saturation level in the core decreases with increasing slip, such that at standstill, only 55%–65% of rated airgap flux remains.
- The slot leakage flux tends to increase with slip (current) and saturates the tooth top unless the slots are open.
- Zig-zag circumferential flux and skewing accentuate the magnetic saturation of teeth top and entire main flux path, respectively, for high currents (above 2–3 times rated current).
- The differential leakage inductance is also reduced when stator (and rotor) current increases as slot, zig-zag, and skewing leakage flux effects increase.
- As the stator (rotor) current increases, the main (magnetising) inductance and leakage inductances are simultaneously influenced by saturation. So leakage and main path saturation are not independent of each other. This is why we use the term: on-load saturation.

As expected, accounting for these complex phenomena simultaneously is not an easy tractable mathematical endeavour. Finite element or even refined analytical methods may be suitable for the scope. Such methods are presented in this chapter after more crude approximations ready for preliminary design are given.

Besides magnetic saturation, skin (frequency) effect influences both the resistances and slot leakage inductances. Again, a simultaneous treatment of both aspects may be practically done only through finite element modelling (FEM).

On the other hand, if slot leakage saturation occurs only on the teeth top and the teeth, additional saturation due to skewing does not influence the flux line distribution within the slot and the two phenomena can be treated separately.

Experience shows that such an approximation is feasible. Skin effect is treated separately for the slot body occupied by a conductor. Its influence on equivalent resistance and slot body leakage geometrical permeance is accounted for by two correction coefficients, K_R and K_X. The slot neck geometry is "corrected" for leakage saturation.

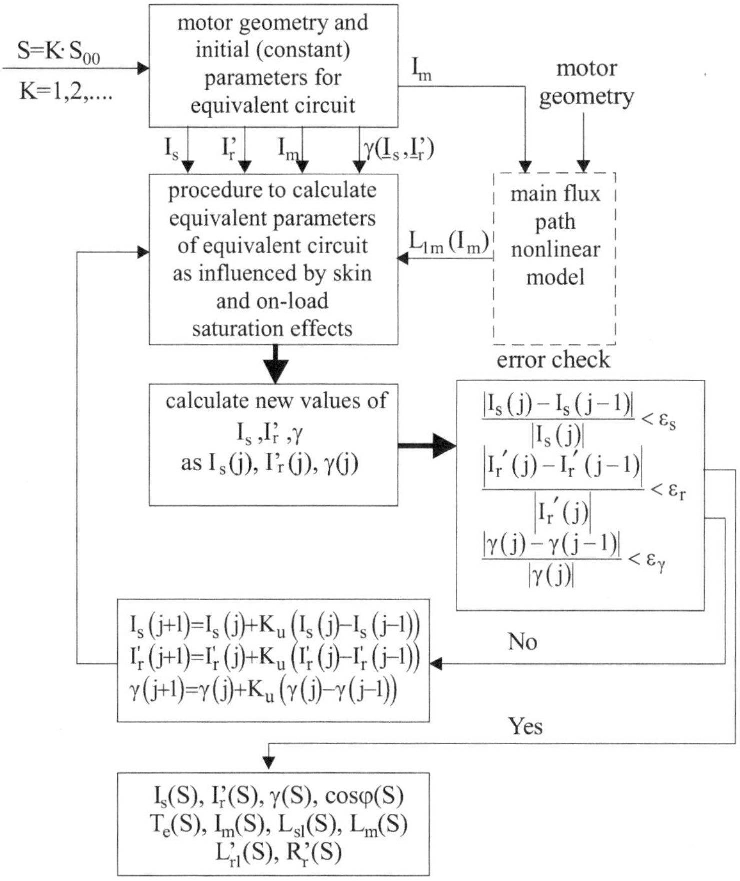

FIGURE 9.4 Iterative algorithm to calculate IM performance and parameters as influenced by skin and on-load saturation effects.

Finally, the on-load saturation effects are treated iteratively for given slip values to find, from the equivalent circuit with variable parameters, the steady-state performance. The above approach may be summarised as in Figure 9.4.

The procedure starts with the equivalent circuit with constant parameters and calculates initial values of stator and rotor currents I_s and I_r' and their phase shift angle γ. Now that we described the whole picture, let us return to its different facets and start with skin effect.

9.2 THE SKIN EFFECT

As already mentioned, skin effects are related to the flux and current density distribution in a conductor (or a group of conductors) flowed by A.C. currents and surrounded by a magnetic core with some airgaps.

Easy-to-use analytical solutions have been found essentially only for rectangular slots, but adaptation for related shapes has also become traditional.

More general shape slots with notable skin effect have been treated so far through equivalent multiple circuits after slicing the conductor(s) in slots in a few elements.

A refined slicing of conductor into many sections may be solved only numerically but within a short computation time. Finally, FEM may also be used to account for skin effect. First, we will summarise some standard results for rectangular slots.

9.2.1 Single Conductor in Rectangular Slot

Rectangular slots are typical for the stator of large induction machines (IMs) and for wound rotors of the same motors. Trapezoidal (and rounded) slots are typical for low-power motors.

The case of a single conductor in slot is (Figure 9.5) typical to single (standard) cage rotors and is commonplace in the literature. The main results are given below.

The standard correction coefficients for resistance and slot leakage inductance K_R and K_X are

$$K_R = \xi \frac{(\sinh 2\xi + \sin 2\xi)}{(\cosh 2\xi - \cos 2\xi)} = \frac{R_{ac}}{R_{dc}}; \quad K_X = \frac{3}{2\xi} \frac{(\sinh 2\xi - \sin 2\xi)}{(\cosh 2\xi - \cos 2\xi)} = \frac{(L_{sls})_{ac}}{(L_{sls})_{dc}} \tag{9.1}$$

with

$$\xi = \beta h_s = \frac{h_s}{\delta_{Al}}; \quad \beta = \frac{1}{\delta_{Al}} = \sqrt{\frac{S\omega_1 \mu_0 \sigma_{Al}}{2} \frac{b_c}{b_s}}; \quad \sigma_{Al} - \text{electrical conductivity} \tag{9.2}$$

The slip S signifies that in this case, the rotor (or secondary) of the IM is considered.

Figure 9.5 depicts K_R and K_x as functions of ξ, which, in fact, represents the ratio between the conductor height and the field penetration depth δ_{Al} in the conductor for given frequency $S\omega_1$. With one conductor in the slot, the skin effects, as reflected in K_R and K_x, increase with the slot (conductor) height, h_s, for given slip frequency $S\omega_1$.

This rotor resistance increase, accompanied by slot leakage inductance (reactance) decrease, leads to, both, a lower starting current and a higher starting torque.

This is how the deep bar cage rotor has evolved. To further increase the skin effects, and thus increase starting torque for even lower starting current ($I_{start} = (4.5–5)I_{rated}$), the double-cage rotor was introduced by the turn of 20th century already by Dolivo-Dobrovolsky and later on by Boucherot.

The advent of power electronics, however, has led to low-frequency starts and thus, up to peak torque at start may be obtained with 2.5–3 times rated current. Skin effect in this case is not needed. Reducing skin effect in large induction motors with cage rotors leads to particular slot shapes adequate for variable frequency supply.

9.2.2 Multiple Conductors in Rectangular Slots: Series Connection

Multiple conductors are placed in the stator slots or in the rotor slots of wound rotors (Figure 9.6).

According to Emde and Richter [1,2], which continued the classic work of field [3], the resistance correction coefficient K_{RP} for the pth layer in slot (Figure 9.6) with current I_p, when total current below pth layer is I_u, is

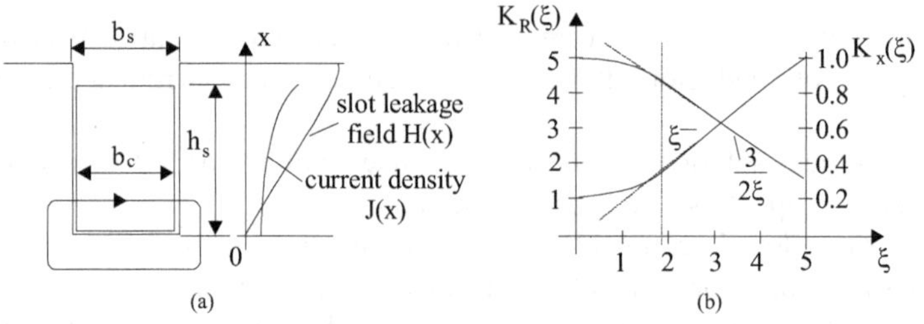

FIGURE 9.5 Rectangular slot (a) slot field (H(x)) and current density (J(x)) distributions and (b) resistance K_R and slot leakage inductance K_X skin effect correction factors.

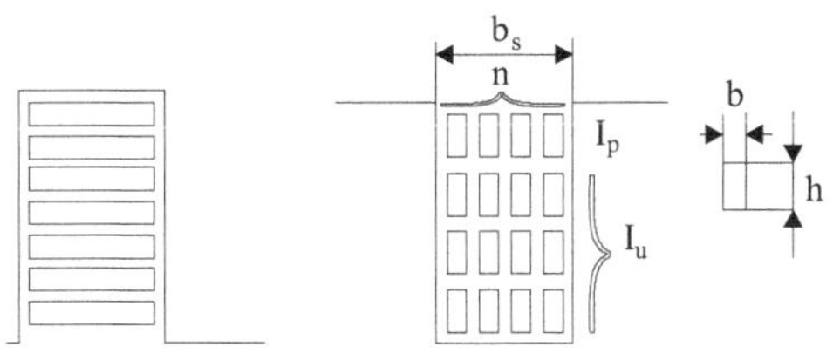

FIGURE 9.6 Multiple conductors in rectangular slots.

$$K_{RP} = \varphi(\xi) + \frac{I_u\left(I_u\cos\gamma + I_p\right)}{I_p^2}\psi(\xi) \tag{9.3}$$

$$\varphi(\xi) = \xi\frac{(\sinh 2\xi + \sin 2\xi)}{(\cosh 2\xi - \cos 2\xi)};\quad \psi(\xi) = 2\xi\frac{(\sinh\xi - \sin\xi)}{(\cosh\xi + \cos\xi)} \tag{9.4}$$

$$\xi = \beta_n h;\quad \beta_n = \sqrt{\frac{S\omega_1\mu_0\sigma_{Al}}{2}\frac{nb}{b_s}}$$

There are n conductors in each layer and γ is the angle between $\underline{I}_p$ and $\underline{I}_u$ phasors.

In two-layer windings with chorded coils, there are slots where the current in all conductors is the same and some in which two phases are located, and thus, the currents are different (or there is a phase shift $\gamma = 60°$).

For the case of $\gamma = 0$ with $I_u = I_p(p - 1)$, Equation (9.3) becomes

$$K_{RP} = \varphi(\xi) + \left(p^2 - p\right)\psi(\xi) \tag{9.5}$$

This shows that the skin effect is not the same in all layers. The average value of K_{RP} for m layers

$$K_{Rm} = \frac{1}{m}\sum_1^m K_{RP}(p) = \varphi(\xi) + \frac{m^2 - 1}{3}\psi(\xi) > 1 \tag{9.6}$$

Based on Ref. [4], for $\gamma \neq 0$ in (9.6), $(m^2 - 1)/3$ is replaced by

$$\frac{m^2(5 + 3\cos\gamma)}{24} - \frac{1}{3} \tag{9.6a}$$

A similar expression is obtained for the slot-body leakage inductance correction K_x [4].

$$K_{xm} = \varphi'(\xi) + \frac{\left(m^2 - 1\right)\psi'(\xi)}{m^2} < 1 \tag{9.7}$$

$$\varphi'(\xi) = \frac{3}{2\xi}\frac{(\sinh 2\xi - \sin 2\xi)}{(\cosh 2\xi - \cos 2\xi)} \tag{9.8}$$

$$\psi'(\xi) = \frac{(\sinh\xi + \sin\xi)}{\xi(\cosh\xi + \cos\xi)} \tag{9.9}$$

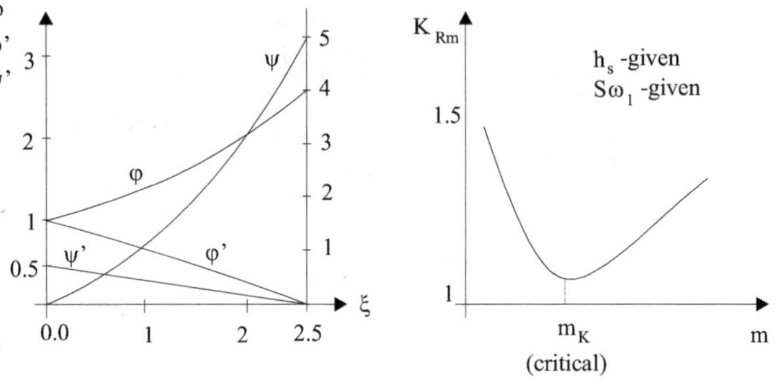

FIGURE 9.7 Helping functions φ, Ψ, φ', Ψ' versus ξ.

Please note that the first terms in K_{Rm} and K_{Rx} are identical to K_R and K_x of (9.1), valid for a single conductor in slot. As expected, K_{Rm} and K_{xm} degenerate into K_R and K_x for one layer (conductor) per slot. The helping functions φ, ψ, φ′, and ψ′ are quite general (Figure 9.7).

For a given slot geometry, increasing the number of conductor layers in slot reduces their height $h = h_s/m$ and thus reduces ξ, which ultimately reduces ψ(ξ) in (9.6). On the other hand, increasing the number of layers, the second term in (9.6) tends to increase.

It is thus evident that there is a critical conductor height h_c for which the resistance correction coefficient is minimum. Reducing the conductor height below h_c does not produce a smaller K_{Rm}.

In large-power or high-speed (frequency), small/medium power machines, this problem of critical conductor height is of great importance to minimise the additional (A.C.) losses in the windings.

A value of $K_{Rm} \approx (1.1–1.2)$ is, in most cases, acceptable. At power grid frequency (50–60 Hz), the stator skin effect resistance correction coefficient is very small (close to 1.0) as long as power is smaller than a few hundred kW.

Inverter-fed IMs, however, show high-frequency current time harmonics for which $K_{Rm\nu}$ may be notable and has to be accounted for.

Example 9.1 Derivation of Resistance and Reactance Corrections

Let us calculate the magnetic field H(x) and current density J(x) in the slot of an IM with m identical conductors (layers) in series making a single-layer winding.

Solution

To solve the problem, we use the field equation in complex numbers for the slot space where, only along slot depth (OX), the magnetic field and current density vary.

$$\frac{\partial^2 \underline{H}(x)}{\partial x^2} = j\frac{b}{b_s}\omega\mu_0\sigma_{Co}\underline{H}(x) \tag{9.10}$$

The solution of (9.10) is

$$\underline{H}(x) = \underline{C}_1 e^{-(1+j)\beta x} + \underline{C}_2 e^{+(1+j)\beta x}; \quad \beta = \sqrt{\frac{b}{b_s}\frac{\omega_1\mu_0\sigma}{2}} \tag{9.11}$$

The boundary conditions are

$$\underline{H}(x_p) \cdot b_s = \underline{I}_s p; \quad x = x_p; \quad x_p = ph$$
$$\underline{H}(x_p - h) \cdot b_s = \underline{I}_s (p-1); \quad x = x_p - h \tag{9.12}$$

From (9.11) and (9.12), we get the expressions of the constants $\underline{C}_1$ and $\underline{C}_2$

$$\underline{C}_1 = \frac{\underline{I}_s}{2b_s \sinh[(1+j)\beta h]}\left[(p-1)e^{(1+j)\beta x_p} - pe^{(1+j)\beta(x_p-h)}\right]$$
$$\underline{C}_2 = \frac{\underline{I}_s}{2b_s \sinh[(1+j)\beta h]}\left[-(p-1)e^{-(1+j)\beta x_p} + pe^{-(1+j)\beta(x_p-h)}\right] \tag{9.13}$$

The current density $\underline{J}(x)$ is

$$\underline{J}(x) = -\frac{b_s}{b}\frac{\partial \underline{H}(x)}{\partial x} = -\frac{b_s}{b}\beta(1+j)\left[\underline{C}_1 e^{-(1+j)\beta x} - \underline{C}_2 e^{(1+j)\beta x}\right] \tag{9.14}$$

For m = 2 conductors in series per slot, the current density distribution (9.14) is as shown qualitatively in Figure 9.8.

The active and reactive powers in the pth conductor, $\underline{S}_p$, is calculated using the Poynting vector [4].

$$\underline{S}_{a.c.} = P_{a.c.} + jQ_{a.c.} = \frac{b_s L}{\sigma_{Co}}\left[\left(\frac{\underline{J}}{2}\frac{\underline{H}^*}{2}\right)_{x=x_p-h} - \left(\frac{\underline{J}}{2}\frac{\underline{H}^*}{2}\right)_{x=x_p}\right] \tag{9.15}$$

Denoting the A.C. resistance and reactance of conductor p by R_{ac} and X_{ac}, we may write

$$P_{ac} = R_{ac} I_s^2 \quad Q_{ac} = X_{ac} I_s^2 \tag{9.16}$$

The D.C. resistance R_{dc} and reactance X_{dc} of conductor p,

$$R_{dc} = \frac{1}{\sigma_{Co}}\frac{L}{hb}; \quad X_{dc} = \omega\mu_0 \frac{b_s L}{3h}; \quad \text{L-stack length} \tag{9.17}$$

The ratios between A.C. and D.C. parameters K_{Rp} and K_{xp} are

$$K_{Rp} = \frac{R_{ac}}{R_{dc}}; \quad K_{xp} = \frac{X_{ac}}{X_{dc}} \tag{9.18}$$

Making use of (9.11) and (9.14) leads to the expressions of K_{Rp} and K_{xp} represented by (9.5) and (9.6).

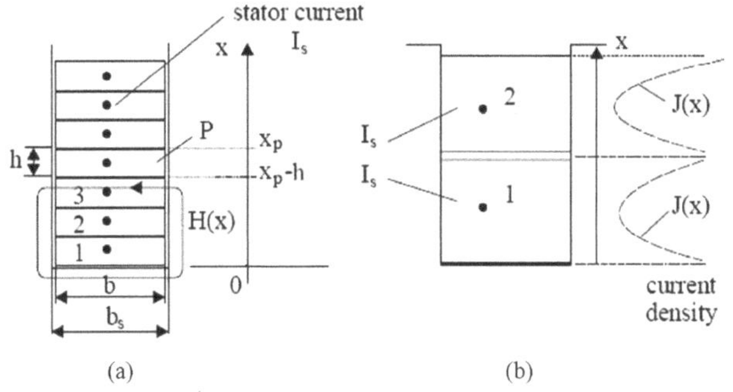

FIGURE 9.8 Stator slot with single coil with m layers (conductors in series) (a) and two conductors in series (b).

9.2.3 Multiple Conductors in Slot: Parallel Connection

Conductors are connected in parallel to handle the phase current, and in such a case, besides the skin effect correction K_{Rm} as described in Section 9.2.2 for series connection, circulating currents will flow between them. Additional losses are produced this way.

When multiple round conductors in parallel are used, their diameter is less than 2.5 (3) mm, and thus, at least for 50 (60) Hz machines, the skin effect may be neglected altogether. In contrast, for medium- and large-power machines, with rectangular shape conductors (Figure 9.9), the skin effect influence has, at least, to be verified. In this case also, the circulating currents' influence is to be considered.

A simplified solution to this problem [5] is obtained by neglecting, for the time being, the skin effect of individual conductors (layers), that is, by assuming a linear leakage flux density distribution along the slot height. Also the inter-turn insulation thickness is neglected.

At the junction between elementary conductors (strands), the average A.C. magnetic flux density $B_{ave} \approx B_m/4$ (Figure 9.11a). The A.C. flux through the cross-section of a strand Φ_{ac} is

$$\Phi_{ac} = B_{ave} h l_{stack} \tag{9.19}$$

The D.C. resistance of a strand R_{dc} is

$$R_{ac} \approx R_{dc} = \frac{1}{\sigma_{Co}} \frac{l_{turn}}{bh} \tag{9.20}$$

Now the voltage induced in a strand turn E_{ac} is

$$E_{ac} = \omega \Phi_{ac} \tag{9.21}$$

So the current in a strand I_{st}, with the leakage inductance of the strand neglected, is

$$I_{st} = E_{ac}/R_{ac} \tag{9.22}$$

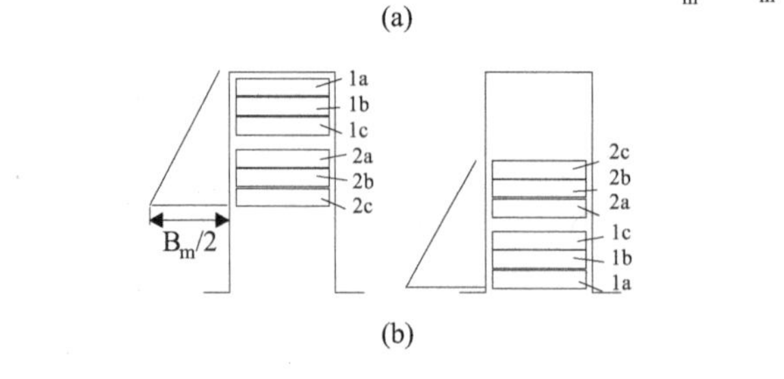

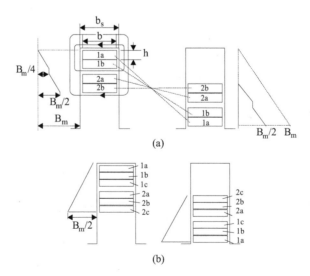

FIGURE 9.9 Slot leakage flux density for coil sides: two turn coils (a) two elementary conductors in parallel (strands) and (b) three elementary conductors in parallel.

The loss in a strand P_{strand} is

$$P_{strand} = \frac{E_{ac}^2}{R_{ac}} = \frac{\omega^2 B_{ave}^2 h^2 l_{stack}^2}{\frac{1}{\sigma_{Co}}\frac{l_{turn}}{bh}} \tag{9.23}$$

As seen from Figure 9.9a, the average flux density B_{ave} is

$$B_{ave} = \frac{B_m}{4} = \frac{\mu_0 n_{coil} I_{phase}(1+\cos\gamma)}{4b_s} \tag{9.24}$$

I_{phase} is the phase current and γ is the angle between the currents in the upper and lower coils. Also, n_{coil} is the number of turns per coil (in our case, $n_{coil} = 2,3$).

The usual D.C. loss in a strand with current (two vertical strands/coil) is

$$P_{dc} = R_{dc}\left(\frac{I_{phase}}{2}\right)^2 \tag{9.25}$$

We may translate the circulating new effect into a resistance additional coefficient, K_{Rad}.

$$K_{Rad} \approx \frac{P_{strand}}{P_{dc}} = \omega^2 \mu_0^2 \sigma_{Co}^2 \frac{b^2 h^4}{b_s^2}\left(\frac{l_{stack}}{l_{turn}}\right)^2 \frac{n_{coil}^2 (1+\cos\gamma)^2}{4} \tag{9.26}$$

Expression (9.26) is strictly valid for two vertical strands in parallel. However, as B_{ave} seems to be the same for other number of strands/turn, Equation (9.26) should be valid in general.

Adding the skin effect coefficient K_{Rm} as already defined to the one due to circulating current between elementary conductors in parallel, we get the total skin effect coefficient $K_{R||}$

$$K_{R||} = K_{Rm}\frac{l_{stack}}{l_{turn}} + K_{Rad} \tag{9.26a}$$

Even with large-power IMs, $K_{R||}$ should be less than 1.25 to 1.3 with $K_{Rad} < 0.1$ for a proper design.

Example 9.2 Skin Effect in Multiple Vertical Conductors in Slot

Let us consider a rather large induction motor with two coils, each made of four elementary conductors in series, respectively, and, of two turns, each of them made of two vertical strands (conductors in parallel) per slot in the stator. The size of the elementary conductor is h·b = 5·20 [mm·mm] and the slot width b_s = 22 mm; the insulation thickness along slot height is neglected. The frequency f_1 = 60 Hz. Let us determine the skin effect in the stack zone for the two cases, if $l_{stack}/l_{turn} = 0.5$.

Solution

As the elementary conductor is the same in both cases, the first skin effect resistance correction coefficient K_{Rm} may be computed first from (9.6) with ξ from (9.4),

$$\xi = \beta_n h; \quad h = 8\text{ mm}$$

$$\beta_n = \sqrt{\frac{\omega_1 \mu_0 \sigma_{Co}}{2}\frac{b}{b_s}} = \sqrt{\frac{2\pi 60 \cdot 1.256 \cdot 10^{-6}}{2 \cdot 1.8 \cdot 10^{-8}}\frac{20}{22}} = 109.32\ \text{m}^{-1}$$

$$\xi = 109.32 \cdot 5 \cdot 10^{-3} = 0.5466$$

The helping functions $\varphi(\xi)$ and $\psi(\xi)$ are (from (9.7)): $\varphi(\xi) = 1.015$, $\psi(\xi) = 0.04$. With m = 8 layers in slot, K_{Rm} (9.6) is

$$K_{Rm} = 1.015 + \frac{8^2 - 1}{3} 0.04 = 1.99$$

Now, for the parallel conductors (two in parallel), the additional resistance correction coefficient K_{Rad} (9.26) for circulating currents is

$$K_{Rad} = \left(1.256 \cdot 10^{-6}\right)^2 (2\pi 60)^2 \frac{1}{\left(1.8 \cdot 10^{-8}\right)^2} \left(\frac{20}{22}\right)^2 \cdot \frac{\left(5 \cdot 10^{-3}\right)^4 \cdot 0.5^2 \cdot 2^2 \cdot (1+1)^2}{4} = 0.3918!$$

The coefficient K_{Rad} refers to the whole conductor (turn) length, that is, it includes the end-turn part of it. In our case, K_{Rm} is too large, to be practical.

9.2.4 The Skin Effect in the End Turns

There is a part of stator and rotor windings that is located outside the lamination stack, mainly in air: the end turns or end rings.

The skin effect for conductors in air is less pronounced than in their portions in slots.

As the machine power or frequency increases, this kind of skin effect is to be considered. In Ref. [6], the resistance correction coefficient K_R for a single round conductor (d_{Co}) is also a function of β in the form (Figure 9.10).

$$\xi = d_{Co} \sqrt{\frac{\omega_1 \mu_0 \sigma_{Co}}{2}} \tag{9.27}$$

On the other hand, a rectangular conductor in air [7] presents the resistance correction coefficient (Figure 9.10) based on the assumption that there are magnetic field lines that follow the conductor periphery.

In general, there are m layers of round or rectangular conductors on top of each other (Figure 9.11).

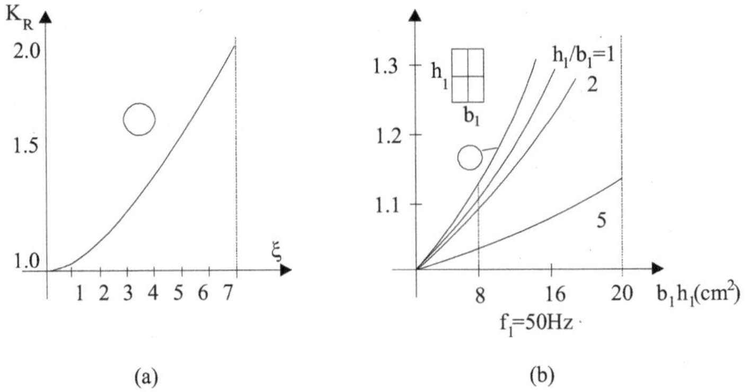

FIGURE 9.10 Skin effect correction factor K_R for a conductor in air: (a) circular cross-section and (b) rectangular cross-section.

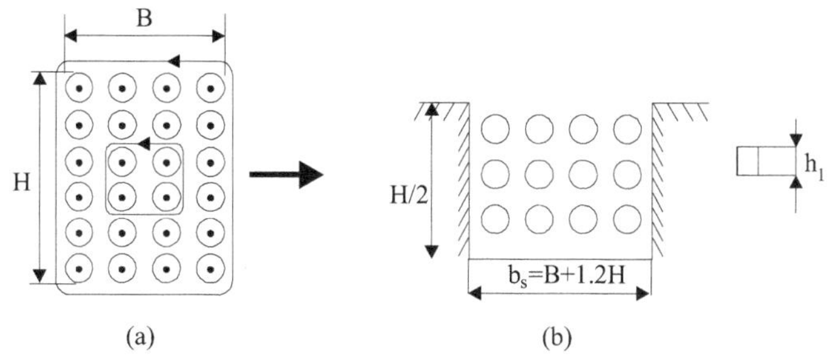

FIGURE 9.11 Four-layer coil in air (a) and its upper part placed in an equivalent (fictitious) slot (b).

Now the value of ξ is

$$\xi = d_{Co}\sqrt{\frac{\omega_1\mu_0\sigma}{2}\frac{B}{B+1.2H}} \quad \text{for round conductors}$$
$$\xi = h_1\sqrt{\frac{\omega_1\mu_0\sigma}{2}\frac{B}{B+1.2H}} \quad \text{for rectangular conductors} \tag{9.28}$$

As the skin effect is to be reduced, ξ should be made smaller than 1.0 by design. And, in this case, for rectangular conductors displaced in m layers [2], the correction coefficient K_{Rme} is

$$K_{Rme} = 1 + \frac{\left(m^2 - 0.8\right)}{36}\xi^4\pi \tag{9.29}$$

For a bundle of Z round conductors [8], K_{Rme} is

$$K_{Rme} = 1 + 0.005\cdot Z\cdot(d/cm)^4(f/50\ Hz)^2 \tag{9.29a}$$

The skin effect in the end rings of rotors may be treated as a single rectangular conductor in air. For small IMs, however, the skin effect in the end rings may be neglected. In large IMs, a more complete solution is needed. This aspect will be treated later in this chapter.

For the IM in Example 9.2, with m = 4, $\xi = 0.5466$, the skin effect in the end turns K_{Rme} (9.29) is

$$K_{Rme} = 1 + \frac{4^2 - 0.8}{36}0.5466^2 = 1.0377!$$

As expected, $K_{Rme} \ll K_{Rm}$ corresponding to the conductors in slot. The total skin effect resistance correction coefficient K_{Rt} is

$$K_{Rt} = \frac{K_{Rm}l_{stack} + K_{Rme}\left(l_{coil} - l_{stack}\right)}{l_{coil}} + K_{Rad} \tag{9.30}$$

For the case in Example 9.2,

$$K_{Rt} = \frac{1.99 + 1.0377\left(1 - \frac{1}{1.5}\right)}{1.5} + 0.3918 = 1.5572 + 0.3918 = 1.949$$

for two conductors in parallel, and K_{Rt} = 1.5572, for all conductors in series.

9.3 SKIN EFFECTS BY THE MULTILAYER APPROACH

For slots of more general shape, adopted to exploit the beneficial effects of rotor cages, a simplified solution is obtained by dividing the rotor bar into n layers of height h_t and width b_j (Figure 9.12). The method originates in [1].

For the pth layer, Faraday's law yields

$$R_p \underline{I}_p - R_{p+1}\underline{I}_{p+1} = -jS\omega_1 \Delta\underline{\Phi}_p \quad (9.31)$$

$$R_p = \frac{1}{\sigma_{Al}}\frac{l_{stack}}{b_p h_t};\quad R_{p+1} = \frac{1}{\sigma_{Al}}\frac{l_{stack}}{b_{p+1} h_t};\quad \Delta\Phi_p = \frac{\mu_0 l_{stack} h_t}{b_p}\sum_{j=1}^{p} I_j \quad (9.32)$$

R_p and R_{p+1} represent the resistances of pth and (p+1)th layer and L_p the inductance of pth layer.

$$L_p = \frac{\mu_0 l_{stack} h_t}{b_p} \quad (9.33)$$

With (9.33), Equation (9.31) becomes

$$I_{p+1} = \frac{R_p}{R_{p+1}} I_p + j\frac{S\omega_1 L_p}{R_{p+1}}\sum_{j=1}^{p} I_j \quad (9.34)$$

Let us consider p = 1, 2 in (9.34)

$$\underline{I}_2 = \frac{R_1}{R_2}\underline{I}_1 + j\frac{S\omega_1 L_1}{R_2}\underline{I}_1 \quad (9.35)$$

$$\underline{I}_3 = \frac{R_2}{R_3}\underline{I}_2 + j\frac{S\omega_1 L_2}{R_3}\left(\underline{I}_1 + \underline{I}_2\right) \quad (9.36)$$

If we assign a value to I_1 in relation to total current I_b, say,

$$\left(\underline{I}_1\right)_{initial} = \frac{I_b}{n}, \quad (9.37)$$

we may use Equations (9.34) through (9.36) to determine the current in all layers. Finally,

$$(I_b)' = \left|\sum_{j=1}^{n} I_j\right| \quad (9.38)$$

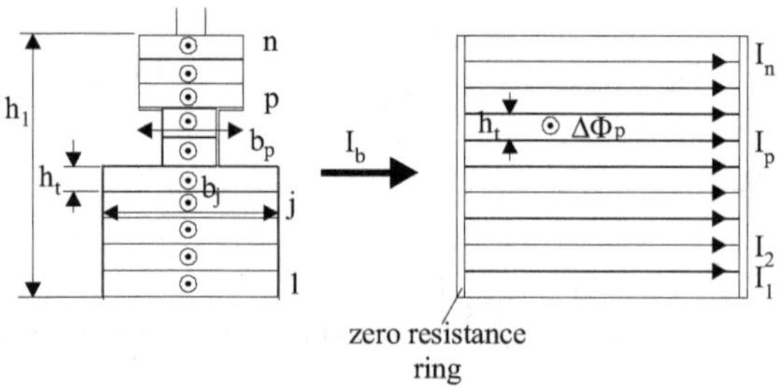

FIGURE 9.12 More general shape rotor bars.

As expected, I_b and I'_b will be different. Consequently, the currents in all layers will be multiplied by I_b/I'_b to obtain their real values. On the other hand, Equations (9.35) and (9.36) lead to the equivalent circuit in Figure 9.13.

Once the layer currents $\underline{I}_1, \ldots, \underline{I}_n$ are known, the total losses in the bar are

$$P_{ac} = \sum_{j=1}^{n} |I_j|^2 R_j \tag{9.39}$$

In a similar manner, the magnetic energy in the slot W_{mac} is

$$W_{mac} = \frac{1}{2}\sum_{j=1}^{n} L_j \left|\sum_{1}^{j} I_k\right|^2 \tag{9.40}$$

The D.C. power loss in the slot (for given total bar current) P_{dc} is

$$P_{dc} = \sum I_{jdc}^2 R_j; \quad I_{jdc} = \frac{I'_b}{A_{bar}} h_t b_j \tag{9.41}$$

Also the D.C. magnetic energy in the slot

$$W_{mdc} = \frac{1}{2}\sum_{j=1}^{n} L_j \left|\sum_{1}^{j} I_{kdc}\right|^2 \tag{9.42}$$

Now the skin effect resistance and inductance correction coefficients K_R and K_x are

$$K_R = \frac{P_{ac}}{P_{dc}} = \frac{\sum_{j=1}^{n} I_j^2 R_j}{\sum_{j=1}^{n} I_{jdc}^2 R_j} \tag{9.43}$$

$$K_x = \frac{\sum_{j=1}^{n} L_j \left|\sum_{k=1}^{j} I_k\right|^2}{\sum_{j=1}^{n} L_j \left|\sum_{k=1}^{j} I_{kdc}\right|^2} \tag{9.44}$$

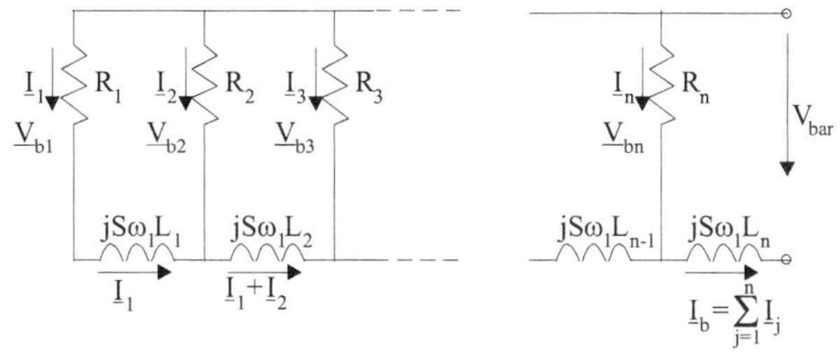

FIGURE 9.13 Equivalent circuit for skin effect evaluation.

Example 9.3 Skin Effects Calculation for General Shape Rotor Bars

Let us consider a deep bar of the shape in Figure 9.12 with the dimensions as in Figure 9.12a. Let us divide the bar in only six layers, each 5 mm high (h_t = 5 mm) and calculate the skin effects for S = 1 and f_1 = 60 Hz.

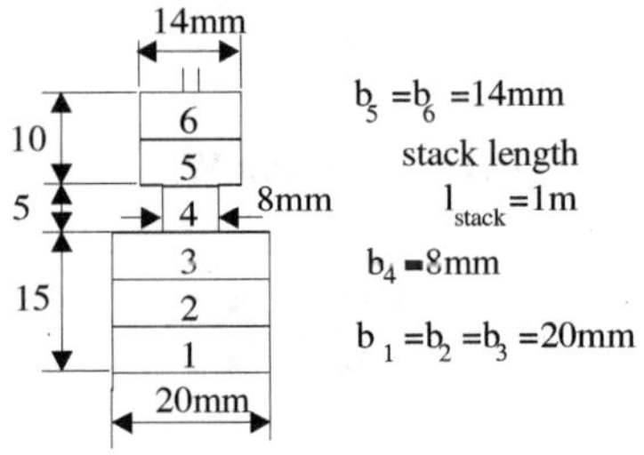

FIGURE 9.12A Deep bar geometry.

Solution

From Figure 9.12a, the layer resistances and inductances are (9.32–9.33).

$$R_1 = R_2 = R_3 = \frac{1}{\sigma_{Al}}\frac{l_{stack}}{b_1 h_t} = \frac{1}{3\cdot 10^7}\frac{1}{20\cdot 10^{-3}\cdot 5\cdot 10^{-3}} = 0.333\cdot 10^{-3}\ \Omega$$

$$R_4 = \frac{1}{\sigma_{Al}}\frac{l_{stack}}{b_4 h_t} = \frac{1}{3\cdot 10^7}\frac{1}{8\cdot 10^{-3}\cdot 5\cdot 10^{-3}} = 0.833\cdot 10^{-3}\ \Omega$$

$$R_5 = R_6 = \frac{1}{\sigma_{Al}}\frac{l_{stack}}{b_5 h_t} = \frac{1}{3\cdot 10^7}\frac{1}{14\cdot 10^{-3}\cdot 5\cdot 10^{-3}} = 0.476\cdot 10^{-3}\ \Omega$$

From (9.33),

$$L_1 = L_2 = L_3 = \frac{\mu_0 l_{stack} h_t}{b_1} = 1.256\cdot 10^{-6}\cdot 1\cdot\frac{0.005}{0.020} = 0.314\cdot 10^{-6}\ \text{H}$$

$$L_4 = \frac{\mu_0 l_{stack} h_t}{b_4} = 1.256\cdot 10^{-6}\cdot\frac{0.005}{0.008} = 0.785\cdot 10^{-6}\ \text{H}$$

$$L_5 = L_6 = \frac{\mu_0 l_{stack} h_t}{b_5} = 1.256\cdot 10^{-6}\cdot\frac{0.005}{0.014} = 0.44857\cdot 10^{-6}\ \text{H}$$

Let us now consider that the bar current is I_b = 3600 A and $I_1 = I_b/n = I_b/6$ = 600 A.

Now $\underline{I}_2$ (in the second layer from slot bottom) is

$$\underline{I}_2 = \frac{R_1}{R_2}\underline{I}_1 + j\frac{S\omega_1 L_1}{R_2}\underline{I}_1 = 600 + j\cdot 1\cdot 2\pi 60\cdot\frac{0.314\cdot 10^{-6}}{0.333\cdot 10^{-3}}\cdot 600 = 600 + j213.18\ \text{A}$$

$$|\underline{I}_2| = 634.74\ \text{A}$$

$$\underline{I}_3 = \frac{R_2}{R_3}\underline{I}_2 + j\frac{S\omega_1 L_2}{R_3}(\underline{I}_1 + \underline{I}_2) = 600 + j213.18$$

$$+ j\cdot 1\cdot 2\pi 60\cdot\frac{0.314\cdot 10^{-6}}{0.333\cdot 10^{-3}}\cdot(600 + 600 + j213.18) = 524.25 + j640\ \text{A}$$

$$|\underline{I}_3| = 827.3 \text{ A}$$

$$\underline{I}_4 = \frac{R_3}{R_4}\underline{I}_3 + j\frac{S\omega_1 L_3}{R_4}\left(\underline{I}_1 + \underline{I}_2 + \underline{I}_3\right) = \frac{0.333\cdot 10^{-3}}{0.833\cdot 10^{-3}}(524.25 + j426.26)$$

$$+ j\cdot 1\cdot 2\pi 60\cdot\frac{0.314\cdot 10^{-6}}{0.833\cdot 10^{-3}}\cdot(600 + 600 + j213.18 + 524 + j640)$$

$$= 55.5 + j490.2 \text{ A}$$

$$|\underline{I}_4| = 493.27 \text{ A}$$

$$\underline{I}_5 = \frac{R_4}{R_5}\underline{I}_4 + j\frac{S\omega_1 L_4}{R_5}\left(\underline{I}_1 + \underline{I}_2 + \underline{I}_3 + \underline{I}_4\right) = \frac{0.833\cdot 10^{-3}}{0.476\cdot 10^{-3}}(55.5 + j490.2)$$

$$+ j\cdot 1\cdot 2\pi 60\cdot\frac{0.785\cdot 10^{-6}}{0.4485\cdot 10^{-3}}\cdot(1779.75 + j1342)$$

$$= -712.088 + j2030.4 \text{ A}$$

$$|\underline{I}_5| = 2151.2\text{A}$$

$$\underline{I}_6 = \frac{R_5}{R_6}\underline{I}_5 + j\frac{S\omega_1 L_5}{R_6}\left(\underline{I}_1 + \underline{I}_2 + \underline{I}_3 + \underline{I}_4 + \underline{I}_5\right) = \frac{0.476\cdot 10^{-3}}{0.476\cdot 10^{-3}}(-712.088 + j2030.4)$$

$$+ j\cdot 1\cdot 2\pi 60\cdot\frac{0.74857\cdot 10^{-6}}{0.476\cdot 10^{-3}}\cdot(-712.088 + j2030.4 + 1724.25 + j853)$$

$$= -1735.75 + j2389.4 \text{ A}$$

$$|\underline{I}_6| = 2953.31 \text{ A}$$

The total current is

$$\underline{I}'_b = \underline{I}_1 + \underline{I}_2 + \underline{I}_3 + \underline{I}_4 + \underline{I}_5 + \underline{I}_6 = -712.088 + j2030.4 - 1735.75 + j2389$$

$$= -2447.75 + j4419.4 \text{ A}$$

$$I'_b \approx 5050 \text{ A}$$

The A.C. power in the bar is

$$P_{ac} = \sum_{j=1}^{n} I_j^2 R_j = 0.333\cdot 10^{-3}\left(600^2 + 636.74^2 + 827.30^2\right)$$

$$+ 0.833\cdot 10^{-3}\cdot 493.27^2 + 0.476\cdot 10^{-3}\left(2151.2^2 + 2955.31^2\right)$$

$$= 482 + 202.68 + 6360 = 7044.68 \text{ W}$$

The D.C. current distribution in the six layer is uniform, therefore,

$$I_{1dc} = I_{2dc} = I_{3dc} = \frac{I'_b}{A_{bar}} b_1 h_t = \frac{5050\cdot 20.5}{(14.10 + 8.5 + 15.20)} = 1052.08 \text{ A}$$

$$I_{4dc} = \frac{I_b'}{A_{bar}} b_4 h_t = \frac{5050 \cdot 8.5}{480} = 420.83\ \text{A}$$

$$I_{5dc} = I_{6dc} = \frac{I_b'}{A_{bar}} b_5 h_t = \frac{5050 \cdot 14.5}{480} = 736.458\ \text{A}$$

$$P_{dc} = \sum_{j=1}^{n} I_{jdc}^2 R_j = 0.333 \cdot 10^{-3} \cdot 3 \cdot 1052.08^2$$

$$+ 0.833 \cdot 10^{-3} \cdot 420.83^2 + 0.476 \cdot 10^{-3} \cdot 2 \cdot 736.458^2 = 1768.81\ \text{W}$$

and the skin effect resistance correction factor K_R is

$$K_R = \frac{P_{ac}}{P_{dc}} = \frac{7044.68}{1768.81} = 3.9827!$$

The magnetic energy ratio K_x is

$$K_x = \frac{A}{B}$$

$$A = L_1\left(|\underline{I}_1|^2 + |\underline{I}_1 + \underline{I}_2|^2 + |\underline{I}_1 + \underline{I}_2 + \underline{I}_3|^2\right) + L_4|\underline{I}_1 + \underline{I}_2 + \underline{I}_3|^2$$

$$+ L_5|\underline{I}_1 + \underline{I}_2 + \underline{I}_3 + \underline{I}_4 + \underline{I}_5|^2 + L_5|\underline{I}_1 + \underline{I}_2 + \underline{I}_3 + \underline{I}_4 + \underline{I}_5 + \underline{I}_6|^2$$

$$B = L_1\left[I_{1dc}^2 + (I_{1dc} + I_{2dc})^2 + (I_{1dc} + I_{2dc} + I_{3dc})^2\right]$$

$$+ L_4(I_{1dc} + I_{2dc} + I_{3dc} + I_{4dc})^2 + L_5(I_b' - I_6)^2 + L_5 I_b'^2$$

$$A = 0.314 \cdot 10^{-6}\left(600^2 + 1.217^2 \cdot 10^6 + 1.9237^2 \cdot 10^6\right) + 0.785 \cdot 10^{-6} 2.229^2 \cdot 10^6$$

$$+ 0.44857 \cdot 10^{-6}\left(3.0554^2 \cdot 10^6 + 5.050^2 \cdot 10^6\right)$$

$$B = 0.314 \cdot 10^{-6}\left(1.052^2 \cdot 10^6 + 2.104^2 \cdot 10^6 + 3.156^2 \cdot 10^6\right)$$

$$+ 0.785 \cdot 10^{-6} 3.576^2 \cdot 10^6 + 0.44857 \cdot 10^{-6}\left(4.3128^2 \cdot 10^6 + 5.05^2 \cdot 10^6\right)$$

$$K_x = \frac{21.267}{34.685} = 0.613!$$

The inductance coefficient refers only to the slot body (filled with conductor) and not to the slot neck, if any.

A few remarks are in order.

- The distribution of current in the various layers is non-uniform when the skin effect occurs.
- Not only the amplitude but also the phase angle of bar current in various layers varies due to skin effect (Figure 9.14).
- At S = 1 (f_1 = 60 Hz), most of the current occurs in the upper part of the slot.
- The equivalent circuit model can be easily put into computer form once the layer geometries – h_t (height) and b_j (width) – are given. For various practical slots, special subroutines may provide b_j and h_t when the number of layers is given.
- To treat a double cage by this method, we only have to consider the current zero in the empty slot layers between the upper and lower cage (Figure 9.15).

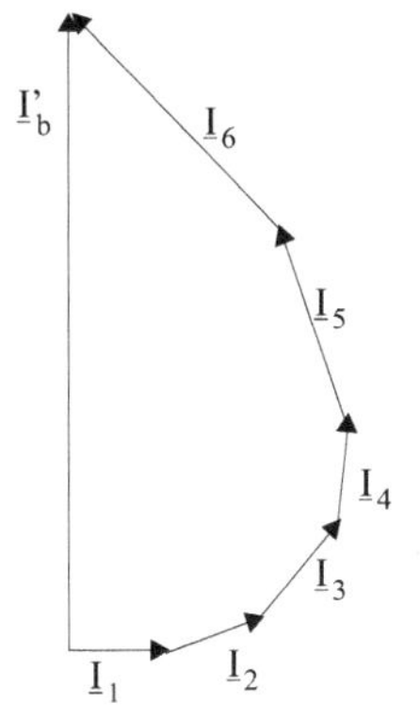

FIGURE 9.14 Layer currents and the bar current with skin effect.

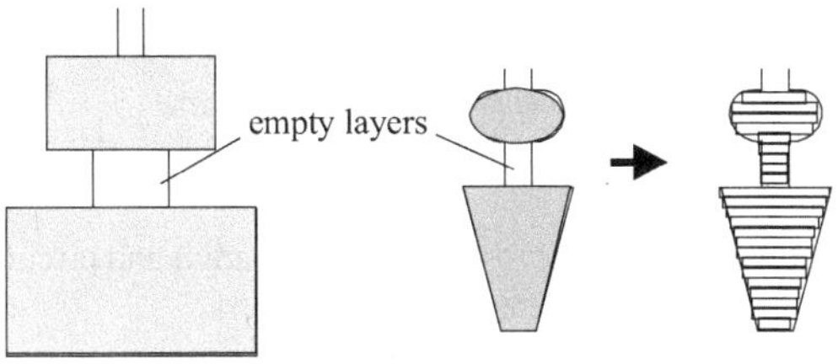

FIGURE 9.15 Treating skin effect with equivalent circuit (or multilayer) method.

Now that both K_R and K_x are known, the bar resistance and slot body leakage geometrical specific permeance λ_{sbody} is modified to account for skin effect.

$$\left(\lambda_{sbody}\right)_{ac} = \left(\lambda_{sbody}\right)_{dc} K_x \tag{9.45}$$

From D.C. magnetic energy W_{mdc} (9.42), we write

$$L_{dc} = \frac{1}{2}\frac{W_{mdc}}{I_b'^2} = \mu_0 l_{stack}\left(\lambda_{sbody}\right)_{dc} \tag{9.46}$$

The slot neck geometrical specific permeance is still to be added to account for the respective slot leakage flux. This slot neck geometrical specific permeance is to be corrected for leakage flux saturation later in this chapter.

9.4 SKIN EFFECT IN THE END RINGS VIA THE MULTILAYER APPROACH

As the end rings are placed in air, although rather close to the motor laminated stack, the skin effect in them is routinely neglected. However, there are applications where the value of slip goes above unity (S = up to 3.0 in standard elevator drives) or the slip frequency is large as in high-frequency (high-speed) motors to be started at rated frequency (400 Hz in avionics).

For such cases, the multilayer approach may be extended to end rings. To do so, we introduce radial and circumferential layers in the end rings (Figure 9.16) as shown in Ref. [7].

In all the layers, the current density is considered uniform. It means that their radial dimension has to be less than the depth of field penetration in aluminium. The currents in neighbouring slots are considered phase shifted by $2\pi p_1/N_r$ radians (N_r – number of rotor slots).

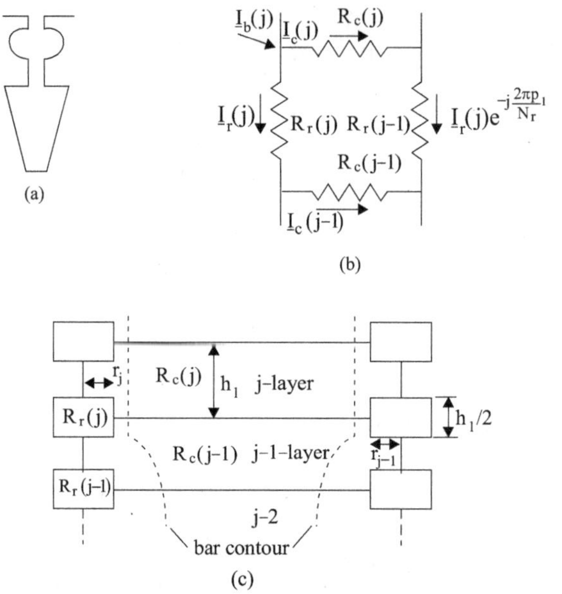

FIGURE 9.16 Bar-end ring transition (a) slot cross-section, (b) radial and circumferential layer end ring currents, and (c) geometry of radial and circumferential end ring layers.

The relationship between bar and end ring layer currents (Figure 9.16) is

$$I_{b(j)} + \underline{I}_{r(j+1)} + \underline{I}_{r(j)} e^{j\frac{2\pi p_1}{N_r}} = \underline{I}_{r(j)} + \underline{I}_{c(j)} \tag{9.47}$$

$$R_{c(j)}\underline{I}_{c(j)} + R_{r(j)}\underline{I}_{r(j)} e^{-j\frac{2\pi p_1}{N_r}} = R_{c(j-1)}\underline{I}_{c(j-1)} + R_{r(j)}\underline{I}_{r(j)} \tag{9.48}$$

The circumferential extension of the radial layer r_j is assigned a value at start. Now if we add the equations for the bar layer currents, we may solve the system of equations. As long as the radial currents increase, γ_j is increased in the next iteration cycle until sufficient convergence is met. Some results, after [2], are given in Figure 9.17.

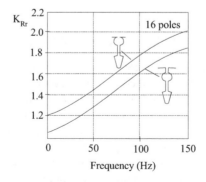

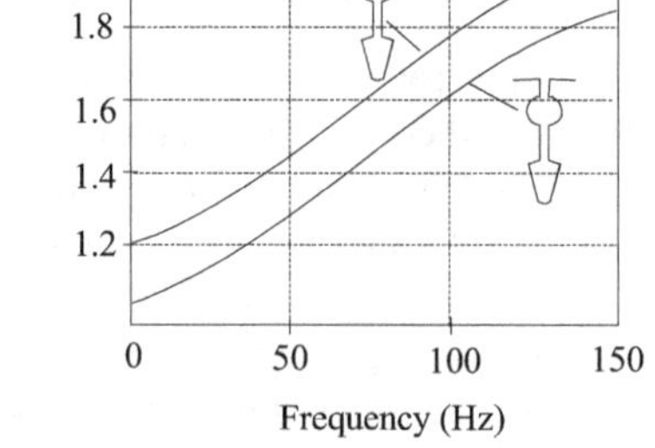

FIGURE 9.17 End ring skin effect resistance coefficient K_{Rr}.

As the slot total height is rather large (above 25 mm), the end ring skin effect is large, especially for rotor frequencies above 50 (60) Hz. In fact, a notable part of this resistance rise is due to the radial ring currents which tend to distribute the bar currents, gathered towards the slot opening, into most of the end ring cross-section.

9.5 THE DOUBLE CAGE BEHAVES LIKE A DEEP BAR CAGE

In some applications, very high starting torque – $T_{start}/T_{rated} \geq 2.0$ – is required. In such cases, a double cage is used. It has been proved that it behaves like a deep bar cage, but it produces even higher starting torque at lower starting current. For the case when skin effect can be neglected in both cages, let us consider a double cage as configured in Figure 9.18 [9].

The equivalent single bar circuit is given in Figure 9.18b. For the common ring of the two cages,

$$R_r = R_{ring}, R_{bs} = R_{bs\ upper\ bar}, R_{bw} = R_{bw\ lower\ bar} \tag{9.49}$$

For separate rings,

$$R_r = 0, R_{bs} = R_{bs} + R^e_{rings}, R_{bw} = R_{bw} + R^e_{ringw} \tag{9.50}$$

The ring segments are included into the bar resistance after approximate reduction as shown in Chapter 6. The value of L_{ring} is the common ring inductance or is zero for separate rings. Also for both cases, $L_e(\Phi_e)$ refers to the slot neck flux.

$$L_e(\Phi_e) = \mu_0 l_{stack} \frac{h_4}{a_4} \tag{9.51}$$

We may add into L_e the differential leakage inductance of rotor.

The start (upper) and work (lower) cage inductances L_{bs} and L_{bw} include the end ring inductances only for separate rings. Otherwise, the bar inductances are

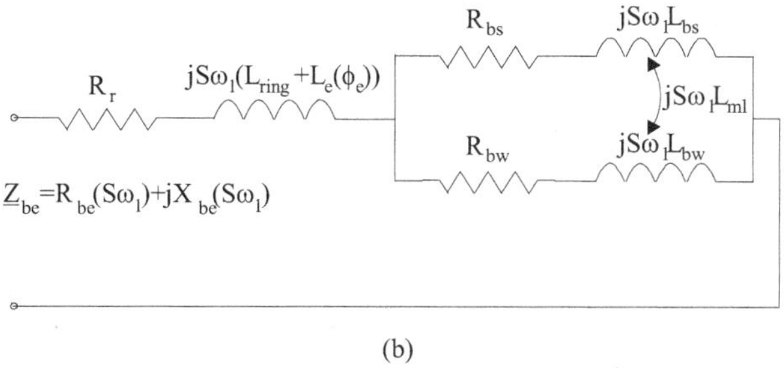

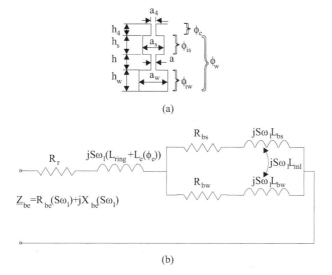

FIGURE 9.18 Double-cage rectangular-shape geometry (a) and equivalent circuit (b).

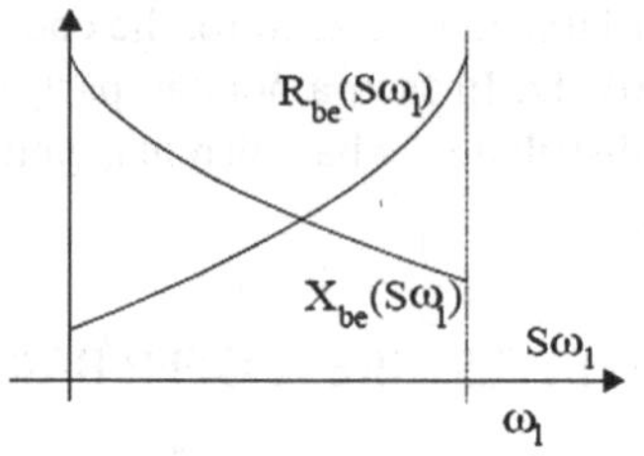

FIGURE 9.19 Equivalent parameters of double cage versus slip frequency.

$$L_{bs} = \mu_0 l_{stack} \frac{h_s}{3a_s}$$
$$L_{bw} = \mu_0 l_{stack} \left(\frac{h_w}{3a_w} + \frac{h}{a} + \frac{h_s}{a_s} \right) \tag{9.52}$$

There is also a flux common to the two cages represented by the flux in the starting cage [3].

$$L_{ml} = \mu_0 l_{stack} \frac{h_s}{2a_s} \tag{9.53}$$

In general, L_{ml} is neglected though it is not a problem to consider it in solving the equivalent circuit in Figure 9.18. It is evident (Figure 9.18a) that the starting (upper) cage has a large resistance (R_{bs}) and a small slot leakage inductance L_{bs}, while for the working cage, the opposite is true.

Consequently, at high slip frequency, the rotor current resides mainly in the upper (starting) cage, while, at low slip frequency, the current flows mainly into the working (lower) cage. Thus, both R_{be} and X_{be} vary with slip frequency as they do in a deep bar single cage (Figure 9.19).

9.6 LEAKAGE FLUX PATH SATURATION – A SIMPLIFIED APPROACH

Leakage flux path saturation occurs mainly in the slot neck zone for semiclosed slots for currents above 2–3 times the rated current or in the rotor slot iron bridges for closed slots even well below the rated current (Figure 9.20).

Consequently,

$$a'_{os} = a_{os} + \frac{(\tau_{ss} - a_{os})}{\mu_s}\mu_0; \quad \tau_{ss} = \frac{\pi D_i}{N_s}$$
$$a'_{or} = a_{or} + \frac{(\tau_{sr} - a_{or})}{\mu_r}\mu_0; \quad \tau_{sr} = \frac{\pi (D_i - 2g)}{N_s} \tag{9.54}$$
$$a'_{or} = \frac{b_{or}\mu_0}{\mu_{br}}, \quad \text{for closed slots}$$

The slot neck geometrical permeances will be changed to a'_{os}/h_{os}, a'_{or}/h_{or}, or a'_{or}/h_{or} dependent on stator (rotor) current.

With n_s the number of turns (conductors) per slot and I_s and I_b the stator and rotor currents, the Ampere's law on Γ_s and Γ_r trajectories in Figure 9.20 yields

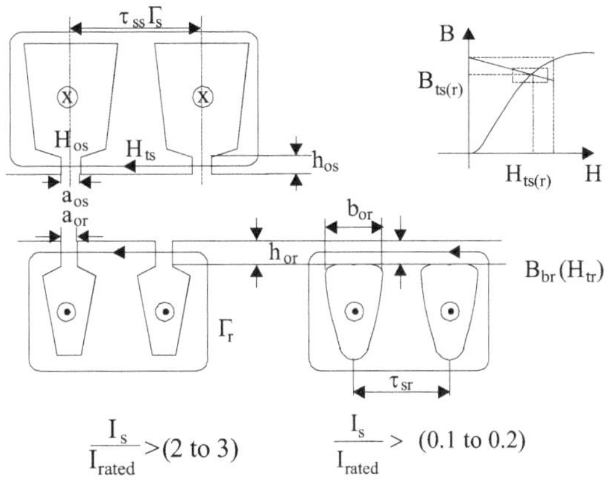

FIGURE 9.20 Leakage flux path saturation conditions.

$$
\begin{aligned}
&H_{ts}(\tau_{ss} - a_{os}) + H_{os}a_{os} \approx n_s I_s \sqrt{2}; \quad \mu_s H_{ts} = B_{ts} = \mu_0 H_{os} \\
&H_{tr}(\tau_{sr} - a_{or}) + H_{or}a_{or} \approx I_b \sqrt{2}; \quad \mu_r H_{tr} = B_{tr} = \mu_0 H_{or} \\
&H_{tr} b_{or} = I_b \sqrt{2}; \quad \mu_{br} = \frac{B_{tb}}{H_{tb}}
\end{aligned}
\tag{9.55}
$$

The relationship between the equivalent rotor current I_r' (reduced to the stator) and I_b is (Chapter 8)

$$
I_b = I_r' \frac{6p_1 q n_s K_{w1}}{N_r} = I_r' n_s K_{w1} \frac{N_s}{N_r} \tag{9.56}
$$

N_s – number of stator slots; K_{w1} – stator winding factor.

When the stator and rotor currents I_s and I_r' are assigned pertinent values, iteratively, using the lamination magnetisation curves, Equations (9.55) may be solved (Figure 9.20) to find the iron permeabilities of teeth tops or closed rotor slot bridges. Finally, from (9.54), the corrected slot openings are found.

With these values, the stator and rotor parameters (resistances and leakage inductances) as influenced by the skin effect (in the slot body zone) and leakage saturation (in the slot neck permeance) are recalculated. Continuing with these values, from the equivalent circuit, new values of stator and rotor currents I_s, I_r' are calculated for given stator and voltage, frequency, and slip.

The iteration cycles continue until sufficient convergence is obtained for stator current.

Example 9.4 Slot Leakage Specific Permeance Calculation

An induction motor has semiclosed rectangular slots whose geometry is shown in Figure 9.21. The current design density for rated current is $j_{Co} = 6.5\ \text{A/mm}^2$ and the slot fill factor $K_{fill} = 0.45$. The starting current is 4.5 times rated current. Let us calculate the slot leakage specific permeance at rated current and at start.

Solution

The slot mmf $n_s I_s$ is obtained from

$$
n_s I_s = (h_s a_s) K_{fill} j_{Co} = 10 \cdot 35 \cdot 0.45 \cdot 6.5 = 1023.75 \text{ Ampereturns}
$$

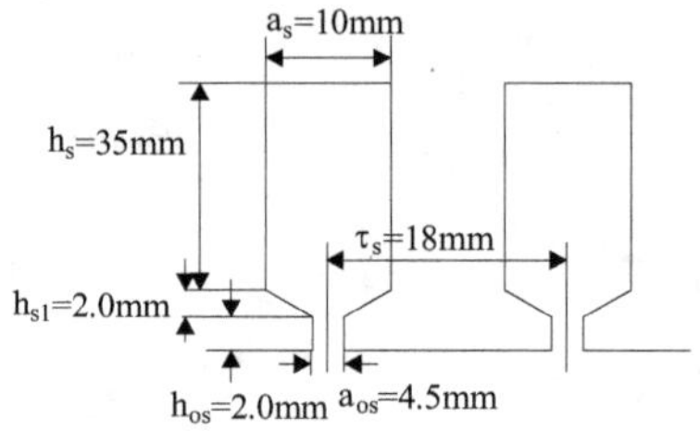

FIGURE 9.21 Slot geometry.

Making use of (9.54),

$$\left[H_{ts}(18-4.5)+\frac{B_{ts}}{\mu_0}4.5\right]\cdot 10^{-3}=1023.75\sqrt{2}$$

$$B_{ts}=\left(1023.75\sqrt{2}-H_{ts}\cdot 13.5\cdot 10^{-3}\right)\frac{1.256\cdot 10^{-6}}{4.5\cdot 10^{-3}}=0.4029-H_{ts}3\mu_0$$

It is very clear that for rated current, $B_{ts} < 0.4029$, so the tooth heads are not saturated. In (9.54), $a'_{os} = a_{os}$ and the slot-specific geometrical permeance λ_s is

$$\lambda_s=\frac{h_s}{3a_s}+\frac{2h_{s1}}{a_{os}+a_s}+\frac{h_{os}}{a_{os}}=\frac{35}{3\cdot 10}+\frac{2\cdot 2.5}{4.5+10}+\frac{2}{4.5}=1.9559$$

For the starting conditions,

$$\left[H_{tstart}(18-4.5)\cdot 10^{-3}+\frac{B_{ts}}{\mu_0}4.5\right]\cdot 10^{-3}=6142.5\sqrt{2}$$

$$B_{tstart}=2.41-H_{tstart}3\mu_0$$

Now we need the lamination magnetisation curve. If we do so, we get $B_{tstart} \approx 2.16$ T, $H_{tstart} \approx 70{,}000$ A/m. The iron permeability μ_s is

$$\frac{\mu_s}{\mu_0}=\frac{B_{tstart}}{H_{tstart}}\mu_0^{-1}=\frac{2.16}{70{,}000\cdot 1.256\cdot 10^{-6}}=24.56$$

Consequently, $a'_{os}=a_{os}+\frac{\left(\tau_s-a_{os}\right)}{\mu_s}\mu_0=4.5+\frac{(18-4.5)}{24.56}=5.05\text{ mm}$

The slot geometric specific permeance is, at start,

$$\lambda'_s=\frac{h_s}{3a_s}+\frac{2h_{s1}}{a'_{os}+a_s}+\frac{h_{os}}{a'_{os}}=\frac{35}{3\cdot 10}+\frac{2\cdot 2.5}{5.05+10}+\frac{2}{5.05}=1.895$$

The leakage saturation of tooth heads at start is not very important in our case. In reality, it could be more important for semiclosed stator (rotor) slots.

Example 9.5 The Rotor Bridge Flux Density

Let us consider a rotor bar with the geometry as shown in Figure 9.22. The bar current at rated current density $j_{Alr} = 6.0$ A/mm^2 is

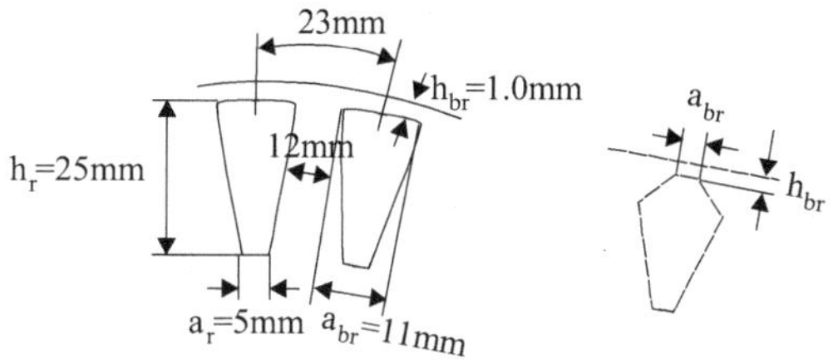

FIGURE 9.22 Closed rotor slots.

$$I_{br} = j_{Alr} h_r \frac{(a_r + b_r)}{2} = 25\frac{(11+5)}{2} \cdot 6 = 1200\text{A}$$

Calculate the rotor bridge flux density for 10%, 20%, 30%, 50%, and 100% of rated current and the corresponding slot geometrical specific permeance.

Solution

Let us first introduce a typical magnetisation curve.

B(T)	0.05	0.1	0.15	0.2	0.25	0.3	0.35	0.4	0.45	0.5
H(A/m)	22.8	35	45	49	57	65	70	76	83	90
B(T)	0.55	0.6	0.65	0.7	0.75	0.8	0.85	0.9	0.95	1
H(A/m)	98	106	115	124	135	148	162	177	198	220

B(T)	1.05	1.1	1.15	1.2	1.25	1.3	1.35	1.4	1.45	1.5
H(A/m)	237	273	310	356	417	482	585	760	1050	1340
B(T)	1.55	1.6	1.65	1.7	1.75	1.8	1.85	1.9	1.95	2.0
H(A/m)	1760	2460	3460	4800	6160	8270	11,170	15,220	22,000	34,000

Neglecting the tooth top saturation and including only the bridge zone, the Ampere's law yields

$$H_{br} a_{br} = I_b; \quad H_{br} = \frac{I_b}{a_{br}}; \quad a_{br} = 11\text{ mm}$$

For various levels of bar current, H_{br} and B_{br} become

I_b (A)	120	240	360	600	1200
H_{br}	10,909	21,818	32,727	54,545	109,090
B_{br}	1.84	1.95	1.98	2.1	2.2
$\mu_{rel} = \frac{\mu_{br}}{\mu_0}$	134.28	71.16	48.165	30.653	16.056
$a'_{or} = \frac{a_{br}}{\mu_{rel}}$ (mm)	0.0819	0.1546	0.2284	0.3588	0.6851
$\lambda_{slot} = \frac{2h_r}{3(a_r + a_{br})} + \frac{h_{br}}{a'_{or}}$	14.293	7.51	5.42	3.8287	2.5013

Note that at low current levels, the slot geometrical specific permeance is unusually high. This is so because the iron bridge is long (a_{br} = 11 mm); so H_{br} is small at small currents.

Also, the iron bridge is 1 mm thick in our case, while 0.5–0.6 mm would produce better results (lower λ_{slot}). The slot height is quite large (h_r = 25 mm), so significant skin effect is expected at high slips. We investigated here only the currents below the rated one so $Sf_1 < 3$ Hz in general. Should the skin effect occur, it would have reduced the first term in λ_{slot} (slot body specific geometrical permeance) by the factor K_x.

As a conclusion to this numerical example, we may infer that a thinner bridge with a smaller length would saturate sooner (at smaller currents below the rated one), producing finally a lower slot leakage permeance λ_{slot}. The advantages of closed slots are related to lower stray load losses and lower noise.

9.7 LEAKAGE SATURATION AND SKIN EFFECTS – A COMPREHENSIVE ANALYTICAL APPROACH

Magnetic saturation in IMs occurs in the main path at low slip frequencies and moderate currents unless closed rotor slots are used when their iron bridges saturate the leakage path.

In contrast, for high slip frequencies and high currents, the leakage flux paths saturate as the main flux decreases with slip frequency for constant stator voltage and frequency.

The presence of slot openings, slot skewing, and winding distribution in slots makes the problem of accounting for magnetic saturation, in the presence of rotor skin effects, when the IM is voltage fed, a very difficult task.

Ultimately, a 3D finite element approach would solve the problem but for prohibitive computation time.

Less computation-intensive analytical solutions have been gradually introduced [10–16]. However, only the starting conditions have been treated (with $I_s = I_r'$ known), with skin effect neglected. Main flux path saturation level, skewing, zig-zag leakage flux have all to be considered for a more precise assessment of induction motor parameters for large currents.

In what follows, an attempt is made to allow for magnetic saturation in the main and leakage path (with skin effect accounted for by appropriate correction coefficients K_R and K_x) for given slip, frequency, voltage, motor geometry, and no-load curve, by iteratively recalculating the stator magnetisation and rotor currents until sufficient convergence is obtained. So, in fact, the saturation and skin effects are explored from no load to motor stall conditions.

The computation algorithm presupposes that the motor geometry and the no-load curve, as $L_m(I_m)$, magnetising inductance versus no-load current are known based on analytical or FEM modelling or from experiments.

The leakage inductances are initially considered unsaturated (as derived in Chapter 6) with their standard expressions. The influence of skin effect on slot body-specific geometric permeance and on resistances is already accounted for in these parameters.

Consequently, given the slip value, the equivalent circuit with the variable parameter $L_m(I_m)$, Figure 9.23, may be solved.

The phasor diagram corresponding to the equivalent circuit in Figure 9.23 is shown in Figure 9.24.

From the equivalent circuit, after a few iterations, based on the $L_m(I_m)$ saturation function, we can calculate I_s, φ_s, I_r', φ_r, I_m, and φ_m.

As expected, at higher slip frequency ($S\omega_1$) and high currents, the main flux ψ_m decreases for constant voltage V_s and stator frequency ω_1. At start ($S = 1$, rated voltage and frequency), due to the large voltage drop over ($R_s + j\omega_1 L_{sl}$), the main flux Ψ_m decreases to 50%–60% of its value at no load.

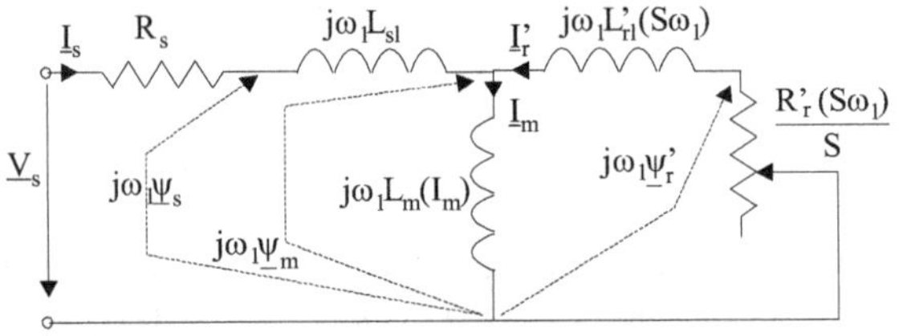

FIGURE 9.23 Equivalent circuit with main flux path saturation and skin effects accounted for.

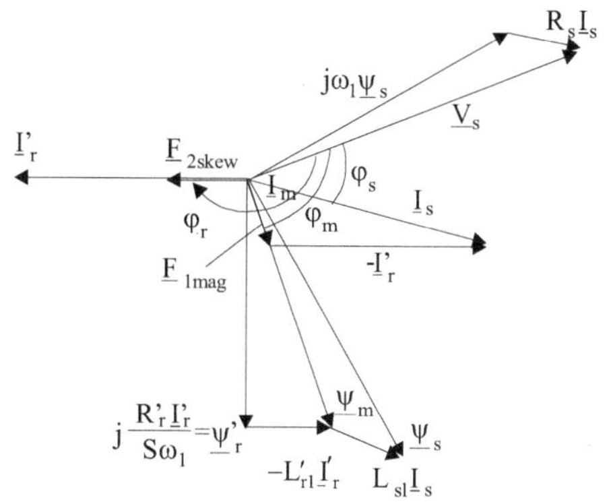

FIGURE 9.24 The phasor diagram.

This ψ_m decreasing with slip and current rise led to the idea of neglecting the main flux saturation level when leakage flux path saturation was approached. However, at a more intrusive view, in the teeth top main flux (Ψ_m), slot neck leakage flux (Φ_{st1}, Φ_{st2}), zig-zag leakage flux (Φ_{z1}, Φ_{z2}), the so-called skewing flux (Φ_{sk1}, Φ_{sk2}), and the differential leakage flux (Φ_{d1}, Φ_{d2}), all meet to produce the resultant flux and saturation level (Figure 9.25). It is true that these fluxes have, at a certain point of time, their maxima at different positions along rotor periphery.

In essence, the main flux phase angle is φ_m, while Φ_{st1}, Φ_{z1}, Φ_{d1}, Φ_{st2}, Φ_{z2}, and Φ_{d2} have the phase angles φ_s and φ_r, while Φ_{sk1} and Φ_{sk2} are related to rotor current uncompensated mmf and, thus, is linked to φ_r again.

As the slip varies, not only I_s, I_r', and I_m vary but also their phase angles: especially φ_s and φ_r', with φ_m about the same.

The differential leakage fluxes (Φ_{d1} and Φ_{d2}) are not shown in Figure 9.25 though they also act within airgap and teeth top zones, AB and DE.

To make the problem easily amenable to computation, it is assumed that when saturation occurs in some part of the machine, the ratios between various flux contributions to total flux remain as they were before saturation.

The case of closed rotor slots is to be treated separately. Let us now define first the initial expressions of various fluxes per tooth and their geometrical specific leakage permeances.

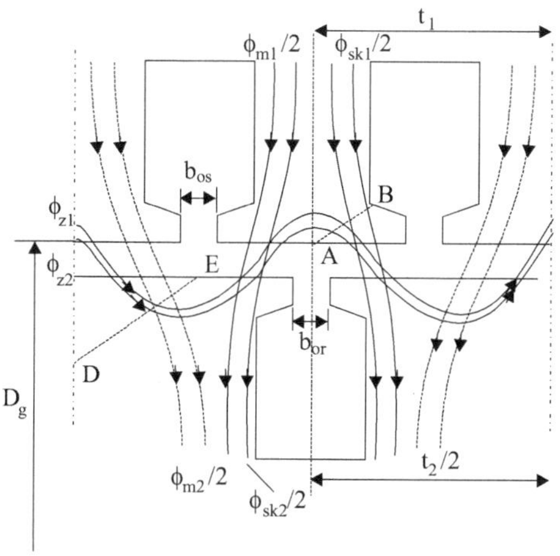

FIGURE 9.25 Computational flux lines.

The main flux per tooth Φ_{m1} is

$$\Phi_{m1} = \frac{\psi_m \cdot 2p_1}{W_1 K_{w1} N_s}; \quad \psi_m = L_m(I_m) I_m \sqrt{2} \tag{9.57}$$

On the other hand, noticing that the stator slot mmf $AT_{1m} = n_s I_m \sqrt{2}$, the main flux per stator tooth is

$$\Phi_{m1} = AT_{1m} \mu_0 l_{stack} \lambda_m; \quad \Phi_{m2} = \Phi_{m1} \frac{N_s}{N_r} \tag{9.58}$$

where λ_m is the equivalent main flux geometrical specific permeance.

In a similar manner, the zig-zag fluxes Φ_{z1} and Φ_{z2} are

$$\begin{aligned} \Phi_{z1} &= AT_1 \mu_0 l_{stack} \lambda_{z1} \\ \Phi_{z2} &= AT_2 \mu_0 l_{stack} \lambda_{z2} \\ AT_1 &= n_s I_s \sqrt{2}; \; AT_2 = n_2 I_2' \sqrt{2}; \end{aligned} \tag{9.59}$$

From (9.56) and (9.59), n_r is

$$n_r = \frac{I_b \sqrt{2}}{I_r' \sqrt{2}} = n_s K_{w1} \frac{N_s}{N_r} \tag{9.60}$$

The unsaturated zig-zag specific permeances λ_{z1} and λ_{z2} are [5]

$$\lambda_{z1} = \frac{K_{w1}{}^2 \pi D_g}{12 N_s g}(1.2a_1 - 0.2) \tag{9.61}$$

$$\lambda_{z2} = \frac{K_{w1}{}^2 \pi D_g}{12 N_r g}(1.2a_2 - 0.2) \tag{9.62}$$

$$a_1 = \frac{t_1(5g + b_{os}) - b_{or}{}^2}{t_1(5g + b_{os})} \tag{9.63}$$

$$a_2 = \frac{t_2(5g + b_{or}) - b_{os}{}^2}{t_2(5g + b_{or})} \tag{9.64}$$

The slot neck fluxes Φ_{st1} and Φ_{st2} are straightforward.

$$\Phi_{st1} = AT_1 \mu_0 l_{stack} \lambda_{st1}; \quad \lambda_{st1} = \frac{h_{os}}{b_{os}} \tag{9.65}$$

$$\Phi_{st2} = AT_2 \mu_0 l_{stack} \lambda_{st2}; \quad \lambda_{st2} = \frac{h_{or}}{b_{or}} \tag{9.66}$$

The differential leakage, specific permeance $\lambda_{d1,2}$ may be derived from a comparison between magnetisation harmonic inductances $L_{m\nu}$ and the leakage inductance expressions L_{sl} and L_{rl}' (Chapter 7).

$$L_{m\nu} = \frac{6\mu_0}{\pi^2} \frac{(\tau/\nu) l_{stack} (W_1 K_{w\nu})^2}{(p_1 \nu) g K_c (1 + K_{steeth})} \tag{9.67}$$

$$L_{sl} = 2\mu_0 \frac{W_1^2}{p_1 q}\left(\lambda_{ss} + \lambda_{z1} + \lambda_{endcon1} + \lambda_{st1} + \lambda_{d1}\right) \tag{9.68}$$

$$L_{rl} = 2\mu_0 \frac{W_1^2}{p_1 q}\left(\lambda_{sr} + \lambda_{z2} + \lambda_{endcon2} + \lambda_{d2} + \lambda_{st2} + \lambda_{sk2}\right) \tag{9.69}$$

$$\lambda_{d1} = \sum_{v>1}^{\infty} 3\frac{K_{wv}^2}{gK_c v^2}\frac{q_1\tau}{\left(1+K_{steeth}\right)} \approx \lambda_{m1}\left(\tau_{d1} - \Delta\tau_{d1}\right)$$

$$\tau_{d1} = \pi^2\frac{\left(10q_1^2+2\right)}{27}\sin^2\left(\frac{30°}{q_1}\right) - 1; \quad \Delta\tau_{d1} = K_{z1}\frac{1}{9q_1^2}$$

$$\lambda_{d2} = \lambda_{m1}\frac{N_s}{N_r}\left(\tau_{d2} - \Delta\tau_{d2}\right); \quad \Delta\tau_{d2} = K_{z2}\left(\frac{2p_1}{N_r}\right)^2 \tag{9.70}$$

$$\tau_{d2} = \frac{\left(\frac{\pi p_1}{N_r}\right)^2}{\sin^2\left(\frac{\pi p_1}{N_r}\right)}\frac{\left(\frac{\pi p_1 \cdot skew}{N_r}\right)^2}{\sin^2\left(\frac{\pi p_1 \cdot skew}{N_r}\right)} - 1$$

The coefficients K_{z1} and K_{z2} are found from Figure 9.26.

The slot-specific geometric permeances λ_{ss} and λ_{sr}, for rectangular slots, with rotor skin effect, are

$$\lambda_{ss} = \frac{h_s}{3b_s} + \lambda_{st1} \tag{9.71}$$

$$\lambda_{sr} = \frac{h_r}{3b_r} + \lambda_{st2}; \quad K_x - \text{skin effect correction} \tag{9.72}$$

The most important "ingredient" in leakage inductance saturation is, for large values of stator and rotor currents, the so-called skewing (uncompensated) mmf.

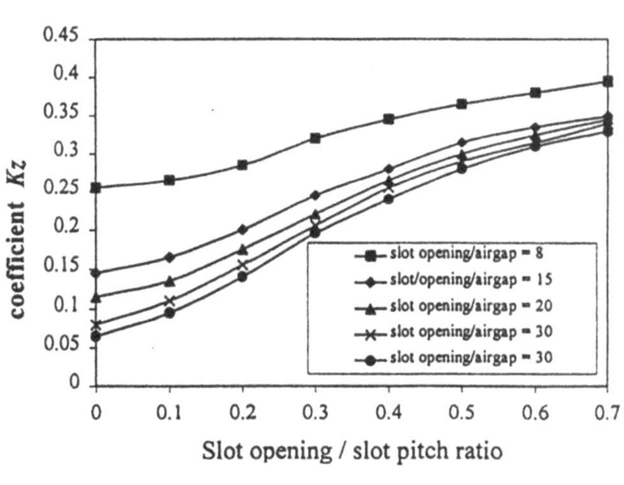

FIGURE 9.26 Correction coefficient for differential leakage damping.

9.7.1 The Skewing mmf

The rotor (stator) slot skewing performed to reduce the first slot harmonic torque (in general) also has some side effects such as a slight reduction in the magnetisation inductance L_m, accompanied by an additional leakage inductance component, L'_{rskew}. We decide to "attach" this new leakage inductance to the rotor as it "disappears" at zero rotor current.

To account for L_m alteration due to skewing, we use the standard skewing factor K_{skew}^{conv}.

$$K_{skew}^{conv} = \frac{\sin\left(\frac{\pi}{2}\text{skew}\Big/m_1q_1\right)}{\frac{\pi}{2}\text{skew}\Big/m_1q_1}; \quad K_{skew} \approx 0.98-1.0 \tag{9.73}$$

and include a rotor current linear dependence to make sure that the influence disappears at zero rotor current,

$$K_{skew} = 1-\left(1-K_{skew}^{conv}\right)\frac{I'_r}{I_s} \tag{9.74}$$

Skew – is the skewing in stator slot pitch counts.

To account for the skewing effect, we first consider the stator and rotor mmf fundamentals for a skewed rotor.

As the current in the bar is the same all along the stack length (with interbar currents neglected), the rotor mmf phasing varies along OY, beside OX and time (Figure 9.27); so the resulting mmf fundamental is

$$F(x,y,t) = F_{1m}\sin\left(\frac{\pi}{\tau}x-2\pi f_1 t\right)+F_{2m}\sin\left(\frac{\pi}{\tau}x-2\pi f_1 t-\left(\varphi_s-\varphi_r\right)-\frac{\pi}{m_1q_1}\text{skew}\frac{y}{l_{stack}}\right)$$

$$-\frac{l_{stack}}{2} \le y \le \frac{l_{stack}}{2} \tag{9.75}$$

We may consider this as made of two terms, the magnetisation mmf F_{1mag} and the skewing mmf F_{2skew}.

$$F(x,y,t) = F_{1mag} + F_{2skew} \tag{9.76}$$

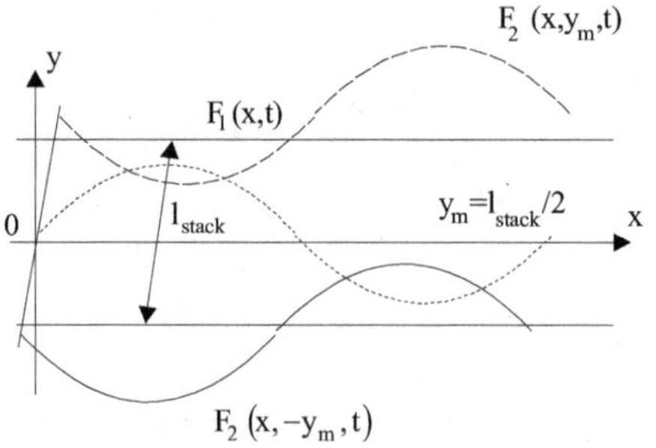

FIGURE 9.27 Phasing of stator and rotor mmfs for skewed rotor slots.

$$F_{1mag} = F_{1m} \sin\left(\frac{\pi}{\tau}x - 2\pi f_1 t\right)$$

$$+F_{2m} \cos\left(\frac{\pi}{m_1 q_1}\text{skew}\frac{y}{l_{stack}}\right)\cdot \sin\left(\frac{\pi}{\tau}x - 2\pi f_1 t - (\varphi_s - \varphi_r)\right) \quad (9.77)$$

$$F_{2skew} = -F_{2m} \sin\left(\frac{\pi}{m_1 q_1}\text{skew}\frac{y}{l_{stack}}\right)\cdot \cos\left(\frac{\pi}{\tau}x - 2\pi f_1 t - (\varphi_s - \varphi_r)\right) \quad (9.78)$$

F_{1mag} is only slightly influenced by skewing (9.77) as also reflected in (9.73).

On the other hand, the skewing (uncompensated) mmf varies with rotor current (F_{2m}) and with the axial position y. It tends to be small even at rated current in amplitude (9.78), but it gets rather high at large rotor currents. Also notice that the phase angle of F_{1mag} (φ_m) and F_{2skew} (φ_r) is different by slightly more than 90° (Figure 9.24), above rated current,

$$\varphi_{skew} = \varphi_r - \varphi_m \quad (9.79)$$

We may now define a skewing magnetisation current I_{mskew}

$$I_{mskew} = \pm I_m \frac{AT_2}{AT_{1m}} \sin\left(\pi \cdot p_1 \cdot \text{skew}/N_s\right) \quad (9.80)$$

Example 9.6 The Skewing mmf

Let us consider an induction motor with the following data: no-load rated current: I_m = 30% of rated current; starting current: 600% of rated current; the number of poles: $2p_1 = 4$, skew = 1 stator slot pitch; number of stator slots $N_s = 36$. Assessment of the maximum skewing magnetisation current I_{mskew}/I_m for starting conditions is required.

Solution

Making use of (9.80), the skewing magnetisation current for $SKW_1 = \pm$skew.

$$\frac{I_{mskew}}{I_m} = \pm\frac{AT_2}{AT_{1m}}\sin\left(\frac{\pi 2p_1(\text{skew}/2)}{N_s}\right) = \pm\left(\frac{I_r'}{I_m}\right)\sin\left(\frac{\pi\cdot 4\cdot 1/2}{36}\right)$$

$$\approx \pm\left(\frac{I_{start}}{I_m}\right)\cdot 0.15643 = \pm\frac{600\%}{30\%}\cdot 0.15643 = \pm 3.1286!$$

This shows that, at start, the maximum flux density in the airgap produced by the "skewing magnetisation current" in the extreme axial segments of the stack will be maximum and much higher than the main flux density in the airgap, which is dominant below rated current when I_{mskew} becomes negligible (Figure 9.28).

Airgap flux densities in excess of 1.0 T are to be expected at start and thus heavy saturation all over magnetic circuit should occur.

Let us remember that in these conditions, the magnetisation current is reduced such that the airgap flux density of main flux is (0.5–0.6) of its rated value, and its maximum is phase shifted with respect to the skewing flux by more than 90° (Figure 9.24). Consequently, at start (high currents), the total airgap flux density is dominated by the skewing mmf contribution! This phenomenon has been documented through FEM [17].

Dividing the stator stack in (N_{seg} + 1) axial segments, we have,

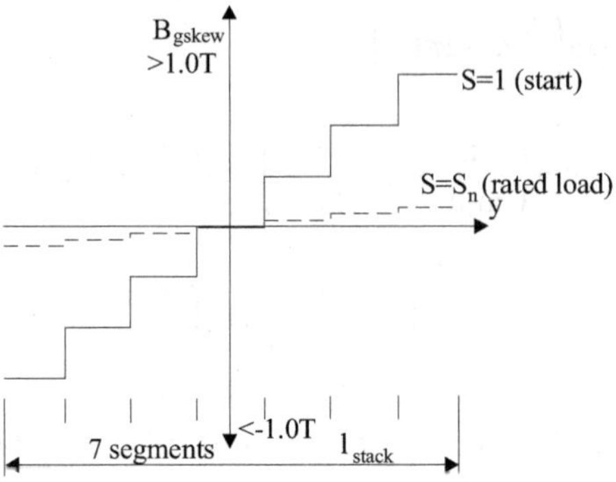

FIGURE 9.28 Skewing magnetisation current produced airgap flux density.

$$SK_{w1} = 0.0, \frac{skew}{N_{seg}}, \frac{skew}{2} \tag{9.81}$$

Now, based on I_{mskew}, from the magnetisation curve ($L_m(I_m)$), we may calculate the magnetisation inductance for skewing magnetisation current $L_m(I_{mskew})$. Finally, we may define a specific geometric permeance for skewing λ_{sk1}.

$$\lambda_{sk1} = \frac{L_m(I_{mskew})}{N_s K_{w1} W_1 n_s \mu_0 l_{stack}} \tag{9.82}$$

W_1 – turns/phase.

λ_{sk1} depends on the segment considered (6–10 segments are enough) and on the level of rotor current.

Now the skewing flux/tooth in the stator and rotor is

$$\Phi_{sk1} = AT_2 \sin\left(\frac{\pi 2p_1 SKW1}{N_s}\right)\mu_0 l_{stack}\lambda_{sk1} \tag{9.83}$$

$$\Phi_{sk2} = \Phi_{sk1}\frac{N_s}{N_r} \tag{9.84}$$

In (9.83), the value of Φ_{sk1} is calculated as if the value of segmented skewing is valid all along the stack length. This means that all calculations will be made n_{seg} times and then average values will be used to calculate the final values of leakage inductances.

Now if the skewing flux occurs both in the stator and the rotor, when the leakage inductance is calculated, the rotor one will include the skewing permeance λ_{sk2}.

$$\lambda_{sk2} = \lambda_{sk1}\frac{N_s}{N_r} \tag{9.85}$$

9.7.2 Flux in the Cross-Section Marked by AB (Figure 9.25)

The total flux through AB, Φ_{AB} is

$$\Phi_{AB} = \frac{\Phi_{m1}}{2} + \frac{\Phi_{sk1}}{2} + \Phi_{z1} + \Phi_{z2} + \Phi_{st1} \tag{9.86}$$

Let us denote

$$C_{m1} = \frac{\Phi_{m1}}{2\Phi_{AB}} \quad C_{sk1} = \frac{\Phi_{sk1}}{2\Phi_{AB}} \quad C_{z1} = \frac{\Phi_{z1}}{\Phi_{AB}} \tag{9.87}$$

$$C_{st1} = \frac{\Phi_{st1}}{\Phi_{AB}} \quad C_{z2} = \frac{\Phi_{z2}}{\Phi_{AB}} \tag{9.88}$$

These ratios, calculated before slot neck saturation is considered, are also taken as valid for total leakage saturation condition.

There are two different situations: one with the stator tooth saturating first and the second with the saturated rotor teeth.

9.7.3 THE STATOR TOOTH TOP SATURATES FIRST

Making use of Ampere's and flux laws, the slot neck zone is considered (Figure 9.29).

$$(t_1 - b_{os})H_1 + b_{os}H_{10} = AT_1 \tag{9.89}$$

The total flux through AB is

$$\Phi_{ABsat} = AB \cdot l_{stack} \cdot B_1 \tag{9.90}$$

On the other end, with C_{st1} known from unsaturated conditions, the slot neck flux Φ_{st1} is

$$\Phi_{st1} = C_{st1}\Phi_{ABsat} = \mu_0 H_{10}h_{os}l_{stack} \tag{9.91}$$

From (9.89)–(9.91), we finally obtain

$$B_1 = \mu_0 h_{os} \frac{|AT_1 - t_1'H_1|}{b_{os}C_{st1}AB}; \quad t_1' = t_1 - b_{os} \tag{9.92}$$

Equation (9.92), corroborated with the lamination $B_1(H_1)$ magnetisation curve, leads to B_1 and H_1 (Figure 9.30).

An iterative procedure as in Figure 9.30 may be used to solve (9.92) for B_1 and H_1. Finally, from (9.90), the saturated value of $(\Phi_{AB})_{sat}$ is obtained.

Further on, new values of various teeth flux components are obtained.

$$\left(\frac{\Phi_{m1}}{2}\right)_{sat} = C_{m1}(\Phi_{AB})_{sat}$$

$$(\Phi_{sk1})_{sat} = C_{sk1}(\Phi_{AB})_{sat}$$

$$(\Phi_{z1})_{sat} = C_{z1}(\Phi_{AB})_{sat} \tag{9.93}$$

$$(\Phi_{st1})_{sat} = C_{st1}(\Phi_{AB})_{sat}$$

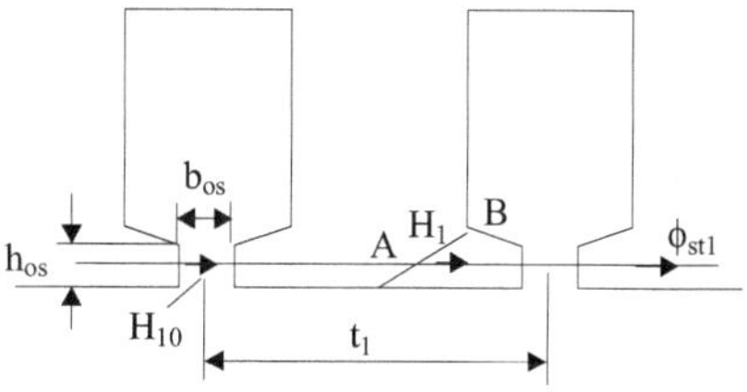

FIGURE 9.29   Stator slotting geometry.

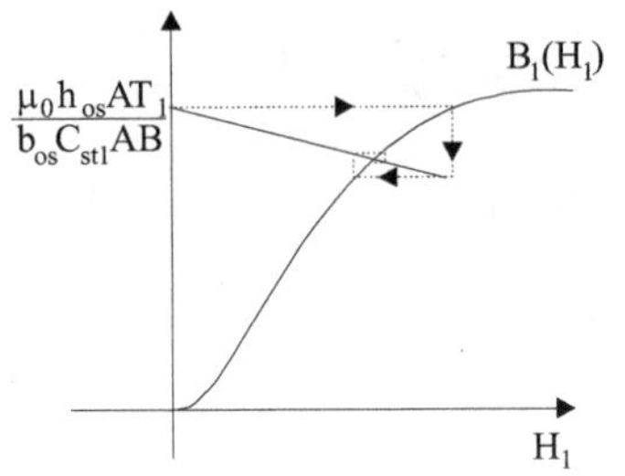

FIGURE 9.30 Tooth top flux density.

The new (saturated) values of stator leakage permeance components are

$$\left(\lambda_{st1}\right)_{sat} = \frac{\left(\Phi_{st1}\right)_{sat}}{AT_1 \mu_0 l_{stack}}$$
$$\left(\lambda_{z1}\right)_{sat} = \frac{\left(\Phi_{z1}\right)_{sat}}{AT_1 \mu_0 l_{stack}} \tag{9.94}$$

The end connection and differential geometrical permeances λ_{endcon} and λ_{d1} maintain, in (9.68), their unsaturated values, while $(\lambda_{st1})_{sat}$ and $(\lambda_{z1})_{sat}$ (via (9.94)) enter (9.68), and thus, the saturated value of stator leakage inductance L_{sl} for given stator and rotor currents and slip is obtained.

We have to continue with the rotor. There may be two different situations.

9.7.4 Unsaturated Rotor Tooth Top

In this case, the saturation flux from the stator tooth is "translated" into the rotor by the ratio of slot numbers N_s/N_r.

$$\left(\lambda_{st2}\right)_{sat} = \frac{\left(\Phi_{st1}\right)_{sat} \frac{N_s}{N_r}}{AT_1 \mu_0 l_{stack}}$$
$$\left(\lambda_{z2}\right)_{sat} = \frac{\left(\Phi_{z1}\right)_{sat} \frac{N_s}{N_r}}{AT_1 \mu_0 l_{stack}} \tag{9.95}$$
$$\left(\lambda_{sk2}\right)_{sat} = \lambda_{sk1} \frac{N_s}{N_r}$$

In Equation (9.95), it is recognised that the skewing flux "saturation" is not much influenced by the tooth top flux zone saturation as it represents a small length flux path in comparison with the main flux path available to skewing flux.

By now, with (9.69), the saturated rotor leakage inductance L'_{rl} may be calculated. With these new, saturated, leakage inductances, new stator, magnetisation, and rotor currents are calculated.

The process is repeated until sufficient convergence is obtained with respect to the two leakage inductances. A new value of slip is then chosen and the computation cycle is repeated.

9.7.5 Saturated Rotor Tooth Tip

In this case, after the AB zone was explored as above, the DE cross-section is investigated (Figure 9.31). Again the Ampere's law across tooth tip and slot neck yields

$$(t_2 - b_{or})H_2 + b_{or}H_{20} = AT_2 \tag{9.96}$$

The flux law results in

$$\Phi_{DE} = \Phi_1 + (\Phi_{st2})_{sat}; \; B_2 = \mu_0 H_{20} \tag{9.97}$$

where

$$\Phi_1 = \Phi_{ABsat}\frac{N_s}{N_r} \tag{9.98}$$

Also,

$$(\Phi_{st2})_{sat} = DE \cdot l_{stack} \cdot B_2(H_2); \quad t_2' = t_2 - b_{or} \tag{9.99}$$

$$(\Phi_{st2})_{sat} = \mu_0 H_{20} h_{os} l_{stack} = \mu_0 h_{or}\frac{(AT_2 - t_2' H_2)}{b_{or}} l_{stack} \tag{9.100}$$

Finally,

$$B_2(H_2) = \frac{1}{l_{stack}DE}\left[\Phi_{ABsat}\frac{N_s}{N_r} + \mu_0 h_{or}\frac{(AT_2 - t_2' H_2)}{b_{or}} l_{stack}\right] \tag{9.101}$$

Equation (9.101), together with the rotor lamination magnetisation curve, $B_2(H_2)$, is solved iteratively (as illustrated in Figure 9.30 for the stator).

Then,

$$(\lambda_{st2})_{sat} = \frac{(\Phi_{st2})_{sat}\lambda_{st2}}{\Phi_{st2}}; \quad \lambda_{st2} = \frac{h_{or}}{b_{or}} \tag{9.102}$$

$$\Phi_{st2} = \mu_0 AT_2 l_{stack} \lambda_{st2} \tag{9.103}$$

The other specific geometrical permeances in the rotor leakage inductance $(\lambda_{z2})_{sat}$ and $(\lambda_{sk2})_{sat}$ are calculated as in (9.95). Now we calculate the saturated value of L_{rl}' in (9.69). Then, from the equivalent circuit, new values of currents are calculated and a new iteration is initiated until sufficient convergence is reached.

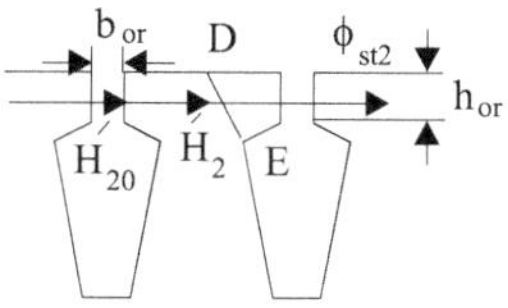

FIGURE 9.31 Rotor slot neck flux Φ_{st2}.

9.7.6 The Case of Closed Rotor Slots

Closed rotor slots (Figure 9.32) are used to reduce starting current, noise, vibration, and windage friction losses though the breakdown and starting torque are smaller and so is the rated power factor.

The slot iron bridge region is discretised in 2n + 1 zones of constant (but distinct) permeability μ_j. A value of n = (4–6) suffices. The Ampere's and flux laws for this region provide the following equations:

$$b_0\left(H_0 + 2\sum_1^n H_j\right) = AT_2 \tag{9.104}$$

$$B_0h_0 = B_1h_1 = B_2h_2 = \ldots = B_nh_n \tag{9.105}$$

Given an initial value for B_0, for given AT_2, B_1, …, B_n are determined from (9.105). From the lamination magnetisation curve, H_0, H_1, …, H_n are found. A new value of AT_2 is calculated, then compared with the given one and a new B_0 is chosen. The iteration cycle is revisited until sufficient convergence is obtained.

Finally, the slot bridge geometrical specific permeance $(\lambda_{st2})_{sat}$ is

$$(\lambda_{st2})_{sat} = \frac{1}{\frac{b_0}{h_0}\left(\frac{B_0}{\mu_0 H_0}\right)^{-1} + \frac{2b_0}{h_1}\left(\frac{B_1}{\mu_0 H_1}\right)^{-1} + \ldots + \frac{2b_n}{h_n}\left(\frac{B_n}{\mu_0 H_n}\right)^{-1}} \tag{9.106}$$

All the other permeances are calculated as in previous paragraphs.

9.7.7 The Algorithm

The algorithm for the computation of permeance in the presence of magnetic saturation and skin effects, as unfolded gradually so far, can be synthesised as the following:

- Load initial data (geometry, winding data, name-plate data, etc.)
- Load magnetisation file: B(H)
- Load magnetisation inductance function $L_m(I_m)$ – from FEM, design, or tests
- Compute differential leakage specific permeances
- For $S = S_{min}$, S_{step}, and S_{max}

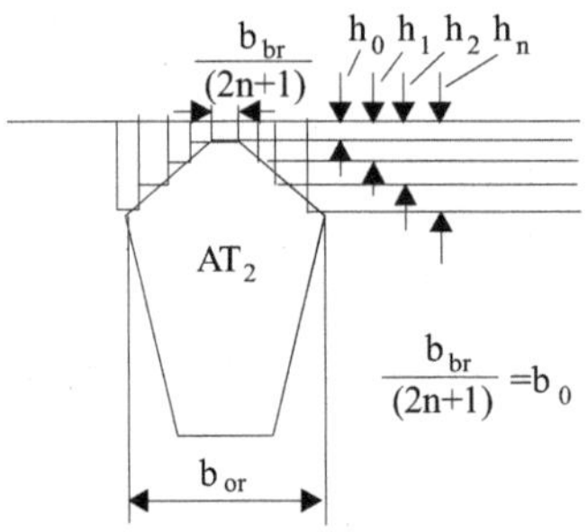

FIGURE 9.32 Closed rotor slot bridge discretisation.

```
    for SKW1 = 0:skew/nseg:skew
       compute unsaturated specific permeances
       while magnetisation current error < 2%
              compute induction motor model using nonlinear Lm(Im) function
       end;
       Save unsaturated parameters
       1. if case = closed rotor slots;
              compute closed rotor case;
       else
              while H1 error < 2%;
                     compute stator saturation case;
              end;
              compute unsaturated rotor case;
              if case = saturated rotor;
                     while H2 error < 2%;
                            compute saturated rotor case
                     end;
              end.
               Compute saturated parameters; compare them with those of
               previous cycle and goto 1. Return to 1 until sufficient
               convergence in Lsl and Lrl' is reached.
       end.
    end (axial segments cycle)
 end (slip cycle)
    Save saturated parameters
    for SKW1 = 0:skew/nseg:skew;
          Compute saturated parameters average values; plot the IM
          parameters (segment number, slip);
          Compute IM equivalent circuit using an average value for each
          parameter (for the nseg, axial segments);
          Compute stator, rotor, magnetisation current, torque, power
          factor versus slip;
          Plot the results
 end.
```

Applying this algorithm for S = 1 an 1.1 kW, two-pole, three-phase IM with skewed rotor, the results shown in Figure 9.33 have been obtained [18]. Heavy saturation occurs at high voltage levels. The model seems to produce satisfactory results.

9.8 THE FEM APPROACH

The FEM has been so far proved successful in predicting the main path flux saturation or the magnetisation curve $L_m(I_m)$. It has also been used to calculate the general shape deep bar [19] and double cage bar [20] resistance and leakage inductance for given currents and slip frequency (Figure 9.34).

The essence of the procedure is to take into consideration only a slot sector and apply symmetry rules (Figure 9.34b).

The stator is considered slotless and an infinitely thin stator current mmf, placed along AB, produces the main flux while the stator core has infinite permeability. Moreover, when the stator (and rotor) mmf has time harmonics, for inverter-fed situations, they may be included, but the saturation background is "provided" by the fundamental to simplify the computation process.

Still the zig-zag and skewing flux contributions to load saturation (in the teeth tips) are not accounted for.

The entire machine, with slotting on both sides of the airgap, and the rotor divided into a few axial slices to account for skewing may be approached by modified 2D-FEM. But the computation

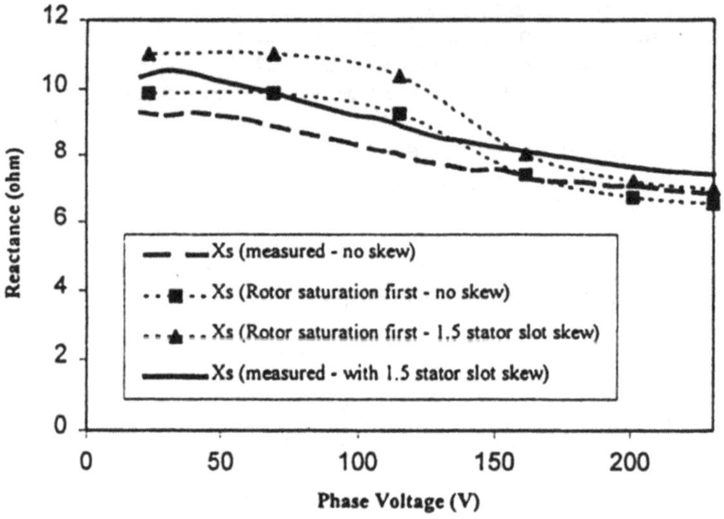

FIGURE 9.33 Leakage reactance for a skewed rotor versus voltage at stall.

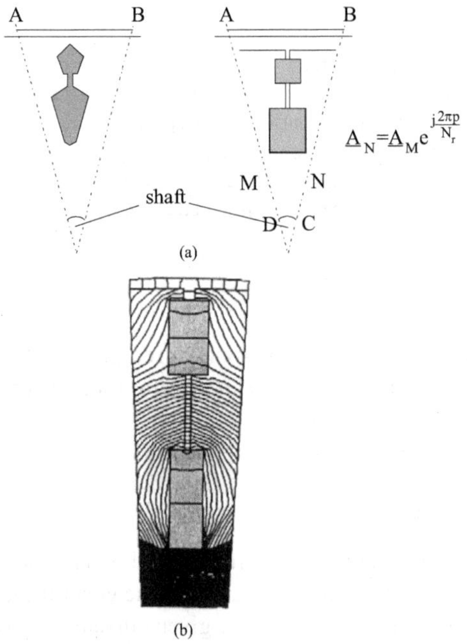

FIGURE 9.34 (a) Deep bar and double-cage rotor geometry and (b) FEM flux lines.

process has to be repeated until sufficient convergence in saturation, then in stator and rotor resistances and leakage inductances, is obtained for given voltage, frequency, and slip.

This enterprise requires a great amount of computation time, but it may be produced more quickly as PCs are getting stronger every day.

High fundamental frequency (above 500 Hz), rich in harmonics stator voltage, and current in high-speed (above 15,000 rpm with four poles) IM drives experience large skin effect in the stator conductors, also. The level of increasing skin effect resistance depends on the current frequencies involved, conductor diameter, and type of their connection:

- Single wire series connected winding (single layer)
- Multiple elementary conductors in parallel with no stranding (single layer)
- A 180° stranding (multiple conductors in parallel, two layers)
- Litz wire or Roebel bar case.

Based on Ref. [6] and examples shown previously, in this chapter, analytical formula for R_{ac} and L_{ac} may be obtained for all the cases mentioned above.

A typical result for a single-layer winding, fundamental frequency: $f_1 = 2333\,Hz$, and various elementary conductors diameters, in parallel, arranged in 18 layers in a single turn (and layer) coil is shown in Figure 9.35.

A comparison between ideal twist, 180° twist, and no twist winding for the first harmonic (2,333 Hz) is shown in Figure 9.35.

The 180° stranding suffices.

The skin effect problem in the stator conductors may also be approached by FEM.

For a two-layer winding, three turns/coil, 18 elementary conductors in parallel, and an equivalent circuit as in Figure 9.36 is introduced.

For diametrical coils, only one slot should be studied through FEM as the currents in the two coils are in phase.

For ideal twist, all elementary conductors in Figure 9.37 may be connected in series for the purpose of skin effect consideration by FEM.

For a 180° twist, the placement of the 6 turns of 18 conductors in parallel (each) is assumed as shown in Figure 9.38, which corresponds to Figure 9.37 indexing: T (turn), L (layer). With random windings, such placement of conductors in slot is not known exactly.

The flux lines in the slot for max coil current and zero coil current are shown in Figure 9.39.

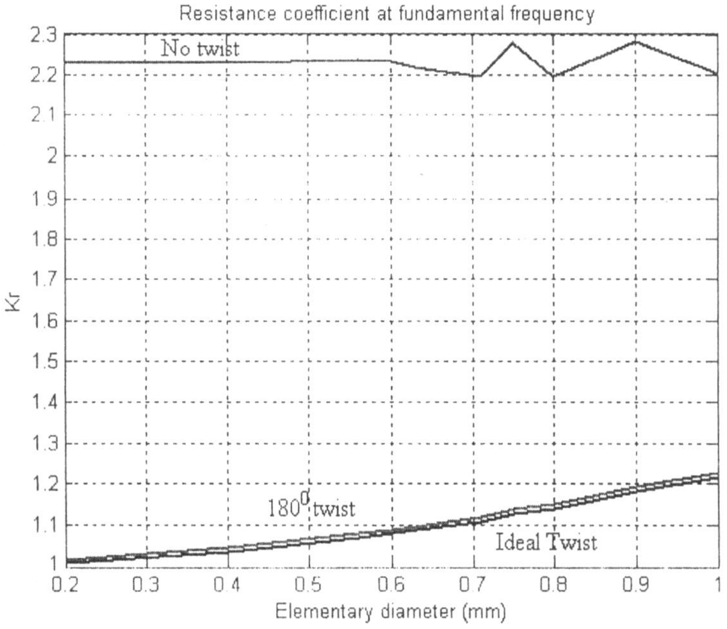

FIGURE 9.35 Resistance A.C. factor K_r for 2333 Hz with 18-layer single-turn coil without twisting.

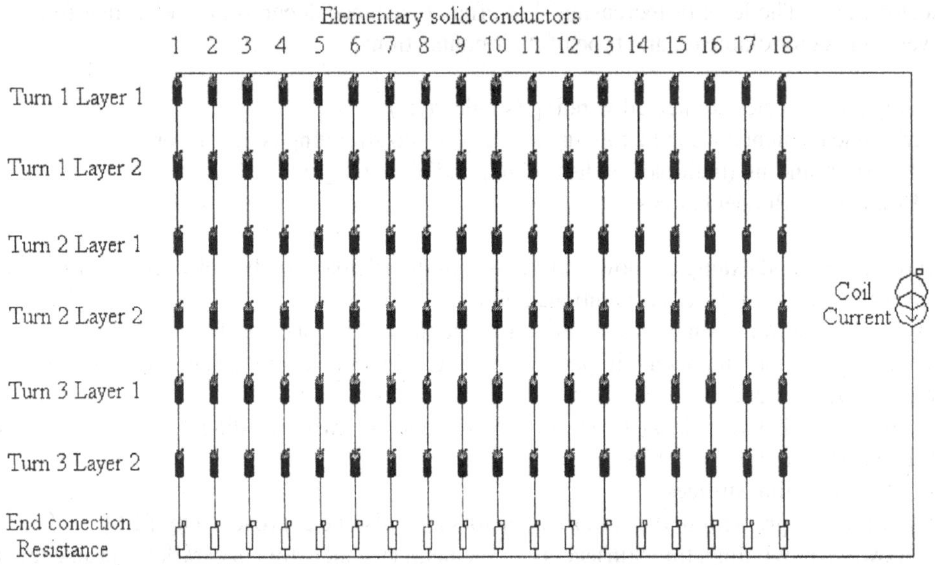

FIGURE 9.36 Two coils, three turns/coil, 18 elementary conductors in parallel.

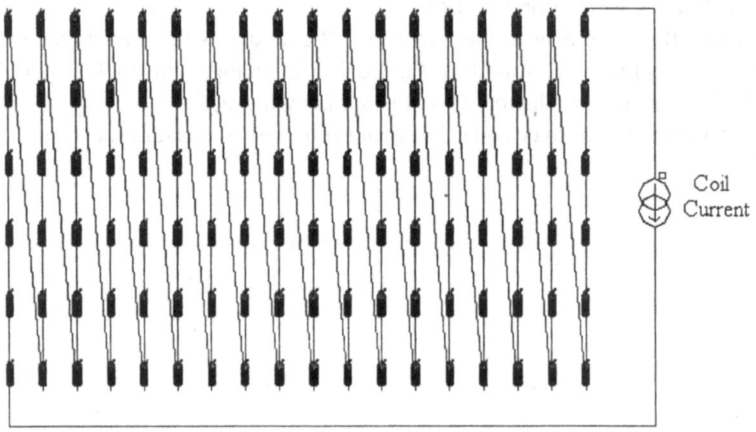

FIGURE 9.37 Series connection or ideal stranding case.

The existence of slot leakage field at zero total coil current is due to eddy currents induced by skin effect in neighbouring paralleled elementary conductors.

The amplitude and phase of current in elementary conductors for 100 A coil current are shown in Figure 9.40.

The different phases of currents in various elementary conductors in parallel explain clearly while even at zero total coil current, there is current in various conductors while the slightly non-uniform current level indicates mild skin effect with practical 180° stranding (only at the coil end!).

It has been proved that the pertinent analytical formula for skin effect (A.C. resistance) coefficient fits rather well to FEM results, at least for elementary conductor diameters below 1 mm diameter, up to 25 kHz.

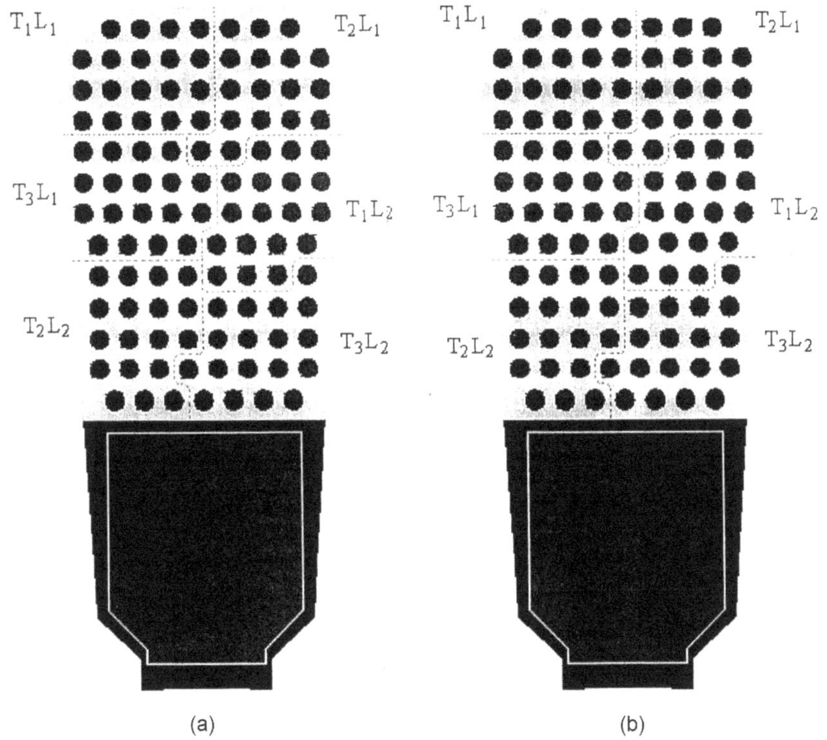

FIGURE 9.38 Eighteen elementary conductors in parallel, three turns/coil, two coils (a) no stranding and (b) 180° stranding.

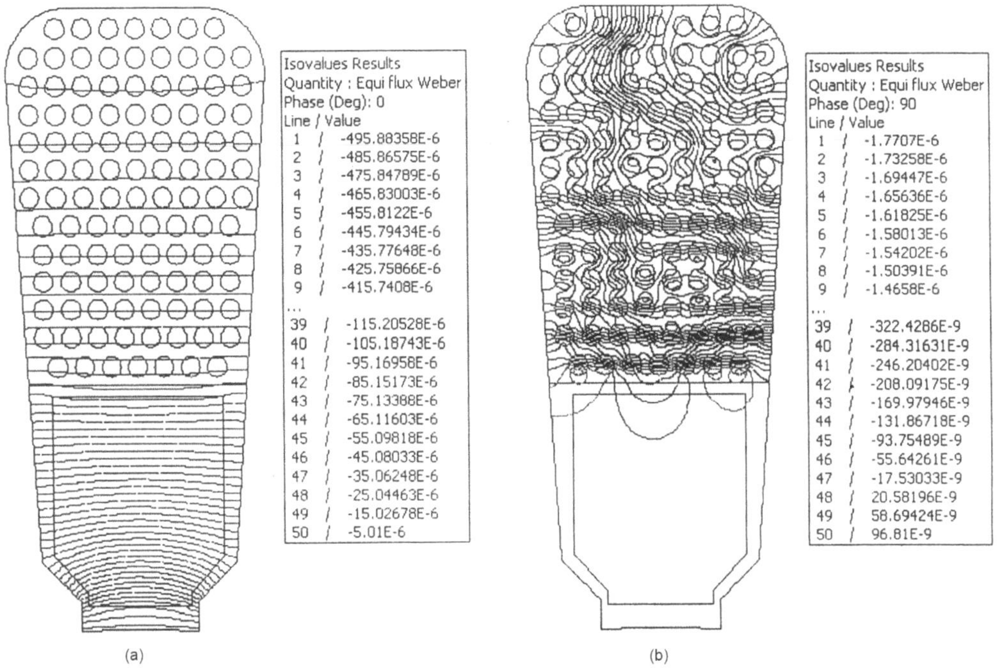

FIGURE 9.39 Flux lines for 180° stranding/18 conductors in parallel, two coils, three turns/coil, at 2333 Hz, conductor diameter 0.63 mm. (a) max coil current and (b) zero coil current.

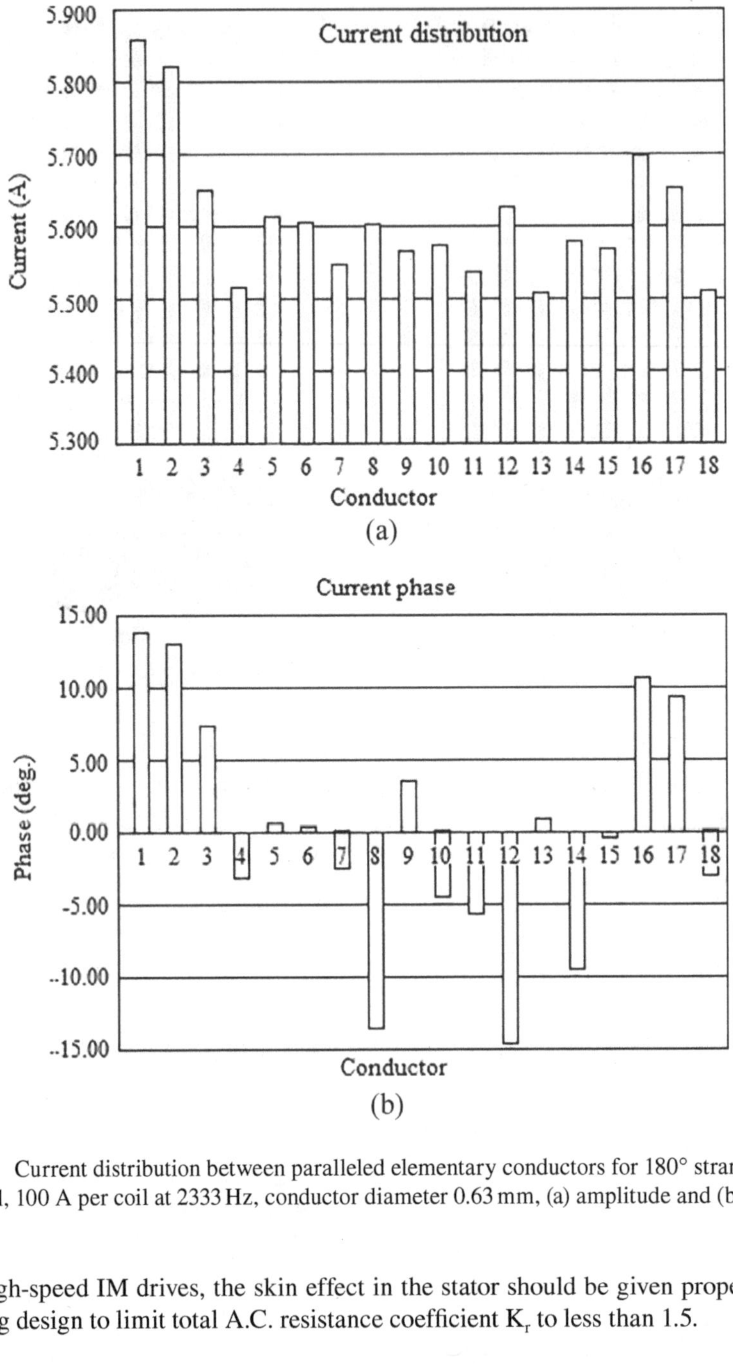

FIGURE 9.40 Current distribution between paralleled elementary conductors for 180° stranding, two coils, three turns/coil, 100 A per coil at 2333 Hz, conductor diameter 0.63 mm, (a) amplitude and (b) phase.

In very high-speed IM drives, the skin effect in the stator should be given proper attention for stator winding design to limit total A.C. resistance coefficient K_r to less than 1.5.

9.9 MAGNETIC SATURATION EFFECTS ON CURRENT/SLIP AND TORQUE/SLIP CURVES

As discussed in this chapter, magnetic saturation affects the main flux paths and the stator and rotor slot leakage flux paths.

The simultaneous treatment of the two magnetic saturation effects may be done by time-domain FEMs, at the price of large computation effort. Here we will tackle a few specific effects of magnetic saturation influence on the torque versus slip and stator current versus slip curves, so important in line-start IM characterisation/design. In future paragraphs, the influence of leakage saturation in closed rotor slots and, respectively, in the solid iron of high-speed IM drives will be treated in some detail.

To treat both magnetic saturation and skin effect, two double-cage rotor models are adopted (Figure 9.41) as in EMTP and, respectively, PSCAD software.

Reactances X_m, X_{1d}, X_{2d}, and X_{12} may all be affected by magnetic saturation of main and leakage flux paths.

A general approximation of reactance dependence on its current is proposed [21]:

$$X(I) = \frac{X_{UNSAT} - X_{SAT}}{\left(1 + \left(\frac{I}{I_{SAT}}\right)^{p}\right)^{1/p}} + X_{SAT} \tag{9.107}$$

X_{UNSAT}, X_{SAT}, I_{SAT}, and p are all coefficients to be determined by a regression method from test results and then used to curve-fit the torque-slip and current-slip curves, for different voltages, all based on experiments [21]. The parameters may be estimated from no-load tests and then used for torque-slip and current-slip curves verification. The results of the investigation in Ref. [21] may be summarised as

- For a fixed value of voltage (but considering saturation level), all dual-cage rotor models yield good torque-slip and current-slip curves fitting.
- If only one (or more) reactances have to account for magnetic saturation, then, if magnetisation reactance is chosen, the latter has a small influence on measured torque and current with slip.
- But if only the stator leakage reactance is considered non-linear (with both DC1 and DC2 models), the prediction of torque-speed and current-speed curves is satisfactory at all voltages up to the rated one.
- For notably higher than rated voltage value (as possible in self-excited induction generators), the saturation has to be treated more thoroughly.
- The correct prediction of torque-slip and current-slip curves is essential in assessing the starting torque, peak torque, and finally, the power factor and efficiency under various loads.
- Sample results [21] of torque-slip, current-slip, and short-circuit impedance-stator current for two IMs (2.2 kW, 50 Hz, rated speeds: 1450 rpm, respectively, 1430 rpm) when only the stator leakage reactance is considered saturation dependent, as shown in Figure 9.42, show remarkable consistency.

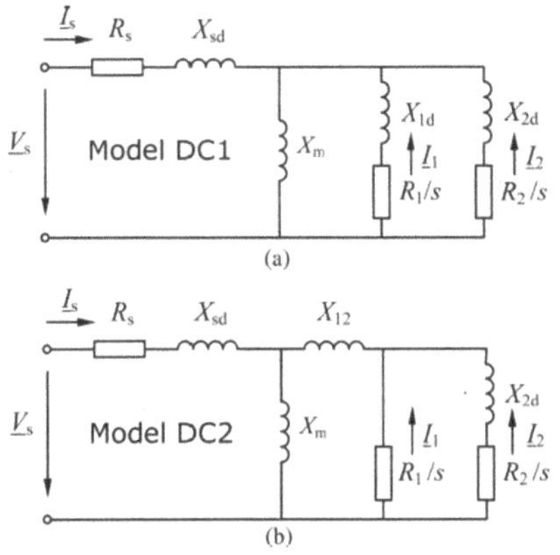

FIGURE 9.41 Steady-state phase equivalent circuit of IM with double cage: (a) model DC1 and (b) model DC2.

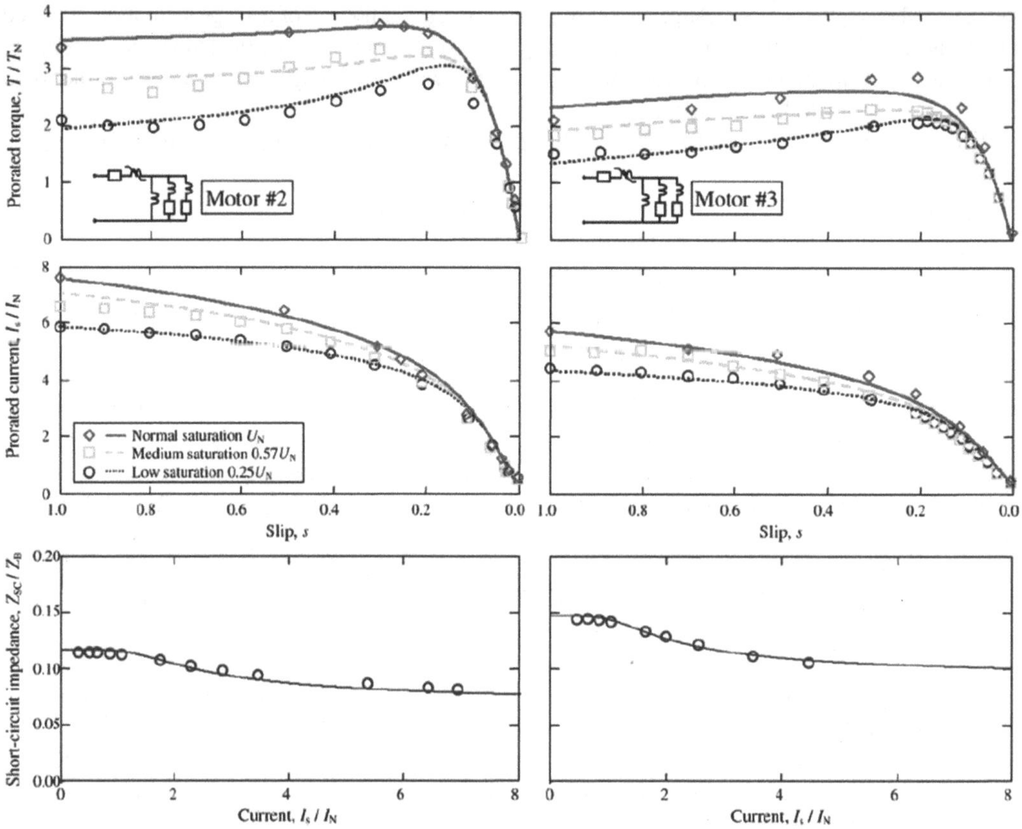

FIGURE 9.42 IMs 2#3 (model DC1): predicted versus measured torque-slip, current-slip, and short-circuit impedance-stator current with only stator leakage reactance X_{1d} considered non-linear (after Ref. [21]).

The prorated calculated values for three different values of torque and stator current have been obtained with torque proportional to voltage squared and current proportional to voltage.

All in all for slips above the rated one, main flux saturation – up to rated voltage – does not place an important role in torque–slip and current–slip curves, but the stator leakage reactance does; the rotor leakage reactance is less important, as long as the rotor slots are not closed.

9.10 ROTOR SLOT LEAKAGE REACTANCE SATURATION EFFECTS

Closed stator and rotor slots are adapted in special applications (when rotor are immersed in fluids), though they are known to reduce starting torque, peak torque, and finally efficiency despite the tendency of reducing rotor surface core losses, noise, and vibration.

It has been proven that the saturation of flux bridges above the rotor (stator) closed slots [8] – Figure 9.43, has to be considered in calculating the torque-speed curve of a high-speed induction motor fed from an inverter up to 30,000 rpm.

It turns out that only by adding an additional emf V_0 in the rotor part of equivalent circuit (Figure 9.44a) [8,21], $V_0 \approx 5.66 \cdot k_{w1} \cdot N_{ph1} \cdot f_1 \cdot \Delta\lambda_{max}$; $\Delta\lambda_{max} \approx B_{sat} \cdot l_{stack} \cdot h_b \cdot 0.95$, where $B_{sat} \approx 2$ T, h_b – bridge height of closed rotor slot, k_{w1} – winding factor, will the torque–speed curve be predicted correctly (Figure 9.44b).

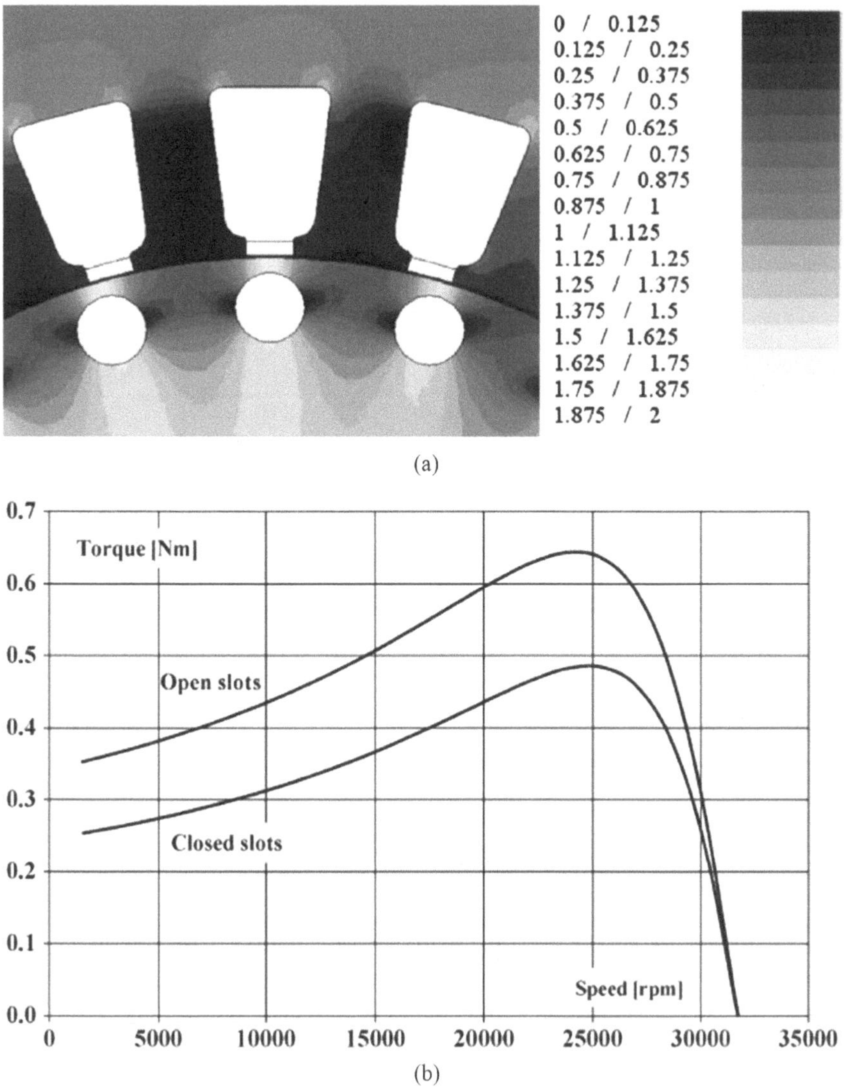

FIGURE 9.43 Flux distribution for closed rotor round slots (a) and the torque-speed curves with open and closed rotor slots (b) (after Ref. [8]).

Note: The presence of closed slots also in the rotor would imply probably the addition of another emf V_{01} to account for the saturated flux bridge below the closed stator slots.

It has also been proved that there are additional time harmonics both on no load and on load with closed rotor slots [8,22].

Low-order current harmonics due to the fifth and seventh space mmf stator harmonics in single-layer windings have been documented as a result of heavy saturation of flux bridges above the closed rotor slots [23].

So, designing with closed rotor slots though it reduces noise and vibration leads to peak and starting torque reduction and to low-order current harmonics due to the heavy saturation of flux bridges of closed rotor slots.

A non-linear FEM analysis of the parameters of closed slot IMs is available in Ref. [24].

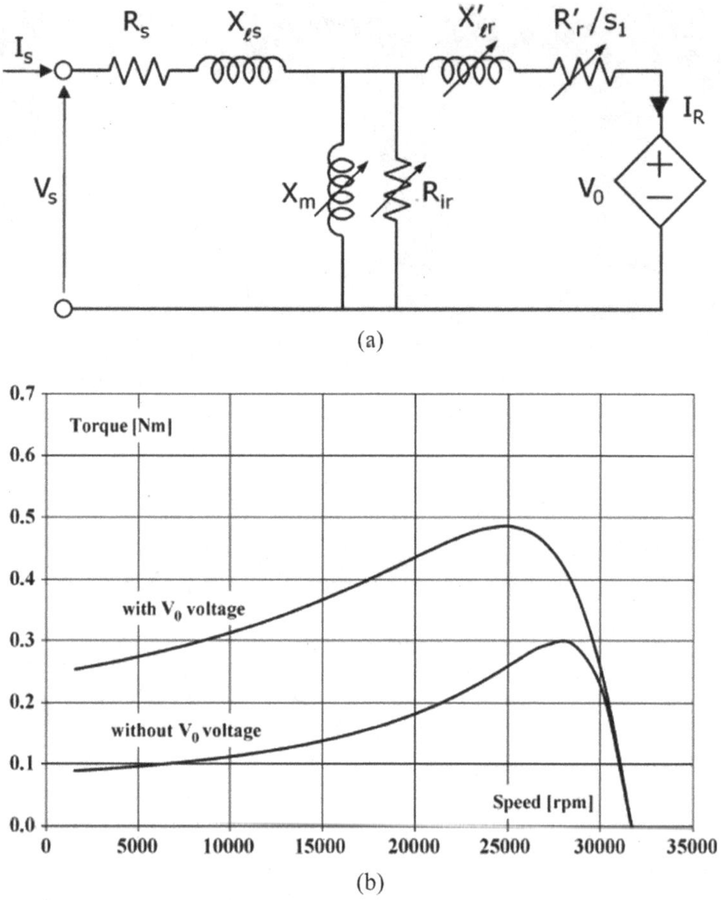

FIGURE 9.44 Equivalent circuit of IM with V_0 for closed rotor slots (a) and torque-speed curves without and with V_0 consideration (b) (after Ref. [8]).

9.11 SOLID ROTOR EFFECTS

Solid rotors with and without copper bars have been proposed for high-speed IMs due to their mechanical and thermal ruggedness after they have been introduced in single-sided linear induction motors (SLIMs).

Both skin effect and magnetic saturation occur in solid rotors of IMs. A time-domain 2(3)D FEM would be adequate to treat this complex situation, at the price of heavy computation effort.

A non-linear multilayer – quasi-2D analytical theory was developed for SLIMs with solid iron secondary [25] to save computer time. It could be used here too, but it is not available yet. Ref. [25] introduces a linear multilayer analytical 2D theory to rapidly calculate the harmonic solid rotor eddy current stationed close to the rotor surface in a heavily saturated (by the fundamental flux density) thin thickness where a constant permeability may be admitted. Sample results are shown in Figure 9.45a and b [26].

They show that there is an optimal copper coating thickness for the particular design (p = 1, stator outer radius = 75 mm, rotor outer radius = 35 mm, airgap = 2 mm, stack length = 140 mm, and f = 800 Hz), and that the harmonics eddy current rotor losses have to be carefully assessed and reduced to yield satisfactory efficiency.

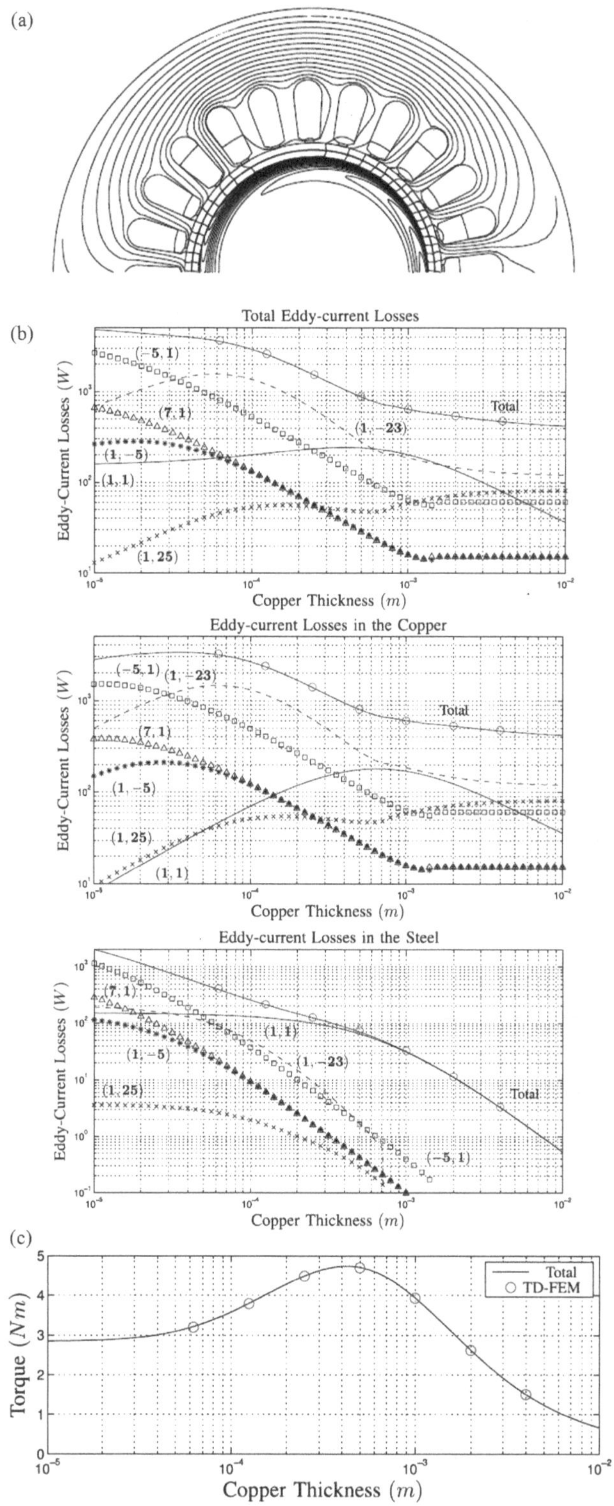

FIGURE 9.45 Field lines for the fundamental time harmonic (η = 1) at 800 Hz (a), (η–ν) harmonic copper and iron rotor losses versus copper coating thickness (b), and total average torque versus copper thickness (c) (after Ref. [26]); ν – space harmonic order.

9.12 STANDARDISED LINE-START INDUCTION MOTORS

As already documented in previous paragraphs and in Chapter 7, the IM steady-state performance is illustrated by torque, current, efficiency, and power factor versus slip for given voltage and frequency.

These characteristics may be influenced in many ways. Among them, the influences of magnetic saturation and skin effects are paramount.

It is the locked-rotor (starting) torque, breakdown torque, pull-up torque, starting current, rated slip, rated efficiency, and rated power factor that interest both, manufacturers and users.

In an effort to put some order into this pursuit, NEMA has defined five designs for IM with cage rotors. They are distinguishable basically by the torque/speed curve (Figure 9.46).

Design A

The starting in-rush current is larger than for general use design B motors. However, starting torque and pull-up and breakdown torques are larger than for design B.

While design A allows for larger starting current, larger starting torque is produced.

These characteristics are obtained with a lower leakage inductance, mainly in the rotor and notable skin effect. Mildly deep rotor bars will do it, in general.

Design B

Design B motors are designed for given maximum lock-rotor current and minimum breakdown and locked-rotor and pull-up torques to make sure the typical load torque/speed curve is exceeded for all slips. They are called general-purpose induction motors and their rated slip varies in general from 0.5 to 3 (maximum 5)%, depending on power and speed.

A normal cage with moderate skin effect is likely to produce design B characteristics.

Design C

Design C motors exhibit a very large locked-rotor torque (larger than 200%, in general) at the expense of lower breakdown torque and larger rated slip – lower efficiency and power factor – than design A & B. Also lower starting currents are typical (I_{start} < 550%). Applications with very high break-away torque such as conveyors are typical for Design C.

Rather deep bars (of rather complex shape) or double cages are required on the rotor to produce C designs.

Design D

Design D induction motors are characterised by a high breakdown slip and large rated slip at very high starting torque and lower starting current (I_{start} < 450%).

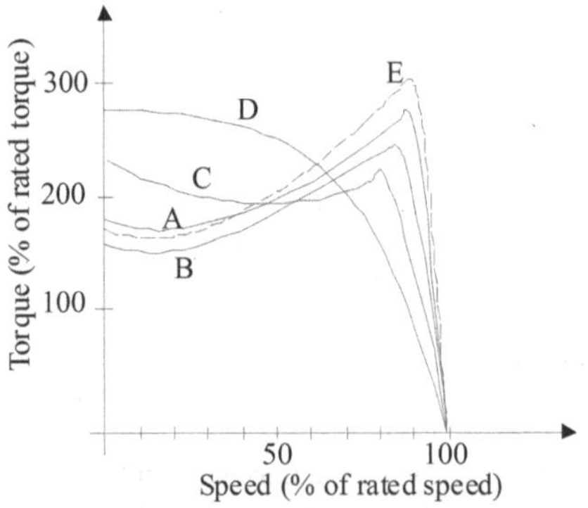

FIGURE 9.46 Torque/speed curve designs.

As they are above the torque/speed curve C (Figure 9.41), it means that the motor will accelerate a given load torque/speed plant the fastest. Also the total energy losses during acceleration will be the lowest (see Chapter 8, the high rotor resistance case).

The heat to be evacuated over the load acceleration process will be less stringent. Solid rotors, made of iron without or with axial slits or slots with copper bars are candidates for design D motors. They are designated to applications where frequent accelerations are more important than running at rated load. Punch presses are typical applications for D designs as well as high-speed applications, too.

Design E (withdrawn in USA in the year 2000)

Design E induction motors are characterised as high-efficiency motors and their efficiency is 1%–4% higher than the B designs for the same power and speed. This superior performance is paid for by larger volume (and initial cost) and higher starting current (up to 30%, in general).

Lower current density, lower skin effect, and lower leakage inductances are typical for design E motors. It has been demonstrated that even a 1% increase in efficiency in a kW range motor has the payback time in energy savings, in a rather loaded machine, of less than 3 years. As the machine life is in the range of more than 10–15 years, it pays off to invest more in design E (larger in volume) induction motor in many applications.

The increase in the starting current, however, will impose stronger local power grids which will tend to increase slightly the abovementioned 3-year payback time for E designs. Design E was withdrawn in the USA in the year 2000, but high-efficiency motors are fabricated routinely in EU. At least for variable speed drives, design E should be pursued aggressively.

A rapid prediction procedure for IM performance combining analytical and FEM analysis is presented in Ref. [27].

9.13 SUMMARY

- As the slip (speed) varies for constant voltage and frequency, the IM parameters: magnetisation inductance (L_m), leakage inductances (L_{sl}, L'_{rl}), and rotor resistance R'_{rl} vary.
- In general, the magnetisation flux decreases with slip to 55%–65% of its full load value at stall.
- As the slip frequency (slip) increases, the rotor cage resistance and slot body leakage inductance vary due to skin effect: the first one increases and the second one decreases.
- Moreover, at high slips (and currents), the leakage flux paths saturate to produce a reduction of stator and rotor leakage inductance.
- In rotors with skewed slots, there is an uncompensated skewing rotor mmf which varies from zero in the axial centre of the stack to maximum positive and negative values at stack axial ends.
- This mmf acts as a kind of independent magnetisation current whose maximum is phase shifted by an angle slightly over 90° unless the rotor current is notably smaller than the rated value.
- At stall, the so-called skewing magnetisation current I_{mskew} may reach values 3 times the rated magnetisation current of the motor in the axial zones close to the axial stack end.
- As the main flux is smaller than usual at stall, the level of saturation seems to be decided mainly by the skewing magnetisation current.
- By slicing the stack axially and using the machine magnetisation curve $L_m(I_m)$, the value of $L_m(I_{mskew})$ in each segment is calculated. Severe saturation of the core occurs in the marginal axial segments. The skewing introduces an additional notable component in the rotor leakage inductance. It also slightly reduces the magnetisation inductance.
- By including the main flux, slot neck flux, zigzag flux, differential leakage flux, and skewing flux per tooth in the stator and rotor, for each axial segment of stack, the stator and rotor saturated leakage inductances are calculated for given stator, rotor and magnetisation current and slip frequency.

- The skin effect is treated separately for stator windings with multiple conductors in slots and for rotor cages with one bar per slot. Standard correction coefficients for bar resistance and leakage inductance are derived. For the general shape rotor bar (including the double cage), the multiple layer approach is extended both for the bar and the end ring.
- The end ring skin effect is notable, especially for S ≥ 1, in applications such as freight elevators.
- Skin and leakage saturation effects in the rotor cage are beneficial for increasing the starting torque for less starting current. However, the slot depth tends to be high which leads to a rather high rotor leakage inductance and thus a smaller breakdown torque and a larger rated slip. Consequently, the efficiency and power factor at full load are slightly lower.
- Skin effect is to be avoided in inverter-fed induction motors as it only increases the winding losses. However, it appears for current time harmonics anyway. Leakage saturation may occur only in servo drives with large transient (or short duration) torque requirements where the currents are large.
- In closed-slot cage rotor motors, the rotor slot iron bridge saturates at values of rotor current notably less than the rated value.
- Skin and leakage saturation effects may not be neglected in line-start motors when assessing the starting, pull-up, or even breakdown torque. The larger the motor power, the more severe is this phenomenon.
- Skin and leakage saturation effects may be controlled to advantage by slot geometry optimisation. This way, the four NEMA designs (A, B, C, D) have been born.
- While skin and leakage (on load) saturation effects can be treated with modified 2D and 3D FEMs [20,28,29], the computation time is still prohibitive for routine calculations. Rather sophisticated analytical tools should be used first, with FEM applied later for final refinements.

REFERENCES

1. Emde, About Current Redistribution, EUM, 1922, pp. 301 (in German).
2. R. Richter, *Electrical Machines*, Vol. 1, Verlag Birkhäuser, Basel, 1951 (in German), pp. 233–243.
3. A. B. Field, Eddy Currents in Conductors Placed in Slots, AIEE, 1905, pp. 659.
4. I. B. Danilevici, V. V. Dombrovski, E. I. Kazovski, *A.C. Machine Parameters*, Science Publishers, St. Petersburg, 1965 (in Russian).
5. P. L. Cochran, *Polyphase Induction Motors*, Marcel Dekker Inc. New York, 1989, pp. 295–312.
6. K. Vogt, *Electrical Machines*, 4th edition, Verlag, Berlin, 1988 (in German), p. 315.
7. J. Martinez – Roman, L. Serrano – Iribarnegaray, Torque speed characteristic of elevator deep rotor bar and double stator winding asynchronous machines. Modelling and measurement, *Record of ICEM*, 1998, Istanbul, Turkey, September, 1998, Vol. 1, pp. 1314–1319.
8. A. Boglietti, R. I. Bojoi, A. Cavagnino, P. Guglielmi, A. Miotto, Analysis and modelling of rotor slot enclosure effects in high speed induction motors, *IEEE Transactions on Industry Applications*, Vol. 48, No. 4, 2012, pp. 1279–1287.
9. R. Richter, *Electric Machines*, Vol. IV, Induction Machines, Verlag Birkhäuser, Bassel (Stuttgart, 1954 in German).
10. P. D. Agarwal, P. L. Alger, Saturation factors for leakage reactance of induction motors, *Transactions of the American Institute of Electrical Engineers. Part III: Power Apparatus and Systems*, Vol. 79, No. 3, 1961, pp. 1037–1042.
11. G. Angst, Saturation factors for leakage reactance of induction motors with skewed rotors, *IEEE Transactions on Power Apparatus and Systems*, 1963, Vol. 82, No. 68 (October), pp. 716–725.
12. B. J. Chalmers, R. Dodgson, Saturated leakage reactances of cage induction motors, *Proceedings of the Institution of Electrical Engineers*, August 1969, Vol. 116, No. 8, pp. 1395–1404.

13. G. J. Rogers, D. S. Benaragama, An induction motor model with deep-bar effect and leakage inductance saturation, *Archiv Für Elektrotechnik*, Vol. 60, 1978, pp. 193–201.
14. M. Akbaba, S. O. Fakhleo, Saturation effects in three-phase induction motors, *EMPS*, Vol. 12, 1987, pp. 179–193.
15. Y. Ning, C. Zhong, K. Shao, Saturation effects in three-phase induction motors, *Record of ICEM*, 1994, Paris, France, D1 section.
16. P. Lagonotte, H. Al Miah, M. Poloujadoff, Modelling and identification of parameters under motor and generator conditions, *EMPS*, Vol. 27, No. 2, 1999, pp. 107–121.
17. B.-Il. Kown, B.-T. Kim, Ch.-S. Jun, Analysis of Axially Nonuniform Distribution in 3 Phase Induction Motor Considering Skew Effect, *IEEE Transactions on Magnetics*, Vol. 35, No. 3, 1999, pp. 1298–1301.
18. I. Boldea, D. G. Dorrel, C. B. Rasmussen, T. J. E. Miller, Leakage reactance saturation in induction motors, *Record of ICEM*, 2000, Vol. 1, pp. 203–207.
19. S. Williamson, M. J. Robinson, Calculation of Cage Induction Motor Equivalent Circuit Parameters Using Finite Elements, *IEE Proceedings B (Electric Power Applications)*, Vol. 138, 1991, pp. 264–276.
20. S. Williamson and D. G. Gersh, Finite Element Calculation of Double Cage Rotor Equivalent Circuit Parameters, *IEEE Transactions on Energy Conversion*, Vol. 11, No. 1, 1996, pp. 41–48.
21. L. Monjo, F. Corcoles, J. Pedra, Saturation effects on torque-slip and current-slip curves of squirrel – cage induction motors, *IEEE Transactions on Energy Conversion*, Vol. 28, No. 1, 2013, pp. 243–254.
22. I. Boldea, S.A. Nasar, *Induction Machine Design Handbook*, 2nd edition, CRC Press, Boca Raton, FL, 2010, p. 685.
23. G-Yu Zhou, J-X Shen, Current harmonics in induction machine with closed-slot rotor, *IEEE Transactions on Industry Applications*, Vol. 53, No. 1, 2017, pp. 134–142.
24. I. Torac, L. N. Tutelea, I. Boldea, Closed slot induction machine inductances and rotor resistance characterization versus current and frequency via FEM, *2017 International Conference on OPTIM and ACEMP*, Brasov, Romania, pp. 17028099, 2017.
25. I. Boldea, M. Babescu, Multilayer approach to the analysis of single-sided linear induction motors, *Proceedings of the Institution of Electrical Engineers*, Vol. 125, No. 4, 1978, pp. 283–287.
26. V. Räisänen, S. Suuriniemi, S. Kurz, L. Kettunen, Rapid computation of harmonic eddy – current losses in high speed induction machines, *IEEE Transactions on Energy Conversion*, Vol. 28, No. 3, 2013, pp. 782–790.
27. L. Alberti, N. Bianchi, S. Bolognani, A Very Rapid Prediction of IM Performance Combining Analytical and Finite Element Analysis, *IEEE Transactions on Industry Applications*, Vol. 44, No. 5, 2008, pp. 1505–1512.
28. M. Sfaxi, F. Bouillault, M. Gabsi, J. F. Rialland, Numerical method for determination the stray losses of conductors in armature slots, *Record of ICEM*, 1998, Istanbul, Turkey, Vol. 3, pp. 1815–1820.
29. P. Zhou, J. Gilmore, Z. Badic, Z. J. Csendes, Finite Element Analysis of Induction Motors Based on Computing Detailed Equivalent Circuit Parameters, *IEEE Transactions on Magnetics*, Vol. 34, No. 5, 1998, pp. 3499–3502.

10 Airgap Field Space Harmonics, Parasitic Torques, Radial Forces, and Noise Basics

The airgap field distribution in induction machines (IMs) is influenced by stator and rotor mmf distributions and by magnetic saturation in the stator and rotor teeth and yokes (back cores).

Previous chapters introduced the mmf harmonics but were restricted to the performance of its fundamental. Slot openings were considered but only in a global way, through an apparent increase of airgap by the Carter coefficient.

We first considered magnetic saturation of the main flux path through its influence on the airgap flux density fundamental. Later on, a more advanced model is introduced (AIM) to calculate the airgap flux density harmonics due to magnetic saturation of main flux path (especially the third harmonic).

However, as shown later in this chapter, slot leakage saturation, rotor static and dynamic eccentricity together with slot openings, and mmf step harmonics produce a multitude of airgap flux density space harmonics. Their consequences are parasitic torques, radial uncompensated forces, and harmonics core and winding losses [1–3]. The harmonic losses will be treated in the next chapter.

In what follows we will gradually use complex analytical tools to reveal various airgap flux density harmonics and their parasitic torques and forces. Such a treatment is very intuitive, but it is merely qualitative and leads to rules for a good design. Only finite element modelling (FEM) – 2D and 3D – could depict the extraordinary involved nature of airgap flux distribution in IMs under various factors of influence, to a good precision, but at the expense of much larger computing time and in an intuitiveless way. For refined investigation, FEM is, however, "the way".

10.1 STATOR MMF PRODUCED AIRGAP FLUX HARMONICS

As already shown in Chapter 4 (Equation 4.17), the stator mmf per pole-stepped waveform may be decomposed in harmonics as

$$F_1(x,t) = \frac{3W_1I_1\sqrt{2}}{\pi p_1}\left[K_{w1}\cos\left(\frac{\pi}{\tau}x-\omega_1 t\right)+\frac{K_{w5}}{5}\cos\left(\frac{5\pi}{\tau}x+\omega_1 t\right)\right.$$
$$\left.+\frac{K_{w7}}{7}\cos\left(\frac{7\pi}{\tau}x-\omega_1 t\right)+\frac{K_{w11}}{11}\cos\left(\frac{11\pi}{\tau}x+\omega_1 t\right)+\frac{K_{w13}}{13}\cos\left(\frac{13\pi}{\tau}x-\omega_1 t\right)\right] \quad (10.1)$$

where $K_{w\nu}$ is the winding factor for the νth harmonic,

$$K_{w\nu} = K_{q\nu}K_{y\nu};\ K_{q\nu} = \frac{\sin\frac{\nu\pi}{6}}{q\sin\frac{\nu\pi}{6q}};\quad K_{y\nu} = \sin\frac{\nu\pi y}{2\tau} \quad (10.2)$$

In the absence of slotting, but allowing for it globally through Carter's coefficient and for magnetic circuit saturation by an equivalent saturation factor $K_{s\nu}$, the airgap field distribution is

$$B_{g1}(\theta,t)=\frac{3\mu_0 W_1 I_1\sqrt{2}}{\pi p_1 g K_c}\left[\frac{K_{w1}}{1+K_{s1}}\cos(\theta-\omega_1 t)+\frac{K_{w5}}{5(1+K_{s5})}\cos(5\theta+\omega_1 t)\right.$$

$$+\frac{K_{w7}}{7(1+K_{s7})}\cos(7\theta-\omega_1 t)+\frac{K_{w11}}{11(1+K_{s11})}\cos(11\theta+\omega_1 t)$$

$$\left.+\frac{K_{w13}}{13(1+K_{s13})}\cos(13\theta-\omega_1 t)\right];\quad \theta=\frac{\pi}{\tau}x \tag{10.3}$$

In general, the magnetic field path length in iron is shorter as the harmonics order gets higher (or its wavelength gets smaller). $K_{s\nu}$ is expected to decrease with ν increasing.

Also, as already shown in Chapter 4, but easy to check through (10.2), for all harmonics of the order ν,

$$\nu=C_1\frac{N_s}{p_1}\pm 1 \tag{10.4}$$

the distribution factor is the same as for the fundamental.

For three-phase symmetrical windings (with integer q slots/pole/phase), even order space harmonics are zero and multiples of three harmonics are zero for star connection of phases. So, in fact,

$$\nu=6C_1\pm 1 \tag{10.5}$$

As shown in Chapter 4 (Equations 4.17–4.19), harmonics of 5th, 11th, 17th, … order travel backwards and those of 7th, 13th, 19th, … order travel forwards – see Equation (10.1).

The synchronous speed of these harmonics ω_ν is

$$\omega_\nu=\frac{d\theta_\nu}{dt}=\frac{\omega_1}{\nu} \tag{10.6}$$

In a similar way, we may calculate the mmf and the airgap field flux density of a cage winding.

10.2 AIRGAP FIELD OF A SQUIRREL-CAGE WINDING

A symmetric (healthy) squirrel-cage winding may be replaced by an equivalent multiphase winding with N_r phases, ½ turns/phase, and unity winding factor. In this case, its airgap flux density is

$$B_{g2}(\theta,t)=\frac{N_r\mu_0 I_b\sqrt{2}}{\pi p_1 g K_c}\sum_{\mu=1}^{\infty}\frac{\cos(\mu\theta\mp\omega_1 t-\phi_{12\nu})}{\mu(1+K_{s\nu})}=\frac{\mu_0 F_2(t)}{gK_c} \tag{10.7}$$

The harmonics order μ, which produces nonzero mmf amplitudes, follows from the applications of expressions of band factors K_{BI} and K_{BII} for $m=N_r$:

$$\mu=C_2\frac{N_r}{p_1}\pm 1 \tag{10.8}$$

Now we have to consider that, in reality, both stator and rotor mmfs contribute to the magnetic field in the airgap and, if saturation occurs, superposition of effects is not allowed. So either a single saturation coefficient is used (say $K_{s\nu}=K_{s1}$ for the fundamental) or saturation is neglected ($K_{s\nu}=0$).

We have already shown in Chapter 9 that the rotor slot skewing leads to variation of airgap flux density along the axial direction due to uncompensated skewing rotor mmf. While we investigated this latter aspect for the fundamental of mmf, it also applies for the harmonics. Such remarks show that the above analytical results should be considered merely as qualitative.

10.3 AIRGAP PERMEANCE HARMONICS

Let us first remember that even the step harmonics of the mmf are due to the placement of windings in infinitely thin slots. However, the slot openings introduce a kind of variation of airgap with position. Consequently, the airgap conductance, considered as only the inverse of the airgap, is

$$\frac{1}{g(\theta)} = f(\theta) \tag{10.9}$$

Therefore, the airgap change $\Delta(\theta)$ is

$$\Delta(\theta) = \frac{1}{f(\theta)} - g \tag{10.10}$$

With stator and rotor slotting,

$$g(\theta) = g + \Delta_1(\theta) + \Delta_2(\theta - \theta_r) = \frac{1}{f_1(\theta)} + \frac{1}{f_2(\theta - \theta_r)} - g \tag{10.11}$$

$\Delta_1(\theta)$ and $\Delta_2(\theta - \theta_r)$ represent the influence of stator and rotor slot openings alone on the airgap function.

As $f_1(\theta)$ and $f_2(\theta - \theta_r)$ are periodic functions whose period is the stator (rotor) slot pitch, they may be decomposed in harmonics:

$$\begin{aligned} f_1(\theta) &= a_0 - \sum_{\nu=1}^{\infty} a_\nu \cos \nu N_s \theta \\ f_2(\theta - \theta_r) &= b_0 - \sum_{\nu=1}^{\infty} b_\nu \cos \nu N_r (\theta - \theta_r) \end{aligned} \tag{10.12}$$

Now, if we use the conformal transformation for airgap field distribution in the presence of infinitely deep, separate, slots – essentially Carter's method – we obtain [1]

$$a_\nu, b_\nu = \frac{\beta}{g} F_\nu\left(\frac{b_{os,r}}{t_{s,r}}\right) \tag{10.13}$$

$$F_\nu\left(\frac{b_{os,r}}{t_{s,r}}\right) = \frac{1}{\nu}\frac{4}{\pi}\left[0.5 + \frac{\left(\frac{\nu b_{os,r}}{t_{s,r}}\right)^2}{0.18 - 2\left(\frac{\nu b_{os,r}}{t_{s,r}}\right)^2}\right]\sin\left(1.6\frac{\pi \nu b_{os,r}}{t_{s,r}}\right) \tag{10.14}$$

β is [2], as in Table 10.1.

TABLE 10.1
$\beta(b_{os,r}/g)$

$b_{os,r}/g$	0	0.5	1.0	1.5	2.0	3.0	4.0	5.0
β	0.0	0.0149	0.0528	0.1	0.1464	0.2226	0.2764	0.3143
$b_{os,r}/g$	6.0	7.0	8.0	10.0	12.0	40.0	∞	
β	0.3419	0.3626	0.3787	0.4019	0.4179	0.4750	0.5	

Equations (10.13 and 10.14) are valid for $\nu = 1, 2, \ldots$. On the other hand, as expected,

$$a_0 \approx \frac{1}{K_{c1}g} \qquad b_0 \approx \frac{1}{K_{c2}g} \tag{10.15}$$

where K_{c1} and K_{c2} are Carter's coefficients for the stator and rotor slotting, respectively, acting separately.

Finally, with a good approximation, the inversed airgap function $1/g(\theta, \theta_r)$ is

$$\lambda_g(\theta,\theta_r) = \frac{1}{g(\theta,\theta_r)} \approx \frac{1}{g}\left\{\frac{1}{K_{c1}K_{c2}} - \frac{a_1}{K_{c2}}\cos N_s\frac{\theta}{p_1} - \frac{b_1}{K_{c1}}\cos N_r\frac{(\theta-\theta_r)}{p_1}\right.$$

$$\left. + a_1 b_1\left[\cos\left(\frac{(N_s+N_r)}{p_1}\theta - \frac{N_r}{p_1}\theta_r\right) + \cos\left(\frac{(N_s-N_r)}{p_1}\theta + \frac{N_r}{p_1}\theta_r\right)\right]\right\} \tag{10.16}$$

As expected, the average value of $\lambda_g(\theta, \theta_r)$ is

$$\left(\lambda_g(\theta,\theta_r)\right)_{\text{average}} = \frac{1}{K_{c1}K_{c2}g} = \frac{1}{K_c g} \tag{10.17}$$

θ and θ_r – stator and rotor electrical angles.

We should notice that the inversed airgap function (or airgap conductance) λ_g has harmonics related directly to the number of stator and rotor slots and their geometry.

10.4 LEAKAGE SATURATION INFLUENCE ON AIRGAP PERMEANCE

As already discussed in Chapter 9, for semiopen or semiclosed stator (rotor) slots at high currents in the rotor (and stator), the teeth heads get saturated. To account for this, the slot openings are increased. Considering a sinusoidal stator mmf and only stator slotting, the slot opening increased by leakage saturation b'_{os} varies with position, being maximum when the mmf is maximum (Figure 10.1).

We might extract the fundamental of $b_{os,r}(\theta)$ function

$$b'_{os,r} \approx b^0{}_{os,r} - b''_{os,r}\cos(2\theta - \omega_1 t); \quad \theta\text{-electric angle} \tag{10.18}$$

The leakage slot saturation introduces a 2p_1 pole pair harmonic in the airgap permeance. This is translated into a variation of a_0, b_0.

$$a_0, b_0 = a_0^0, b_0^0 - a'_0, b'_0\sin(2\theta - \omega_1 t) \tag{10.19}$$

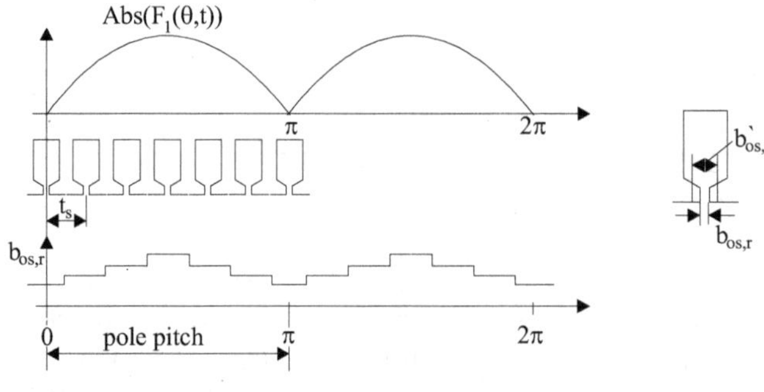

FIGURE 10.1 Slot opening $b_{os,r}$ variation due to slot leakage saturation.

a_0^0, b_0^0 – are the new average values of $a_0(\theta)$ and $b_0(\theta)$ with leakage saturation accounted for.

This harmonic, however, travels at synchronous speed of the fundamental wave of mmf. Its influence is notable only at high currents.

Example 10.1 Airgap Conductance Harmonics

Let us consider only the stator slotting with $b_{os}/t_s = 0.2$ and $b_{os}/g = 4$ which are quite practical value and calculate the airgap conductance harmonics.

Solution

Equations (10.13 and 10.14) are to be used with β from Table 10.1, β(4) = 0.2764, with

$$K_{c1} \approx \frac{t_s}{t_s - 1.6\beta b_{os}} = \frac{1}{1 - 1.6 \cdot 0.2764 \cdot 0.2} = 1.097$$

$$a_0 = \frac{1}{g}\frac{1}{K_{c1}} = \frac{1}{g}\frac{1}{1.097} = \frac{0.91155}{g}$$

$$a_1 = \frac{1}{g}0.2764\frac{4}{\pi}\left[0.5 + \frac{(1 \cdot 0.2)^2}{0.78 - 2 \cdot (0.2)^2}\right]\sin(1.6\pi \cdot 1 \cdot 0.2) = \frac{1}{g}0.1542$$

$$a_2 = \frac{1}{g}0.2764\frac{4}{2\pi}\left[0.5 + \frac{(2 \cdot 0.2)^2}{0.78 - 2 \cdot (2 \cdot 0.2)^2}\right]\sin(1.6\pi \cdot 2 \cdot 0.2) = \frac{1}{g}0.04886$$

The airgap conductance $\lambda_1(\theta)$ is

$$\lambda_1(\theta) = \frac{1}{g(\theta)} = \frac{1}{g}\left[0.91155 - 0.1542\cos(N_s\theta) - 0.04886\cos(2N_s\theta)\right]$$

If for the ν th harmonics the "sine" term in (10.14) is zero, so is the ν th airgap conductance harmonic. This happens if

$$1.6\pi\nu\frac{b_{os,r}}{t_{s,r}} = \pi; \quad \nu\frac{b_{os,r}}{t_{s,r}} = 0.625$$

Only the first 1, 2 harmonics are considered, so only with open slots the above condition may be met approximately.

10.5 MAIN FLUX SATURATION INFLUENCE ON AIRGAP PERMEANCE

In Chapter 5, an iterative analytical model (AIM) to calculate the main flux distribution accounting for magnetic saturation was introduced. Finally, we showed the way to use AIM to derive the airgap flux harmonics due to main flux path (teeth or yokes) saturation. A third harmonic was particularly visible in the airgap flux density.

As expected, this is the result of a virtual second harmonic in the airgap conductance interacting with the mmf fundamental, but it is the resultant (magnetising) mmf and not stator (or rotor) mmfs alone.

The two second-order airgap conductance harmonics due to leakage slot and main flux path saturation are phase shifted as their originating mmfs are by the angle $\varphi_m - \varphi_1$ for the stator and $\varphi_m - \varphi_2$ for the rotor.

φ_1, φ_2, and φ_m are the stator, rotor, and magnetisation mmf phase current phase shift angle with respect to stator phase voltage:

$$\lambda_s(\theta_s) = \frac{1}{gK_c}\left[\frac{1}{1+K_{s1}} + \frac{1}{1+K_{s2}}\sin\left(2\theta - \omega_1 t - (\phi_m - \phi_1)\right)\right] \tag{10.20}$$

The saturation coefficients K_{s1} and K_{s2} result from the main airgap field distribution decomposition into first and third harmonics.

There should not be much influence (amplification) between the two second-order saturation-caused airgap conductance harmonics unless the rotor is skewed and its rotor currents are large (>3 to 4 times rated current), when both the skewing rotor mmf and rotor slot mmf are responsible for large main and leakage flux levels, especially towards the axial ends of stator stack.

10.6 THE HARMONICS-RICH AIRGAP FLUX DENSITY

It has been shown [1] that, in general, the airgap flux density $B_g(\theta,t)$ is

$$B_g(\theta, t) = \mu_0 \lambda_{1,2}(\theta) F_v \cos(v\theta \mp \omega t - \phi_v) \tag{10.21}$$

where F_v is the amplitude of the mmf harmonic considered and $\lambda(\theta)$ is the inversed airgap (airgap conductance) function. $\lambda(\theta)$ may be considered as containing harmonics due to slot openings, leakage, or main flux path saturation. However, using superposition in the presence of magnetic saturation is not correct in principle, so mere qualitative results are expected by such a method.

10.7 THE ECCENTRICITY INFLUENCE ON AIRGAP MAGNETIC PERMEANCE

In rotary machines, the rotor is hardly located symmetrically in the airgap, either due to rotor (stator) un-roundedness, bearing eccentric support, or shaft bending.

A one-sided magnetic force (uncompensated magnetic pull or UMP) is the main result of such a situation. This force tends to further increase the eccentricity and produces vibrations, noise, and increases the critical rotor speed.

When the rotor is positioned off centre to the stator bore, according to Figure 10.2, the airgap at angle θ_m is

$$g(\theta_m) = R_s - R_r - e\cos\theta_m = g - e\cos\theta_m \tag{10.22}$$

where g is the average airgap (with zero eccentricity: e = 0.0).

The airgap magnetic conductance $\lambda(\theta_m)$ is

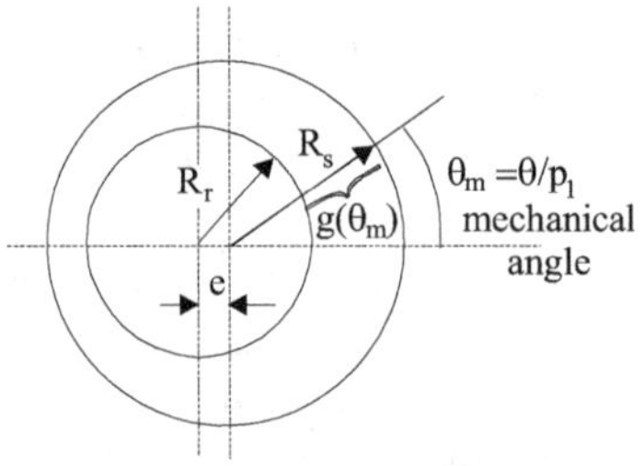

FIGURE 10.2 Rotor eccentricity R_s – stator radius and R_r – rotor radius.

$$\lambda(\theta_m) = \frac{1}{g(\theta_m)} = \frac{1}{g(1-\varepsilon\cos\theta_m)}; \quad \varepsilon = \frac{e}{g} \tag{10.23}$$

Now (10.23) may be easily decomposed into harmonics to obtain

$$\lambda(\theta_m) = \frac{1}{g}(c_0 + c_1\cos\theta_m + \ldots) \tag{10.24}$$

with

$$c_0 = \frac{1}{\sqrt{1-\varepsilon^2}}; \quad c_1 = \frac{2(c_0 - 1)}{\varepsilon} \tag{10.25}$$

Only the first geometrical harmonic (notice that the period here is the entire circumference: $\theta_m = \theta/p_1$) is hereby considered.

The eccentricity is static if the angle θ_m is independent of rotor position, that is, if the rotor revolves around its axis, but this axis is shifted with respect to stator axis by e.

In contrast, the eccentricity is dynamic if θ_m is dependent on rotor motion.

$$\theta_m = \theta_m - \frac{\omega_r t}{p_1} = \theta_m - \frac{\omega_1(1-S)t}{p_1} \tag{10.26}$$

It corresponds to the case when the axis of rotor revolution coincides with the stator axis, but the rotor axis of symmetry is shifted.

Now using Equation (10.21) to calculate the airgap flux density produced by the mmf fundamental as influenced only by the rotor eccentricity, static and dynamic, we obtain

$$B_g(\theta,t) = \mu_0 F_1\cos(\theta-\omega_1 t)\frac{1}{g}\left[c_0 + c_1\cos\left(\frac{\theta}{p_1} - \phi_s\right) + c_1'\cos\left(\frac{\theta-\omega_1(1-S)t}{p_1} - \phi_d\right)\right] \tag{10.27}$$

As seen from (10.27) for $p_1 = 1$ (2 pole machines), the eccentricity produces two homopolar flux densities, $B_{gh}(t)$,

$$B_{gh}(t) = \frac{\left[\frac{\mu_0 F_1 c_1}{g}\cos(\omega_1 t - \phi_s) + \frac{\mu_0 F_1 c_1'}{g}\cos(S\omega_1 t - \phi_d)\right]}{1 + c_h} \tag{10.28}$$

These "homopolar" components close their flux lines axially through the stator frame, then radially through the end frame, bearings, and axially through the shaft (Figure 10.3).

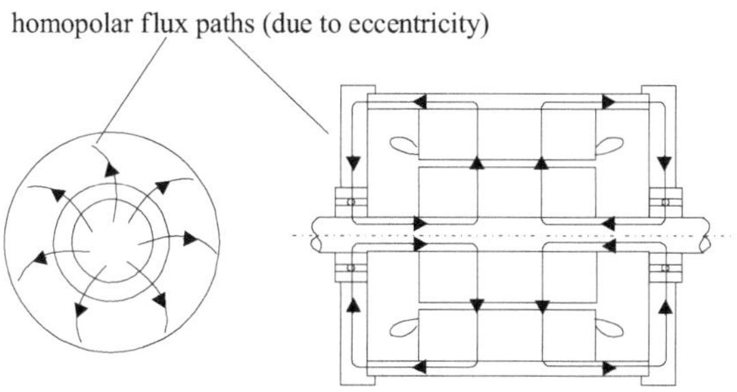

FIGURE 10.3 Homopolar flux due to rotor eccentricity.

The factor c_h accounts for the magnetic reluctance of axial path and end frames and may have a strong influence on the homopolar flux. For nonmagnetic frames and (or) insulated bearings, c_h is large, while for magnetic steel frames, c_h is smaller. Anyway, c_h should be much larger than unity at least for the static eccentricity component because its depth of penetration (at ω_1), in the frame, bearings, and shaft, is small. For the dynamic component, c_h is expected to be smaller as the depth of penetration in iron (at $S\omega_1$) is larger.

The "homopolar flux" may produce A.C. voltage along the shaft length and, consequently, shaft and bearing currents, thus contributing to bearing deterioration.

10.8 INTERACTIONS OF MMF (OR STEP) HARMONICS AND AIRGAP MAGNETIC PERMEANCE HARMONICS

It is now evident that various airgap flux density harmonics may be calculated using (10.21) with the airgap magnetic conductance $\lambda_{1,2}(\theta)$ either from (10.16) to account for slot openings, with a_0, b_0 from (10.19) for slot leakage saturation, or with $\lambda_s(\theta_s)$ from (10.20) for main flux path saturation, or $\lambda(\theta_m)$ from (10.24) for eccentricity.

$$B_g(x,t) = \frac{\mu_0}{g}\Big[\underset{(10.1)}{F_1(\theta,t)+F_2(\theta,t)}\Big]\left[\underset{(10.16)\text{with}(10.19)}{\lambda_{1,2}(\theta)} + \underset{(10.20)}{\lambda_s(\theta)} + \underset{(10.24)}{\lambda\left(\frac{\theta}{p_1}\right)}\right] \tag{10.29}$$

As expected, there will be a very large number of airgap flux density harmonics and its complete exhibition and analysis is beyond our scope here.

However, we noticed that slot openings produce harmonics whose order is a multiple of the number of slots (10.12). The stator (rotor) mmfs may produce harmonics of the same order either as sourced in the mmf or from the interaction with the first airgap magnetic conductance harmonic (10.16).

Let us consider an example where only the stator slot opening first harmonic is considered in (10.16).

$$B_{g\nu}(\theta,t) \approx \frac{\mu_0 F_{1\nu}}{g}\cos(\nu\theta' - \omega_1 t)\left[\frac{1}{K_{c1}} + a_1\cos\frac{N_s\theta'}{p_1}\right] \tag{10.30}$$

with

$$\theta' = \theta + \frac{\pi}{N_s}p_1 \tag{10.31}$$

θ' takes care of the fact that the axis of airgap magnetic conductance falls in a slot axis for coil chordings of 0, 2, and 4 slot pitches. For odd slot pitch coil chordings, $\theta' = \theta$. The first step (mmf) harmonic which might be considered has the order

$$\nu = c_1\frac{N_s}{p_1} \pm 1 = \frac{N_s}{p_1} \pm 1 \tag{10.32}$$

Writing (10.30) into the form

$$B_{g\nu}(\theta,t) = \frac{\mu_0 F_{1\nu}}{gK_{c1}}\cos\left[\nu\left(\theta+\frac{\pi P_1}{N_s}\right)-\omega_1 t\right]$$

$$+\frac{a_1\mu_0 F_{1\nu}}{2}\cos\left[\nu\left(\theta+\frac{\pi P_1}{N_s}\right)+\frac{N_s\theta}{P_1}-\omega_1 t\right]+\cos\left[\nu\left(\theta+\frac{\pi P_1}{N_s}\right)-\frac{N_s\theta}{P_1}-\omega_1 t\right] \tag{10.33}$$

For $\nu = \frac{N_s}{P_1} + 1$, the argument of the third term of (10.33) becomes

$$\left[\left(\frac{N_s}{p_1} + 1\right)\left(\theta + \frac{\pi p_1}{N_s}\right) - \pi - \omega_1 t\right]$$

But this way, it is the opposite of the first term argument, so the first step, mmf harmonic $\frac{N_s}{p_1} + 1$, and the first-slot opening harmonics subtract each other. The opposite is true for $\nu = \frac{N_s}{p_1} - 1$ as they are added. So the slot openings may amplify or attenuate the effect of the step harmonics of order $\nu = \frac{N_s}{p_1} \mp 1$, respectively.

Other effects such as differential leakage fields affected by slot openings have been investigated in Chapter 6 when the differential leakage inductance has been calculated. Also we did not discuss yet on the currents induced by the flux harmonics in the rotor and stator conductors.

In what follows, some attention will be paid to the main effects of airgap flux and mmf harmonics: parasitic torques and radial forces.

10.9 PARASITIC TORQUES

Not long after the cage-rotor induction motors have reached industrial use, it was discovered that a small change in the number of stator or rotor slots prevented the motor to start from any rotor position or the motor became too noisy to be usable. After Georges (1896), Punga (1912), Krondl, Lund, Heller, Alger, and Jordan presented detailed theories about additional (parasitic) asynchronous torques and Dreyfus (1924) derived the conditions for the manifestation of parasitic synchronous torques.

10.9.1 When Do Asynchronous Parasitic Torques Occur?

Asynchronous parasitic torques occur when a harmonic ν of the stator mmf (or its airgap flux density) produces currents in the rotor cage whose mmf harmonic has the same order ν.

The synchronous speed of these harmonics $\omega_{1\nu}$ (in electrical terms) is

$$\omega_{1\nu} = \frac{\omega_1}{\nu} \tag{10.34}$$

The stator mmf harmonics have orders like: ν = −5, +7, −11, +13, −17, +19, …, in general, $6c_1 \pm 1$, while the stator slotting introduces harmonics of the order $\frac{c_1 N_s}{p_1} \pm 1$. The higher the harmonic order, the lower its mmf amplitude. The slip S_ν of the νth harmonic is

$$S_\nu = \frac{\omega_{1\nu} - \omega_r}{\omega_{1\nu}} = 1 - \nu(1 - S) \tag{10.35}$$

For synchronism $S_\nu = 0$ and, thus, the slip for the synchronism of harmonic ν is

$$S = 1 - \frac{1}{\nu} \tag{10.36}$$

For the first mmf harmonic (ν = −5), the synchronism occurs at

$$S_5 = 1 - \frac{1}{(-5)} = \frac{6}{5} = 1.2 \tag{10.37}$$

For the seventh harmonic ($\nu = +7$),

$$S_7 = 1 - \frac{1}{(+7)} = \frac{6}{7} \tag{10.38}$$

All the other mmf harmonics have their synchronism at slips S_ν

$$|S_5| > |S_\nu| > S_7 \tag{10.39}$$

In a first approximation, the slip at synchronism for all harmonics is $S \approx 1$. That is, only close to motor stall, asynchronous parasitic torques occur.

As the same stator current is at the origin of both the stator mmf fundamental and harmonics, the steady-state equivalent circuit may be extended in series to include the asynchronous parasitic torques (Figure 10.4).

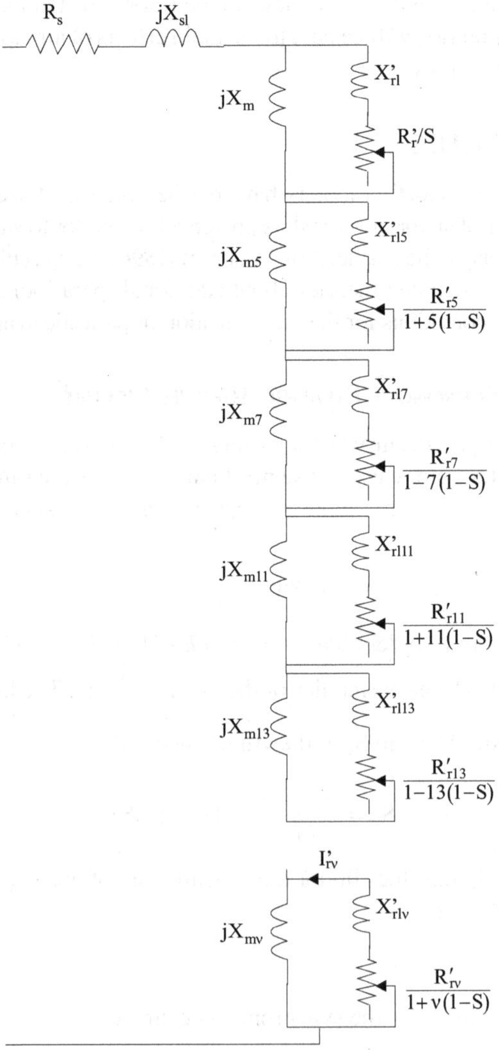

FIGURE 10.4 Equivalent circuit including asynchronous parasitic torques due to mmf harmonics (phase band and slot driven).

The mmf harmonics, whose order is lower than the first-slot harmonic $\nu_{s\min} = \frac{N_s}{p_1} \pm 1$, are called phase-belt harmonics. Their order is: −5, +7, …. They are all indicated in Figure 10.4 and considered to be mmf harmonics.

One problem is to define the parameters in the equivalent circuit. First of all, the magnetising inductances X_{m5}, X_{m7}, … are in fact the "components" of the up-to-now called differential leakage inductance.

$$L_{mv} = \frac{6\mu_0\tau L}{\pi^2 p_1 g K_c (1+K_{st})}\left(\frac{W_1 K_{wv}}{\nu}\right)^2 \tag{10.40}$$

The saturation coefficient K_{st} in (10.40) refers to the teeth zone only as the harmonics wavelength is smaller than that of the fundamental, and therefore, the flux paths close within the airgap and the stator and rotor teeth/slot zone.

The slip $S_v \approx 1$, and thus, the slip frequency for the harmonics $S_v\omega_1 \approx \omega_1$. Consequently, the rotor cage manifests a notable skin effect towards harmonics, much as for short-circuit conditions. Consequently, $R'_{rv} \approx (R'_r)_{start}$, $L'_{rlv} \approx (L'_{rl})_{start}$.

As these harmonics act around S = 1, their torque can be calculated as in a machine with given current $I_s \approx I_{start}$.

$$T_{ev} \approx \frac{3\cdot p_1\cdot\nu\cdot R'_{rstart}}{S_v\omega_1} I_{start}^2 \frac{L_{mv}^2}{\left[(L'_{rl})_{start} + L_{mv}\right]^2 + \left(\frac{R'_{rstart}}{S_v\omega_1}\right)^2} \tag{10.41}$$

In (10.41), the starting current I_{start} has been calculated from the complete equivalent circuit where, in fact, all magnetisation harmonic inductances L_{mv} have been lumped into $X_{sl} = \omega_1 L_{sl}$ as a differential inductance like if $S_v = \infty$. The error is not large.

Now (10.41) reflects a torque/speed curve similar to that of the fundamental (Figure 10.5). The difference lies in the rather high current, but the factor K_{wv}/ν is rather small and overcompensates this situation, leading to a small harmonic torque in well-designed machines.

We may dwell a little on (10.41) noticing that the inductance L_{mv} is of the same order (or smaller) than $(L'_{rl})_{start}$ for some harmonics. In these cases, the coupling between stator and rotor windings is small and so is the parasitic torque.

Only if $(L'_{rl})_{start} < L_{mv}$, a notable parasitic torque is expected. On the other hand, to reduce the first parasitic asynchronous torques T_{e5} or T_{e7}, chording of stator coils is used to make $K_{w5} = 0$.

$$K_{w5} = \frac{\sin\frac{5\pi}{6}}{q\sin\left(\frac{5\pi}{6q}\right)}\sin\frac{5\pi y}{2\tau} \approx 0; \quad \frac{y}{\tau} = \frac{4}{5} \tag{10.42}$$

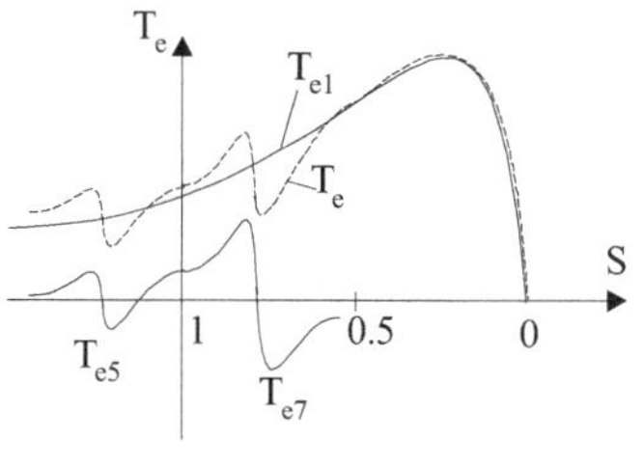

FIGURE 10.5 Torque/slip curve with 5th and 7th harmonic asynchronous parasitic torques.

As $y/\tau = 4/5$ is not feasible for all values of q, the closest values are $y/\tau = 5/6, 7/9, 10/12$, and $12/15$. As can be seen for $q = 5$, the ratio $12/15$ fulfills exactly condition (10.42). In a similar way, the 7th harmonic may be cancelled instead, if so desired.

The first-slot harmonics torque $\left(\nu_{s\min} = \frac{N_s}{p_1} \pm 1\right)$ may be cancelled by using skewing.

$$K_{w\nu_{s\min}} = \frac{\sin\frac{\nu_{s\min}\pi}{6}}{q\sin\left(\frac{\nu_{s\min}\pi}{6q}\right)} \sin\frac{\nu_{s\min}\pi y}{2\tau} \cdot \frac{\sin \nu_{s\min}\frac{c}{\tau}\frac{\pi}{2}}{\nu_{s\min}\frac{c}{\tau}\frac{\pi}{2}} \approx 0 \tag{10.43}$$

From (10.43),

$$\nu_{s\min}\frac{c}{\tau}\frac{\pi}{2} = K\pi; \quad \frac{c}{\tau} = \frac{2K}{\nu_{s\min}} = \frac{2K}{\frac{N_s}{p_1} \pm 1} \tag{10.44}$$

In general, the skewing c/τ corresponds to 0.5 to 2 slot pitches of the part (stator or rotor) with less slots.

For $N_s = 36$, $p_1 = 2$, $m = 3$, and $q = N_s/2p_1m = 36/(2\cdot2\cdot3) = 3$: $c/\tau = 2/17(19) \approx 1/9$. This, in fact, means a skewing of one slot pitch as a pole has $q\cdot m = 3\cdot3 = 9$ slot pitches.

10.9.2 Synchronous Parasitic Torques

Synchronous torques occur through the interaction of two harmonics of the same order (one of the stator and one of the rotor) but originating from different stator harmonics. This is not the case, in general, for wound rotor whose harmonics do not adapt to stator ones but produce about the same spectrum of harmonics for same number of phases.

For cage rotors, however, the situation is different. Let us first calculate the synchronous speed of harmonic ν_1 of the stator.

$$n_{1\nu_1} = \frac{f_1}{p_1\nu_1} \tag{10.45}$$

The relationship between a stator harmonic ν and the rotor harmonic ν' produced by it in the rotor is

$$\nu - \nu' = c_2\frac{N_r}{p_1} \tag{10.46}$$

This is easy to accept as it suggests that the difference in the number of periods of the two harmonics is a multiple of the number of rotor slots N_r.

Now the speed of rotor harmonics ν' produced by the stator harmonic ν with respect to rotor, $n_{2\nu\nu'}$, is

$$n_{2\nu\nu'} = \frac{S_\nu f_1}{p_1\nu'} = \frac{n_1}{\nu'}[1 - \nu(1 - S)] \tag{10.47}$$

With respect to stator, $n_{1\nu\nu'}$ is

$$n_{1\nu\nu'} = n + n_{2\nu\nu'} = \frac{f_1}{p_1\nu'}[1 + (\nu - \nu')(1 - S)] \tag{10.48}$$

With (10.46), Equation (10.48) becomes

$$n_{1\nu\nu'} = \frac{f_1}{P_1\nu'}\left[1 + c_2\frac{N_r}{P_1}(1-S)\right] \quad (10.49)$$

Synchronous parasitic torques occur when two harmonic field speeds $n_{1\nu_1}$ and $n_{1\nu\nu'}$ are equal to each other.

$$n_{1\nu_1} = n_{1\nu\nu'} \quad (10.50)$$

or

$$\frac{1}{\nu_1} = \frac{1}{\nu'}\left[1 + c_2\frac{N_r}{P_1}(1-S)\right] \quad (10.51)$$

It should be noticed that ν_1 and ν' may be either positive (forward) or negative (backward) waves. There are two cases when such torques occur.

$$\begin{aligned} \nu_1 &= +\nu' \\ \nu_1 &= -\nu' \end{aligned} \quad (10.52)$$

When $\nu_1 = +\nu'$, it is mandatory (from (10.51)) to have S = 1 or zero speed.

On the other hand, when $\nu_1 = -\nu'$,

$$(S)_{\nu_1=+\nu'} = 1; \quad \text{standstill} \quad (10.53)$$

$$(S)_{\nu_1=-\nu'} = 1 + \frac{2p_1}{c_2 N_r} <> 1 \quad (10.54)$$

As seen from (10.54), synchronous torques at nonzero speeds, defined by the slip $(S)_{\nu_1=-\nu'}$, occur close to standstill as $N_r >> 2p_1$.

Example 10.2. Synchronous Parasitic Torques

For the unusual case of a three-phase IM with $N_s = 36$ stator slots and $N_r = 16$ rotor slots, $2p_1 = 4$ poles, and frequency $f_1 = 60$ Hz, let us calculate the speed of the first synchronous parasitic torques as a result of the interaction of phase belt and step (slot) harmonics.

Solution

First, the stator and rotor slotting-caused harmonics have the orders

$$\nu_{1s} = c_1\frac{N_s}{p_1} + 1; \qquad \nu'_{f,b} = c_2\frac{N_r}{p_1} + 1; \qquad c_1 c_2 <> 0 \quad (10.55)$$

The first stator slot harmonics ν_{1s} occur for $c_1 = 1$.

$$\nu_{1f} = +1\frac{36}{2} + 1 = 19; \quad \nu_{1b} = -1\frac{36}{2} + 1 = -17 \quad (10.56)$$

On the other hand, a harmonic in the rotor $\nu'_f = +17 = -\nu_{1b}$ is obtained from (10.55) for $c_2 = 2$.

$$\nu'_f = +2\frac{16}{2} + 1 = +17 \quad (10.57)$$

The slip S for which the synchronous torque occurs (10.54),

$$S_{(17)} = 1 + \frac{2\cdot 2}{2\cdot 16} = \frac{9}{8}; \quad n_{17} = \frac{f_1}{P_1}\left(1 - S_{(17)}\right) = \frac{60}{2}\left(1 - \frac{9}{8}\right)\cdot 60 = -255 \text{ rpm} \tag{10.58}$$

Now, if we consider the first (phase belt) stator mmf harmonic ($6c_1 \pm 1$), $\nu_f = +7$, and $\nu_b = -5$ from (10.55), we discover that for $c_2 = -1$.

$$\nu_b' = -1\frac{16}{2} + 1 = -7 \tag{10.59}$$

So the first rotor slot harmonic interacts with the seventh-order (phase belt) stator mmf harmonic to produce a synchronous torque at the slip.

$$S_{(7)} = 1 + \frac{2\cdot 2}{(-1)\cdot 16} = \frac{3}{4}; \quad n_7 = \frac{f_1}{P_1}\left(1 - S_{(7)}\right) = \frac{60}{2}\left(1 - \frac{3}{4}\right)\cdot 60 = 450 \text{ rpm} \tag{10.60}$$

These two torques occur superposed on the torque/speed curve, as discontinuities at corresponding speeds and at any other speeds, their average values are zero (Figure 10.6).

To avoid synchronous torques, the conditions (10.52) have to be avoided for harmonics orders related to the first rotor slot harmonics (at least) and stator mmf phase-belt harmonics.

$$6c_1 + 1 \neq c_2\frac{N_r}{p_1} + 1 \tag{10.61}$$

$$c_1 = \pm 1, \pm 2 \text{ and } c_2 = \pm 1, \pm 2$$

For stator and rotor slot harmonics,

$$c_1\frac{N_s}{p_1} + 1 \neq c_2\frac{N_r}{p_1} + 1 \tag{10.62}$$

Large synchronous torque due to slot/slot harmonics may occur for $c_1 = c_2 = \pm 1$ when

$$N_s - N_r = 0; \quad \nu_1 = \nu_2' \tag{10.63}$$

$$N_s - N_r = 2p_1; \quad \nu_1 = -\nu_2' \tag{10.64}$$

When $N_s = N_r$, the synchronous torque occurs at zero speed and makes the motor less likely to start without hesitation from some rotor positions. This situation has to be avoided in any case, so $N_s \neq N_r$.

Also from (10.61), it follows that (c_1, $c_2 > 0$), for zero speed,

$$c_2 N_r \neq 6c_1 p_1 \tag{10.65}$$

Condition (10.64) refers to first-slot harmonic synchronous torques occurring at nonzero speed:

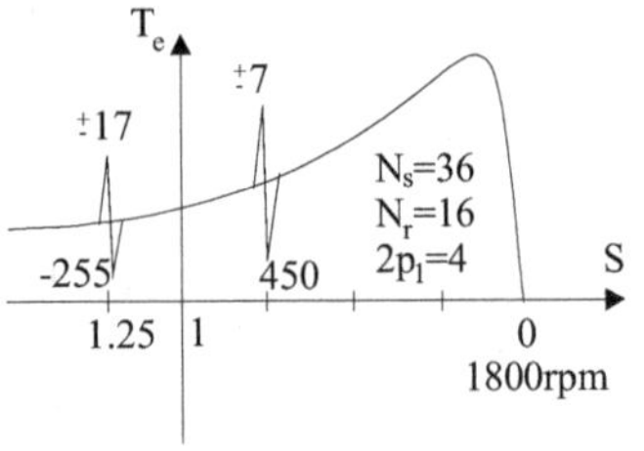

FIGURE 10.6 Torque/speed curve with parasitic synchronous torques.

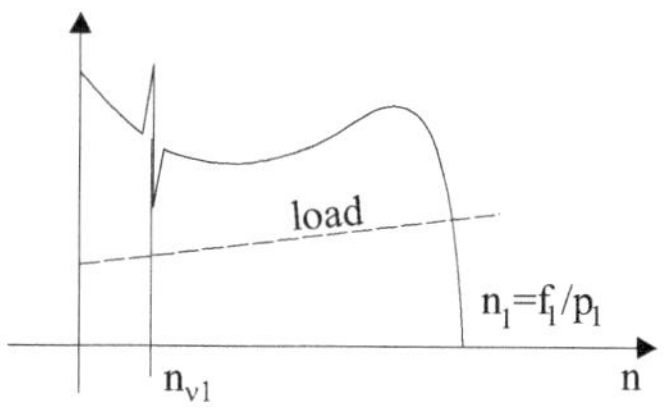

FIGURE 10.7 Synchronous torque on descending torque/speed curve.

$$n_{v_1=v_2'} = \frac{f_1}{p_1}(1-S) = \frac{f_1}{p_1}\frac{2p_1}{c_2N_r} = \frac{-2f_1}{c_2N_r} \tag{10.66}$$

In our case, $c_2 = \pm 1$. Equation (10.66) can also be written in stator harmonic terms as

$$n_{v_1=-v_2'} = \frac{f_1}{p_1 v_1} = \frac{f_1}{p_1}\frac{1}{\left(c_1\frac{Z_1}{p_1}+1\right)} = \frac{f_1}{c_1 Z_1 + p_1} \tag{10.67}$$

with $c_1 = \pm 1$.

The synchronous torque occurs at the speed $n_{v_1=-v_2'}$ if there is a geometrical phase shift angle γ_{geom} between stator and rotor mmf harmonics v_1 and $-v_2'$, respectively. The torque should have the expression T_{sv_1}

$$T_{sv_1} = \left(T_{sv_1}\right)_{max} \sin\left(c_1 Z_1 + p_1\right)\cdot\gamma_{geom} \tag{10.68}$$

The presence of synchronous torques may cause motor locking at the respective speed. This latter event may appear more frequently with IMs with descending torque/speed curve (Design C, D), Figure 10.7.

So far, we considered mmf harmonics and slot harmonics as a source of synchronous torques. However, there are other sources for them such as the airgap magnetic conductance harmonics due to slot leakage saturation.

10.9.3 Leakage Saturation Influence on Synchronous Torques

The airgap field produced by the first rotor mmf slot harmonic $v' = \pm\frac{N_r}{p_1}+1$ due to the airgap magnetic conductance fluctuation caused by leakage saturation (second term in (10.19)) is

$$B_{gv'} = \mu_0 F_{2v_m}\cdot\cos\left[\left(\pm\frac{N_r}{p_1}+1\right)\theta \mp \left(\omega_1 \pm N_r\Omega_r\right)t\right]\cdot\frac{a_0'}{g}\sin\left(2\theta-\omega_1 t\right) \tag{10.69}$$

If B_{gv} contains a harmonic term multiplied by $\cos(v\theta \mp \omega_1 t)$, for any $v = \pm 6c_1 + 1$ that corresponds to the stator mmf harmonics, a synchronous torque condition is again obtained. Decomposing (10.69) into two terms, we find that only if

$$\frac{N_r}{p_1} \pm 3 \neq (6c_1 \pm 1) \tag{10.70}$$

such torques are avoided. Condition (10.70) becomes

$$N_r \neq 2p_1(3c_1 \mp 1);\quad c_1 = 1,2,3,\ldots \tag{10.71}$$

TABLE 10.2
Starting Torque (Maximum and Minimum Value)

Motor Power (kW) $f_1 = 50$ Hz			Starting Torque (Nm)	
$2p_1 = 2$ Poles	N_s	N_r	Max	Min
0.55	24	18	1.788	0
0.85	24	22	0.75	0.635
1.4	24	26	0.847	0.776
2	24	28	2	1.7647
3	24	30	5.929	2.306

The largest such torque occurs in interaction with the first stator slot mmf harmonic $\nu_1 = \frac{N_s}{p_1} \pm 1$ unless

$$\frac{N_s}{p_1} \pm 1 \neq \frac{N_r}{p_1} \pm 3 \tag{10.72}$$

So,

$$N_r \neq N_s \mp 4p_1 \tag{10.73}$$

As an evidence of the influence of synchronous torque at standstill with rotor bars skewed by one stator slot pitch, some experimental results are given in Table 10.2 [1].

As Table 10.2 shows, the skewing of rotor bars cannot prevent the occurrence of synchronous torque, and a small change in the number of rotor slots drastically changes the average starting torque. The starting torque variation (with rotor position) is a clear indication of the presence of synchronous torque at stall.

The worst situation occurs with $N_r = 18 = 6c_1p_1$ (10.65). The speed for this synchronous torque comes from (10.69) as

$$\pm 3\omega_1 + N_r\Omega_r = \pm\omega_1 \tag{10.74}$$

or

$$\Omega_r = \pm\frac{4\omega}{N_r}; \quad n = \frac{4f_1}{N_r} \tag{10.75}$$

Calculating the synchronous torques (even at zero speed) analytically is a difficult task. It may be performed by calculating the stator mmf harmonics airgap field interaction with the pertinent rotor current mmf harmonic.

After [1], the maximum tangential force $f_{\theta max}$ per unit rotor area due to synchronous torques at standstill should be

$$f_{\theta max} \approx \frac{\mu_0 \left(n_s I_s \sqrt{2}\right)^2 d}{4\gamma_{slot}} \tag{10.76}$$

where $n_s I_s \sqrt{2}$ is the peak slot mmf, γ_{slot} is the stator slot angle, and d is the greatest common divisor of N_s and N_r.

From this point of view, the smaller the d, the better. The number d = 1 if Ns and Nr are prime numbers and it is maximum for $N_s = N_r = d$.

With $N_s = 6p_1q_1$, for q integer, (that is, an even number) with N_r an odd number, with d = 1, the smallest variation of starting torque with rotor position is obtained.

However, with N_r an odd number, the IM tends to be noisy, so most IMs have an even number of rotor slots, but with d = 2 if possible.

10.9.4 The Secondary Armature Reaction

So far, in discussing the synchronous torques, we considered only the reaction of rotor currents on the fields of the fundamental stator current. However, in some cases, harmonic rotor currents can induce in the stator windings additional currents of a frequency different from the power grid frequency. These additional stator currents may influence the IM starting properties and add to the losses in an IM [4,5]. These currents are related to the term of secondary armature reaction [4].

With delta stator connection, the secondary armature reaction currents in the primary windings find a good path to flow, especially for multiples of three order harmonics. Parallel paths in the stator winding also favour this phenomenon by its circulating currents.

For wound (slip ring) rotor IMs, the field harmonics are not influenced by delta connection or the parallel path [6]. Let us consider the first step (slot) stator harmonic $\frac{N_s}{p_1} \pm 1$ whose current $I_{1\left(\frac{N_s}{p_1}\pm 1\right)}$ produces a rotor current $I_{2\left(\frac{N_s}{p_1}\pm 1\right)}$.

$$I_{2\left(\frac{N_s}{p_1}\pm 1\right)} \approx \frac{I_{1\left(\frac{N_s}{p_1}\pm 1\right)}}{1+\tau_{d\left(\frac{N_s}{p_1}\pm 1\right)}} \tag{10.77}$$

τ_d is the differential leakage coefficient for the rotor cage [1].

$$\tau_{d\left(\frac{N_s}{p_1}\pm 1\right)} = \left[\frac{\pi(N_s \pm p_1)}{N_r}\right]^2 \cdot \frac{1}{\sin^2\left[\frac{\pi(N_s \pm p_1)}{N_r}\right]} - 1 \tag{10.78}$$

As known, N_s and N_r are numbers rather close to each other, so $\tau_{d\left(\frac{N_s}{p_1}\pm 1\right)}$ might be a large number and thus, the rotor harmonic current is small. A strong secondary armature reaction may change this situation drastically. To demonstrate this, let us notice that $I_{2\left(\frac{N_s}{p_1}\pm 1\right)}$ can now create harmonics of order ν' (10.46), for $c_2 = 1$ and $\nu = \left(\frac{N_s}{p_1} \pm 1\right)$

$$\nu' = \frac{N_r}{P_1} \pm \left(\frac{N_s}{P_1} \pm 1\right) \tag{10.79}$$

The ν' rotor harmonic mmf $F_{\nu'}$ has the amplitude

$$F_{\nu'} = F_{2\left(\frac{N_s}{p_1}\pm 1\right)} \cdot \frac{N_s \pm p_1}{N_r \pm N_s \pm p_1} \tag{10.80}$$

If these harmonics induce notable currents in the stator windings, they will, in fact, reduce the differential leakage coefficient from $\tau_{d\left(\frac{N_s}{P_1}\pm 1\right)}$ to $\tau'_{d\left(\frac{N_s}{P_1}\pm 1\right)}$ [1].

$$\tau'_{d\left(\frac{N_s}{P_1}\pm1\right)} \approx \tau_{d\left(\frac{N_s}{P_1}\pm1\right)} - \left(\frac{N_s \pm P_1}{N_r - (N_s \pm P_1)}\right)^2 \quad (10.81)$$

Now the rotor current $I_{2\left(\frac{N_s}{P_1}\pm1\right)}$ of (10.77) will become $I'_{2\left(\frac{N_s}{P_1}\pm1\right)}$.

$$I'_{2\left(\frac{N_s}{P_1}\pm1\right)} = \frac{I_{1\left(\frac{N_s}{P_1}\pm1\right)}}{\tau'_{d\left(\frac{N_s}{P_1}\pm1\right)} + 1} \quad (10.82)$$

The increase of rotor current $I'_{2\left(\frac{N_s}{P_1}\pm1\right)}$ is expected to increase both asynchronous and synchronous torques related to this harmonic.

In a delta connection, such a situation takes place. For a delta connection, the rotor current (mmf) harmonics

$$\nu' = \frac{N_r}{p_1} - \left(\frac{N_s}{p_1} \pm 1\right) = 3c \quad c = 1,3,5,\ldots \quad (10.83)$$

can flow freely because their induced voltages in the three stator phases are in phase (multiple of three harmonics). They are likely to produce notable secondary armature reaction.

To avoid this undesirable phenomenon,

$$|N_s - N_r| \neq (3c \mp 1)p_1 = 2p_1, 4p_1, \ldots \quad (10.84)$$

An induction motor with $2p_1 = 6$, $N_s = 36$, and $N_r = 28$ slots does not satisfy (10.84), so it is prone to a less favourable torque-speed curve in Δ connection than it is in star connection.

In parallel path stator windings, circulating harmonics current may be responsible for notable secondary armature reaction with star or delta winding connection.

These circulating currents may be avoided if all parallel paths of the stator winding are at any moment at the same position with respect to rotor slotting.

However, for an even number of current paths, the stator current harmonics of the order $\frac{N_r}{p_1} - \frac{N_s}{p_1} - 1$ or $\frac{N_r}{p_1} - \frac{N_s}{p_1} + 1$ may flow within the winding if the two numbers are multiples of each other. The simplest case occurs if

$$|N_s - N_r| = 3P_1 \quad (10.85)$$

When (10.85) is fulfilled, care must be exercised in using parallel path stator windings.

We should also notice again that the magnetic saturation of stator and rotor teeth leads to a further reduction of differential leakage by 40%–70%.

10.9.5 Notable Differences between Theoretical and Experimental Torque/Speed Curves

By now, the industry has accumulated enormous data on the torque/speed curves of IM of all power ranges.

In general, it was noticed that there are notable differences between theory and tests especially in the braking regime ($S > 1$). A substantial rise of torque in the braking regime is noticed. In contrast, a smaller reduction of torque in the large slip motoring regime appears frequently.

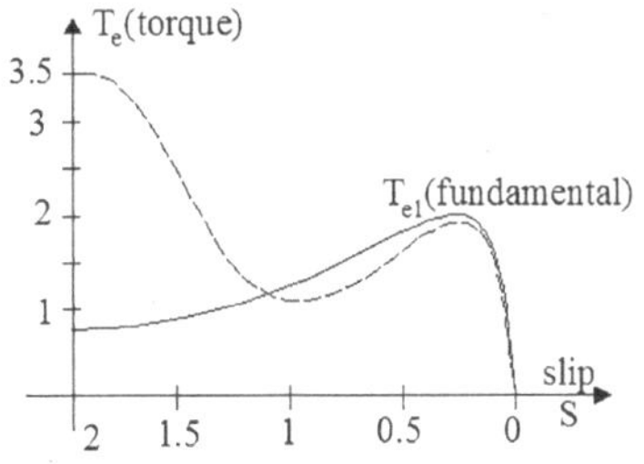

FIGURE 10.8 Torque-slip curve (relative values) for $N_s = 36$, $N_r = 28$, $2p_1 = 4$, skewed rotor (one slot pitch), and insulated copper bars.

A similar effect is produced by transverse (lamination) rotor currents between skewed rotor bars.

However, in this latter case, there should be no difference between slip ring and cage rotor, which is not the case in practice.

Finally, these discrepancies between theory and practice occur even for insulated copper bar cage rotors where the transverse (lamination) rotor currents cannot occur (Figure 10.8) [1].

What may cause such a large departure of experiments from theory? Although a general, simple, answer to this question is still not available, it seems intuitive to assume that the airgap magnetic conductance harmonics due to slot openings and the tooth head (leakage) saturation are the main causes of spectacular braking torque augmentation and notable reductions in high-slip motoring torque.

This is true when the contact resistance of rotor bars (practically insulated rotor bars) is high and, thus, the transverse (lamination) currents in skewed rotors are small.

Slot openings have been shown to considerably increase the parasitic torques. (Remember we showed earlier in this chapter that the first-slot (step) stator mmf harmonic $\nu_{s\,min} = \frac{N_s}{p_1} - 1$ is amplified by the influence of the first airgap magnetic conductance – due to slot openings – (10.33)).

Now the leakage path saturation-produced airgap magnetic conductance (10.11) may add further to amplify the parasitic torques.

For skewed rotors, at high slips, where parasitic torques occur, main path flux saturation due to the uncompensated (skewing) rotor mmf may lead to a fundamental torque increase in the braking region (due to higher current, for lower leakage inductance) and to the amplification of parasitic torque through main path saturation-caused airgap conductance harmonics (10.20). And, finally, rotor eccentricity may introduce its input to this braking torque escalation.

We have to stress again that leakage flux path saturation and skin effects alone, also cause important increases in rotor resistance, decreases in motor reactance, and thus produce also more fundamental torque.

In fact, the extreme case in Figure 10.8 with a steady increase of torque with slip (for $S > 1$) seems to be a mixture of such intricated contributions. However, the minimum torque around zero speed shows the importance of parasitic torques as the skin effect and leakage saturation depend on $\sqrt{S}$; so they change slower with slip than Figure 10.7 suggests.

10.9.6 A Case Study: $N_s/N_R = 36/28$, $2p_1 = 4$, $y/\tau = 1$, and 7/9; $m = 3$ [7]

As there is no general answer, we will summarise the case of a three-phase induction motor with $P_n = 11$ kW, $N_s/N_r = 36/28$, $2p_1 = 4$, and $y/\tau = 9/9$ and 7/9 (skewed rotor by one stator slot pitch).

For this machine, and $y/\tau = 9/9$, with straight slots, condition (10.71) is not satisfied as

$$N_r = 28 = 2p_1(3c \mp 1) = 2 \cdot 2(3c + 1); \quad \text{for } c = 2 \tag{10.86}$$

This means that a synchronous torque occurs at the speed

$$n = \frac{4f_1}{N_r} = \frac{4 \cdot 50 \cdot 60}{28} = 215\,\text{rpm} \tag{10.87}$$

Other synchronous, but smaller, torques may be detected. A very large synchronous torque has been measured at this speed (more than 14 Nm, larger than the peak torque for 27% voltage). This leads to a negative total torque at 215 rpm; successful acceleration to full speed is not always guaranteed. It is also shown that as the voltage increases and saturation takes place, the parasitic synchronous torque does not increase significantly over this value.

Also, a motor of 70 kW, $N_s/N_r = 48/42$, $2p_1=4$, and $y/\tau = 10/12$ with Δ connection and four parallel current paths has been investigated. As expected, a strong synchronous torque has been spotted at standstill because condition (10.65) is not met as

$$N_r = 6\frac{c_1}{c_2}p_1 = 6\frac{7}{2}\cdot 2 = 42 \tag{10.88}$$

The parasitic synchronous torque at stall is amplified by Δ connection with four paths in parallel stator winding as $N_s - N_r = 48 - 42 = 6 = 3p_1 = 3 \cdot 2$.

The situation is so severe that, for 44% voltage and 30% load torque, the motor remains blocked at zero speed.

Strong variations of motor torque with position have been measured and calculated satisfactorily with finite difference method. Saturation consideration seems mandatory for realistic torque calculations during starting transients.

It has also been shown that the amplification of parasitic torque due to slot openings is higher for $N_r > N_s$ because the differential leakage coefficient $\tau_{d\left(\frac{N_s}{p_1}\pm 1\right)}$ is smaller in this case, allowing for large rotor mmf harmonics. From the point of view of parasitic torques, the rotor with $N_r < N_s$ is recommended and may be used without skewing for series-connected stator windings.

For $N_r > N_s$, skewing is recommended unless magnetic wedges in the stator slots are used.

10.9.7 Evaluation of Parasitic Torques by Tests (after Ref. [1])

To assist potential designers of IMs here is a characterisation of torque/speed curve for quite a few stator/rotor slot numbers of practical interest.

For $N_s = 24$ stator slots, $f_1 = 50\,\text{Hz}$, $2p_1 = 4$

- $N_r = 10$; small synchronous torques at n = 150, 300, and 600 rpm and a large one at n = –300 rpm
- $N_r = 16$; unsuitable (high synchronous torque at n = 375 rpm)
- $N_r = 18$; very suitable (low asynchronous torque dips at n = 155 rpm and at –137 rpm)
- $N_r = 19$; unsuitable because of high noise, though only small asynchronous torque dips occur at n = 155 rpm and n = –137 rpm
- $N_r = 20$; greatest synchronous torque of all situations investigated here at n = –300 rpm, unsuitable for reversible operation.
- $N_r = 22$; slight synchronous torques at n = 690, 600, ±550, 214, and –136 rpm
- $N_r = 24$; unsuitable because of high synchronous torque at n = 0 rpm
- $N_r = 26$; larger asynchronous torque deeps at n = 115 and –136 rpm and synchronous one at n = –300 rpm
- $N_r = 27$; high noise
- $N_r = 28$; high synchronous torque dip at n = 214 rpm and slight torque dips at n = 60, –300, and –430 rpm. Asynchronous torque dips at n = 115 and –136 rpm. With skewing, the situation gets better, but at the expense of 20% reduction in fundamental torque.

- $N_r = 30$; typical asynchronous torque dips at n = 115 and −136 rpm, small synchronous ones at n = 600, ±400, ±300, 214, ±200, −65, −100, and −128 rpm. Practically noiseless.
- $N_r = 31$; very noisy.

Number of slots N_s = 36, $f_1 = 50\,Hz$, $2p_1 = 4$

This time, the harmonics asynchronous torques are small. At n = 79–90 rpm, they become noticeable for higher number of rotor slots (44, 48).

- $N_r = 16$; large synchronous torque at n = −188 and +376 rpm
- $N_r = 20$; synchronous torque at n = −376 rpm
- $N_r = 24$; large synchronous torque at n = 0 rpm, impractical
- $N_r = 28$; large synchronous torque at n = −214 rpm, −150, and −430 rpm
- $N_r = 31$; good torque curve but very noisy
- $N_r = 32$; large synchronous torque at n = −188 rpm and a smaller one at n = −375 rpm
- $N_r = 36$; impractical, motor locks at zero speed
- $N_r = 44$; large synchronous torque at n = −136 rpm, smaller ones at n = 275, −40, and −300 rpm
- $N_r = 48$; moderate synchronous torque dip at zero speed

Number of slots N_s = 48, $f_1 = 50\,Hz$, $2p_1 = 4$

- $N_r = 16$; large synchronous torque at n = −375, a smaller one at n= −188 rpm
- $N_r = 20$; large synchronous torque at n = −300 rpm, small ones at n = 600, 375, 150, 60, −65, −136, and −700 rpm
- $N_r = 22$; large synchronous torque dip at n = −136 rpm due to the second rotor slip harmonic $\left(c_2 = 2, n = \frac{2f_1}{c_2 N_r} = \frac{-2\cdot 50\cdot 60}{2\cdot 22} = -136\,\text{rpm}\right)$
- $N_r = 28$; high synchronous torque at n = 214 rpm, smaller ones at n = 375, 150, −65, −105, −136, and −214 rpm
- $N_r = 32$; high synchronous torque at n = −188 rpm, small ones at n = 375, 188, −90, 90, and −300 rpm
- $N_r = 36$; large torque dip at zero speed, not good start
- $N_r = 44$; very large synchronous torque at n = −136 rpm, smaller ones at n = 275, 214, 60, and −300 rpm
- $N_r = 48$; motor locks at zero speed ($N_s = N_r = 48$!), after warming up, however, it starts!

10.10 RADIAL FORCES AND ELECTROMAGNETIC NOISE BASICS

Tangential forces in IMs are producing torques which smoothly drive the IM (or cause torsional vibrations) while radial forces are causing vibration and noise [8–15].

This kind of vibration and noise is produced by electromagnetic forces which vary along the rotor periphery and with time. Using Maxwell tensor, after neglecting the tangential field component in the airgap (as it is much lower than the radial one), the radial stress on the stator is

$$p_r = \frac{B^2(\theta_m, t)}{2\mu_0}; \quad \theta = \theta_m p_1 \tag{10.89}$$

where $B(\theta,t)$ is the resultant airgap flux density.

The expression of $B(\theta_m,t)$ is as derived earlier in this chapter,

$$B(\theta_m, t) = \mu_0 \left(F_1(\theta_m, t) + F_2(\theta_m, t)\right) \cdot \lambda(\theta_m, t) \tag{10.90}$$

with $F_1(\theta_m,t)$ and $F_2(\theta_m,t)$ the stator and rotor mmfs and $\lambda(\theta_m,t)$ the airgap magnetic conductance (inversed airgap length) as influenced by stator and rotor slot openings, leakage, and main path saturation and eccentricity.

$$\lambda(\theta,t)=\lambda_0+\sum_{\sigma=1}^{\infty}\lambda_\sigma\cos(\sigma\theta_m+\phi_\sigma)+\sum_{\rho=1}^{\infty}\lambda_\rho\cos\left(\rho\left(\theta_m-\frac{\omega_r t}{p_1}\right)+\phi_\rho\right)+\ldots \quad (10.91)$$

$$F_1(\theta_m,t)=\sum F_{1\nu}\cos(\nu\theta_m\pm\omega_\nu t) \quad (10.92)$$

$$F_2(\theta_m,t)=\sum F_{2\mu}\cos(\mu\theta_m\pm\omega_\mu t+\phi_\mu) \quad (10.93)$$

In (10.89)–(10.93), we have used the mechanical angle θ_m as the permeance harmonics are directly related to it and not the electrical angle $\theta = p_1\theta_m$. Consequently, now the fundamental, with p_1 pole pairs, will be $\nu_1 = p_1$ and not $\nu = 1$.

Making use of (10.92 and 10.93) in (10.89), we obtain terms of the form

$$p_r = A_r\cos(\gamma\theta_m-\Omega_r t) \quad (10.94)$$

$\gamma = 0,1,2,\ldots$. The frequency of these forces is $f_r = \frac{\Omega_r}{2\pi}$, which is not to be confused to the rotor speed ω_r/p_1.

If the number of pole pairs (γ) is small, stator vibrations may occur due to such forces. As inferred from (10.90), there are terms from same harmonic squared and interaction terms where 1(2) mmf harmonics interact with an airgap magnetic conductance. Such combinations are likely to yield low orders for γ.

For $\gamma = 0$, the stress does not vary along rotor periphery but varies in time producing variable radial pressure on the stator (and rotor).

For $\gamma = 1$, (10.94) yields a single-sided pull on the rotor which circulates with the angular velocity of Ω_r. If Ω_r corresponds to a mechanical resonance of the machine, notable vibration will occur.

Case $\gamma = 1$ results from the interaction of two mechanical harmonics whose order differs by unity.

For $\gamma = 2, 3, 4, \ldots$ stator deflections occur. In general, all terms p_r with $\gamma \le 4$ should be avoided, by adequately choosing the stator and rotor slot number pairs N_s and N_r.

However, as mechanical resonance of stators (without and with frames attached) decreases with motor power increase, it means that a certain N_s and N_r combination may lead to low noise in low-power motors but to larger noise in large-power motors.

Now as in (10.90), the mmf step harmonics and the airgap magnetic conductance occur, we will treat their influence separately to yield some favourable N_s and N_r combinations.

10.10.1 Constant Airgap (No Slotting, No Eccentricity)

In this case, $\lambda(\theta_m,t) = \lambda_0$ and a typical p_r term occurs from two different stator and rotor mmf harmonics.

$$(p_r)_{\gamma=\nu-\mu}=\frac{\mu_0 F_{1\nu}F_{2\mu}}{2g^2}\lambda_0^2\cos\left[(\gamma\mp\mu)\theta_m\pm(\omega_\nu+\omega_\mu)-\phi_\mu\right] \quad (10.95)$$

In this case, $\gamma = \nu - \mu$, that is the difference between stator and rotor mmf mechanical harmonics order.

Now if we are considering the first-slot (step) mmf harmonics ν and μ,

$$\begin{aligned} \nu &= N_s \pm p_1 \\ \mu &= N_r \pm p_1 \end{aligned} \tag{10.96}$$

It follows that to avoid such forces with $\gamma = 0, 1, 3, 4$,

$$\begin{aligned} |N_s - N_r| &\neq 0, 1, 2, 3, 4 \\ |N_s - N_r| &\neq 2p_1, 2p_1 \pm 1, 2p_1 \pm 2, 2p_1 \pm 3, 2p_1 \pm 4 \end{aligned} \tag{10.97}$$

The time variation of all stator space harmonics has the frequency $\omega_\nu = \omega_1 = 2\mu f_1$, while for the rotor,

$$\omega_\mu = \omega_1 \mp cN_r\Omega_r = \omega_1\left[1 \mp c\frac{N_r}{p_1}(1-S)\right] \tag{10.98}$$

Notice that Ω_r is the rotor mechanical speed here.

Consequently, the frequency of such forces is $\omega_\nu \pm \omega_\mu$

$$\omega_\xi = \omega_\nu \pm \omega_\mu = \begin{cases} \omega_1 c\dfrac{N_r}{p_1}(1-S) \\ 2\omega_1 + \omega_1 c\dfrac{N_r}{p_1}(1-S) \end{cases} \tag{10.99}$$

In general, only frequencies ω_ξ low enough ($c = 1$) can cause notable stator vibrations.

The frequency ω_ξ is twice the power grid frequency or zero at zero speed ($S = 1$) and increases with speed.

10.10.2 Influence of Stator/Rotor Slot Openings, Airgap Deflection, and Saturation

As shown earlier in this chapter, the slot openings introduce harmonics in the airgap magnetic conductance which have the same order as the mmf slot (step) harmonics.

Zero-order ($\gamma = 0$) components in radial stress p_r may be obtained when two harmonics of the same order, but different frequency, interact. One springs from the mmf harmonics and one from the airgap magnetic conductance harmonics. To avoid such a situation simply,

$$\begin{aligned} N_s \pm p_1 &\neq N_r \\ N_r \pm p_1 &\neq N_s \end{aligned} \quad \text{or} \quad |N_s - N_r| \neq \pm p_1 \tag{10.100}$$

Also, to avoid first-order stresses for $\gamma = 2$,

$$2(N_r - N_s \mp P_1) \neq 2 \tag{10.101}$$

$$|N_s - N_r| \neq \pm p_1 \pm 1 \tag{10.102}$$

The frequency of such stresses is still calculated from (10.99) as

$$\omega_\xi = \left[\omega_1 \pm \omega_1\frac{N_r}{p_1}(1-S)\right]2 \tag{10.103}$$

The factor 2 in (10.103) above comes from the fact that this stress component springs from a term squared so all the arguments (of space and time) are doubled.

It has been shown that airgap deformation itself, due to such forces, may affect the radial stresses of low order.

In the most serious case, the airgap magnetic conductance becomes

$$\lambda(\theta_m,t)=\frac{1}{g(\theta_m,t)}=\left[a_0+a_1\cos N_s\theta_m+a_2\cos N_r(\theta_m-\omega_r t)\right]\cdot$$
$$\cdot\left[1+B\cos(K\theta_m-\Omega t)\right] \tag{10.104}$$

The last factor is related to airgap deflection. After decomposition of $\lambda(\theta_m,t)$ in simple sin(cos) terms for $K = p_1$, strong airgap magnetic conductance terms may interact with the mmf harmonics of orders p_1, $N_s \pm p_1$, and $N_r \pm p_1$ again. So we have to have

$$\begin{gathered}(N_s\pm p_1)-(N_r\pm p_1)\neq p_1\\ \text{or}\quad |N_s-N_r|\neq p_1\\ \text{and}\quad |N_s-N_r|\neq 3p_1\end{gathered} \tag{10.105}$$

In a similar way, slot leakage flux path saturation, already presented earlier in this chapter with its contribution to the airgap magnetic conductance, yields conditions as above to avoid low-order radial forces.

10.10.3 Influence of Rotor Eccentricity on Noise

We have shown earlier in this chapter that rotor eccentricity – static and dynamic – introduces a two-pole geometrical harmonic in the airgap magnetic conductance. This leads to a $p_1 \pm 1$ order flux density strong components (10.90) in interaction with the mmf fundamentals. These forces vary at twice stator frequency $2f_1$, which is rather low for resonance conditions. In interaction with another harmonic of same order but different frequency, zero-order ($\gamma = 0$) vibrations may occur again. Consider the interference with first mmf slot harmonics $N_s \pm p_1$ and $N_r \pm p_1$

$$\begin{gathered}(N_s\pm P_1)-(N_r\pm P_1)\neq P_1\pm 1\\ \text{or}\quad |N_s-N_r|\neq 3P_1\pm 1\\ |N_s-N_r|\neq P_1\pm 1\end{gathered} \tag{10.106}$$

These criteria, however, have been found earlier in this paragraph. The dynamic eccentricity has also been shown to produce first-order forces ($\gamma = 1$) from the interaction of $p_1 \pm 1$ airgap magnetic conductance harmonic with the fundamental mmf (p_1) as their frequencies are different. Noise is thus produced by the dynamic eccentricity.

10.10.4 Parallel Stator Windings

Parallel path stator windings are used for large-power, low-voltage machines (now favoured in power electronics adjustable drives). As expected, if consecutive poles are making a current path, and some magnetic asymmetry occurs, a two-pole geometric harmonic occurs. This harmonic in the airgap magnetic conductance can produce, at least in interaction with the fundamental mmf, radial stresses with the order $\gamma = p_1 \pm 1$ which might cause vibration and noise.

On the contrary, with two parallel paths, if north poles make one path and south poles the other current path, with two-layer full-pitch windings, no asymmetry will occur. However, in single-layer winding, this is not the case and stator harmonics of low order are produced ($\nu = N_s/2c$). They will never lead to low-order radial stress, because, in general, N_s and N_r are not far from each other in actual motors.

Parallel windings with chorded double-layer windings, however, due to circulating current, lead to additional mmf harmonics and new rules to eliminate noise [1],

$$\begin{aligned} |N_s - N_r| &= \left|\frac{p_1}{a} \pm p_1 \pm 1\right|; \quad \text{a-odd} \\ |N_s - N_r| &= \left|\frac{2p_1}{a} \pm p_1 \pm 1\right|; \quad \text{a-even} \end{aligned} \tag{10.107}$$

where a is the number of current paths.

Finally, for large-power machines, the first-slot (step) harmonic order $N_s \pm p_1$ is too large so its phase-belt harmonics $n = (6c_1 \pm 1)p_1$ are to be considered in (10.107) [1],

$$\begin{aligned} &\left|N_s \pm \frac{p_1}{a} - (6c_1 \pm 1)p_1\right| \neq 1; \quad a > 2, \text{ odd} \\ &\left|N_r \pm \frac{2p_1}{a} - (6c_1 \pm 1)p_1\right| \neq 1; \quad a > 2, \text{ even} \\ &\left|N_r \pm 2p_1 - (6c_1 \pm 1)p_1\right| \neq 1; \quad a = 2 \end{aligned} \tag{10.108}$$

with c_1 a small number $c_1 = 1, 2, \ldots$.

Symmetric (integer q) chorded windings with parallel paths introduce, due to circulating current, a $2p_1$ pole pair mmf harmonic which produces effects that may be avoided by fulfilling (10.108).

10.10.5 Slip-Ring Induction Motors

The mmf harmonics in the rotor mmf are now $\nu = c_2N_r \pm p_1$ only. Also, for integer q_2 (the practical case), the number of slots is $N_r = 2{\cdot}3{\cdot}p_1{\cdot}q_2$. This time, the situation of $N_s = N_r$ is to be avoided; the other conditions are automatically fulfilled in general due to this constraint on N_r. Using the same rationale as for the cage rotor, the radial stress order, for constant airgap, is

$$r = \nu - \mu = (6c_1 \pm 1)p_1 - (6c_2 \pm 1)p_1 \tag{10.109}$$

Therefore "r" may be only an even number.

Now as soon as eccentricity occurs, the absence of armature reaction leads to higher radial uncompensated forces than for cage rotors. The one for $r = p_1 \pm 1$ – that in interaction with the mmf fundamental, which produces a radial force at $2f_1$ frequency, may be objectionable because it is large (no armature reaction) and produces vibrations.

So far, the radial forces have been analysed in the absence of magnetic saturation and during steady state. The complexity of this subject makes experiments in this domain mandatory.

Thorough experiments [8] for a few motors with cage and wound rotors confirm the above analysis in general and reveal the role of magnetic saturation of teeth and yokes in reducing these forces (above 70%–80% voltage). The radial forces due to eccentricity level off or even decrease.

During steady state, the unbalanced magnetic pull due to eccentricity has been experimentally proved much higher for the wound rotor than for the cage rotor! During starting transients, both are, however, only slightly higher than for the wound rotor under steady state. Parallel windings have been proven to reduce the UMP force ($\gamma = 0$).

It seems that even with 20% eccentricity, UMP may surpass the cage rotor mass during transients and the wound rotor mass even under steady state [8].

The effect of radial forces on the IM depends not only on their amplitude and frequency but also on the mechanical resonance frequencies of the stator; so the situation is different for low-, medium-, and large-power motors.

10.10.6 Mechanical Resonance Stator Frequencies

The stator ring resonance frequency for r = 0 (first order) radial forces F_0 is [1]

$$F_0 = \frac{8\cdot 10^2}{R_a}\sqrt{\frac{G_{yoke}}{G_{yoke}+G_{teeth}}} \tag{10.110}$$

with R_a – average stator yoke radius and G_{yoke} and G_{teeth} stator yoke and teeth weight.

For $r \neq 0$,

$$F_r = F_0\frac{1}{2\sqrt{3}}\frac{h_{yoke}}{R_a}\frac{r\left(r^2-1\right)}{\sqrt{r^2+1}}K_r \tag{10.111}$$

h_{yoke} – yoke radial thickness.

$K_r = 1$ for $h_{yoke}/R_a > 0.1$ and $0.66 < K_r < 1$ for $h_{yoke}/R_a < 0.1$.

The actual machine has a rather elastic frame fixture to ground and a less than rigid attachment of stator laminations stack to the frame.

This leads to more than one resonance frequency for a given radial force of order r. A rigid framing and fixture will reduce this resonance "multiplication" effect.

To avoid noise, we have to avoid radial force frequencies to be equal to resonance frequencies. For the step harmonics of mmfs contribution, (10.100) applies.

for r = 0, at S = 0 (zero slip),

$$\frac{F_0}{f_1} \neq c\frac{N_r}{p_1} \quad \text{or} \quad \frac{F_0}{f_1} \neq c\frac{N_r}{p_1} \pm 2; \quad c = 1,2 \tag{10.112}$$

We may proceed, for r = 1, 2, 3, by using F_r instead of F_0 in (10.112) and with c = 1.

In a similar way, we should proceed for the frequency of other radial forces (derived in previous paragraphs).

In small-power machines, such conditions are easy to meet as the stator resonance frequencies are higher than most of the radial stress frequencies. So radial forces of small pole pairs r = 0, 1, 2 are to be checked. For large-power motors, radial forces with r = 3, 4 are more important to account for.

The subject of airgap flux distribution, torque pulsations (or parasitic torques), and radial forces has been approached recently by FEM [9,10] with good results but at the price of prohibitive computation time.

Consequently, it seems that intuitive studies, based on harmonics analysis, are to be used in the preliminary design study and, once the number of stator/rotor slots are defined, FEM is to be used for precision calculation of parasitic torques and radial forces.

10.11 ELECTROMAGNETIC VIBRATION: A PRACTICAL VIEW

As already alluded to earlier in this chapter, electromagnetic vibration in IMs (in all electric machinery) is produced by

- Radial electromagnetic forces

$$F_{rad} \sim B_r^2(\alpha,t)/2\mu_0 \quad \left(N/m^2\right) \tag{10.113}$$

- Tangential electromagnetic forces

$$F_t \sim B_r(\alpha,t)\cdot B_t(\alpha,t)/\mu_0 \quad \left(\mathrm{N/m^2}\right) \tag{10.114}$$

As fundamental B(α, t) varies with frequency ω_1, the radial force varies with $2\omega_1$ and produces vibration at this frequency even in a fault-less IM fed with symmetrical phase voltages.

Higher frequency vibrations produced by the radial force are the result of stator and rotor harmonics interactions at frequencies f_r:

$$f_r = f_1 K N_b (1-S)/p;\ f_r = f_1\left[K N_b (1-S)/p_1 \pm 2\right] \tag{10.115}$$

N_s, N_r – number of stator (rotor) slots; S – slip, k = 1, 2, 3…

The order of radial forces *r* is

$$r = K_1 N_s \pm K_2 N_r \pm 2p; \quad K_{1,2} = 0,1,2,3 \tag{10.116}$$

Another source of vibration, the tangential force Ft, manifests itself in torque pulsations ΔT_e

$$\Delta T_e = \sum T_{es}(\Omega_v)\cos\left(\Omega_v t - \Psi_{t\Omega_v}\right) \tag{10.117}$$

Torque pulsations in line-start IMs are provoked primarily by asymmetrical stator voltages at $2\omega_1$ frequency (due to inverse component) but they may also be the result of asymmetrical rotor currents (broken bars), at $2S\omega_1$ frequency. Source voltage time harmonics of 5th and 7th order produce torque pulsations at $6\omega_1$ frequency. Finally, PWM converters in faulty conditions might produce D.C. voltages whose additional D.C. currents interact with the main (fundamental) flux to produce torque ripple at frequency ω_1.

Static eccentricity was also claimed to produce vibration at $2\omega_1$ frequency, but this is not clear yet [16] as it also depends on the mechanical resonance frequency. Dynamic eccentricity may produce vibrations around at *1 x running speed* sidebands [17].

Ref. [17] shows a row of experiments that certify the electromagnetic vibrations at $2S\omega_1$ as caused by broken bars, those at $2\omega_1$ as caused by voltage imbalance, and at ω_1 as produced by the faulty inverter D.C. voltage components.

Ref. [18], however, points out about the overlapping of $2S\omega_1$ torque oscillations due to broken bars and those at Ω_r/r_t in IMs with speed reduction gears with r_t transmission ratio (especially with $r_t > 10$) and suggests that the current signature analysis (CSA) for diagnosis fails in this case but the active and, respectively, reactive instantaneous power oscillations can discriminate between the two sources of torque pulsations in such cases.

In general, CSA and vibration analysis have been in general used to the diagnosis of electric machine internal or external faults. It seems that active and reactive instantaneous power pulsations analysis is another synthetic way in monitoring and diagnosis of IM drives.

10.12 SUMMARY

- The fundamental component of resultant airgap flux density distribution in interaction with the fundamental of stator (or rotor) mmf produces the fundamental (working) electromagnetic torque in the IM.
- The placement of windings in slots (even in infinitely thin slots) leads to a stepped-like waveform of stator (rotor) mmfs, which exhibit space harmonics besides the fundamental wave (with p_1 pole pairs). These are called mmf harmonics (or step harmonics); their lower orders are called phase-belt harmonics ν = 5, 7, 11, 13, … to the first "slot" harmonic order $\nu_{s\,min} = \frac{N_s}{p_1} \pm 1$. In general, $\nu = (6c_1 \pm 1)$.

- A wound rotor mmf has its own harmonic content $\mu = (6c_2 \pm 1)$.
- A cage rotor, however, adapts its mmf harmonic content (μ) to that of the stator such that

$$\nu - \mu = \frac{c_2 N_r}{p_1}$$

- ν and μ are here electric harmonics, so $\nu = \mu = 1$ means the fundamental (working) wave.
- The mechanical harmonics ν_m and μ_m are obtained if we multiply ν and μ by the number of pole pairs

$$\nu_m = p_1 \nu; \quad \mu_m = p_1 \mu;$$

- In the study of parasitic torques, it seems better to use electric harmonics, while, for radial forces, mechanical harmonics are favoured. As the literature uses both concepts, we thought it is appropriate to use them both in a single chapter.
- The second source of airgap field distribution is the magnetic specific airgap conductance (inversed airgap function) $\lambda_{1,2}(\theta_m,t)$ variation with stator or/and rotor position. In fact, quite a few phenomenon are accounted for by decomposing $\lambda_{1,2}(\theta_m,t)$ into a constant component and various harmonics. They are outlined below.
- The slot openings, both on the stator and rotor, introduce harmonics in $\lambda_{1,2}(\theta_m,t)$ of the orders c_1N_s/p_1, c_2N_r/p_1, and $(c_1N_s \pm c_2N_r)\cdot p_1$, with $c_1 = c_2 = 1$ for the first harmonics.
- At high currents, the teeth heads saturate due to large leakage fluxes through slot necks. This effect is similar to a fictitious increase of slot openings, variable with slot position with respect to stator (or rotor) mmf maximum. A $\lambda_{1,2}(\theta_m,t)$ second-order harmonic (4 p_1 poles) travelling at the frequency of the fundamental occurs mainly due to this aspect.
- The main flux path saturation produces a similar effect, but its second-order harmonic ($4p_1$ poles) is in phase with the magnetisation mmf and not with stator or rotor mmfs separately!
- The second permeance harmonic leads to a third harmonic in the airgap flux density.
- The rotor eccentricity, static and dynamic, produces mainly a two-pole harmonic in the airgap conductance. For a two-pole machine, in interaction with the fundamental mmf, a homopolar flux density is produced. This flux is closing axially through the frame, bearings, and shaft, producing an A.C. shaft voltage and bearing current of frequency f_1 (for static eccentricity) and Sf_1 (for the dynamic eccentricity).
- The interaction between mmf and airgap magnetic conductance harmonics is producing a multitude of harmonics in the airgap flux distribution.
- Stator and rotor mmf harmonics of same order and same stator origin produce asynchronous parasitic torques. Practically, all stator mmf harmonics produce asynchronous torques in cage rotors as the latter adapts to the stator mmf harmonics. The no-load speed of asynchronous torque is $\omega_r = \omega_1/\nu$.
- As $\nu \geq 5$ in symmetric (integer q) stator windings, the slip S for all asynchronous torques is close to unity. So they all occur around standstill.
- The rotor cage, as a short-circuited multiphase winding, may attenuate asynchronous torques.
- Chording is used to reduce the first asynchronous parasitic torques (for $\nu = -5$ or $\nu = +7$); skewing is used to cancel the first-slot mmf harmonic $\nu = 6q_1 \pm 1$ (q_1 – slots/pole/phase).
- Synchronous torques are produced at some constant specific speeds where two harmonics, of same order $\nu_1 = \pm\nu'$ but of different origin, interact.
- Synchronous torques may occur at standstill if $\nu_1 = +\nu'$ and at

$$S = 1 + \frac{2p_1}{c_2 N_r}; \quad -1 \geq C_2 \geq 1$$

for $\nu_1 = -\nu'$.

- Various airgap magnetic conductance harmonics and stator and rotor mmf harmonics may interact in a cage-rotor IM in many ways to produce synchronous torques. Many stator/rotor slot number N_s/N_r combinations are to be avoided to eliminate most important synchronous torques. The main benefit is that the machine will not lock into such speed and will accelerate quickly to the pertinent load speed.
- The harmonics currents induced in the rotor cage by various sources may induce certain current harmonics in the stator, especially for Δ connection or for parallel stator windings (a >1).
- This phenomenon, called secondary armature reaction, reduces the differential leakage coefficient τ_d of the first-slot harmonic $\nu_{smin} = N_s/p_1 \pm 1$ and thus, in fact, increases its corresponding (originating) rotor current. Such an augmentation may lead to the amplification of some synchronous torques.
- The stator harmonics currents circulating between phases (in Δ connection) or in between current paths (in parallel windings), whose order is multiple of three may be avoided by forbidding some N_s and N_r combinations $(N_s - N_r) \neq 2p_1, 4p_1, \ldots$.
- Also, if the stator current paths are in the same position with respect to rotor slotting, no such circulating currents occur.
- Notable differences, between the linear theory and tests, have been encountered, with large torque amplifications in the braking region (S > 1). The main cause of this phenomenon seems to be magnetic saturation.
- Slot opening presence also tends to amplify synchronous torques. This tendency is smaller if $N_r < N_s$; even straight rotor slots (no skewing) may be adopted. Attention to noise for no skewing!
- As a result of numerous investigations, theoretical and experimental, clear recommendations of safe stator/rotor slot combinations are given in some design books. Attention is to be paid to the fact that, as noise is concerned, low-power machines and large-power machines behave differently.
- Radial forces are somehow easier to calculate directly by Maxwell's stress method from various airgap flux harmonics.
- Radial stress (force per unit stator area) p_r is a wave with a certain order r and a certain electrical frequency Ω_r. Only r = 0, 1, 2, 3, 4 cases are important.
- Investigating again the numerous N_s and N_r combinations' contributions to the mechanical stress components coming from the mmfs and airgap magnetic conductance harmonics, new stator/rotor slot number N_s/N_r restrictions are developed.
- Slip ring IMs behave differently as they have "clear cut" stator and rotor mmf harmonics. Also damping of some radial stress component through induced rotor current is absent in wound rotors.
- Especially, radial forces due to rotor eccentricity are much larger in slip ring rotor than in cage rotor with identical stators, during steady state. The eccentric radial stress during starting transients are, however, about the same.
- The circulating current of parallel winding might, in this case, reduces some radial stresses. By increasing the rotor current, they reduce the resultant flux density in the airgap which, squared, produces the radial stress.
- The effect of radial stress in terms of stator vibration amplitude manifests itself predominantly with zero-, first-, and second-order stress harmonics (r = 0, 1, 2).
- For large-power machines, the mechanical resonance frequency is smaller, and thus, the higher-order radial stress r = 3, 4 are to be avoided if low noise machines are to be built.
- For precision calculation of parasitic torques and radial forces, after preliminary design rules have been applied, FEM is to be used, though at the price of still very large computation time.
- Intricate modified winding function approach-based methods have been successfully introduced recently to account for stator and rotor slotting, combined static-dynamic and axial eccentricity, in IM behaviour [12–14].

- Vibration and noise is a field in itself with a rich literature and standardisation [11,19]
- The vibrations in a large doubly fed induction generator (DFIG) is pertinently investigated in Ref. [20].

REFERENCES

1. B. Heller, V. Hamata, *Harmonics Effects in Induction Motors*, Chapter 6, Elsevier, Amsterdam, The Netherlands, 1977.
2. R. Richter, *Electric Machines*, 2nd edition, Vol. 1, p. 173, Verlag Birkhäuser, Basel, 1951 (in German).
3. K. Vogt, *Electric Machines–Design of Rotary Electric Motors*, Chapter 10, VEB Verlag Technik, Berlin, 1988 (in German).
4. K. Oberretl, New knowledge on parasitic torques in cage rotorinduction motors, *Buletin Oerlikon*, Vol. 348, 1962, pp. 130–155.
5. K. Oberretl, The theory of harmonic fields of induction motors considering the influence of rotor currents on additional stator harmonic currents in windings with parallel paths, *Archiv für Electrotechnik*, Vol. 49, 1965, pp. 343–364 (in German).
6. K. Oberretl, Field harmonics theory of slip ring motor taking multiple reaction into account, *Proceedings of the IEEE*, No. 8, 1970, pp. 1667–1674.
7. K. Oberretl, Parasitic synchronous and pendulation torques in induction machines; the influence of transients and saturation, part II – III, *Archiv für Electrotechnik*, Vol. 77, 1994, pp. 1–11, pp. 277–288 (in German).
8. D. G. Dorrel, Experimental behaviour of unbalanced magnetic pull in 3-phase induction motors with eccentric rotors and the relationship with teeth saturation, *IEEE Transactions*, Vol. EC – 14, No. 3, 1999, pp. 304–309.
9. J. F. Bangura, N. A. Dermerdash, Simulation of inverter – Fed induction motor drives with pulse-width-modulation by a time-stepping FEM model–flux linkage- based space model, *IEEE Transactions*, Vol. EC – 14, No. 3, 1999, pp. 518–525.
10. A. Arkkio, O. Lingrea, Unbalanced magnetic pull in a high speed induction motor with an eccentric rotor, *Record of ICEM*, 1994, Paris, France, Vol. 1, pp. 53–58.
11. S. J. Yang, A. J. Ellison, *Machinery Noise Measurement*, Clarendon Press, Oxford, 1985.
12. S. Nandi, R. M. Bharadwaj, H. A. Toliyat, Performance analysis of a three-phase IM under mixed eccentricity condition, *IEEE Transactions*, Vol. EC – 17, No. 3, 2002, pp. 392–399.
13. S. Nandi, Modeling of IM including stator and rotor slot effects, *IEEE Transactions*, Vol. 40, No. 4, 2004, pp. 1058–1065.
14. X. Li, Q. Wu, S. Nandi, Performance analysis of a three-phase IM with inclined static excentricity, *IEEE Transactions*, Vol. IA-43, No. 2, 2007, pp. 531–541.
15. P. L. Timar, *Noise and Vibration of Electrical Machines, Technical Publishers*, Budapest, Hungary, 1986.
16. R. Mistry, W. R. Finley, S. Kreitzer, Induction motor vibrations, *IEEE IA Magazine*, Vol. 16, No. 6, 2010, pp. 37–46.
17. M. Tsypkin, The origin of the electromagnetic vibration of induction motors operating in modern industry: Practical experience – Analysis and diagnostics, *IEEE Transactions*, Vol. IA-53, No. 2, 2017, pp. 1669–1676.
18. M. Drif, H. Kim, J. Kim, S. B. Lee, A.J.M. Cardoso, Active and reactive power spectra-based detection and separation of rotor faults and low frequency load torque oscillations, *IEEE Transactions*, Vol. IA-53, No. 3, 2017, pp. 2702–2710.
19. J. Martinez, A. Belahcen, A. Muetze, Analysis of the vibration magnitude of an induction motor with different numbers of broken bars, *IEEE Transactions*, Vol. IA-53, No. 3, 2017, pp. 2711–2720.
20. T. Lugand, A. Schwery, Comparison between the salient-pole synchronous machine and the doubly – Fed induction machine with regard to electromagnetic parasitic force and stator vibrations, *IEEE Transactions*, Vol. IA-55, No. 6, 2017, pp. 5284–5294.

11 Losses in Induction Machines

Losses in induction machines (IMs) occur in windings, magnetic cores, besides mechanical friction and windage losses. They determine the efficiency of energy conversion in the machine and the cooling system that is required to keep the temperatures under control.

In the design stages, it is natural to try to calculate the various types of losses as precisely as possible. After the machine is manufactured, the losses have to be determined by tests. Loss segregation has become a standard method to determine the various components of losses, because such an approach does not require to shaft-load the machine. Consequently, the labour and energy costs for testings are low.

On the other hand, when prototyping or for more demanding applications, it is required to validate the design calculations and the loss segregation method. The input–output method has become standard for the scope. It is argued that, for high efficiency machines, measuring of the input and output P_{in} and P_{out} to determine losses Σp on load

$$\Sigma p = P_{in} - P_{out} \quad (11.1)$$

requires high precision measurements. This is true, as for a 90% efficiency machine, a 1% error in P_{in} and P_{out} leads to a 10% error in the losses.

However, by now, less than 0.1% to 0.2% error in power measurements is available so this objection has been reduced considerably.

On the other hand, shaft-loading the IM requires a dynamometer, takes time, and energy. Still, as soon as 1912 [1], it was noticed that there was notable difference between the total losses determined from the loss segregation method (no-load + short-circuit tests) and direct load tests. This difference is called "stray load losses". The dispute on the origin of "stray load losses" and how to measure them correctly is still on today, after numerous attempts made so far [2–8].

To reconcile such differences, standards have been proposed. In the USA (IEEE Standard 112B), the combined loss segregation and input–output tests are used to calculate a posteriori for each motor type the "stray load losses" and thus guarantee the efficiency.

In many other standards, the "stray load losses" are assigned 0.5% or 1% of rated power despite the fact that most measurements done suggest notably higher values.

The use of static power converters to feed IMs for variable speed drives complicates the situation even more, as the voltage time harmonics are producing additional winding and core losses in the IM.

Faced with such a situation, we decided to retain only the components of losses which proved notable (greater than (3% to 5%) of total losses) and explore their computation one by one by analytical methods.

Further on, numerical, finite element, loss calculation results are given.

11.1 LOSS CLASSIFICATIONS

The first classification of losses, based on their location in the IM, includes

- Winding losses – stator and rotor
- Core losses – stator and rotor
- Friction and windage losses – rotor.

Electromagnetic losses include only winding and core losses.

A classification of electromagnetic losses by origin would include

- Fundamental losses
  - Fundamental winding losses (in the stator and rotor)
  - Fundamental core losses (in the stator).
- Space harmonics losses
  - Space harmonic winding losses (in the rotor)
  - Space harmonic core losses (stator and rotor).
- Time harmonic losses
  - Time harmonic winding losses (stator and rotor)
  - Time harmonic core losses (stator and rotor).

Time harmonics are to be considered especially when the IM is static converter fed, and thus, the voltage time harmonics content depends on the type of the converter and the pulse width modulation (PWM) used with it. The space harmonics in the stator (rotor) mmf and the airgap field are related directly to mmf space harmonics, airgap permeance harmonics due to slot openings, leakage, or main path saturation.

All these harmonics produce additional (stray) core and winding losses called

- Surface core losses (mainly on the rotor)
- Tooth flux pulsation core losses (in the stator and rotor teeth)
- Tooth flux pulsation cage-current losses (in the rotor cage).

Load, coil chording, and the rotor bar-tooth contact electric resistance (interbar currents) influence all these stray losses.

No-load tests include all the above components but at zero fundamental rotor current. These components will be calculated first; then corrections will be applied to compute some components on load.

11.2 FUNDAMENTAL ELECTROMAGNETIC LOSSES

Fundamental electromagnetic losses refer to core loss due to space fundamental airgap flux density – essentially stator based – and time fundamental conductor losses in the stator and rotor cage or winding.

The fundamental core losses contain the hysteresis and eddy current losses in the stator teeth and core:

$$P_{Fe1} \approx C_h\left(\frac{f_1}{50}\right)\left[\left(\frac{B_{1ts}}{1}\right)^n G_{teeth} + \left(\frac{B_{1cs}}{1}\right)^n G_{core}\right] + C_e\left(\frac{f_1}{50}\right)^2\left[\left(\frac{B_{1ts}}{1}\right)^2 G_{teeth} + \left(\frac{B_{1cs}}{1}\right)^2 G_{core}\right] \quad (11.2)$$

where C_h [W/kg], C_e [W/kg] at 50 Hz and 1 Tesla, and n = (1.7–2.0) are material coefficients for hysteresis and eddy currents, dependent on the lamination hysteresis cycle shape, electrical resistivity, and lamination thickness, G_{teeth} and G_{core}, the teeth and back core weights, and B_{1ts} and B_{1cs} – teeth and core fundamental flux density values.

At any instant in time, the flux density is different in different locations and, in some regions around tooth bottom, the flux density changes direction, that is, it becomes rotating.

Hysteresis losses are known to be different with alternative and rotating, respectively, fields. In rotating fields, hysteresis losses peak at around 1.4 to 1.6 T, while they increase steadily for alternative fields.

Moreover, the mechanical machining of stator bore (when stamping is used to produce slots) is known to increase core losses by, sometimes, 40%–60%.

The above remarks show that the calculation of fundamental core losses is not a trivial task. Even when FEM is used to obtain the flux distribution, formulas like (11.2) are used to calculate core losses in each element, so some of the errors listed above still hold. The winding (conductor) fundamental losses are

$$P_{co} = 3R_s I_{1s}^2 + 3R_r I_{1r}^2 \tag{11.3}$$

The stator and rotor resistances R_s and R_r' are dependent on skin effect. In this sense, $R_s(f_1)$ and $R_r'(Sf_1)$ depend on f_1 and S. The depth of field penetration in copper $\delta_{Co}(f_1)$ is

$$\delta_{Co}(f_1) = \sqrt{\frac{2}{\mu_0 2\pi f_1 \sigma_{Co}}} = \sqrt{\frac{2}{1.256\cdot 10^{-6} 2\pi 60\left(\frac{f_1}{60}\right)\cdot 5\cdot 10^7}} = 0.94\cdot 10^{-2}\sqrt{\frac{60}{f_1}}\,\text{m} \tag{11.4}$$

If either the elementary conductor height d_{Co} or the fundamental frequency f_1 is large, whenever

$$\delta_{Co} > \frac{d_{Co}}{2} \tag{11.5}$$

the skin effect is to be considered. As the stator has many layers of conductors in slot even for $\delta_{Co} \approx d_{Co}/2$, there may be some skin effect (resistance increase). This phenomenon was treated in detail in Chapter 9.

In a similar way, the situation stands for the wound rotor at large values of slip S. The rotor cage is a particular case for one conductor per slot. Again, Chapter 9 treated this phenomenon in detail. For rated slip, however, skin effect in the rotor cage is, in general, negligible.

In skewed cage rotors with uninsulated bars (cast aluminium bars), there are interbar currents (Figure 11.1a).

The treatment of various space harmonic losses at no-load follows.

Depending on the relative value of the transverse (contact) resistance R_q and skewing c, the influence of interbar currents on fundamental rotor conductor losses will be notable or small.

The interbar currents influence also depends on the fundamental frequency as for $f_1 = 500\,\text{Hz}$, for example, the skin effect notably increases the rotor cage resistance even at rated slip.

On the other hand, skewing leaves an uncompensated rotor mmf (under load) which modifies the airgap flux density distribution along the stack length (Figure 11.1b).

As the flux density squared enters the core loss formula, it is obvious that the total fundamental core loss will change due to skewing.

As skewing and interbar currents also influence the space harmonics losses, we will treat their influence once for all harmonics. The fundamental will then become a particular case.

Fundamental core losses seem impossible to segregate in a special test which would hold the right flux distribution and frequency. However, a standstill test at rated rotor (slip) frequency $f_2 = Sf_1$ would, in fact, yield the actual value of $R_r'(S_n f_1)$. The same test at various frequencies would produce precise results on conductor fundamental losses in the rotor. The stator fundamental conductor losses might be segregated from a standstill A.C. single-phase test with all phases in series at rated frequency as, in this case, the core fundamental and additional losses may be neglected by comparison.

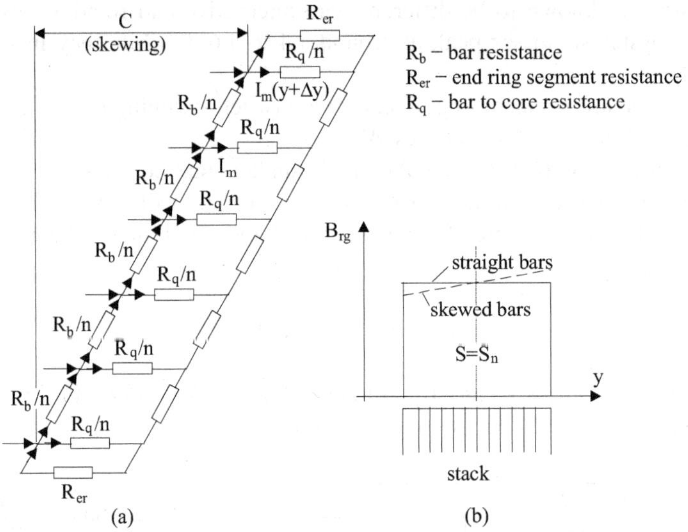

FIGURE 11.1 Interbar currents ($I_m(Y)$) in a skewed cage rotor (a) and the resultant airgap flux density along stack length (b).

11.3 NO-LOAD SPACE HARMONICS (STRAY NO-LOAD) LOSSES IN NONSKEWED IMS

Let us remember that airgap field space harmonics produce on no load, in nonskewed IMs, the following types of losses:

- Surface core losses (rotor and stator)
- Tooth flux pulsation core losses (rotor and stator)
- Tooth flux pulsation cage losses (rotor).

The interbar currents produced by the space harmonics are negligible in nonskewed machines if the rotor end ring (bar) resistance is very small ($R_{cr}/R_b < 0.15$).

11.3.1 No-Load Surface Core Losses

As already documented in Chapter 10, dedicated to airgap field harmonics, the stator mmf space harmonics (due to the very placement of coils in slots) as well the slot openings produce airgap flux density harmonics. Further on, main flux path heavy saturation may create third flux harmonics in the airgap.

It has been shown in Chapter 10 that the mmf harmonics and the first slot opening harmonics with a number of pole pairs $\nu_s = N_s \pm p_1$ are attenuating and augmenting each other, respectively, in the airgap flux density.

For these mmf harmonics, the winding factor is equal to that of the fundamental. This is why they are the most important, especially in windings with chorded coils where the 5th, 7th, 11th, and 17th stator mmf harmonics are reduced considerably.

Let us now consider the fundamental stator mmf airgap field as modulated by stator slotting openings (Figure 11.2).

$$B_{N_s \pm p_1}(\theta_m, t) = \frac{\mu_0 F_{1m} a_1}{a_0} \cos(\omega_1 t - p_1\theta_m) \cos N_s\theta_m \tag{11.6}$$

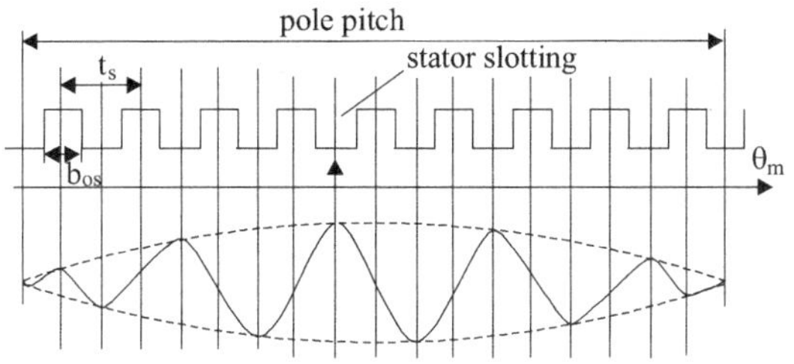

FIGURE 11.2 First slot opening (airgap permeance) airgap flux density harmonics.

This represents two waves, one with $N_s + p_1$ pole pairs and the other with $N_s - p_1$ pole pairs, that is, exactly, the first airgap magnetic conductance harmonic.

Let us consider that the rotor slot openings are small or that closed rotor slots are used. In this case, the rotor surface is flat.

The rotor laminations have a certain electrical conductivity, but they are insulated from each other axially by an adequate coating.

For frequencies of 1200 Hz, characteristic to $N_s \pm p_1$ harmonics, the depth of field penetration in silicon steel for $\mu_{Fe} = 200\,\mu_0$ is (from 11.4) $\delta \approx 0.4$ mm for 0.6 mm thick laminations. This means that the skin effect is not significant.

Therefore, we may neglect the rotor lamination-induced current reaction to the stator field. That is, the airgap field harmonics (11.6) penetrate the rotor without being disturbed by induced rotor surface eddy currents. The eddy currents along axial direction are neglected, since the laminations are insulated from each other.

But now, let us consider a general harmonic of stator mmf produced field B_{vg}.

$$B_{vg} = B_v \cos\left(\frac{v\pi x_m}{p_1\tau} - S_v\omega t\right) = B_v \cos\left(\frac{v x_m}{R} - S_v\omega t\right) \tag{11.7}$$

R – rotor radius; τ – pole pitch of the fundamental.

The slip for the vth mmf harmonic, S_v, is

$$(S_v)_{S=0} = \left(1 - \frac{v}{p_1}(1 - S)\right)_{S=0} = 1 - \frac{v}{p_1} \tag{11.8}$$

For constant iron permeability μ, the field equations in the rotor iron are

$$\mathrm{rot}B_v = 0, \quad \mathrm{div}B_v = 0, \quad B_z = 0 \tag{11.9}$$

This leads to

$$\frac{\partial^2 B_x}{\partial x^2} + \frac{\partial^2 B_x}{\partial y^2} = 0 \qquad \frac{\partial^2 B_y}{\partial x^2} + \frac{\partial^2 B_y}{\partial y^2} = 0 \tag{11.10}$$

In iron, these flux density components decrease along y (inside the rotor) direction

$$B_y = f_y(y)\cos\left(\frac{v x_m}{R} - S_v\omega t\right) \tag{11.11}$$

According to $\frac{\partial B_x}{\partial x} + \frac{\partial B_y}{\partial y} = 0$,

$$B_x = \left(\frac{\frac{\partial f_y(y)}{\partial y}}{\frac{\nu}{R}} \right) \sin\left(\frac{\nu x_m}{R} - S_\nu \omega t \right) \tag{11.12}$$

From (11.10),

$$\frac{\partial^2 f_y(y)}{\partial y^2} - \left(\frac{\nu}{R} \right)^2 f_y(y) = 0 \tag{11.13}$$

or

$$f_y(y) = B_\nu e^{\frac{\nu y}{R}} \tag{11.14}$$

Equation (11.14) retains one term because y < 0 inside the rotor and $f_y(y)$ should reach zero for $y \approx -\infty$.

The resultant flux density amplitude in the rotor iron $B_{r\nu}$ is

$$B_{r\nu} = B_\nu e^{\frac{\nu y}{R}} \tag{11.15}$$

But the Faraday law yields

$$\text{rot}\bar{E} = \text{rot}\left(\frac{1}{\sigma_{Fe}} \bar{J} \right) = -\frac{\partial \bar{B}}{dt} \tag{11.16}$$

In our case, the flux density in iron has two components B_x and B_y, so

$$\begin{aligned} \frac{\partial J_y}{\partial z} &= -\sigma_{Fe} \frac{\partial B_x}{\partial t} \\ \frac{\partial J_x}{\partial z} &= -\sigma_{Fe} \frac{\partial B_y}{\partial t} \end{aligned} \tag{11.17}$$

The induced current components J_x and J_y are thus

$$\begin{aligned} J_x &= -\sigma_{Fe} S_\nu \omega B_\nu e^{\frac{\nu y}{R}} z \cos\left(\frac{\nu x_m}{R} - S_\nu \omega t \right) \\ J_y &= -\sigma_{Fe} S_\nu \omega B_\nu e^{\frac{\nu y}{R}} z \sin\left(\frac{\nu x_m}{R} - S_\nu \omega t \right) \end{aligned} \tag{11.18}$$

The resultant current density amplitude $J_{r\nu}$ is

$$J_{r\nu} = \sqrt{J_x^2 + J_y^2} = \sigma_{Fe} S_\nu \omega B_\nu e^{\frac{\nu y}{R}} z \tag{11.19}$$

The losses per one lamination (thickness d_{Fe}) are

$$P_{lam\nu} = \frac{1}{\sigma_{Fe}} \int_{-d_{Fe}/2}^{d_{Fe}/2} \int_{y=0}^{y=-\infty} \int_{x=0}^{2\pi R} J_{r\nu}^2 \, dx \, dy \, dz \tag{11.20}$$

For the complete rotor of axial length l_{stack},

$$P_{0v} = \frac{\sigma_{Fe}}{24} B_v^2 (S_v \omega)^2 d_{Fe}^2 \frac{R}{v} 2\pi R l_{stack} \tag{11.21}$$

Now, if we consider the first and second airgap magnetic conductance harmonic (inversed airgap function) with $a_{1,2}$ (Chapter 10),

$$a_{1,2} = \frac{\beta}{g} F_{1,2}\left(\frac{b_{os}}{t_s}\right) \tag{11.22}$$

$\beta(b_{os,r}/g)$ and $F_{1,2}$ $(b_{os,r}/t_{s,r})$ are to be found from Table 10.1 and (10.14) in Chapter 10 and

$$B_v = B_{g1} \frac{a_{1,2}}{a_0}; \quad a_0 = \frac{1}{K_c g} \tag{11.23}$$

where B_{g1} is the fundamental airgap flux density with

$$v = N_s \pm p_1 \quad \text{and} \quad S_v = 1 - \frac{(N_s \pm p_1)}{p_1} \approx \frac{N_s}{p_1} \tag{11.24}$$

The no-load rotor surface losses P_{ov} are thus

$$(P_{0v})_{N_s \pm P_1} \approx 2 \frac{\sigma_{Fe}}{24} B_{g1}^2 \left(\frac{N_s}{p_1}\right)^2 (2\pi f_1)^2 d_{Fe}^2 \frac{Rp_1}{N_s} 2\pi R l_{stack} \left[\frac{a_1^2 + 2a_2^2}{a_0^2}\right] \tag{11.25}$$

Example 11.1 Rotor Surface Losses Calculation

Let us consider an IM with open stator slots and $2R = 0.38$ m, $N_s = 48$ slots, $t_s = 25$ mm, $b_{os} = 14$ mm, $g = 1.2$ mm, $d_{Fe} = 0.5$ mm, $B_{g1} = 0.69$ T, $2p_1 = 4$, $n_0 = 1500$ rpm, $N_r = 72$, $t_r = 16.6$ mm, $b_{or} = 6$ mm, and $\sigma_{Fe} = 10^8/45(\Omega m)^{-1}$. To determine the rotor surface losses per unit area, we first have to determine a_1, a_2, and a_0 from (11.22). For $b_{os}/g = 14/1.2 = 11.66$ from Table 10.1, $\beta = 0.4$. From Equation (10.14), $F_1(14/25) = 1.02$ and $F_2(0.56) = 0.10$.

Also, $a_0 = \frac{1}{K_{c1,2}g}$; K_c (from Equations (5.3–5.5)) is $K_{c1,2} = 1.85$ (due to open stator slots).

Now the rotor surface losses can be calculated from (11.25).

$$\frac{P_{0v}}{2\pi R l_{stack}} = \frac{2 \cdot 2.22 \cdot 10^6}{24} \cdot 0.69^2 \cdot \left(\frac{48}{2}\right) (2\pi 60)^2 \cdot 0.5^2 \cdot 10^{-6} \cdot 0.19$$

$$\cdot \left[\frac{(0.4 \cdot 1.02)^2 + 2(0.4 \cdot 0.1)^2}{1.85^2}\right] = 8245.68 \text{ W/m}^2$$

As expected, with semiclosed slots, a_1 and a_2 would become much smaller; also, the Carter coefficient decreases. Consequently, the rotor surface losses will be much smaller. Also, increasing the airgap has the same effect, however, at the price of larger no-load current and lower power factor of the machine. The stator surface losses produced by the rotor slotting may be calculated in a similar way by replacing $F_1(b_{os}/t_s)$, $F_2(b_{os}/t_s)$, $\beta(b_{os}/g)$ with $\beta(b_{or}/g)$, $F_1(b_{or}/t_s)$, and $F_2(b_{or}/t_s)$ and N_s/p_1 with N_r/p_1.

As in general, the rotor slots are semiclosed, $b_{or} \ll b_{os}$, the stator surface losses are notably smaller than those of the rotor; they are, in general, neglected.

11.3.2 No-Load Tooth Flux Pulsation Losses

As already documented in the previous paragraph, the stator (and rotor) slot openings produce variation in the airgap flux density distribution (Figure 11.3).

In essence, the total flux in a stator and rotor tooth varies with rotor position due to stator and rotor slot openings only in the case when the number of stator and rotor slots are different from each other. This is, however, the case, as $N_s \neq N_r$ at least to avoid large synchronous parasitic torques at zero speed (as demonstrated in Chapter 10).

The stator tooth flux pulsates due to rotor slot openings with the frequency $f_{PS} = N_r f \frac{(1-S)}{p_1} = N_r \frac{f}{p_1}$, for S = 0 (no load). The flux variation coefficient K_ϕ is

$$K_\phi = \frac{\phi_{max} - \phi_{min}}{2\phi_0} \tag{11.26}$$

The coefficient K_ϕ, as derived when the Carter coefficient was calculated in Ref. [6]

$$K_\phi = \frac{\gamma_2 g}{2t_s} \tag{11.27}$$

with γ_2 as

$$\gamma_2 = \frac{\left(\frac{b_{or}}{g}\right)^2}{5 + \frac{b_{or}}{g}} \tag{11.28}$$

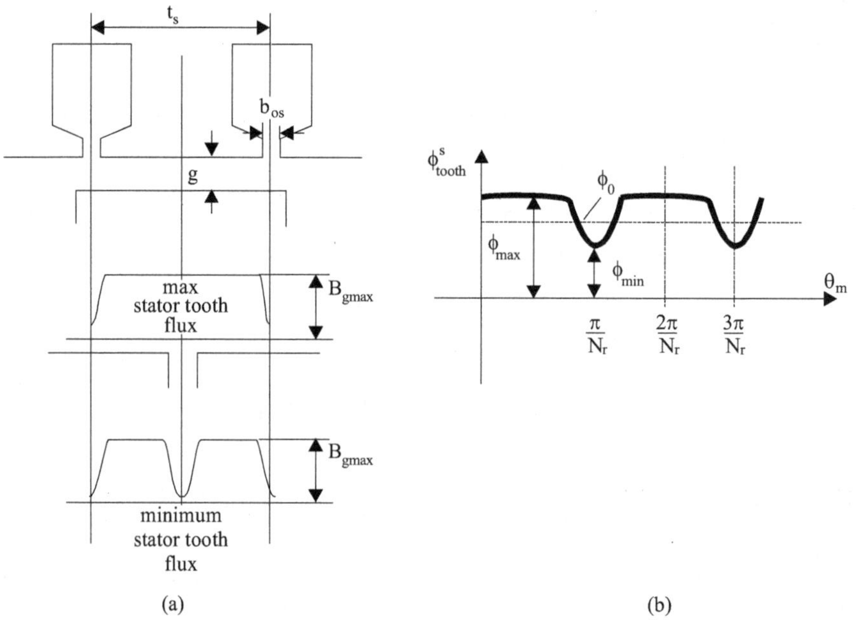

FIGURE 11.3 Airgap flux density as influenced by rotor and stator slotting (a) and stator tooth flux versus rotor position (b).

Now, by denoting the average flux density in a stator tooth with B_{ots}, the flux density pulsation B_{ps} in the stator tooth is

$$B_{Ps} = B_{ots}\frac{\gamma_2 g}{2t_s} \quad (11.29)$$

$$B_{ots} = B_{g0}\frac{t_s}{t_s - b_{os}} \quad (11.30)$$

B_{g0} – airgap flux density fundamental.

We are now in the classical case of an iron region with an A.C. magnetic flux density B_{Ps}, at frequency $f_{Ps} = N_r f/p_1$. As $N_r f/p_1$ is a rather high frequency, eddy current losses prevail, so

$$P_{Ps0} \approx C_{ep}\left(\frac{B_{Ps}}{1}\right)^2\left(\frac{f_{Ps}}{50}\right)^2 G_{steeth}; \quad G_{steeth} - \text{stator teeth weight} \quad (11.31)$$

In (11.31), C_{ep} represents the core losses at 1 T and 50 Hz. It could have been for 1 T and 60 Hz as well.

Intuitively, the magnetic saturation of main flux path places the pulsation flux on a local hysteresis loop with lower (differential) permeability, so saturation is expected to reduce the flux pulsation in the teeth. However, Equation (11.31) proves satisfactory even in the presence of saturation. Similar tooth flux pulsation core losses occur in the rotor due to stator slotting.

Similar formulas as above are valid.

$$P_{Pr0} = C_{ep}\left(\frac{B_{Pr}}{1}\right)^2\left(\frac{f_{Pr}}{50}\right)^2 G_{teeth}; \quad f_{Pr} = N_s\frac{f}{P_1}$$

$$B_{Pr} = B_{otr}\frac{\gamma_1 g}{2t_r}; \qquad B_{otr} = B_{g0}\frac{t_r}{t_r - b_{or}} \quad (11.32)$$

$$\gamma_1 = \frac{\left(\frac{b_{os}}{g}\right)^2}{5 + \frac{b_{os}}{g}}$$

As expected, C_{ep} – power losses per Kg at 1 T and 50 Hz slightly changes when the frequency increases, as it does at $N_s\frac{f}{p_1}$ or $N_r\frac{f}{p_1}$. $C_{ep} = C_e K$, with K = 1.1–1.2 as an empirical coefficient.

Example 11.2 Tooth Flux Pulsation Losses Calculation

For the motor in Example 10.1, let us calculate the tooth flux pulsation losses per Kg of teeth in the stator and rotor.

Solution

In essence, we have to determine the flux density pulsations B_{Ps} and B_{Pr}, then f_{Ps} and f_{Pr}, and apply Equation (11.31).

From (11.26) and (11.32),

$$\gamma_2 = \frac{\left(\frac{b_{or}}{g}\right)^2}{5 + \frac{b_{or}}{g}} = \frac{\left(\frac{6}{1.2}\right)^2}{5 + \frac{6}{1.2}} = 2.5$$

$$\gamma_1 = \frac{\left(\frac{b_{os}}{g}\right)^2}{5+\frac{b_{os}}{g}} = \frac{\left(\frac{14}{1.2}\right)^2}{5+\frac{14}{1.2}} = 8.166$$

$$B_{Ps} = B_{g0}\frac{t_s}{t_s - b_{or}}\frac{\gamma_2 g}{2t_s} = \frac{0.69\cdot 2.5\cdot 1.2}{2(25-14)} = 0.094\ \text{T}$$

$$B_{Pr} = B_{g0}\frac{t_r}{t_r - b_{or}}\frac{\gamma_1 g}{2t_r} = \frac{0.69\cdot 8.166\cdot 1.2}{2(16.6-6)} = 0.3189\ \text{T}$$

With $C_{ep} = 3.6$ W/Kg at 1 T and 50 Hz,

$$P_{Ps} = C_{ep}\left(\frac{B_{Ps}}{1}\right)^2\left(\frac{f_{Pr}}{50}\right)^2 = 3.6\left(\frac{0.094}{1}\right)^2\left(\frac{72\cdot 60}{2\cdot 50}\right)^2 = 59.3643\ \text{W/Kg}$$

$$P_{Pr} = C_{ep}\left(\frac{B_{Pr}}{1}\right)^2\left(\frac{f_{Ps}}{50}\right)^2 = 3.6\left(\frac{0.3189}{1}\right)^2\left(\frac{48\cdot 60}{2\cdot 50}\right)^2 = 303.66\ \text{W/Kg}$$

The open slots in the stator produce large rotor tooth flux pulsation no-load specific losses (P_{Pr}).

The values just obtained, even for straight rotor (and stator) slots, are too large.

Intuitively, we feel that at least the rotor tooth flux pulsations will be notably reduced by the currents induced in the cage by them. At the expense of no-load circulating cage-current losses, the rotor flux pulsations are reduced.

Not so for the stator unless parallel windings are used where circulating currents would play the same role as circulating cage currents.

When such a correction is done, the value of B_{Pr} is reduced by a subunitary coefficient [4],

$$B'_{Pr} = B_{Pr}\cdot K_{tk}\cdot \sin\left(\frac{Kt_r p_1}{2R}\right) = K_{dk}B_{Pr};\quad K = \frac{N_s}{P_1} \tag{11.33}$$

$$K_{tr} = 1 - \frac{L_{mk}+L_{2dk}}{L_{mk}+L_{2dk}+L_{slot}+L_{mk}\left(\frac{1}{K_{skewk}{}^2-1}\right)} \tag{11.34}$$

where L_{mk} is the magnetising inductance for harmonic K, L_{2dk} the differential leakage inductance of Kth harmonic, K_{skewk} the skewing factor for harmonic K, and L_{slot} – rotor slot leakage inductance.

$$L_{mk} \approx L_m\frac{1}{K^2} \tag{11.35}$$

$$L_{dk} \approx K_{dlk}\cdot \tau_{2dk}\cdot L_{mk} \tag{11.36}$$

$K_{dlk} = \tanh\dfrac{\left(\dfrac{d_{Fe}}{2\delta_{Fe}}\right)}{\dfrac{d_{Fe}}{2\delta_{Fe}}}$; d_{Fe} – lamination thickness;

$$\delta_{Fe} = \sqrt{\frac{2}{\mu_{Fediff}S_k\omega\sigma_{Fe}}} \tag{11.37}$$

where μ_{Fediff} is the iron differential permeability for given saturation level of fundamental.

$$S_k\omega = 2\pi f_k = 2\pi\frac{N_s}{p_1} \tag{11.38}$$

$$\tau_{2dk} = \left(\frac{\frac{\pi p_1 K}{R}}{\sin\left(\frac{\pi p_1 K}{R}\right)} \right)^2 - 1 \tag{11.39}$$

$$K_{skewk} = \frac{\sin\left(\frac{KP_1c}{R}\right)}{\frac{KP_1c}{R}} \tag{11.40}$$

Skewing is reducing the cage circulating current reaction and thus increasing the value of K_{tk} and, consequently, the teeth flux pulsation core losses stay undamped.

Typical values for K_{dk} – the total damping factor due to cage circulating currents – would be in the range $K_{dk} = 1$ to 0.05, depending on rotor number slots, skewing, etc.

A skewing of one stator slot pitch is said to reduce the circulating cage reaction to almost zero ($K_{dk} \approx 1$), so the rotor tooth flux pulsation losses stay high. For straight rotor slots with $K_{dk} \approx 0.1$–0.2, the rotor tooth flux pulsation core losses are reduced to small relative values but still have to be checked.

11.3.3 No-Load Tooth Flux Pulsation Cage Losses

Now that the rotor tooth flux pulsation, as attenuated by the corresponding induced bar currents, is known, we may consider the rotor bar mesh in Figure 11.4.

$$\Delta\phi_{rk} = K_{dk} B_{prk} (t_r - b_{or}) l_{stack} \tag{11.41}$$

In our case, $K = \frac{N_s}{p_1}$. For this harmonic, the rotor bar resistance is increased due to skin effect by $K_{Rk}\left(f_k = \frac{N_s}{p_1} f_1\right)$.

The skin effect in the end ring is much smaller, so the ring resistance may be neglected. If we neglect also the leakage reactance in comparison with rotor bar resistance, the equation of the mesh circuit in Figure 11.4 becomes

$$2\pi f K \Delta\phi_{rk} = R_b K_{Rk} \left(I_{bk}^{(1)} - I_{bk}^{(2)}\right) \tag{11.42}$$

The currents I_{bk} in the neighbouring bars (1) and (2) are phase shifted by an angle $2\pi \frac{N_s}{N_r}$, so

$$I_{bk}^{(1)} - I_{bk}^{(2)} = 2 I_{bk} \sin \pi \frac{N_s}{N_r} \tag{11.43}$$

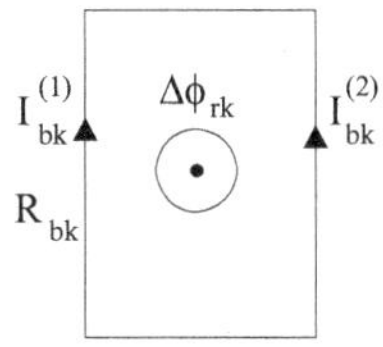

FIGURE 11.4 Rotor mesh with two bars.

So, from (11.41) and (11.42),

$$I_{bk} = \frac{\Delta\phi_{rk} 2\pi f K}{2\left(\sin\frac{N_s}{N_r}\right) R_b K_{Rk}} \tag{11.44}$$

Finally, for all bars, the cage losses for harmonic K are

$$P_{ocagek} = N_r R_b K_{Rk} \left(\frac{I_{bk}}{\sqrt{2}}\right)^2 \tag{11.45}$$

or

$$P_{ocagek} \approx \frac{N_r}{8} \frac{(\Delta\phi_{rk})^2 (2\pi f)^2 \left(\frac{N_s}{P_1}\right)^2}{\sin^2\left(\pi\frac{N_s}{N_r}\right) R_b K_{Rk}} \tag{11.46}$$

Expression (11.46) is valid for straight rotor slots. In a skewed rotor, the rotor tooth flux pulsation is reduced (per stack length) and, thus, the corresponding no-load cage losses are also reduced. They should tend to zero for one stator slot pitch skewing.

Example 11.3 The Calculation of the Cage Losses

Consider the motor in Examples 11.1 and 11.2 with the stack length $l_{stack} = 0.4\,m$, rotor bar cross-section $A_{bar} = 250\,mm^2$, and $N_s = 48$, $N_r = 72$, and $2p_1 = 4$. The rotor bar skin effect coefficient for $f_k = \frac{N_s}{P_1} f = \frac{48}{2} \cdot 60 = 1440\,Hz$ is $K_{Rk} = 15$. Let us calculate the cage losses due to rotor tooth flux pulsations if the attenuating factor of $K_{dk} = 0.05$.

Solution

With $K_{dk} = 0.05$ and $B_{pr} = 0.3189$, from Example 11.2, we may calculate $\Delta\phi_k$ (from 11.41).

$$\Delta\phi_{rk} = K_{dk} B_{prk} (t_r - b_{or}) l_{stack} = 0.05 \cdot 0.3189 \cdot (16.6 - 6) \cdot 10^{-3} \cdot 0.4 = 0.067 \cdot 10^{-3}\ Wb$$

The dc bar resistance R_b is

$$R_b \approx \frac{l_{stack}}{A_{bar}} \frac{1}{\sigma_{Al}} = \frac{0.4}{250 \cdot 10^{-6}} \frac{1}{3.0 \cdot 10^7} = 0.533 \cdot 10^4 \Omega$$

So the bar current $(I_{bk})_{K=\frac{N_r}{P_1}}$ is (11.43)

$$I_{bk} = \frac{0.067 \cdot 10^{-3} \cdot 2\pi 60 \cdot \frac{48}{2}}{2 \sin\pi\frac{48}{72} \cdot 0.533 \cdot 10^{-4} \cdot 15} = 443\ A$$

This is indeed a large value, but let us remember that the rotor bars are many ($N_r >> N_s$) and the stator slots are open, with a small airgap.

Now the total rotor flux pulsation losses P_{ocagek} are (11.45)

$$P_{0cagek} = 72 \cdot \left(\frac{443}{\sqrt{2}}\right)^2 \cdot 0.533 \cdot 10^{-4} \cdot 15 = 5682.2\ \text{W}$$

Let us compare this with the potential power of the IM under consideration ($f_x = 3 \cdot 10^4$ N/cm² – specific force)

$$P_n \approx f_x \cdot 2\pi R \cdot L \cdot R \cdot \frac{2\pi f}{P_1} = 3 \cdot 10^4 \cdot 2\pi \cdot 0.19^2 \cdot 0.4 \cdot \frac{2\pi 50}{2} \approx 427\ \text{kW}$$

In such a case, the no-load cage losses would represent more than 1% of all power.

11.4 LOAD SPACE HARMONICS (STRAY LOAD) LOSSES IN NONSKEWED IMS

Slot opening and mmf space harmonics act together to produce space harmonics load (stray load) losses in the presence of larger stator and rotor currents.

In general, the no-load stray losses are augmented by load, component by component. The mmf space harmonics of the stator, for integral-slot windings, is

$$\nu = (6c \pm 1) p_1 \tag{11.47}$$

The slot opening (airgap magnetic conductance) harmonics are

$$\nu_s = c_1 N_s \pm p_1 \tag{11.48}$$

As expected, they overlap. The first slot harmonics ($c_1 = 1$) $\nu_{smin} = N_s \pm p_1$ are most important as their winding factor is the same as for the fundamental.

The mmf harmonic F_ν amplitude is

$$F_\nu = F_1 \frac{K_{w\nu}}{K_{w1}} \frac{p_1}{\nu} \tag{11.49}$$

where K_{w1} and $K_{w\nu}$ are the winding factors of the fundamental and harmonic ν, respectively.

If the number of slots per pole and phase (q) is not very large, the first slot (opening) harmonics $N_s \pm p_1$ produce the largest harmonics field in the airgap.

Only for full-pitch windings, the phase-belt mmf harmonics ($\nu = 5, 7, 11, 13$) may produce worthy of consideration airgap fields because their winding factors are not small enough (as they are in chorded coil windings). However, even in such cases, the losses produced by the phase-belt mmf harmonics may be neglected by comparison with the first slot opening harmonics $N_s \pm p_1$ [6].

Now, for these first slot harmonics, it has been shown in Chapter 10 that their mmf companion harmonics of same order produce an increase for $N_s - p_1$ and a decrease for $N_s + p_1$ [6]

$$\xi_{(N_s - p_1)0} = \left(1 + \frac{N_s - p_1}{p_1} \frac{a_1}{2a_0}\right) \frac{1}{K_c} > 1 \tag{11.50}$$

$$\xi_{(N_s + p_1)0} = \left(1 - \frac{N_s + p_1}{p_1} \frac{a_1}{2a_0}\right) \frac{1}{K_c} < 1$$

where a_1, a_0, and K_s have been defined earlier in this chapter for convenience.

For load conditions, the amplification factors $\xi_{(N_s-p_1)0}$ and $\xi_{(N_s+p_1)0}$ are to be replaced by [6]

$$\xi_{(N_s-p_1)n} = \frac{1}{K_c}\sqrt{1+2(\sin\varphi_n)\frac{I_0}{I_n}\frac{N_s-p_1}{p_1}\frac{a_1}{2a_0}+\left(\frac{I_0}{I_n}\frac{N_s-p_1}{p_1}\frac{a_1}{2a_0}\right)^2}$$
$$\xi_{(N_s+p_1)n} = \frac{1}{K_c}\sqrt{1-2(\sin\varphi_n)\frac{I_0}{I_n}\frac{N_s+p_1}{p_1}\frac{a_1}{2a_0}+\left(\frac{I_0}{I_n}\frac{N_s+p_1}{p_1}\frac{a_1}{2a_0}\right)^2} \quad (11.51)$$

I_0 is the no-load current, I_n the current under load, φ_n power factor angle on load.

From now on, using the above amplification factors, we will calculate the correction coefficients for load to multiply the various no-load stray losses and thus find the stray load losses.

Based on the fact that losses are proportional to harmonic flux densities squared, for $N_s \pm P_1$ slot harmonics, we obtain,

- Rotor surface losses on load
  The ratio between load P_{0n} and no-load P_{00} surface losses C_{loads} is

$$P_{0n} = P_{00}\cdot C_{loads}; \quad C_{loads} > 1 \quad (11.52)$$

$$C_{loads} = \left(\frac{I_n}{I_0}\right)^2 \frac{\left[\left(\frac{p_1}{N_s-p_1}\right)^2 \xi^2_{(N_s-p_1)n} + \left(\frac{p_1}{N_s+p_1}\right)^2 \xi^2_{(N_s+p_1)n}\right]}{\left[\left(\frac{p_1}{N_s-p_1}\right)^2 \xi^2_{(N_s-p_1)0} + \left(\frac{p_1}{N_s+p_1}\right)^2 \xi^2_{(N_s+p_1)0}\right]} \quad (11.53)$$

- Tooth flux pulsation load core losses
  For both stator and rotor, the no-load surfaces P_{sp0} and P_{rp0} are to be augmented by amplification coefficients similar to C_{load},

$$P_{spn} = P_{sp0}\cdot C_{loadr}$$
$$P_{rpn} = P_{rp0}\cdot C_{loads} \quad (11.54)$$

  where C_{loadr} is calculated from (11.53) with N_r instead of N_s.
- Tooth flux pulsation cage losses on load
  For chorded coil windings, the same reasoning as above is used.

$$P_{ncage} = P_{0cage}\cdot C_{loadr} \quad (11.55)$$

For full-pitch windings, the fifth and seventh (phase-belt) mmf harmonics losses are to be added.

The cage equivalent current I_{rv} produced by a harmonic ν is

$$I_{rv} \sim \frac{F_v v}{1+\tau_v} = F_1\frac{K_{wv}}{K_{w1}}\frac{p_1}{1+\tau_v} \quad (11.56)$$

τ_v is the leakage (approximately differential leakage) coefficient for harmonic ν ($\tau_v \approx \tau_{dv}$).

The cage losses,

$$P_{rv} \sim I_{rv}^2 R_b K_{Rv} \quad (11.57)$$

K_{Rv} is, again, the skin effect coefficient for frequency f_v.

$$f_v = S_v f_1;\quad S_v = 1 - \frac{v}{p_1}(1 - S) \tag{11.58}$$

The differential leakage coefficient for the rotor (Chapter 6)

$$\tau_{dv} = \left(\frac{\pi v}{N_r}\right)^2 \frac{1}{\sin^2\left(\frac{\pi v}{N_r}\right)^2} - 1 \tag{11.59}$$

Equation (11.57) may be applied both for 5th and 7th ($5p_1$, $7p_1$), phase belt, mmf, harmonics, and for the first slot opening harmonics $N_s \pm p_1$.

Adding up all these four terms for load conditions [6] yields

$$P_{ncage} \approx P_{0cage} \cdot C_{loadr}\left[1 + \frac{2\left(\frac{K_{w5}}{K_{w1}}\right)^2 \frac{(1+\tau_{dN_s})^2}{(1+\tau_{d6})^2}\sqrt{\frac{6p_1}{N_s}}}{\xi_{(N_s-P_1)n}{}^2 + \xi_{(N_s+P_1)n}{}^2}\right] \tag{11.60}$$

$$(1+\tau_{d6}) = \frac{\left(\frac{6\pi p_1}{N_r}\right)^2}{\sin^2\left(\frac{6\pi p_1}{N_r}\right)};\quad (1+\tau_{dN_s}) = \frac{\left(\frac{\pi N_s}{N_r}\right)^2}{\sin^2\left(\frac{\pi N_s}{N_r}\right)} \tag{11.61}$$

Magnetic saturation may reduce the second term in (11.60) by as much as 60%–80%. We may use (11.61) even for chorded coil windings, but, as K_{w5} is almost zero, the factor in straight parenthesis is reduced to almost unity.

Example 11.4 The Calculation of the Load Stray Losses

Let us calculate the stray load loss amplification with respect to no load for a motor with nonskewed insulated bars and full-pitch stator winding; open stator slots, $a_0 = 0.67/g$, $a_1 = 0.43/g$, $I_0/I_n = 0.4$, $\cos\varphi_n = 0.86$, $N_s = 48$, $N_r = 40$, $2p_1 = 4$, $\beta = 0.41$, and $K_c = 1.8$.

Solution

We have to calculate first from (11.50) the amplification factors for the airgap field $N_s \pm P_1$ harmonics by the same order mmf harmonics.

$$\xi_{(N_s-p_1)0} = \frac{1}{K_c}\left(1 + \frac{N_s - p_1}{p_1}\frac{a_1}{2a_0}\right) = \frac{1}{K_c}\left(1 + \frac{(48-2)}{2}\cdot\frac{0.43}{2\cdot 0.67}\right) = \frac{8.4}{K_c}$$

$$\xi_{(N_s-P_1)0} = \frac{1}{K_c}\left(1 + \frac{N_s - P_1}{P_1}\frac{a_1}{2a_0}\right) = \frac{1}{K_c}\left(1 + \frac{(48-2)}{2}\cdot\frac{0.43}{2\cdot 0.67}\right) = \frac{8.4}{K_c}$$

Now, the same factors under load are found from (11.51).

$$\xi_{(N_s-p_1)n} = \frac{1}{K_c}\sqrt{1 + 2\cdot 0.53\cdot 0.4\cdot\frac{48-2}{2}\cdot\frac{0.43}{2\cdot 0.67} + \left(0.4\cdot\frac{48-2}{2}\cdot\frac{0.43}{2\cdot 0.67}\right)^2} = \frac{3.6}{K_c}$$

$$\xi_{(N_s+p_1)n} = \frac{1}{K_c}\sqrt{1-2\cdot 0.53\cdot 0.4\cdot\frac{48+2}{2}\cdot\frac{0.43}{2\cdot 0.67}+\left(0.4\cdot\frac{48+2}{2}\cdot\frac{0.43}{2\cdot 0.67}\right)^2} = \frac{2.8}{K_c}$$

The stray loss load amplification factor C_{loads} is (11.53)

$$C_{loads} \approx \left(\frac{I_n}{I_0}\right)^2\left[\frac{\xi^2_{(N_s-p_1)n}+\xi^2_{(N_s+p_1)n}}{\xi^2_{(N_s-p_1)0}+\xi^2_{(N_s+p_1)0}}\right] = \left(\frac{1}{0.4}\right)^2\left[\frac{3.60^2+2.80^2}{8.4^2+7^2}\right] = 1.0873!$$

So the rotor surface and the rotor tooth flux pulsation (stray) load losses are increased at full load only by 8.73% with respect to no load.

Let us now explore what happens to the no-load rotor tooth pulsation (stray) cage losses under load. This time we have to use (11.60)

$$\frac{P_{ncage}}{P_{0cage}} = C_{load}\cdot\left[1+\frac{2K_s\left(\frac{K_{w5}}{K_{w1}}\right)^2\frac{(1+\tau_{dN_s})^2}{(1+\tau_{d6})^2}\sqrt{\frac{6p_1}{N_s}}}{\xi^2_{(N_s-p_1)n}+\xi^2_{(N_s+p_1)n}}\right]$$

with K_s a saturation factor ($K_s = 0.2$), $K_{w5} = 0.2$, and $K_{w1} = 0.965$. Also τ_{d48} and τ_{d6} from (11.61) are $\tau_{d48} = 42$ and $\tau_{d6} = 0.37$. Finally,

$$\frac{P_{ncage}}{P_{0cage}} = 1.0873\cdot\left[1+\frac{2\cdot 0.2\left(\frac{0.2}{0.965}\right)^2\frac{(1+42)^2}{(1+0.37)^2}\sqrt{\frac{6\cdot 2}{48}}}{3.60^2+2.80^2}\right] = 1.43!$$

As expected, the load stray losses in the insulated nonskewed cage are notably larger for no-load conditions.

11.5 FLUX PULSATION (STRAY) LOSSES IN SKEWED INSULATED BARS

When the rotor slots are skewed and the rotor bars are insulated from rotor core, no interbar currents flow between neighbouring bars through the iron core.

In this case, the cage no-load stray losses P_{0cage} due to first slot (opening) harmonics $N_s \pm p_1$ are corrected as [6]

$$P_{0cages} = \frac{P_{0cage}}{2}\left[\left(\frac{\sin\frac{\pi c}{t_s N_s}(N_s+p_1)}{\frac{\pi c}{t_s N_s}(N_s+p_1)}\right)^2+\left(\frac{\sin\frac{\pi c}{t_s N_s}(N_s-p_1)}{\frac{\pi c}{t_s N_s}(N_s-p_1)}\right)^2\right] \tag{11.62}$$

where c/t_s is the skewing in stator slot pitch t_s units.

When skewing equals one stator slot pitch ($c/t_s = 1$), Equation (11.62) becomes

$$\left(P_{0cages}\right)_{c/t_s=1} \approx P_{0cage}\cdot\left(\frac{p_1}{N_s}\right)^2 \tag{11.63}$$

Consequently, in general, skewing reduces the stray losses in the rotor insulated cage (on no-load and on load), as a bonus from (11.60). For one stator pitch skewing, this reduction is spectacular.

Two things we have to remember here.

- When stray cage losses are almost zero, the rotor flux pulsation core losses are not attenuated and are likely to be large for skewed rotors with insulated bars.
- Insulated bars are made, in general, of brass or copper.

For the vast majority of small- and medium-power IMs, cast aluminium uninsulated bar cages are used. Interbar current losses are expected and they tend to be augmented by skewing.

We will now treat this problem separately.

11.6 INTERBAR CURRENT LOSSES IN UNINSULATED SKEWED ROTOR CAGES

For cage rotor IMs with skewed slots (bars) and uninsulated bars, transverse (cross-path) or interbar additional currents through rotor iron core between adjacent bars occur.

Measurements suggest that the cross-path or transverse impedance Z_d is, in fact, a resistance R_d up to at least f = 1 kHz. Also, the contact resistance between the rotor bar and rotor teeth is much larger than the cross-path iron core resistance. This resistance tends to increase with the frequency of the harmonic considered and it depends on the manufacturing technology of the cast aluminium cage rotor. To have a reliable value of R_d measurements are mandatory.

To calculate the interbar currents (and losses), a few analytical procedures have been put forward [9,10,6]. While Refs. [9,10] do not refer specially to the first slot opening harmonics, Ref. [6] ignores the end ring resistance.

In what follows, we present a generalisation of Refs. [9,11] to include the first slot opening harmonics and the end ring resistance.

Let us consider the rotor stack divided into many segments with the interbar currents lumped into definite resistances (Figure 11.5).

The skewing angle γ is

$$\tan \gamma = \frac{c}{l_{stack}} \tag{11.64}$$

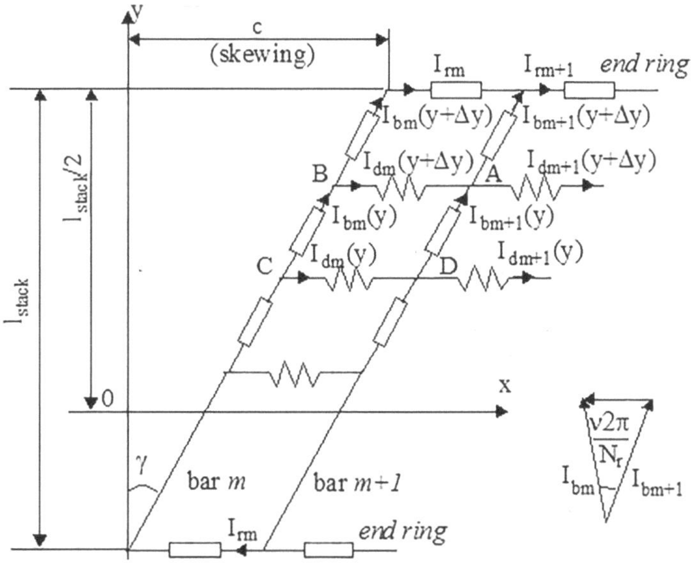

FIGURE 11.5 Bar and interbar currents in a rotor cage with skewed slots.

What airgap field harmonics are likely to produce interbar currents? In principle, all space harmonics, including the fundamental, do so. Additionally, time harmonics have a similar effect. However, for chorded coil stator windings, the first slot opening harmonic $N_s - p_1$ augments the same stator mmf harmonic; $N_s + p_1$ harmonic is attenuated in a similar way as shown in previous paragraphs.

Only for full-pitch stator windings, the losses produced by the first two 5th and 7th phase-belt stator mmf harmonics are to be added. (For chorded coil windings, their winding factors, and consequently their amplitudes, are negligible).

At 50 (60) Hz apparently, the fundamental component losses in the cross-path resistance for skewed uninsulated bar rotor cages may be neglected. However, this is not so in high- (speed) frequency IMs (say above 300 Hz fundamental frequency). Also, inverter-fed IMs show current and flux time harmonics, so additional interbar current losses due to the space fundamental and time harmonics are to be considered. Later in this chapter, we will return to this subject.

For the time being, let us consider only one stator frequency f with a general space harmonic ν with its slip S_ν.

Thus, even the case of time harmonics can be dealt with, one by one, changing only f, but for $\nu = p_1$ (pole pairs), the fundamental.

The relationship between adjacent bar and cross-path resistance currents on Figure 11.5 are

$$
\begin{aligned}
I_{bm}(y) - I_{bm+1}(y) &= 2je^{-\frac{j\pi\nu}{N_r}} I_{bm}(y)\sin\frac{\pi\nu}{N_r} \\
I_{dm}(y) - I_{dm+1}(y) &= 2je^{-\frac{j\pi\nu}{N_r}} I_{dm}(y)\sin\frac{\pi\nu}{N_r} \\
I_{rm} - I_{rm+1} &= 2je^{-\frac{j\pi\nu}{N_r}} I_{rm}\sin\frac{\pi\nu}{N_r}
\end{aligned}
\tag{11.65}
$$

Kirchhoff's first law in node A yields

$$
I_{dm}(y+\Delta y) - I_{dm+1}(y+\Delta y) \approx \frac{I_{bm}(y+\Delta y) - I_{bm+1}(y)}{\Delta y} \tag{11.66}
$$

Note that the cross-path currents I_{dm} and I_{dm+1} refer to unity stack length.

With (11.65) in (11.66), we obtain

$$
I_{dm}(y) \approx -j\frac{e^{-\frac{j\pi\nu}{N_r}}}{2\sin\frac{\pi\nu}{N_r}} \cdot \frac{d}{dy} I_{bm}(y) \tag{11.67}
$$

Also, for the ring current I_{rm},

$$
I_{rm} - I_{rm+1} = \left(I_{bm+1}\right)_{y=\frac{l_{stack}}{2}} \tag{11.68}
$$

with (11.65), Equation (11.68) becomes

$$
I_{rm} = j\frac{e^{-\frac{j\pi\nu}{N_r}}}{2\sin\frac{\pi\nu}{N_r}} \left(I_{bm}\right)_{y=\frac{l_{stack}}{2}} \tag{11.69}
$$

The second Kirchhoff's law along the closed path ABCD yields

$$R_q l_{stack}\left(I_{dm}(y+\Delta y)-I_{dm}(y)\right)+\left(R_{bv}+jS_v X_{0v}(1+\tau_{dv})\right)$$
$$\cdot\left(I_{bm}(y)-I_{bm+1}(y)\right)\cdot\frac{\Delta y}{l_{stack}}=E_m^v\cdot e^{-\frac{j\gamma v y}{l_{stack}}}\cdot\frac{\Delta y}{l_{stack}} \qquad (11.70)$$

E_m^v represents the stator current produced induced voltage in the contour ABCD. X_{0v} represents the airgap reactance for harmonic v seen from the rotor side and τ_{dv} is the differential leakage coefficient for the rotor (11.59).

Making use of (11.67) into (11.70) yields

$$R_{de}l_{stack}^2\frac{d^2 I_{bm}(y)}{dy^2}-Z_{be}I_{bm}(y)=E_m^v e^{-j\frac{\gamma v y}{l_{stack}}} \qquad (11.71)$$

The boundary conditions at $y=\pm l_{stack}/2$ are

$$\left(I_{rm}\right)_{y=\pm l_{stack}/2}\cdot Z_{er}=\mp\left(I_{dm}\right)_{y=\pm l_{stack}/2}R_d l_{stack};\quad R_{de}=\frac{R_d}{4\sin^2\frac{v\pi}{N_r}} \qquad (11.72)$$

Z_{er} represents the impedance of end ring segment between adjacent bars.

$$\underline{Z}_{er}=R_{erv}+jX_{erv};\quad Z_{ere}=\frac{\underline{Z}_{er}}{4\sin^2\frac{v\pi}{N_r}};\quad \underline{Z}_{be}=R_{be}+jX_{be} \qquad (11.73)$$

R_{erv}, X_{erv} are the resistance and reactance of the end ring segment. Z_{be} is bar equivalent impedance.

Solving the differential Equation (11.71) with (11.72) yields [10]

$$I_{bm}(y)=\frac{-E_{mv}}{\left(\underline{Z}_{be}+v^2\gamma^2 R_{de}l_{stack}\right)}$$
$$\cdot\left[\left(\cos\frac{\gamma v y}{l_{stack}}-j\sin\frac{\gamma v y}{l_{stack}}\right)-A\cosh\frac{y}{l}\sqrt{\frac{Z_{be}}{R_{de}}}-B\sinh\frac{y}{l}\sqrt{\frac{Z_{be}}{R_{de}}}\right] \qquad (11.74)$$

The formula of the cross-path current $I_{dm}(y)$ is obtained from (11.67) with (11.74).

Also,

$$A=\frac{Z_{ere}\cos\left(\frac{v\gamma}{2}\right)-v\gamma R_{de}\sin\left(\frac{v\gamma}{2}\right)}{Z_{ere}\cosh\left(\frac{1}{2}\sqrt{\frac{Z_{be}}{R_{de}}}\right)+R_{de}\sqrt{\frac{Z_{be}}{R_{de}}}\sinh\left(\frac{1}{2}\sqrt{\frac{Z_{be}}{R_{de}}}\right)} \qquad (11.75)$$

$$B=-j\frac{Z_{ere}\sin\left(\frac{v\gamma}{2}\right)-v\gamma R_{de}\cos\left(\frac{v\gamma}{2}\right)}{Z_{ere}\sinh\left(\frac{1}{2}\sqrt{\frac{Z_{be}}{R_{de}}}\right)+R_{de}\sqrt{\frac{Z_{be}}{R_{de}}}\cosh\left(\frac{1}{2}\sqrt{\frac{Z_{be}}{R_{de}}}\right)} \qquad (11.76)$$

When the skewing is zero ($\gamma = 0$), B = 0 but A is not zero and, thus, the bar current in (11.74) still varies along y (stack length) and thus some cross-path currents still occur. However, as the end ring impedance Z_{ere} tends to be small with respect to R_{de}, the cross-path currents (and losses) are small.

However, only for zero skewing and zero end ring impedance, the interbar current (losses) are zero (bar currents are independent of y). Also, as expected, for infinite transverse resistance ($R_{de} \sim \infty$), again the bar current does not depend on y in terms of amplitude. We may now calculate the sum of the bar and interbar (transverse) losses together, P^d_{cagev}.

$$P^d_{cagev} = N_r\left[\int_{-\frac{l_{stack}}{2}}^{\frac{l_{stack}}{2}} \left|I_{bm}{}^2(y)\right| \frac{R_b}{l_{stack}} dy + \int_{\frac{l_{stack}}{2}}^{\frac{l_{stack}}{2}} R_d l_{stack} \left|I_{dm}(y)\right|^2 dy + 2R_{er}\left|I_{rm}\right|^2\right] \tag{11.77}$$

Although (11.77) looks rather cumbersome, it may be worked out rather comfortably with I_{bm} (bar current) from (11.74), I_{dm} from (11.67), and I_{rm} from (11.69).

We still need the expressions of emfs E^{ν}_m which would refer to the entire stack length for a straight rotor bar pair.

$$E^{\nu}_m = \frac{V_{r\nu}\left(B_\nu - B_\nu e^{-j\frac{2\nu\pi}{N_r}}\right) l_{stack}}{\sqrt{2}}; \text{ (RMS value)} \tag{11.78}$$

$V_{r\nu}$ is the field harmonic ν speed with respect to the rotor cage and B_ν the νth airgap harmonic field. For the case of a chorded coil stator winding, only the first stator slot (opening) harmonics $\nu = N_s \pm p_1$ are to be considered.

For this case, accounting for the first airgap magnetic conductance harmonic, the value of B_ν is

$$B_{N_s} = B_{g1}\frac{a_1}{2a_0}; \quad N_s = (N_s \pm p_1) \mp p_1 \tag{11.79}$$

a_1 and a_0 from (11.22 and 11.23).

The speed $V_{r\nu}$ is

$$V_{r\nu} = R\frac{2\pi f}{p_1 60}\left(1 \pm \frac{p_1}{\nu}\right) \tag{11.80}$$

with

$$\nu = N_s \pm p_1 \qquad V_{r\nu} \approx R\frac{2\pi f}{p_1 60} = V_{N_s} \tag{11.81}$$

The airgap reactance for the $\nu = N_s \pm p_1$ harmonics, $X_{0\nu}$, is

$$X_{0N_s} \approx X_{0p}\frac{p_1}{N_s} \tag{11.82}$$

X_{0p}, the airgap reactance for the fundamental, as seen from the rotor bar,

$$X_{0p} = \frac{X_m}{\alpha^2}; \quad \alpha = \frac{4 \cdot 3 \cdot W_1{}^2 K_{w1}{}^2}{N_r} \tag{11.83}$$

where X_m is the main (airgap) reactance for the fundamental (reduced to the stator),

$$X_m = \frac{6\mu_0\omega_1}{\pi^2}\frac{W_1^2K_{w1}^2\tau l_{stack}}{p_1 g K_c(1+K_s)} \quad (11.84)$$

(For the derivation of (11.83), see Chapter 5).

W_1 – turns/phase, K_{w1} – winding factor for the fundamental, K_s – saturation factor, g – airgap, p_1 – pole pairs, and τ – pole pitch of stator winding.

Now that we have all data to calculate the cage and the transverse losses for skewed uninsulated bars, making use of a computer programming environment like Matlab etc., the problem could be solved rather comfortably in terms of numbers.

Still we can hardly draw any design rules this way. Let us simplify the solution (11.74) by considering that the end ring impedance is zero ($Z_{er} = 0$) and that the contact (transverse) resistance is small.

- Small transverse resistance
  Let us consider that R_q is the contact transverse resistance per unit rotor length and it is so small that

$$R_q\left(\frac{\nu\gamma}{R}\right)^2 << 4\sin^2\left(\frac{\nu\pi}{N_r}\right)X_{0\nu}(1+\tau_{d\nu}) \quad (11.85)$$

For this case in Ref. [6], the following solution has been obtained:

$$I_{bm}(y) = \frac{-jE_m^\nu e^{-j\frac{\nu\gamma y}{l_{stack}}}}{X_{0\nu}(1+\tau_{d\nu})}; \quad I_{dm} = \frac{-jE_m^\nu e^{-j\frac{\nu\gamma y}{l_{stack}}}}{2\sin\left(\frac{\nu\pi}{N_r}\right)X_{0\nu}(1+\tau_{d\nu})}\gamma\frac{l_{stack}}{n}\frac{1}{R} \quad (11.86)$$

n – the number of stack axial segments.

The transverse losses P_{d0} are

$$P_{d0} \approx \frac{R_q}{l_{stack}}\frac{N_r^3}{4\pi^2}\left(\frac{E_m^\nu}{X_{0\nu}}\right)^2\cdot\frac{1}{1+\tau_{d\nu}}\cdot\left(\frac{c}{R}\right)^2 \quad (11.87)$$

The bar losses P_{b0} are

$$P_{b0} = N_r R_{be}\frac{\frac{E_m^\nu}{l_{stack}}}{X_{0\nu}^2(1+\tau_{d\nu})^2}l_{stack} \quad (11.88)$$

Equations (11.87 and 11.88) lead to remarks such as

- For zero end ring resistance and low transverse resistance R_d, the transverse (interbar) rotor losses P_d are proportional to R_q and to the skewing squared.
- The bar losses are not influenced by skewing (11.88), so skewing is not effective in this case and it shall not be used.

- Large transverse resistance.
  In this case, the opposite of (11.85) is true. The final expressions of transverse and cage losses P_{d0} and P_{b0} are

$$P_{d0} \approx \frac{2\pi^2}{N_r}\frac{\left(\frac{E_m^v}{l_{stack}}\right)^2}{\left(\frac{c}{R}\right)^2} l_{stack}^{\ 3} \cdot \frac{1}{(1+\tau_{dv})} \cdot \frac{\left(1-\frac{1}{3}\sin^2\left(\frac{\nu l_{stack}}{2R}\right)\right)}{\frac{R_q}{2}} \tag{11.89}$$

$$P_{b0} = N_r R_{be}\left(\frac{E_m^v}{X_{0v}(1+\tau_{dv})}\right)^2 K_{skewv}^{\ 2}, \tag{11.90}$$

$$K_{skewv} = \frac{\sin\left(\frac{\gamma\nu l_{stack}}{2R}\right)}{\frac{\gamma\nu l_{stack}}{2R}}$$

  The situation is now different.
- The transverse losses are proportional to the third power of stack length, to the square of inversed skewing factor, and inversely proportional to transverse resistance per unit length $R_q(\Omega m)$.
- The rotor bar losses P_b are also proportional with skewing squared and, thus, by proper skewing for the first slot (opening) harmonics $\nu \approx N_s$, this harmonic bar losses are practically zero; skewing is effective. Also, with long stacks, only chorded windings are practical.
- With long stacks, large transverse resistivities have to be provided by bad contact between bars and tooth walls or by insulated bars. Plotting the transverse losses by the two formulas, (11.87) and (11.89), versus transverse resistance per unit length R_q shows a maximum (Figure 11.6).

So there is a critical transverse resistance R_{q0} for which the transverse losses are maximum. Such conditions are to be avoided.

In Ref. [6], the critical value of R_{q0} is

$$R_{q0} \approx \frac{X_{0v} l_{stack} 4\pi^2 \sqrt{\left[1-\frac{1}{3}\sin^2\left(\frac{\nu l_{stack}}{2R}\right)\right]}}{N_r^{\ 2}\left(\frac{c}{R}\nu\right)^2} \tag{11.91}$$

X_{0v}, R_{be}, Z_{er}, and R_d are in Ω and E_{mv} in Volt.

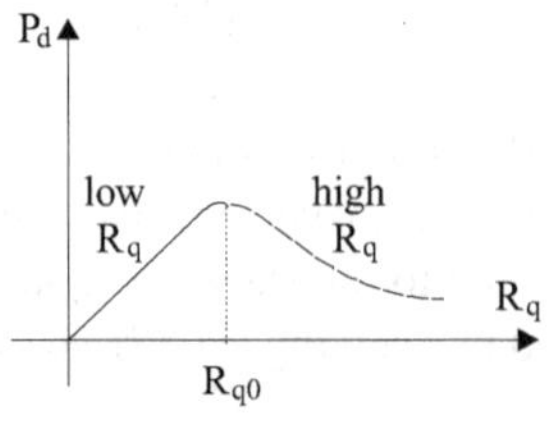

FIGURE 11.6 Transverse losses versus transverse resistance per unit length R_q (Ωm).

Increasing the stack length tends to increase the critical transverse resistivity R_{q0}.

R_{q0} is inversely proportional with skewing. Also, it seems that $(P_b)_{R_{q0}}$ or maximum transverse rotor losses depend on skewing.

11.7 NO-LOAD ROTOR SKEWED UNINSULATED CAGE LOSSES

For small or high transverse resistivity (R_q), expressions (11.87–11.90) yield the computation of transverse and cage no-load stray losses with E_m^{ν} from (11.78) and $\nu = N_s \pm p_1$. It has been shown that for this case, with $\nu \approx N_s$, the transverse losses have a minimum for $0.9 < N_r/N_s < 1.1$.

It means that for low contact (transverse) resistivity, the rotor and stator slot numbers should not be too far away from each other. Especially for $N_r/N_s < 1$, small transverse rotor losses are expected.

11.8 LOAD ROTOR SKEWED UNINSULATED CAGE LOSSES

While under no load, the airgap permeance first harmonic was important as it acted upon the airgap flux distribution, under load, it is the stator mmf harmonics of same order $N_s \pm p_1$ that are important as the stator current increases with load. This is not so for chorded pitch stator windings when the 5th and 7th pole pair phase-belt harmonics are small.

The airgap flux density B_v will be now coming from a different source

$$B_v = \frac{p_1}{\nu}\frac{K_{w\nu}}{K_{w1}}\frac{\mu_0 F_{1p}}{gK_c(1+K_s)} \tag{11.92}$$

F_{1p} is the stator mmf fundamental amplitude.

For full-pitch windings however, the 5th and 7th (phase belt) harmonics are to be considered and the conditions of low or high transverse resistivity (11.85) have to be verified for them as well, with $\nu = 5p_1, 7p_1$.

Transverse cage losses for both smaller or higher transverse resistivity R_q are inversely proportional to differential leakage coefficient $\tau_{d\nu}$ which, for $\nu = N_s$, is

$$\tau_{dN_s} = \frac{\left(\dfrac{\pi N_s}{N_r}\right)^2}{\sin^2\left(\dfrac{\pi N_s}{N_r}\right)^2} - 1 \tag{11.93}$$

It may be showed that τ_{dN_s} increases when $N_s/N_r > 1$.

Building IMs with $N_s/N_r > 1$ seems very good to reduce the transverse cage losses with skewed uninsulated rotor bars. For such designs, it may be adequate to even use nonskewed rotor slots, when the additional transverse losses are almost zero.

Care must be exercised to check if the parasitic torques are small enough to secure safe starts. We should also notice that skewing leads to a small attenuation of tooth flux pulsation core losses by the rotor cage currents.

In general, the full-load transverse cage loss P_{dn} is related to its value under no-load P_{d0} by the load multiplication factor C_{loads} (11.53) for chorded pitch stator windings:

$$P_{dn} = P_{d0}C_{loads} \tag{11.94}$$

For a full-pitch (single-layer) winding, C_{loads} has to be changed to add the 5th and 7th phase-belt harmonics in a similar way as in (11.60 and 11.61).

$$P_{dn} = P_{d0}C_{loads}\left[1+\frac{2\left(\frac{K_{w5}}{K_{w1}}\right)^2\frac{(1+\tau_{dN_s})}{(1+\tau_{d6})}}{\xi_{(N_s-p_1)n}{}^2+\xi_{(N_s+p_1)n}{}^2}\right] \tag{11.95}$$

with $\xi_{(N_s-p_1)n}$ and $\xi_{(N_s+p_1)n}$ from (11.51).

Example 11.5 The Load to No-Load Transverse Rotor Losses Calculation

For the motor with the data in Example 11.4, let us determine the P_{dn}/P_{d0} (load to no-load transverse rotor losses).

Solution

We are to use (11.95).

From Example 11.4, $\xi_{(N_s-p_1)n} = 3.6/K_c$, $K_c = 1.8$, $K_{w5} = 0.2$, $K_{w1} = 0.965$, $C_{loads} = 1.0873$, $\xi_{(N_s+p_1)n} = 2.8/K_c$, $1+\tau_{dN_s} = 43$, and $1+\tau_{d6} = 1.37$.

We now have all data to calculate

$$\frac{P_{dn}}{P_{d0}} = 1.0873\left[1+\frac{2\left(\frac{0.2}{0.965}\right)^2\frac{43}{1.37}}{\left(\frac{3.6}{1.8}\right)^2+\left(\frac{2.8}{1.8}\right)^2}\right] = 1.54!$$

11.9 RULES TO REDUCE FULL-LOAD STRAY (SPACE HARMONICS) LOSSES

So far, the rotor surface core losses, rotor and stator tooth flux pulsation core losses, and space harmonic cage losses have been included into stray load losses. They all have been calculated for motor no load and then corrected for load conditions by adequate amplification factors.

Insulated and noninsulated, skewed and nonskewed bar rotors have been investigated. Chorded pitch and full-pitch windings cause differences in terms of stray load losses. Other components such as end-connection leakage flux produced losses, which occur in the windings surroundings, have been left out as their study by analytical methods is almost impractical. In high-power machines, such losses are to be considered.

Based on analysis, such as the one corroborated above, with those of [12], we here line up a few rules to reduce full-load stray losses

- Large number of slots per pole and phase, q, if possible, to increase the first phase belt and first slot (opening) harmonics.
- Insulated or large transverse resistance cage bars in long stack skewed rotors to reduce transverse cage losses.
- Skewing is not adequate for low transverse resistance at it does not reduce the stray cage losses while the transverse cage losses are large. Check the tooth flux pulsation core losses in skewed rotors.
- $N_r < N_s$ to reduce the differential leakage coefficient of the first slot (opening) harmonics $N_s \pm p_1$, and thus reduce the transverse cage losses.
- For $N_r < N_s$, skewing may be eliminated after the parasitic torques are checked and found small enough. For q = 1, 2, skewing is mandatory, though.
- Chorded coil windings with $y/\tau \approx 5/6$ are adequate as they reduce the first phase-belt harmonics.

- With full-pitch winding use large number of slot/pole/phase whenever possible.
- Skewing seems efficient for uninsulated rotor bars with high transverse resistance as it reduces the transverse rotor losses.
- With delta connection, $(N_s - N_r) \neq 2p_1, 4p_1, 8p_1$.
- With parallel path winding, the circulating stator currents induced by the bar current mmf harmonics have to be avoided by observing a certain symmetry.
- Use small stator and rotor slot openings, if possible, to reduce the first slot opening flux density harmonics and their losses. Check the starting and peak torque as they tend to decrease due to slot leakage inductance increase.
- Use magnetic wedges for open slots, but check for the additional eddy current losses in them and secure their mechanical ruggedness.
- Increase the airgap, but maintain good power factor and efficiency.
- Use sharp tools and annealed lamination sheets (especially for low-power motors) to reduce surface core losses.
- Return rotor surface to prevent laminations to be short-circuited and thus reduce rotor surface core losses.
- As storing the motor with cast aluminium (uninsulated) cage rotors after fabrication leads to a marked increase of rotor bar to slot wall contact electrical resistivity, a reduction of stray losses of 40% to 60% may be obtained after 6 months of storage.

11.10 HIGH-FREQUENCY TIME HARMONICS LOSSES

High-frequency time harmonics in the supply voltage of IMs may occur either because the IM itself is fed from a PWM static power converter for variable speed or because, in the local power grid, some other power electronics devices produce voltage time harmonics at IM terminals [14].

For voltage-source static power converters, the time harmonics frequency content and distribution depends on the PWM strategy and the carrier ratio c_r ($2c_r$ switchings per fundamental period). For high-performance symmetric regular sampled, asymmetric regular sampled, and optimal regular sampled PWM strategies, the main voltage harmonics are $(c_r \pm 2)f_1$ and $(2c_r \pm 1)f_1$, respectively [13], Figure 11.7.

It seems that accounting for time harmonics losses at frequencies close to carrier frequency suffices. As of today, the carrier ratio c_r varies from 20 to more than 200. Smaller values relate to larger powers.

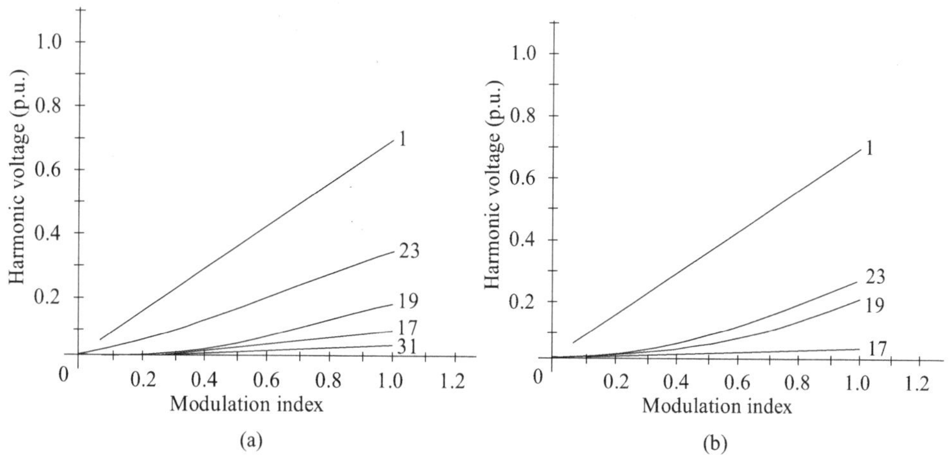

FIGURE 11.7 Harmonic voltages with modulation index for $c_r = 21$ (a) symmetric regular sampled PWM and (b) asymmetric regular sampled PWM.

Switching frequencies of up to 20 kHz are typical for low-power induction motors fed from IGBT voltage-source converters (more than 30 kHz for silicon-carbide converters (SIC)). Exploring the conductor and core losses up to such large frequencies becomes necessary.

High carrier frequencies tend to reduce the current harmonics and thus reduce the conductor losses associated with them, but the higher frequency flux harmonics may lead to larger core loss. On the other hand, the commutation losses in the PWM converter increase with carrier frequency.

The optimum carrier frequency depends on the motor and the PWM converter itself. The 20 kHz is typical for hard-switched PWM converters. Higher frequencies are practical for soft-switched (resonance) or SIC converters. We will explore, first rather qualitatively and then quantitatively, the high-frequency time harmonics losses in the stator and rotor conductors and cores.

At high carrier frequency, the skin effect, both in the conductors and iron cores, may not be neglected.

11.10.1 Conductor Losses

The variation of resistance R and leakage L_l inductance for conductors in slots with frequency, as studied in Chapter 9, is at first rapid, being proportional to f^2. As the frequency increases further, the field penetration depth gets smaller than the conductor height and the rate of change of R and L_l decreases to become proportional to $f^{1/2}$ (Figure 11.8).

For the end connections (or end rings), the skin effect is less pronounced, but it may be calculated with similar formulas provided virtual larger slots are defined (Chapter 9, Figure 9.10).

For high frequencies, the equivalent circuit of the IM may be simplified by eliminating the magnetisation branch (Figure 11.9).

Notice that the slip $S_v \approx 1$.

In general, the reactances prevail at high frequencies

$$I_v \approx \frac{V_v}{2\pi f_v L_\sigma(f_v)}; \quad L_\sigma(f_v) = L_{sl}(f_v) + L_{rl}(f_v) \tag{11.95a}$$

The conductor losses are then

$$P_{con} = 3I_v^2\left(R_s(f_v) + R_r(f_v)\right) \approx \frac{3V_v^2}{\left(2\pi f_v L_\sigma(f_v)\right)^2}\left(R_s(f_v) + R_r(f_v)\right) \tag{11.96}$$

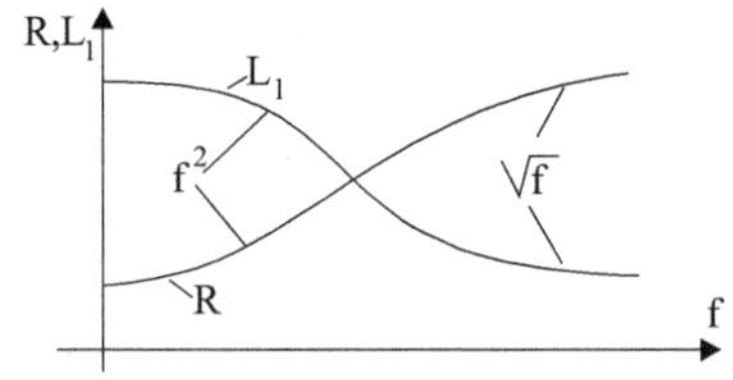

FIGURE 11.8 R & L_l variation with frequency.

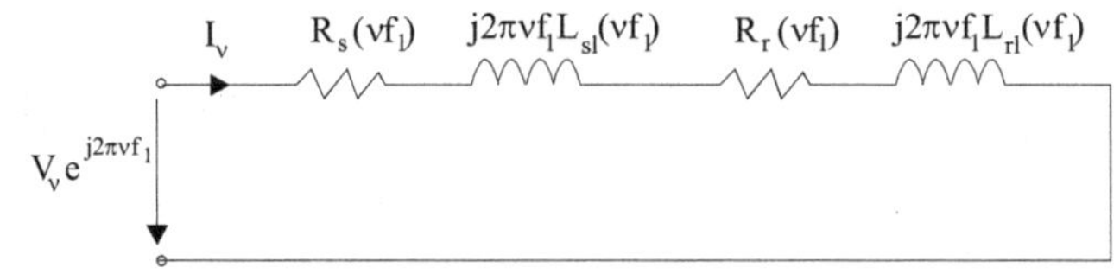

FIGURE 11.9 Equivalent circuit for time harmonic v.

For a given (fixed) value of the current harmonic I_v, the conductor losses will increase steadily with frequency. This case would be typical for current control.

For voltage control, when voltage harmonics V_v are given, however, (11.96) shows that it is possible to have conductor losses decreasing with frequency if the decrease of leakage inductance L_σ (f_v) with f_v and the increase factors of $\left(R_s(f_v)+R_r(f_v)\right)$ with f_v are less than proportional to f_v^2.

Measurements have shown that the leakage inductance decreases to 0.5 to 0.3 of the rated frequency (60 Hz) value at 20 kHz.

So, in general,

$$L_\sigma(f_v) \approx K_L f_v^{-0.16} \tag{11.97}$$

For high frequencies, R_r variation with frequency is in the $f_v^{0.5}$ range, approximately, so the rotor conductor losses are

$$P_{conr} \approx \frac{3V_v^2}{(2\pi f_v K_L)^2 f_v^{-0.32}} K_R f_v^{0.5} \sim \frac{V_h^2}{f_v^{1.18}} \tag{11.98}$$

The rotor conductor losses drop notably as the time harmonic frequency increases.

The situation in the stator is different as there are many conductors in every slot (at least in small power induction machines).

So the skin effect for R_s will remain f_v dependent in the initial stages ($K_{Rs} = C_{Rs} \cdot f_v$)

$$P_{cons} \sim V_v^2 f_v^{0.32} \tag{11.99}$$

The stator conductor losses tend to increase slightly with frequency. For large-power IMs (MW range), the stator conductor skin effect is stronger and the situation comes closer to that of the rotor: the stator conductor losses slightly decrease at higher frequencies. We should also mention that the equivalent carrier frequency in large power is 1 to 3 kHz.

For low- and medium-power motors, as the carrier frequency reaches high levels (20 kHz or more), the skin effect in the stator conductors enters the $f_v^{0.5}$ domain and the stator conductor losses, for given harmonic voltage, behave like the rotor cage losses (decrease slightly with frequency (11.98)). This situation occurs when the penetration depth becomes smaller than conductor height.

11.10.2 Core Losses

Predicting the core loss at high frequencies is difficult because the flux penetration depth in lamination becomes comparable with (or smaller than) the lamination thickness.

The leakage flux paths may then prevail, and thus, the reaction of core eddy currents may set up significant reaction fields.

The field penetration depth in laminations, δ_{Fe}, is

$$\delta_{Fe} = \sqrt{\frac{1}{\pi f_v \sigma_{Fe} \mu_{Fe}}} \tag{11.100}$$

σ_{Fe} – iron electrical conductivity; μ_{Fe} – iron magnetic permeability.

For f = 60 Hz, $\sigma_{Fe} = 2 \cdot 10^6\ (\Omega m)^{-1}$, $\mu_{Fe} = 800\ \mu_0$, and δ_{Fe} = 1.63 mm, while at 20 kHz, δ_{Fe} = 0.062 mm. In contrast, for copper, $(\delta_{Co})_{60Hz}$ = 9.31 mm and $(\delta_{Co})_{20kHz}$ = 0.51 mm and in aluminium $(\delta_{Al})_{60Hz}$ = 13.4 mm and $(\delta_{Al})_{20kHz}$ = 0.73 mm.

The penetration depth at 800 μ_0 and 20 kHz, δ_{Fe} = 0.062 mm, shows the importance of skin effect in laminations. To explore the dependence of core losses on frequency, let us distinguish three cases.

- Case 1 – no lamination skin effect: $\delta_{Fe} >> d$
This case corresponds to low-frequency time harmonics. Both hysteresis and eddy current losses are to be considered.

$$P_{Fe} = \left(K'_{hl} B_v{}^n f_v + K'_{el} B_v{}^2 f_v{}^2\right) A_l l \tag{11.101}$$

where B_v is the harmonic flux density:

$$B_v = \frac{\varphi_v}{A_l} \approx \frac{V_v}{2\pi f_v A_l} \tag{11.102}$$

A_l is the effective area of the leakage flux path and l is its length.
With (11.102), (11.101) becomes

$$P_{Fe} = \left(K_{hl} V_v^n f_v^{1-n} + K_{el} V_v{}^2\right) A_l l \tag{11.103}$$

Since n > 1, the hysteresis losses decrease with frequency while the eddy current losses stay constant. P_{Fe} is almost constant in these conditions.
- Case 2 – slight lamination skin effect, $\delta_{Fe} \approx d$
When $\delta_{Fe} \approx d$, the frequency f_v is already high, and thus, $\delta_{Al} < d_{Al}$ and a severe skin effect in the rotor slot occurs. Consequently, the rotor leakage flux is concentrated close to the rotor surface. The "volume" where the core losses occur in the rotor decreases. In general then, the core losses tend to decrease slowly and level out at high frequencies.
- Case 3 – strong lamination skin effect, $\delta_{Fe} < d$
With large enough frequencies, the lamination skin depth $\delta_{Fe} < d$, and thus, the magnetic field is confined to a skin depth layer around the stator slot walls and on the rotor surface (Figure 11.10).
The conventional picture of rotor leakage flux paths around the rotor slot bottom is not valid in this case anymore.
The area of leakage flux is now, for the stator, $A_l = l_f\, \delta_{Fe}$, with l_f the length of the meander zone around the stator slot.

$$l_f = (2h_s + b_s) N_s \tag{11.104}$$

Now the flux density B_v is

$$B_v = \frac{K_v V_v}{f_v l_f \delta_{Fe}} \sim K f_v^{-\frac{1}{2}} \tag{11.105}$$

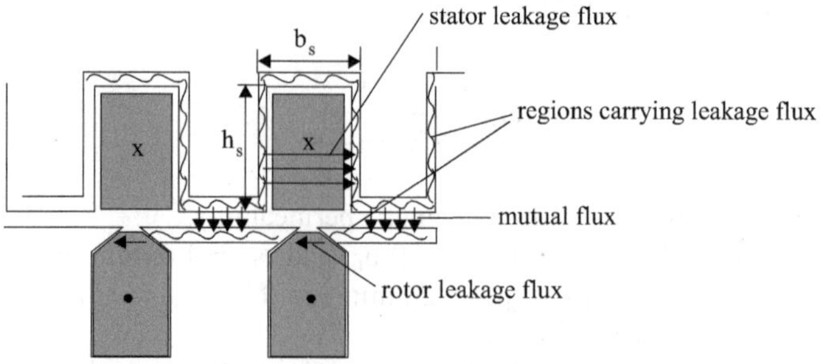

FIGURE 11.10 Leakage flux paths at high frequency.

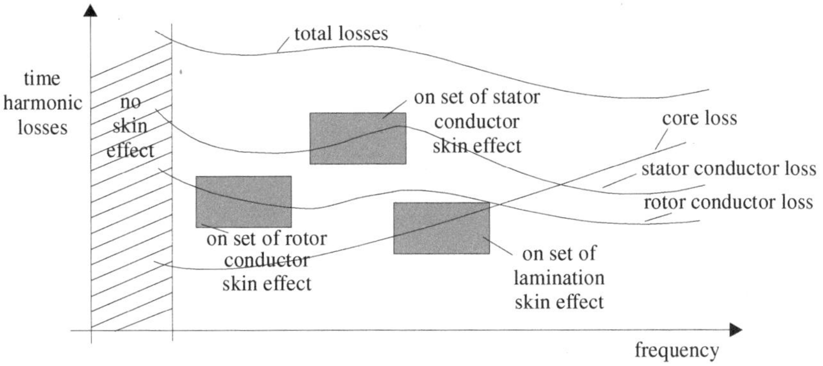

FIGURE 11.11 Time harmonic IM losses with frequency (constant time harmonic voltage)/(after Ref. [14]).

Consequently, the core losses (with hysteresis losses neglected) are

$$P_{Fee} = K_e B_v{}^2 f_v{}^2 l_f l_{stack} \delta_{Fe} \sim K_1 V_v{}^2 f_v{}^{\frac{1}{2}} \tag{11.106}$$

A slow steady-state growth of core losses at high frequencies is thus expected.

11.10.3 Total Time Harmonics Losses

As the discussion above indicates, for a given harmonic voltage, above certain frequency, the conductor losses tend to decrease as $f_v^{-1.2}$ (11.98) while the core losses (11.106) increase.

Consequently, the total time harmonics may level out above a certain frequency (Figure 11.11). Even a minimum may be observed. At what frequency such a minimum occurs depends on the machine design, power, and PWM frequency spectrum.

11.11 COMPUTATION OF TIME-HARMONICS CONDUCTOR LOSSES

As Equation (11.96) shows, to compute the time-harmonics conductor losses, the variation of stator and rotor leakage inductances and resistances with frequency due to skin effect is needed. This problem has been treated in detail, for a single, unassigned, frequency in Chapter 9, both for the stator and the rotor.

Here we simplify the correction factors in Chapter 9 to make them easier to use and interpret.

In essence, with different skin effect in the slot and end-connection (end ring) zones, the stator resistance $R_s(f_v)$ is

$$R_s(f_v) = R_{sdc}\left[K_{Rss}(f_v)\frac{l_{stack}}{l_{coil}} + K_{Rse}(f_v)\frac{l_{endcon}}{l_{coil}}\right] \tag{11.107}$$

l_{coil} – coil length, l_{endcon} – coil end-connection length.

$K_{Rss} > 1$ is the resistance skin effect correction coefficient for the slot zone and K_{Rse} corresponds to the end connection. A similar expression is valid for the stator phase leakage inductance.

$$L_{sl}(f_v) = L_{sldc}\left[K_{Xlss}(f_v)\frac{l_{stack}}{l_{coil}} + K_{Xlse}(f_v)\frac{l_{endcon}}{l_{coil}}\right] \tag{11.108}$$

K_{Xlss} and K_{Xlse} are leakage inductance correction factors.

Now with m layers of conductors per slot from Chapter 9, Equation (9.6) yields

$$K_{Rss} = \varphi(\xi) + \frac{m^2-1}{3}\psi(\xi) > 1 \tag{11.109}$$

$$\xi = \beta_n h, \quad \beta_n = \sqrt{\pi f_v \mu_0 \sigma_{Co} \frac{nb}{b_s}}$$

h – conductor height, n – conductors per layer, b – conductor width (or diameter), and b_s – slot width. Equation (11.109) is strictly valid for rectangular slots (Figure 11.12). More evolved methods, as described in Chapter 9, may be used for general shape slots

The same formula (11.109) may be used for end connection coefficient K_{Rse} but with b_{se} instead of b_s, h_{se} instead of h_s, and m/2 instead of m.

In a similar way, the reactance correction coefficient K_{Xss} is (Equation (9.7))

$$K_{Xss} = \frac{\varphi'(\xi) + \left(m^2-1\right)\psi'(\xi)}{m^2} < 1 \tag{11.110}$$

Also, from (9.4) and (9.8), (9.9)

$$\varphi(\xi) = \xi\frac{\left(\sinh(2\xi)+\sin(2\xi)\right)}{\left(\cosh(2\xi)-\cos(2\xi)\right)}; \quad \psi(\xi) = 2\xi\frac{\left(\sinh(\xi)-\sin(\xi)\right)}{\left(\cosh(\xi)+\cos(\xi)\right)} \tag{11.111}$$

$$\varphi'(\xi) = \frac{3}{2\xi}\frac{\left(\sinh(2\xi)-\sin(2\xi)\right)}{\left(\cosh(2\xi)-\cos(2\xi)\right)}; \quad \psi'(\xi) = \frac{\left(\sinh(\xi)+\sin(\xi)\right)}{\xi\left(\cosh(\xi)+\cos(\xi)\right)} \tag{11.112}$$

As the number of layers in the virtual slot of end connection is m/2 and the slot width is much larger than for the actual slot, $b_{se} - b_s \approx 1.2h_s \approx (4\text{–}6)b_s$, the skin effect coefficients for the end connection zone are smaller than for the slot zone but still, at high frequencies, they are to be considered.

For the cage-rotor resistance R_r and leakage inductance L_{rl}, expressions similar to (11.107–11.108) are valid.

$$R_r\left(f_v\right) = R_{bdc}K_{Rb}\left(f_v\right) + R_{erdc}K_{Re}\left(f_v\right) \tag{11.113}$$

$$L_{rl}\left(f_v\right) = L_{bdc}K_{Xb}\left(f_v\right) + L_{erdc}K_{Xe}\left(f_v\right) \tag{11.114}$$

Again, expressions (11.109) and (11.110) are valid, but with m = 1, with $n_b = b_s$, and σ_{Al} in (11.109).

Also, in (11.110), ξ is to be calculated with $h_{se} = h$ and $b_{se} = b_s + K_h h$. In other words, the end ring has been assimilated to an end connection, but the factor K_h should be 1.2 for a distant ring and less than that for an end ring placed close to the rotor stack.

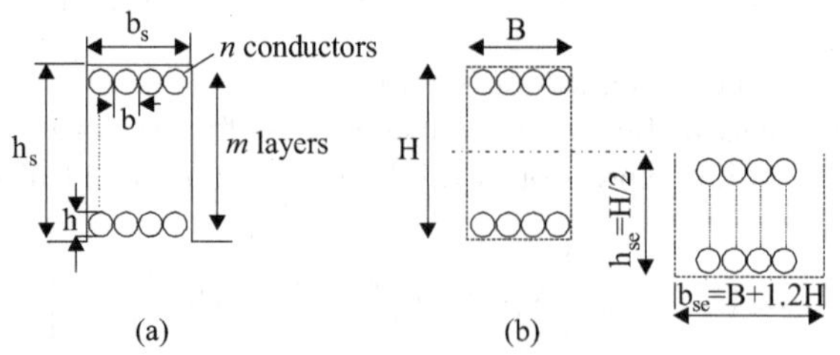

FIGURE 11.12 Stator slots (a) and end connection (b) geometries.

Example 11.6 Time Harmonics Conductor Losses Calculation

Time harmonics conductor losses. A 11 kW, 6 pole, $f_1 = 50\,Hz$, 220 V induction motor is fed from a PWM converter at carrier frequency $f_v = 20\,kHz$. For this frequency, the voltage $V_v = 126\,V$ (see Ref. [14]).

For the rotor, $\xi = 18.38$, $K_{Xb} = 0.0816$, $K_{Rb} = 18.38$, $R_r'(f_v) = 2.2\,\Omega$ (end ring resistance neglected), and $X_{rl}' = 2\pi f_v \times L_{rl}(f_v) = 47.9\,\Omega$ and for the stator, $\xi = 3.38$, $K_{Rss} = K_{Rse} = 79.4$, $R_s(f_v) = 11.9\,\Omega$, $K_{Xss} = K_{Xse} = 0.3$, and $X_{sl} = 2\pi f_v \cdot L_{rl}(f_v) = 63.3\,\Omega$.

The harmonic current I_v is

$$I_v = \frac{I_v}{\sqrt{(R_r' + R_s)^2 + (X_{rl} + X_{sl})^2}} = \frac{126}{\sqrt{(2.2 + 11.9)^2 + (47.9 + 63.3)^2}} \approx 1.12\text{ A}$$

Now the conductor losses at 20 kHz are

$$(P_{con})_{20\,kHz} = 3(R_r + R_s)I_v^2 = 3(2.2 + 11.9)1.12^2 = 53.06\text{ W}$$

We considered above $V_v \approx V_1$, which is not the case in general, although in some PWM strategies, at some modulation factor values, such a situation may occur.

11.12 TIME-HARMONICS INTERBAR ROTOR CURRENT LOSSES

Earlier in this chapter, we have dealt with space harmonic-induced losses in skewed uninsulated bar rotor cages. This phenomenon shows up also due to time harmonics. At the standard fundamental frequency (50 or 60 Hz), the additional transverse rotor losses due to interbar currents are negligible. In high-frequency motors ($f_1 > 300\,Hz$), these losses count [11]. Also, time harmonics are likely to produce transverse losses for skewed uninsulated bar rotor cages.

To calculate such losses, we may use directly the theory developed in Section 11.6, with $S = 1$ and $\nu = 1$ but for given frequency f_v and voltage V_v.

However, to facilitate the computation process, we adopt here the final results of [11] which calculates the rotor resistance and leakage reactance including the skin effect and the transverse (interbar currents) losses

$$R_r^*(f_v) = \left(R_{bdc}K_{Rb}(f_v) + R_{ere}K_{Re}(f_v)\right)\cdot|\alpha_0|\cos\gamma_0 + 2\pi f_v\left(L_{bdc}K_{Xb}(f_v) + L_{ere}K_{XRe}(f_v)\right)\cdot|\alpha_0|\sin\gamma_0 + X_m(f_v)\cdot\frac{\sin\gamma_0}{\eta^2\left|\dot{K}\right|} \quad (11.115)$$

$$X_{rl}^*(f_v) = 2\pi f_v\left(L_{bdc}K_{Xb}(f_v) + L_{ere}K_{XRe}(f_v)\right)\cdot|\alpha_0|\cos\gamma_0 - \left(R_{bdc}K_{Rb}(f_v) + R_{ere}K_{Re}(f_v)\right)\cdot|\alpha_0|\sin\gamma_0 + X_m(f_v)\cdot\left(\frac{\cos\gamma_0}{\eta^2\left|\dot{K}\right|} - 1\right) \quad (11.116)$$

with

$$\alpha_0 = \frac{4\cdot 3\cdot 2\pi f_v K_{w1}^2}{N_r\cdot\dot{K}};\quad \dot{K} = \left|\dot{K}\right|e^{j\gamma_0};\quad \eta = \frac{\sin\left(\frac{\pi}{N_r}\right)}{\frac{\pi}{N_r}} \quad (11.117)$$

$$\dot{K} = \frac{Z_{be} + R_{ere}}{Z_{be} + R_{de} \cdot \left(\gamma^2\right)} \cdot \left[1 - (A+B) \frac{\sinh \frac{1}{2}\left(\sqrt{\frac{Z_{be}}{R_{de}}} + j\gamma\right)}{\left(\sqrt{\frac{Z_{be}}{R_{de}}} + j\gamma\right)} - (A-B) \frac{\sinh \frac{1}{2}\left(\sqrt{\frac{Z_{be}}{R_{de}}} + j\gamma\right)}{\left(\sqrt{\frac{Z_{be}}{R_{de}}} + j\gamma\right)} \right] \quad (11.118)$$

where γ_0 is the skewing angle.

A and B are defined in (11.75) and (11.76). Z_{be} and R_{de} are defined in (11.73) and refer to rotor bar impedance and equivalent rotor end ring, and bar-tooth wall resistance R_{de} is reduced to the rotor bar.

$X_m(f_v)$ is the airgap reactance for frequency f_v and fundamental pole pitch,

$$X_m(f_v) = X_m(f_1) \cdot \frac{f_v}{f_1} \quad (11.119)$$

In Ref. [11], for a 500Hz fundamental, high speed, 2.2kW motor, the variation of rectangular bar rotor resistance $R_r^*(R_{de})$ for various frequencies was found to have a maximum whose position depends only slightly on frequency (Figure 11.13a).

The rotor leakage reactance X_{rl}^* (R_{de}) levels out for all frequencies at high bar-tooth wall contact resistance values (Figure 11.13b).

We should note that Equations (11.115) and (11.116) and Figure 11.13 refer both to skin effect and transverse losses.

As the skin effect at, say, 9500Hz is very strong and only the upper part of rotor bar is active, the value of transverse resistance R_{de} is to be increased in the ratio

$$R_{de} = R_{dedc} \frac{h}{\delta_{Al}(f_v)} \quad (11.120)$$

This, as shown in Figure 11.13, would read values of R_r^* and X_{rl}^* calculated for $(R_{de})_{dc}$ at the abscissa R_{de} for (11.120).

Also, as X_{rl}^* increases with R_{de}, the harmonic current and conductor losses tend to decrease for large R_{de}.

Increasing the frequency tends to push the transverse resistance R_{de} to larger values, beyond the critical value, and thus, lower interbar currents and losses are to be obtained.

However, such a condition has to be verified up to carrier frequency so that a moderate influence of interbar currents is secured.

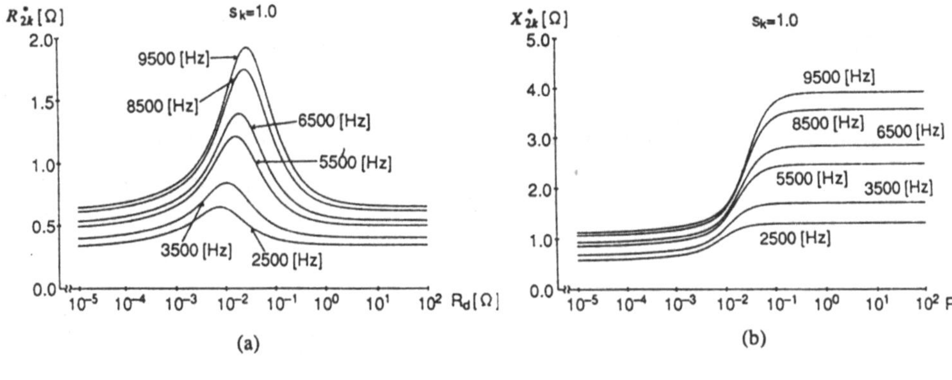

FIGURE 11.13 Rotor cage resistance and leakage reactance versus bar-tooth wall contact resistance R_{de}, for various time harmonics (after Ref. [11]).

11.13 COMPUTATION OF TIME-HARMONIC CORE LOSSES

The computation of core losses for high frequency in IMs is a very difficult task. However, for the case when $\delta_{Fe} < d$ (large skin effect) when all the flux paths are located around stator and rotor slots and on rotor surface in a thin layer (due to airgap flux pulsation caused by stator slot openings), such an attempt may be made easily through analytical methods.

Core losses may occur also due to axial fluxes, in the stack-end laminations because of end-connection high-frequency leakage flux. So we have three types of losses here.

a. Slot wall time harmonic core losses
b. Airgap flux pulsation (zig-zag) time harmonic core losses
c. End connection leakage flux time harmonic core losses

11.13.1 Slot Wall Core Losses

Due to the strong skin effect, the stator and rotor currents are crowded towards the slot opening in each conductor layer (Figure 11.14).

The stator average leakage flux ϕ_{vs} is

$$\phi_{vs} = \frac{1}{W_1}\left[\frac{L_{slslot}(f_v)I_v}{\frac{N_s}{3}}\right]\cdot\frac{1}{\left(\frac{h_s}{\xi}\right)}[\text{Wb/m}] \tag{11.121}$$

where W_1 – turns per phase
$N_s/3$ – slots per stator phase
h_s/ξ – the total height per m skin depths
m conductor layers in slot

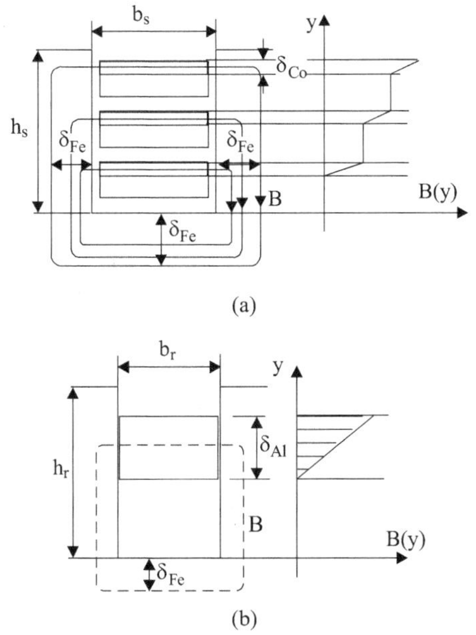

FIGURE 11.14 Stator (a) and rotor (b) slot leakage flux path and flux density distribution.

From Ref. [14], the surface core losses, with skin effect considered for the stator, are

$$P_{sw} = \frac{\phi_{vs}^2}{\delta_{Fe}^3 \sigma_{Fe} \mu_{Fe}^2} \cdot \frac{\left(\sinh\left(\frac{2d}{\delta_{Fe}}\right) - \sin\left(\frac{2d}{\delta_{Fe}}\right)\right)}{\left(\cosh\left(\frac{2d}{\delta_{Fe}}\right) - \cos\left(\frac{2d}{\delta_{Fe}}\right)\right)} \cdot K_{er} N_s l_{stack} (2h_s + b_s)\delta_{Fe} \tag{11.122}$$

K_{er} is the reaction-field factor: $K_{er} \approx \varphi'$ (see (11.112)) for

$$\xi = \frac{2d}{\delta_{Fe}} \tag{11.123}$$

In a similar way, we may proceed for the rotor slot, but here, m = 1 (conductors per slot). The rotor average slot leakage flux ϕ_{vr} is

$$\phi_{vr} = \frac{1}{W_2}\left[\frac{L_{rlslot}(f_v) I_v}{\frac{N_r}{m_2}}\right] \cdot \frac{1}{\left(\frac{h_r}{\xi}\right)} (\text{Wb/m}) \tag{11.124}$$

W_2 – turns per rotor phase and m_2 – rotor equivalent phases. For a cage rotor, we may consider $W_2 = W_1$ and $m_2 = 3$ as $L_{rlslot}(\nu)$ is already reduced to the stator.

The rotor slot wall losses P_{rw} are

$$P_{rw} = \frac{\phi_{vr}^2}{\delta_{Fe}^3 \sigma_{Fe} \mu_{Fe}^2} \cdot \frac{\left(\sinh\left(\frac{2d}{\delta_{Fe}}\right) - \sin\left(\frac{2d}{\delta_{Fe}}\right)\right)}{\left(\cosh\left(\frac{2d}{\delta_{Fe}}\right) - \cos\left(\frac{2d}{\delta_{Fe}}\right)\right)} \cdot K_{er} N_r l_{stack} (2h_r + b_r)\delta_{Fe} \tag{11.125}$$

11.13.2 Zig-Zag Rotor Surface Losses

The time harmonic airgap flux density pulsation on the rotor tooth heads, similar to zig-zag flux, produces rotor surface losses.

The maximum airgap flux density pulsation B_{vK} due to stator slot opening is

$$B_{vK} = \frac{\pi}{4} \frac{\beta}{2-\beta} B_{vg} \tag{11.126}$$

The factor β (b_{os}/g) has been defined in Chapter 5 (Figure 5.4) or Table 10.1. It varies from 0 to 0.41 for b_{os}/g from 0 to 12. The frequency of B_{vK} is f_v ($N_s/p_1 \pm 1$). B_{vg} is the f_v frequency airgap space fundamental flux density.

To calculate B_{vg}, we have to consider that all the airgap flux is converted in leakage rotor flux. So they are equal to each other.

$$\left(\frac{2}{\pi} B_{vg} \tau l_{stack}\right) \frac{W_1 K_{w1}}{\sqrt{2}} = I_v L_{rv} \tag{11.127}$$

Based on (11.126 and 11.127) in Ref. [14], the following result has been obtained for zig-zag rotor surface losses:

$$P_{zr} = 2(\pi D\delta_{Fe})\frac{1}{8}\frac{\pi^2}{16}\left(\frac{\beta}{2-\beta}\right)^2 B_{vg}^{\ 2} l_{stack}^{\ 2}\cdot\frac{K_{er}C_{v1r}}{\delta_{Fe}^{\ 3}\sigma_{Fe}\mu_{Fe}^{\ 2}}\cdot\frac{\left(\sinh\left(\frac{2d}{\delta_{Fe}}\right)-\sin\left(\frac{2d}{\delta_{Fe}}\right)\right)}{\left(\cosh\left(\frac{2d}{\delta_{Fe}}\right)-\cos\left(\frac{2d}{\delta_{Fe}}\right)\right)} \quad (11.128)$$

$$C_{v1r} = \left(\frac{\sin\left(\frac{K\pi K_2}{2R'}\right)}{\left(\frac{K\pi K_2}{2R'}\right)}\right)^2 + \left(\frac{\sin\left(\frac{K\pi K_2}{R'}\right)}{\left(\frac{K\pi K_2}{R'}\right)}\right)^2 - \frac{\sin\left(\frac{K\pi K_2}{2R'}\right)}{\left(\frac{K\pi K_2}{2R'}\right)}\cdot\frac{\sin\left(\frac{K\pi K_2}{R'}\right)}{\left(\frac{K\pi K_2}{R'}\right)}\cdot\cos\frac{K\pi K_2}{2R'} \ll 1 \quad (11.129)$$

$$K = \frac{N_s}{p_1} \pm 1; \quad K_2 = \frac{(t_r - b_{or})}{t_r}; \quad R' = \frac{N_r}{2p_1} \quad (11.130)$$

where t_r – rotor slot pitch; b_{or} – rotor slot opening.

Zig-zag rotor surface losses tend to be negligible, at least in small power motors. End-connection leakage core losses proved negligible in small motors. In high-power motors, only 3D FEM are able to produce trustworthy results.

In fact, it seems practical to calculate only the slot wall time harmonics core losses in the stator and rotor. For the motor, in Example 11.6 in Ref. [14], it was found that $P_{sw} + P'_{rw} = 6.8 + 7.43 = 14.23$ W. This is about 4 times less than conductor losses at 20 kHz.

11.14 LOSS COMPUTATION BY FEM BASICS

The FEM, even in its 2D version, allows for the computation of magnetic field distribution once the stator and rotor currents are known. Under no load, only the stator current waveform fundamental is required. It is thus possible to calculate the additional currents induced in the rotor cage by conjugating field distribution with circuit equations.

The core loss may be calculated so far only from the distribution of field in the machine with zero electrical conductivity in the iron core. This is a strong approximation, especially at high time harmonic frequencies (see previous paragraphs). In Refs. [15,16], such a computation approach is followed for both sinusoidal and PWM voltages. Sample results of losses for a 37 kW motor on no-load are shown in Figure 11.15 [16].

More importantly, the flux density radial (B_r) and tangential (B_t) flux density hodographs in three points a, b, and c (Figure 11.16c) are shown in Figure 11.16a and b [16].

This is proof that the rotor slot opening (point a) and tooth (point b) experience A.C. field while point c (slot bottom) experiences a quasi-travelling field. Although such knowledge is standard, a quantitative proof is presented here.

When the motor is under load, for sinusoidal voltage supply, FEM has been also used to calculate the losses for a skewed bar cage rotor machine. Insulated bars have been used for the computation. This time again, but justified, the iron skin effect was neglected as time harmonics do not exist. Semiempirical loss formulas, as used with analytical models, are still used with FEM.

The influence of skewing in the rotor is considered separately, and then, by using the coupling field-circuit FEM, the stack was sliced into 5–8 axial segments properly shifted to consider the skewing effects [17].

The already documented axial variation of airgap and core flux density due to skewing changes the balance of losses by increasing the stator fundamental and especially the rotor (stray) core losses and decreasing the rotor stray (additional) cage losses.

In low-power induction motors, where conductor losses dominate, skewing tends to reduce total losses on load, while for large machines where core loss is relatively more important, the total losses on load tend to increase slightly due to skewing [17].

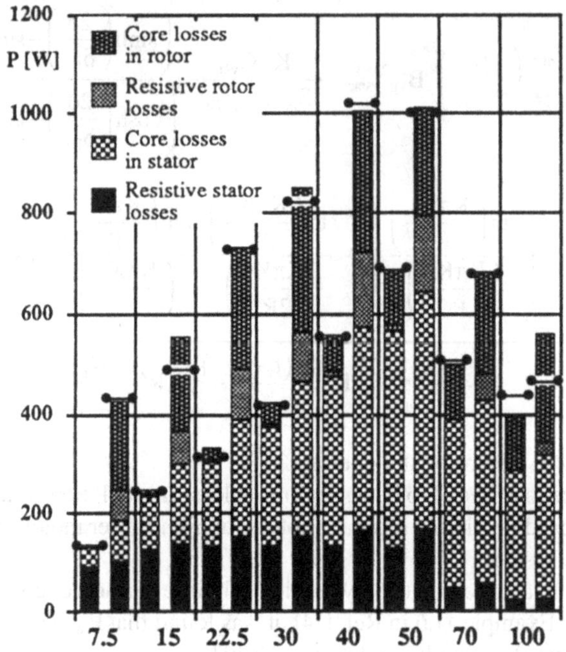

FIGURE 11.15 No-load losses (37 kW IM) at 7.5 Hz, …, 100 Hz fundamental frequency. Left column for sinusoidal voltage and right column for PWM voltage. Measured losses marked by cross bars (after Ref. [16]).

The network – field coupled time – stepping finite element 2D model with axial stack segmentation of skewed-rotor cage IMs has been used to include also the interbar currents [18].

The fact that such complex problems can be solved by quasi-2D FEM today is encouraging as rather reasonable computation time is required. However, all the effects are mixed and no easy way to derive design hints seems in sight.

For other apparently simpler problems, however, such that the usage of magnetic wedges for stator slots in large machines, FEM is extremely useful [19]. A notable reduction of rotor stray core losses is obtained. Consequently, higher efficiency for open stator slots is expected.

A similar situation occurs when the hysteresis losses in a cage IM are calculated by investigating the IM from rated to zero positive and negative slips [20].

A pertinent review of FEM usage in the computation of IM performance (and losses) is given in Refs. [21–24].

11.15 SUMMARY

- Conductor, core, frictional, and windage losses occur in IMs.
- Conductor and core losses constitute electromagnetic losses.
- Electromagnetic losses may be divided into fundamental and harmonic (space and/or time harmonics) losses.
- Fundamental conductor losses depend on skin effect, temperature, and the machine power and specific design.
- Fundamental core losses depend on the airgap flux density, yoke and teeth flux densities, and the supply voltage frequency.
- In a real IM, fundamental electromagnetic losses hardly exist separately.
- Nonfundamental electromagnetic losses are due to space or/and time harmonics.
- The additional (stray) losses, besides the fundamental, are caused by space airgap flux density harmonics or by voltage time harmonics (when PWM converters are used to feed the IM).

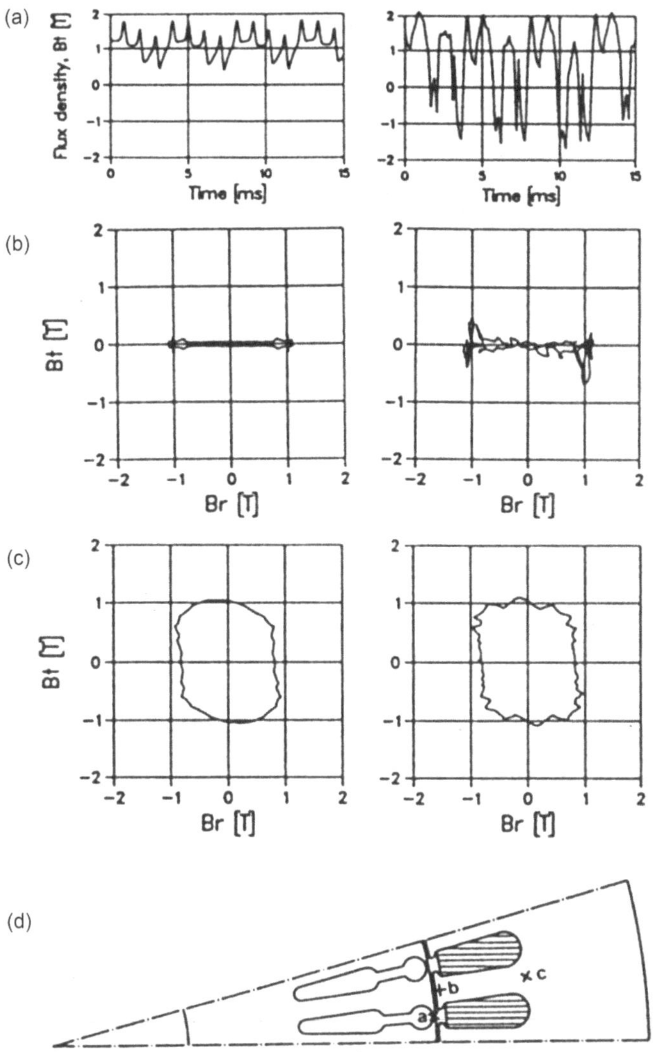

FIGURE 11.16 No-load flux density hodograph in points a, b, and c, at 40 Hz, left – sinusoidal voltage, right – PWM voltage (a–c); abc prints (d) (after Ref. [16]).

- The airgap flux density space harmonics are due to mmf space harmonics, airgap magnetic conductance space harmonics due to slot opening, and leakage or main flux path magnetic saturation.
- No-load stray losses are space harmonics losses at no load. They are mainly surface, tooth flux pulsation, and tooth flux pulsation cage losses.
- For phase-belt (mmf) harmonics, 5, 7, 11, 13 and up to the first slot opening harmonic ($N_s/p_1 \pm 1$), the skin depth in laminations is much larger than lamination thickness. So the reaction of core eddy currents on airgap harmonic field is neglected. This way, the rotor (stator) surface core losses are calculated.
- As the number of stator and rotor slots are different from each other $N_s \neq N_r$, the stator and rotor tooth flux pulsates with N_r and N_s pole pairs, respectively, per revolution. Consequently, tooth flux pulsation core losses occur both in the stator and rotor teeth.
- However, such tooth flux pulsations and losses are attenuated by the corresponding currents induced in the rotor cage.

- For the rotor slot skewing by one stator slot pitch, the reaction of cage currents to rotor tooth flux pulsation is almost zero and, thus, large rotor tooth flux pulsation core losses persist (undamped).
- For straight slot rotors, the rotor tooth flux pulsation produces, as stated above, no-load tooth flux pulsation cage losses. They are not to be neglected, especially in large power motors.
- All space harmonics losses under load are called stray load losses.
- Under load, the no-load stray losses are "amplified". A load amplification factor C_{load} is defined and used to correct the no-load stator and rotor tooth flux pulsation core losses. A distinct load correction factor is used for calculating rotor tooth flux pulsation cage losses. For full-pitch stator windings when calculating stray losses, besides the first slot harmonics $N_s \pm p_1$, the first phase-belt harmonics (5, 7) are contributing to losses. In skewed rotors, the load stray cage losses are also corrected by a special factor.
- Cast aluminium cages are used with low-to-medium power IMs. Their bars are uninsulated from slot walls, and thus, a transverse (cross-path) resistance between bars through iron occurs. In this case, interbar current losses occur especially for skewed rotors. Thus, despite of the fact that with proper skewing the stray cage losses are reduced, notable interbar current losses may occur making the skewing less effective.
- Pertinent work on interbar current effects backed by test results is given in Ref. [20,21].
- There is a critical transverse resistance for which the interbar current losses are maximum. This condition is to be avoided by proper design. Interbar current losses occur both on no load and on load.
- A number of rules to reduce stray losses are presented. The most interesting is $N_s > N_r$ where even straight rotor slots may be used after safe starting is secured.
- With the advent of power electronics, supply-voltage time harmonics occur. With PWM static power converters, around carrier frequency, the highest harmonics occur unless random PWM is used. Frequencies up to 20 kHz or even more occur this way.
- Both stator and rotor conductor losses, due to these voltage time harmonics, are heavily influenced by the skin effect (frequency).
- In general, as the frequency rises over 2–3 kHz, for given time harmonic voltage, the conductor losses decrease slightly with frequency while core losses do increase with frequency.
- The computation of core losses at high time frequencies (up to 20 kHz or more) is made accounting for the skin depth in iron δ_{Fe} as all field occurs around slots and on the rotor surface in a thin layer (δ_{Fe} – thick). This way, slot wall and rotor surface core losses are calculated. Only slot wall core losses are not negligible and they represent about 20% to 30% of all time harmonic losses.
- FEM has been applied recently to calculate all no-load or load losses, thus including implicitly the space harmonics losses. Still, the core losses are determined with analytical expressions where the local flux densities variation in time is considered [22, 23]. For field distribution computation, the laminated core electrical conductivity is considered zero. So the computation of time harmonic high-frequency (20 kHz) core losses including the iron skin depth is not available yet with FEM. The errors vary from 5% to 30%.
- However, the effect of skewing, interbar currents, magnetic wedges, and relative number of slots N_s/N_r has been successfully investigated by field-coupled circuit 2D FEMs for reasonable amounts of computation time [24].
- New progress with 3D FEM investigation of IMs is expected in the near future.
- Measurements of losses will be dealt with separately in the chapter dedicated to IM testing (in Volume 2).

REFERENCES

1. E. M. Olin, Determination of power efficiency of rotating electric machines: Summation of losses versus input – Output method, *AIEE Transaction*, Vol. 31, Part. 2, 1912, pp. 1695–1719.
2. L. Dreyfus, The additional core losses in A.C. synchronous machines, *Elektrotechnik und Maschinenbau*, Vol. 45, 1927, pp. 737–756 (in German).
3. P. L. Alger, G. Angst, E. J. Davies, Stray load losses in polyphase induction machines, *AIEE Transactions*, Vol. 78, 1957, pp. 349–357.
4. N. Christofides, Origin of load losses in induction machines with cast aluminum rotors, *Proceedings of the IEEE*, Vol. 112, 1965, pp. 2317–2332.
5. A. Odok, Stray load losses and stray torques in induction machines, *AIEE Transactions*, Vol. 77, Part 2, 1958, pp. 43–53.
6. B. Heller, V. Hamata, *Harmonic Field Effects in Induction Machines*, Elsevier Scientific, Amsterdam; New York, 1977.
7. A. A. Jimoh, S. R. D. Findlay, M. Poloujadoff, Stray losses in induction machines, Part 1 and 2, *IEEE Transactions*, Vol. PAS – 104, No. 6, 1985, pp. 1500–1512.
8. C. N. Glen, Stray load losses in induction motors: A challenge to academia, *Record of EMD*, 1977, IEE Publication, No. 444, pp. 180–184.
9. A. M. Odok, Stray load losses and stray torques in induction machines, *AIEE Transactions*, Vol. 77, No. 4, 1958, pp. 43–53.
10. R. Weppler, A contribution to the design of induction motors with uninsulated – Cage rotors, *Archiv für Electrotechnik*, Vol. 50, No. 4, 1966, pp. 248–252 (in German).
11. K. Matsuse, T. Hayashida, I. Miki, H. Kubota, Y. Yoshida, Effect of crosspath resistance between adjacent rotor bars on performance of inverter – Fed high speed induction motor, *IEEE Transactions*, Vol. IA-30, No. 3, 1994, pp. 621–627.
12. K. Oberretl, 13 Rules for minimum stray losses in induction machines, Bull – Oerlikon, 1969, No. 389/390, pp. 2–12.
13. J. Singh, Harmonic analysis and loss comparison of microcomputer – Based PWM strategies for induction motor drive, *EMPS Journal*, Vol. 27, No. 10, 1999, pp. 1129–1140.
14. D. W. Novotny, S. A. Nasar, High frequency losses in induction motors, Part II, Contract no. MAG 3-940, Final Report, University of Wisconsin, ECE Dept, 1991.
15. G. Bertotti et al., An improved estimation of core losses in rotating electrical machines, *IEEE Transactions*, Vol. MAG-27, 1991, pp. 5007–5009.
16. A. Arkkio, A. Miemenmaa, Estimation of losses in cage induction motors using FEM, *Record of ICEM*, 1992, Vol. 1, pp. 317–321.
17. C. I. McClay, S. Williamson, The variation of cage motor losses with skew, *Record of IEEE – IAS*, 1998, Vol. 1, pp. 79–86.
18. S. L. Ho, H. L. Li, W. N. Fu, Inclusion of interbar currents in the network field-coupled Time-stepping FEM of skewed rotor induction motors, *IEEE Transactions*, Vol. MAG – 35, No. 5, 1999, pp. 4218–4225.
19. T. J. Flack, S. Williamson, On the possible case of magnetic slot wedges to reduce iron losses in cage motors, *Record of ICEM*, 1998, Vol. 1, pp. 417–422.
20. D. G. Dorrell, T. J. E. Miller, C. B. Rasmussen, Inter-bar currents in induction machines, *IEEE Transactions*, Vol. IA-39, No. 3, 2003, pp. 677–684.
21. R. Carlson, C. A. da Silva, N. Sadowski, Y. Lefevre, M. Lajoie-Mazenc, Analysis of inter-bar currents on the performance of cage induction motors, *IEEE Transactions*, Vol. IA-39, No. 6, 2003, pp. 1674–1680.
22. S. Yarase, H. Kimata, Y. Okazaki, S. Hashi, A simple predicting method for magnetic losses of electrical steel sheets under arbitrary induction waveform, *IEEE Transactions*, Vol. MAG-41, No. 11, 2005, pp. 4365–4367.
23. W. A. Roshen, Magnetic losses for nonsinusoidal waveforms found in A.C. motors, *IEEE Transactions*, Vol. PE-21, No. 4, 2006, pp. 1138–1141.
24. K. Yamazaki, Y. Haruishi, Stray load loss analysis of induction motor comparison of measurement due to IEEE Standard 112 and direct calculation by finite element method, *IEEE Transactions*, Vol. IA-40, No. 2, 2004, pp. 543–549.

12 Thermal Modelling and Cooling

12.1 INTRODUCTION

Besides electromagnetic, mechanical and thermal designs are equally important.

Thermal modelling of an electric machine is in fact more non-linear than electromagnetic modelling. Any electric machine design is highly thermally constrained.

The heat transfer in an induction motor depends on the level and location of losses, machine geometry, and the method of cooling [1].

Electric machines work in environments with temperature that varies, say from –20°C to 50°C, from 20° to 100° in special applications.

The thermal design should make sure that the motor winding temperatures do not exceed the limit for the pertinent insulation class in the worst situation. Heat removal and the temperature distribution within the induction motor are the two major objectives of thermal design. Finding the highest winding temperature spots is crucial to insulation (and machine) working life.

The maximum winding temperature in relation to insulation classes is shown in Table 12.1.

Practice has shown that increasing the winding temperature over the insulation class limit reduces the insulation life L versus its value L_0 at the insulation class temperature (Figure 12.1):

$$\mathrm{Log}\,L \approx a - bT \tag{12.1}$$

It is thus very important to set the maximum winding temperature as a design constraint. The highest temperature spot is located in general in the stator end connections. The rotor cage bars experience a larger temperature, but they are not, in general, insulated from the rotor core. If they are, the maximum (insulation class dependent) rotor cage temperature also has to be observed.

The thermal modelling depends essentially on the cooling approach.

12.2 SOME AIR COOLING METHODS FOR IMS

For induction motors, there are few main classes of cooling systems

- Totally enclosed design with no (zero air speed) ventilation (TENV)
- Drip-proof axial internal cooling
- Drip-proof radial internal cooling
- Drip-proof radial-axial cooling.

TABLE 12.1
Insulation Classes

Insulation Class	Typical Winding Temperature Limit [°C]
Class A	105
Class B	130
Class F	155
Class H	180

HEATING, COOLING, AND VENTILATING

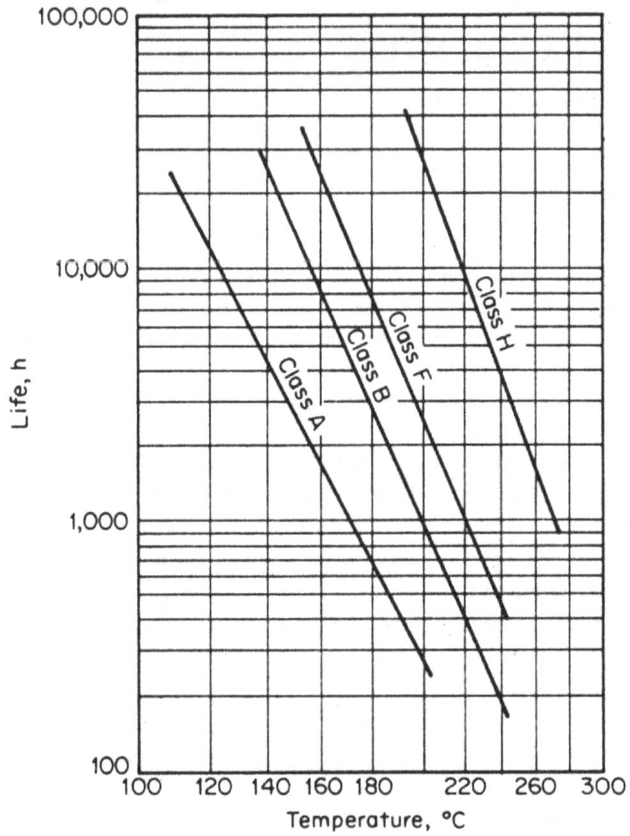

FIGURE 12.1 Insulation life versus temperature rise.

In general, fan air cooling is typical for induction motors. Only for very large powers, a second heat exchange medium (forced air or liquid) is used in the stator to transfer the heat to the ambient temperature.

TENV induction motors are typical for special servos to be mounted on machine tools etc., where limited space is available. It is also common for some static power converter-fed induction machines (IMs) that operate at large loads for extended periods of time at low speeds to have an external ventilator running at constant speed, to maintain high cooling in all conditions.

The totally enclosed motor cooling system with external ventilator only (Figure 12.2b) has been extended lately to hundreds of kW by using finned stator frames.

Radial and radial-axial cooling systems (Figure 12.2c and d) are in favour for medium and large powers.

However, axial cooling with internal ventilator and rotor, stator axial channels in the core, and special rotor slots seem to gain ground for very large powers as it allows for lower rotor diameter, and finally, greater efficiency is obtained, especially with two-pole motors (Figure 12.3) [2].

The rotor slots may be provided with their axial channels to facilitate a kind of direct cooling (Figure 12.3).

The rather complex (anisotropic) structure of the IM for all cooling systems presented in Figures 12.2 and 12.3 suggests that the thermal modelling has to be rather difficult to build.

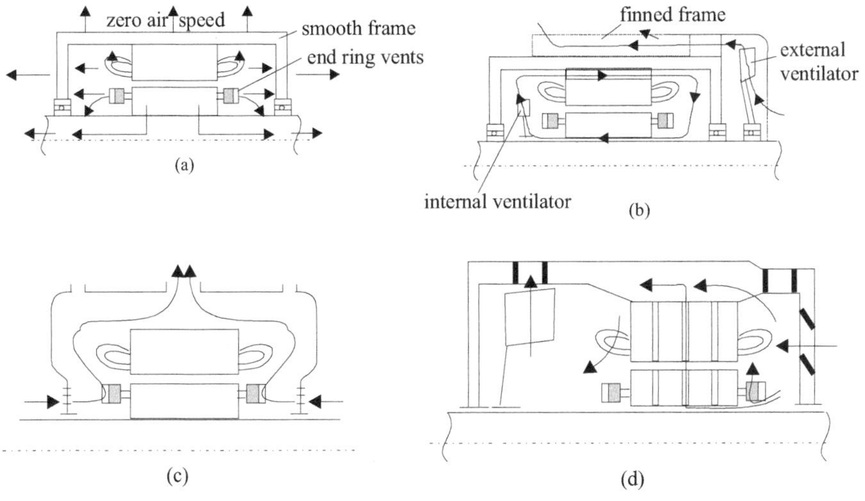

FIGURE 12.2 Cooling methods for IMs: (a) TENV; (b) totally enclosed motor with internal and external ventilator; (c) radially cooled IM; and (d) radial–axial cooling system.

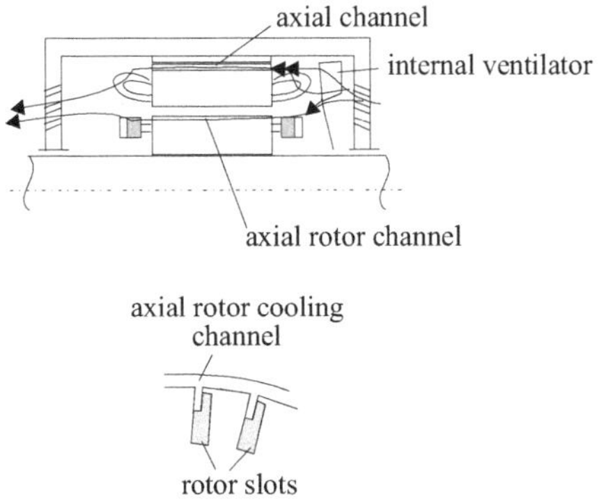

FIGURE 12.3 Axial cooling of large IMs.

There are thermal circuit and distributed (FEM) models. Thermal circuit models are similar to electric circuits, and they may be used both for thermal steady state and transients. They are less precise but easy to handle and require a smaller computation effort. In contrast, distributed (FEM) models are more precise but require large amounts of computation time.

We will first define the elements of thermal circuits based on the three basic methods of heat transfer: conduction, convection, and radiation.

12.3 CONDUCTION HEAT TRANSFER

Heat transfer is related to thermal energy flow from a heat source to a heat sink.

In electric (induction) machines, the thermal energy flows from the windings in slots to laminated core teeth through the conductor insulation and slot line insulation.

On the other hand, part of the thermal energy in the end-connection windings is transferred through thermal conduction through the conductors axially towards the winding part in slots. A similar heat flow through thermal conduction takes place in the rotor cage and end rings.

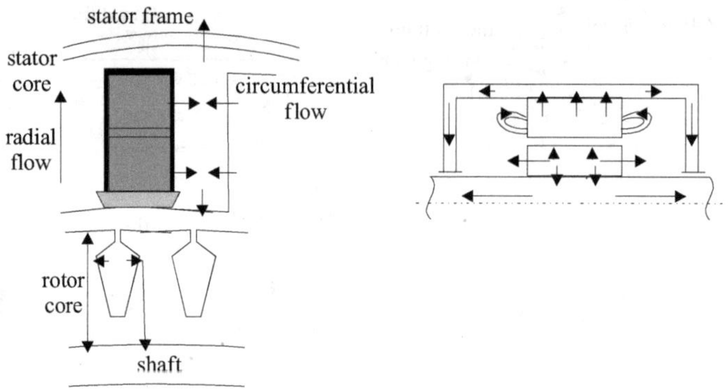

FIGURE 12.4 Heat conduction flow routs in the IM.

There is also thermal conduction from the stator core to the frame through the back core iron region and from the rotor cage to the rotor core, respectively, to shaft and axially along the shaft. Part of the conduction heat flow through slot insulation to core will be directed axially through the laminated core. The presence of lamination insulation layers will make the thermal conduction along the axial direction more difficult. In long stack IMs, axial temperature differentials of a few degrees (less than 10°C in general), (Figure 12.4), occur.

So, to a first approximation, the axial heat flow may be neglected.

Second, after accounting for conduction heat flow from windings in slots to the core teeth, the circumferential symmetry of the machine makes the neglecting of circumferential temperature variation possible.

So we end up with a one-dimensional temperature variation, along the radial direction. For this crude approximation, defining thermal conduction, convection, and radiation and their equivalent circuit becomes a rather simple task.

The Fourier's law of conduction may be written, for steady state, as

$$\nabla(-K\Delta\theta) = q \tag{12.2}$$

where q – heat generation rate per unit volume (W/m^3)

K is thermal conductivity (W/m°C) and θ is local temperature.

For one-dimensional heat conduction, Equation (12.2), with constant thermal conductivity K, becomes

$$-K\frac{\partial^2\theta}{\partial x^2} = q \tag{12.3}$$

A basic heat conduction element (Figure 12.5) shows that power Q transported along distance l of cross-section A is

$$Q \approx q \cdot l \cdot A \tag{12.4}$$

with q, A – constant along distance l.

The thermal conduction resistance R_{con} may be defined as similar to electrical resistance.

$$R_{con} = \frac{1}{KA}[°C/W] \tag{12.5}$$

Temperature takes the place of voltage, and power (losses) replaces the electrical current.

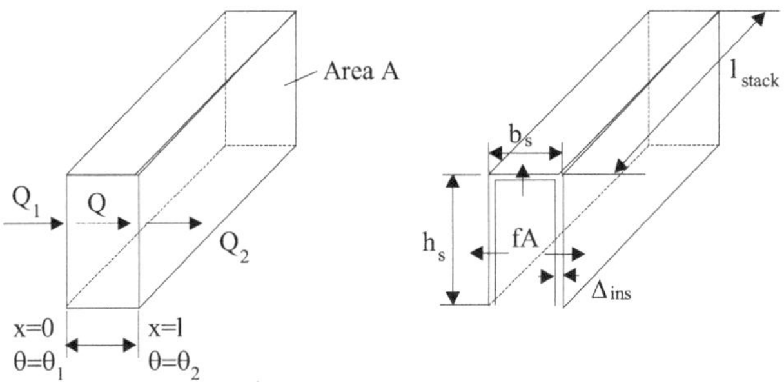

FIGURE 12.5 One-dimensional heat conduction.

For a short l, the Fourier's law in differential form yields

$$f \approx -K\frac{\Delta\theta}{\Delta x}; \quad f\text{ - heat flow density}\left[W/m^2\right] \tag{12.6}$$

If the heat source is in a thin layer,

$$f = \frac{p_{cos}}{A} \tag{12.7}$$

p_{cos} in watts is the electric power producing losses and A the cross-section area.

For the heat conduction through slot insulation Δ_{ins} (total, including all conductor insulation layers from the slot middle (Figure 12.5)), the conduction area A is

$$A = (2h_s + b_s)l_{stack}N_s; \quad N_s - \text{slots/stator} \tag{12.8}$$

The temperature differential between winding in slots and the core teeth $\Delta\theta_{Co}$ is

$$\Delta\theta_{cos} = p_{cos}R_{con}; \quad R_{con} = \frac{\Delta_{ins}}{AK} \tag{12.9}$$

In well-designed IMs, $\Delta\Theta_{cos} < 10°C$, with notably smaller values for small-power induction motors.

The improvements of insulation materials in terms of thermal conductivity and in thickness reduction have been decisive factors in reducing the slot insulation conductor temperature differential. Thermal conductivity varies with temperature and is constant only to a first approximation. Typical values are given in Table 12.2. The low axial thermal conductivity of the laminated cores is evident.

12.4 CONVECTION HEAT TRANSFER

Convection heat transfer takes place between the surface of a solid body (say stator frame) and a fluid (air, for example) by the movement of the fluid.

The temperature of a fluid (air) in contact with a hotter solid body rises and sets a fluid circulation, and thus, heat transfer to the fluid occurs.

The heat flow out of a body by convection is

$$q_{conv} = hA\Delta\theta \tag{12.10}$$

where A is the solid body area in contact with the fluid; $\Delta\theta$ is the temperature differential between the solid body and bulk of the fluid, and h is the convection heat coefficient ($W/m^{2\circ}C$).

TABLE 12.2
Thermal Conductivity

Material	Thermal Conductivity (W/m°C)	Specific Heat Coefficient C_s (J/Kg/°C)
Copper	383	380
Aluminium	204	900
Carbon steel	45	
Motor grade steel	23	500
Si steel lamination		490
• Radial	20–30	
• Axial	2.0	
Mica sheet	043	-
Varnished cambric	2.0	-
Press board Normex	0.13	-

The convection heat transfer coefficient depends on the velocity of the fluid, fluid properties (viscosity, density, and thermal conductivity), the solid body geometry, and orientation. For free convention (zero forced air speed and smooth solid body surface [2]),

$$\begin{aligned} h_{Co} &\approx 2.158(\Delta\theta)^{0.25} \text{ vertical up- } W/\left(m^2{}^\circ C\right) \\ h_{Co} &\approx 0.496(\Delta\theta)^{0.25} \text{ vertical down- } W/\left(m^2{}^\circ C\right) \\ h_{Co} &\approx 0.67(\Delta\theta)^{0.25} \text{ horizontal- } W/\left(m^2{}^\circ C\right) \end{aligned} \tag{12.11}$$

where $\Delta\theta$ is the temperature differential between the solid body and the fluid.

For $\Delta\theta = 20°C$ (stator frame $\theta_1 = 60°C$, ambient temperature $\theta_2 = 40°C$) and vertical – up – surface,

$$h_{Co} = 2.158(60-40)^{0.25} = 4.5\ W/\left(m^2{}^\circ C\right)$$

When air is blown with a speed U along the solid surfaces, the convection heat transfer coefficient h_c is

$$h_C{}^\circ(u) = h_{Co}\left(1+K\sqrt{U}\right) \tag{12.12}$$

with K = 1.3 for perfect air blown surface; K = 1.0 for the winding end-connection surface, K = 0.8 for the active surface of rotor, and K = 0.5 for the external stator frame.

Alternatively,

$$h_C(u) = 1.77\frac{U^{0.75}}{L^{0.25}}\ W/\left(m^2{}^\circ C\right) \quad \text{for } U > 5 \text{ m/s} \tag{12.13}$$

U in m/s and L is the length of surface in m.

For a closed-air blowed surface – inside the machine,

$$h_C{}^c(u) = h_{Co}\left(1+K\sqrt{U}\right)(1-a/2); \quad a = \frac{\theta_{air}}{\theta_a} \tag{12.14}$$

θ_{air} – local air heating; θ_a – heating (temperature) of solid surface.

In general, $\theta_{air} = 35°C–40°C$ while θ_a varies with machine insulation class. So, in general, $a < 1$.

For convection heat transfer coefficient in axial channels of length L, (12.13) is to be used.

In radial cooling channels, $h_c^c(U)$ does not depend on the channel's length but only on speed.

$$h_c^c(U) \approx 23.11U^{0.75}\ W/\left(m^2{}^\circ C\right) \tag{12.15}$$

12.5 HEAT TRANSFER BY RADIATION

Between two bodies at different temperatures, there is a heat transfer by radiation. One body radiates heat and the other absorbs heat. Bodies which do not reflect heat, but absorb it, are called black bodies.

Energy radiated from a body with emissivity ε to black surroundings is

$$q_{rad} = \sigma\varepsilon A\left(\theta_1{}^4 - \theta_2{}^4\right) = \sigma\varepsilon A(\theta_1 + \theta_2)\left(\theta_1{}^2 + \theta_2{}^2\right)(\theta_1 - \theta_2) \tag{12.16}$$

σ – Boltzmann's constant, $\sigma = 5.67 \cdot 10^{-8}\ W/(m^2K^4)$; ε – emissivity, for a black painted body, ε = 0.9; A – radiation area.

In general, for IMs, the radiated energy is much smaller than the energy transferred by convection, except for totally enclosed natural ventilation (TENV) or for class F(H) motor with very hot frame (120°–150°C).

For the case when $\theta_2 = 40°$ and $\theta_1 = 80°C, 90°C, 100°C$, ε = 0.9, $h_{rad} = 7.67, 8.01, 8.52\ W/(m^{2\circ}C)$.

For TENV with $h_{Co} = 4.56\ W/(m^{2\circ}C)$ (convection), the radiation is superior to convection, and thus, it cannot be neglected. The total (equivalent) convection coefficient

$$h_{(c+r)0} = h_{Co} + h_{rad} \geq 12\ W/\left(m^2{}^\circ C\right)$$

The convection and radiation combined coefficients $h_{(c+r)0} \approx 14.2\ W/(m^{2\circ}C)$ for steel unsmoothed frames, $h_{(c+r)0} = 16.7\ W/(m^{2\circ}C)$ for steel smoothed frames, and $h_{(c+r)0} = 13.3\ W/(m^{2\circ}C)$ for copper/aluminium or lacquered or impregnated copper windings.

In practice, for design purposes, this value of h_{Co}, which enters Equations (12.12–12.14), is, in fact, $h_{(c+r)0}$, the combined convection radiation coefficient.

It is well understood that the heat transfer is three dimensional, and as K, h_c, and h_{rad} are not constants, the heat flow, even under thermal steady state, is a very complex problem. Before advancing to more complex aspects of heat flow, let us work out a simple example.

Example 12.1 One-Dimensional Simplified Heat Transfer

In an induction motor with $p_{Co1} = 500$ W, $p_{Co2} = 400$ W, and $p_{iron} = 300$ W, the stator slot perimeter $2h_s + b_s = (50 + 8)$ mm, 36 stator slots, stack length $l_{stack} = 0.15$ m, an external frame diameter $D_e = 0.30$ m, finned area frame (4 to 1 area increase by fins), frame length 0.30 m, let us calculate the winding in slots temperature and the frame temperature, if the air temperature increase around the machine is 10°C over the ambient temperature of 30°C and the slot insulation total thickness is 0.8 mm. The ventilator is used and the end connection/coil length is 0.4.

Solution

First, the temperature differential of the windings in slots has to be calculated. We assume here that all rotor loss heat cross the airgap, and through the stator core, it flows towards the stator frame.

In this case, the stator winding in slot temperature differential is (12.3)

$$\Delta\theta_{cos} = \frac{\Delta_{ins} p_{Co1}\left(1 - \dfrac{l_{endcon}}{l_{coil}}\right)}{K_{ins} N_s (2h_s + b_s) l_{stack}} = \frac{0.8 \cdot 10^{-3} \cdot 500 \cdot 0.6}{2.0 \cdot 36 \cdot 0.058 \cdot 0.15} = 3.83°C$$

Now we consider that stator winding in slot losses, rotor cage losses, and stator core losses produce heat that flows radially through stator core by conduction without temperature differential (infinite conduction!).

Then all these losses are transferred to ambient temperature through the motor frame through combined free convection and radiation.

$$\theta_{core} - \theta_{air} = \frac{q_{total}}{h_{(c+r)0} A_{frame}} = \frac{(500+400+300)}{14.2 \cdot \pi \cdot 0.30 \cdot 0.3 \cdot (4/1)} = 74.758°C$$

with $\theta_{air} = 40°$, $\theta_{ambient} = 30°$, the frame (core) temperature $\theta_{core} = 40 + 74.758 = 114.758°C$, and the winding in slots temperature $\theta_{cos} = \theta_{core} + \Delta\theta_{cos} = 114.758 + 3.83 = 118.58°C$.

In such TENV IMs, the unventilated stator winding end turns are likely to experience the highest temperature spot. However, it is not at all simple to calculate the end-connection temperature distribution.

12.6 HEAT TRANSPORT (THERMAL TRANSIENTS) IN A HOMOGENOUS BODY

Although the IM is not a homogenous body, let us consider the case of a homogenous body – where temperature is the same all over.

The temperature of such a body varies in time if the heat produced inside, by losses in the induction motor, is applied at a certain point in time – as after starting the motor. The heat balance equation is

$$\underset{\substack{\text{losses}\\\text{per unit}\\\text{time in W}}}{P_{loss}} = \underset{\substack{\text{heat accumulation}\\\text{in the body}}}{Mc_t \frac{d(T-T_0)}{dt}} + \underset{\substack{\text{heat transfer from the body}\\\text{through convection, conduction, radiation}}}{A\, h_{\substack{cond\\(conv)}} (T-T_0)} \quad (12.17)$$

M – body mass (in Kg), c_t – specific heat coefficient (J/(Kg·°C))
A – area of heat transfer from (to) the body
h – heat transfer coefficient.
Denoting by

$$C_t = Mc_t \text{ and } R_{\substack{conv\\(rad)}} = \frac{1}{Ah}; \left(R_{cond} = \frac{1}{KA}\right) \quad (12.18)$$

Equation (12.17) becomes

$$P_{loss} = C_t \frac{d(T-T_0)}{dt} + \frac{(T-T_0)}{R_t} \quad (12.19)$$

This is similar to a R_t and C_t parallel electric circuit fed from a current source P_{loss} with a voltage $T - T_0$ (Figure 12.6).

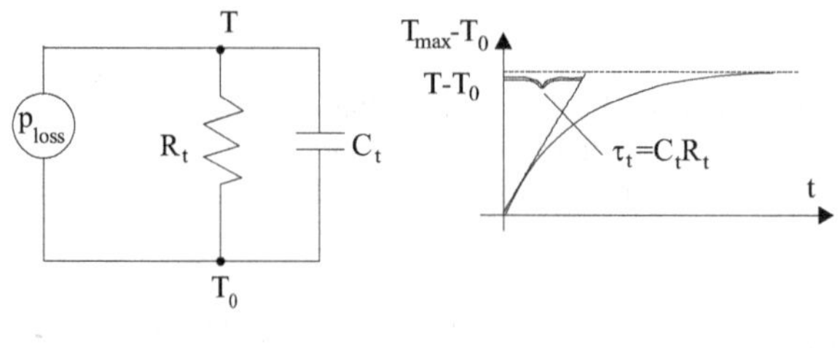

FIGURE 12.6 Equivalent thermal circuit.

For steady state, C_t does not enter Equation (12.17) and the equivalent circuit (Figure 12.6) simplifies.

The solution of this electric circuit is evident.

$$T = (T_{max} - T_0)\left(1 - e^{-\frac{t}{\tau_t}}\right) + T_0\, e^{-\frac{t}{\tau_t}} \tag{12.20}$$

The thermal time constant $\tau_t = C_tR_t$ is very important as it limits the machine working time with a certain level of losses and given cooling conditions. Intermittent operation, however, allows for more losses (more power) for the same given maximum temperature T_{max}.

The thermal time constant increases with machine size and effectivity of the cooling system. A TENV motor is expected to have a smaller thermal time constant than a constant speed ventilator-cooled configuration.

12.7 INDUCTION MOTOR THERMAL TRANSIENTS AT STALL

The IM at stall is characterised by very large conductor losses. Core loss may thus be neglected by comparison. If the motor remains at stall, the temperature of the windings and cores increases in time. There is a maximum winding temperature limit T_{max}^{cooper} given by the insulation class, (155°C for class F) which should not be surpassed. This is so to maintain a reasonable working life for conductor insulation. The machine is designed for lower winding temperatures at full continuous load.

To simplify the problem, let us consider two extreme cases, one with long end-connection stator winding and the other with a long stack and short end connections.

For the first case, we may neglect the heat transfer by conduction to the winding in slots portion. Also, if the motor is totally enclosed, the heat transfer through free convection to the air inside the machine is rather small (because this air gets hot easily). In fact, all the heat produced in the end connection (p_{Coend}) serves to increase the end winding temperature

$$\frac{\Delta\theta_{endcon}}{\Delta t} \approx \frac{p_{Coend}}{C_{endcon}};\ C_{endcon} = M_{endcon}C_{tcopper} \tag{12.21}$$

With $p_{Coend} = 1000$ W, $M_{endcon} = 1$ Kg, $c_{tcopper} = 380$ J/Kg/°C, the winding would heat up 115°C (from 40°C to 155°C) in a time interval Δt (12.21).

$$(\Delta t)_{40\rightarrow155°C} = \frac{115\cdot1\cdot380}{1000} = 43.7 \text{ seconds} \tag{12.22}$$

Now if the machine is already hot at, say, 100°C, $\Delta\theta_{endcon} = 155° - 100° = 55°C$. So the time allowed to keep the machine at stall is reduced to

$$(\Delta t)_{100\rightarrow155°C} = \frac{55\cdot1\cdot380}{1000} = 20.9 \text{ seconds}$$

The equivalent thermal circuit for this oversimplified case is shown in Figure 12.7a.

On the contrary, for long stacks, only the winding losses in slots are considered. However, this time, some heat is accumulated in the core also after the same heat is transferred through thermal conduction through insulation from slot conductors to core.

With $p_{Coslot} = 1000$ W, $M_{slotcopper} = 1$ Kg, $C_{slotcopper} = 380$ J/Kg/°C, insulation thickness 0.3 mm $K_{ins} = 6$ W/m/°C, $c_{tcore} = 490$ J/Kg/°C, slot height $h_s = 20$ mm, slot width $b_s = 8$ mm, slot number $N_s = 36$, $M_{core} = 5$ Kg, and stack length $l_{stack} = 0.1$ m,

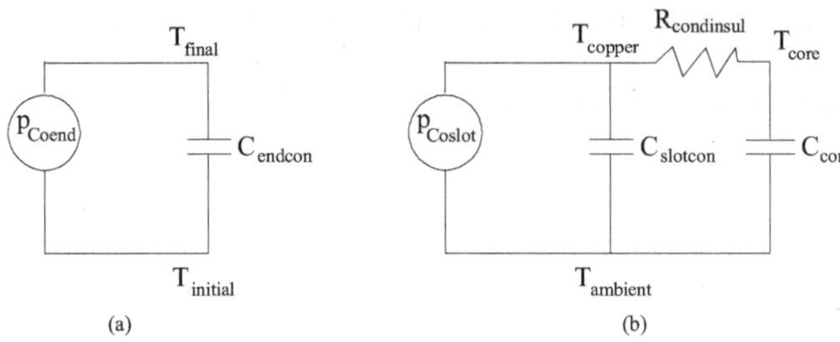

FIGURE 12.7 Simplified thermal equivalent circuits for stator winding temperature rise at stall (a) long end connections and (b) long stacks.

$$R_{condinsul} = \frac{\Delta_{ins}}{(2h_s + b_s)l_{stack}N_sK_{ins}} = \frac{0.3 \cdot 10^{-3}}{(2 \cdot 20 + 8) \cdot 10^{-3} \cdot 0.1 \cdot 36 \cdot 2} = 8.68 \cdot 10^{-4}\ (°C/m) \quad (12.23)$$

$$C_{slotcon} = M_{slotcon}C_{tcopper} = 1 \cdot 380 = 380\ J/°C$$
$$C_{core} = M_{core}C_{tcore} = 5 \cdot 490 = 2450\ J/°C \quad (12.24)$$

The temperature rise in the copper and core versus time (solving the circuit of Figure 12.7b) is

$$T_{copper} - T_{ambient} = P_{Coslot}\left[\frac{t}{C_{slotcon} + C_{core}} + \frac{\tau_t^2}{C_{slotcon}\tau_{con}}\left(1 - e^{-\frac{t}{\tau_t}}\right)\right]$$
$$T_{core} - T_{ambient} = \frac{P_{Coslot}}{C_{slotcon} + C_{core}}\left[t - \tau_t\left(1 - e^{-\frac{t}{\tau_t}}\right)\right] \quad (12.25)$$

with

$$\tau_{con} = C_{slotcon}R_{condinsul}; \quad \tau_t = \frac{C_{slotcon} \cdot C_{core}}{C_{slotcon} + C_{core}}R_{condinsul} \quad (12.26)$$

As expected, the copper temperature rise is larger than core temperature rise. Also, the core accumulates a good part of the winding-produced heat, so the time after which the conductor insulation temperature limit (155°C for class F) is reached at stall is larger than for the end-connection windings.

The thermal time constant τ_t is

$$\tau_t = \frac{380 \cdot 2450}{2450 + 380} \cdot 8.68 \cdot 10^{-4} = 0.2855 \text{ seconds} \quad (12.27)$$

The second term in (12.25) dyes out quickly, so, in fact, only the first, linear, term counts. As $C_{core} \gg C_{slotcon}$, the time to reach the winding insulation temperature limit is increased a few times: for $T_{ambient} = 40°C$ and $T_{copper} = 155°C$ from (12.25).

$$(\Delta t)_{40°C \to 155°C} = \frac{(155 - 40)(380 + 2450)}{1000} = 325.45 \text{ seconds}$$

Consequently, longer stack motors seem advantageous if they are to be used frequently at or near stall at high currents (torques).

Based on dominant heat transfer mode (subtransient, transition, transient, and temperature creep) Ref. [3] calculates the IM transient thermal response at stall in more detail, to characterise start-time ratings.

12.8 INTERMITTENT OPERATION

Intermittent operation with IMs occurs both in line-start constant frequency and voltage and in variable speed drives (variable frequency and voltage).

In most line-start applications, as the voltage and frequency stay constant, the magnetisation current I_m is constant. Also, the rotor circuit is dominated by the rotor resistance term (R_r/S), and thus, the rotor current I_r is 90° ahead of I_m and the torque may be written as

$$T_e \approx 3p_1 L_m I_m I_r = 3p_1 L_m I_m \sqrt{I_s^2 - I_m^2} \tag{12.28}$$

The torque is proportional to the rotor current, and the stator and rotor winding losses and core losses are related to torque by the expression

$$\begin{aligned} P_{dis} &= p_{core} + p_{Costator} + p_{Corotor} \approx 3R_s I_s^2 + 3R_r I_r^2 + p_{core} \\ &= 3R_s\left(I_m^2 + \left(\frac{T_e}{3p_1 L_m I_m}\right)^2\right) + 3R_r\left(\frac{T_e}{3p_1 L_m I_m}\right)^2 + \frac{3(\omega_1 L_m I_m)^2}{R_{mII}} \end{aligned} \tag{12.29}$$

For fractional power (sub-kW) or low-speed ($2p_1 = 10, 12$) motors, I_m (magnetisation current) may reach 70%–80% of rated current I_{sn}, and thus, (12.29) remains a rather complicated expression of torque, with $I_n = \text{const.}$

For medium- and large-power (and $2p_1 = 2, 4, 6$) IMs, in general, $I_m < 30\%\ I_{sn}$ and I_m may be neglected in (12.29), which becomes

$$P_{dis} \approx (p_{core})_{const} + 3(R_s + R_r)\left(\frac{T_e}{3p_1 L_m I_m}\right)^2 \tag{12.30}$$

Electromagnetic losses are proportional to torque squared. For variable speed drives with IMs, the magnetisation current is reduced with torque reduction to cut down (minimise) core and winding losses together.

Thus, (12.29) may be used to obtain $\partial P_{dis}/\partial I_m = 0$ and obtain $I_m(T_e)$ and, again, from (12.29), $P_{dis}(T_e)$. Qualitatively for the two cases, the electromagnetic loss variation with torque is shown in Figure 12.8.

As expected, for an on–off sequence (t_{ON}, t_{OFF}), more than rated (continuous duty) losses are acceptable during the on time. Therefore, motor overloading is permitted. For constant magnetisation current, however, as the losses are proportional to torque squared, the overloading is not very large but still similar to the case of PM motors [4], though magnetisation losses $3R_{ms}I_m^2$ are additional for the IM.

The duty cycle d may be defined as

$$d = \frac{t_{ON}}{t_{ON} + t_{OFF}} \tag{12.31}$$

Complete use of the machine in intermittent operation is made if, at the end of ON time, the rated temperature of windings is reached. Evidently, the average losses during ON time P_{dis} may surpass the rated losses P_{disn} for continuous steady-state operation. By how much, it depends both on the t_{ON} value and the machine equivalent thermal time constant τ_t.

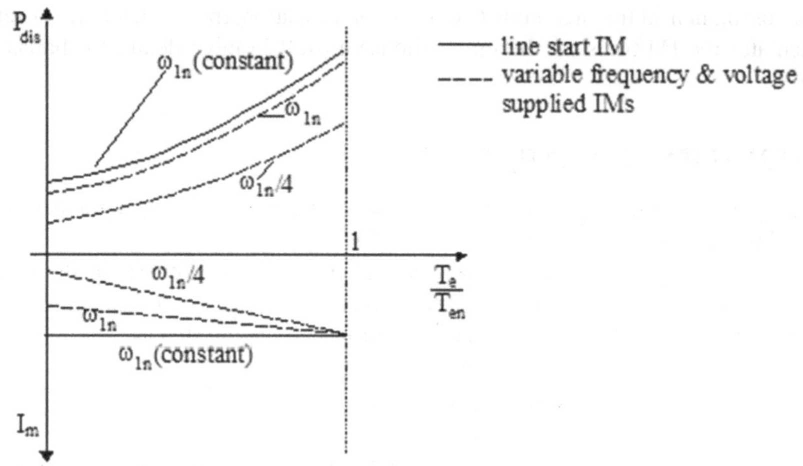

FIGURE 12.8 Electromagnetic losses P_{dis} and magnetisation current I_m versus torque.

$$\tau_t = R_t C_t \tag{12.32}$$

C_t – thermal capacity of winding (J/°C) and R_t – thermal resistance between windings and the surroundings (°C/W). The value of τ_t depends on machine geometry, rated power and speed, and on the cooling system and may run from tens of seconds to tens of minutes or even several hours.

12.9 TEMPERATURE RISE (T_{ON}) AND FALL (T_{OFF}) TIMES

The loss (dissipated power) P_{dis} may be considered approximately proportional to load squared.

$$\frac{P_{dis}}{P_{disn}} \approx \left(\frac{T_e}{T_{en}}\right)^2 = K_{load}^{\ 2} \tag{12.33}$$

The temperature rise, for an equivalent homogeneous body, during t_{ON} time is (12.19)

$$T - T_0 = R_t P_{dis}\left(1 - e^{-\frac{t_{ON}}{\tau_t}}\right) + (T_c - T_0)e^{-\frac{t_{ON}}{\tau_t}} \tag{12.34}$$

where T_c is the initial temperature and T_0 the ambient temperature ($T_{max} - T_0 = R_t \cdot P_{dis}$), with P_{disn} ($K_{load} = 1$) and $T(t_{ON}) = T_{rated}$. Replacing P_{dis} in (12.34) by

$$P_{dis} = P_{disn} \cdot K_{load}^{\ 2} = \frac{(T_{rated} - T_0)}{R_t} \cdot K_{load}^{\ 2} \tag{12.35}$$

with $T_{max} = T_{rated}$,

$$(T_{rated} - T_0)\left[1 - K_{load}^{\ 2}\left(1 - e^{-\frac{t_{ON}}{\tau_t}}\right)\right] = (T_c - T_0)e^{-\frac{t_{ON}}{\tau_t}} \tag{12.36}$$

Equation (12.36) shows the dependence of t_{ON} time such that to reach the rated winding temperature, from an initial temperature T_c for a given overload factor K_{load}. As expected, t_{ON} time decreases with the rise in initial winding temperature T_c.

During t_{OFF} time, the losses are zero, and the initial temperature is $T_c = T_r$. So with $K_{load} = 0$ and $T_c = T_r$, (12.36) becomes

$$T - T_0 = (T_r - T_0) \cdot e^{-\frac{t_{OFF}}{\tau_t}} \tag{12.37}$$

For steady-state intermittent operation, however, the temperature at the end of OFF time is equal, again, to T_c.

$$T_c - T_0 = (T_r - T_0) \cdot e^{-\frac{t_{OFF}}{\tau_t}} \tag{12.38}$$

For given initial (low) T_c, final (high) T_{rated} temperatures, load factor K_{load}, and thermal time constant τ_t, Equations (12.36) and (12.38) allow for the computation of t_{ON} and t_{OFF} times.

Now, introducing the duty cycle $d = \frac{t_{ON}}{t_{ON} + t_{OFF}}$ to eliminate t_{OFF}, from (12.36) and (12.38), we obtain

$$K_{load} = \sqrt{\frac{1 - e^{-\frac{t_{ON}}{d\tau_t}}}{1 - e^{-\frac{t_{ON}}{\tau_t}}}} \tag{12.39}$$

It is to be noted that using (12.36)–(12.38) is most practical when $t_{ON} < \tau_t$ as it is known that the temperature stabilises after 3 to $4\tau_t$.

For example, with $t_{ON} = 0.2\tau_t$ and $d = 25\%$, $K_{load} = 1.743$, and $\tau_t = 75$ minutes, it follows that $t_{ON} = 15$ minutes and $t_{OFF} = 45$ minutes.

For very short on–off cycles ($(t_{ON} + t_{OFF}) < 0.2\tau_t$), we may use Taylor's formula to simplify (12.39) to

$$K_{load} = \sqrt{\frac{1}{d}} \tag{12.40}$$

For short cycles, when the machine is overloaded as in (12.40), the medium loss will be the rated one.

For a single pulse, we may use $d = 0$ in (12.39) to obtain

$$K_{load} = \sqrt{\frac{1}{1 - e^{-\frac{t_{ON}}{\tau_t}}}} \tag{12.41}$$

As expected for one pulse, K_{load} allowed to reach rated temperature for given t_{ON} is larger than for repeated cycles.

With same start and end of cooling period temperature T_c, the t_{ON} and t_{OFF} times are again obtained from (12.39) and (12.38), respectively, even for a single cycle (heat up, cool down).

$$t_{OFF} = -\tau_t \ln\left[K_{load}^2 - \left(K_{load}^2 - 1\right)e^{\frac{t_{ON}}{\tau_t}}\right] \tag{12.42}$$

For a given K_{load} from (12.41), we may calculate t_{ON}/τ, while then from (12.42) t_{OFF} time, for a single steady-state cycle, T_c to T_r to T_c temperature excursion ($T_r > T_c$) is obtained. It is also feasible to set t_{OFF} and, for given K_{load}, to determine from (12.42), t_{ON}.

Rather simple formulas, as presented in this chapter, may serve well in predicting the thermal transients for given overload and intermittent operation.

After this almost oversimplified picture of IM thermal modelling, let us advance one more step further by building more realistic thermal equivalent circuits [5–12].

12.10 MORE REALISTIC THERMAL EQUIVALENT CIRCUITS FOR IMS

Let us consider the overall heating of the stator (or rotor) winding with radial channels. The air speed and temperatures inside the motor are taken as known. (The ventilator design is a separate problem which produces the airflow rate and temperatures of air as its output, for given losses in the machine and its geometry).

A half longitudinal cross-section is shown in Figure 12.9a and b for the stator and the rotor, respectively.

The objective here is to set a more realistic equivalent thermal circuit and explicitate the various thermal resistances R_{t1}, … R_{t9}.

To do so, a few assumptions are made

- The winding end-connection losses do not contribute to the stator (rotor) stack heating
- The end-connection and in-slot winding temperature, respectively, do not vary axially or radially
- The core heat centre is placed $l_s/4$ away from elementary stack radial channel.

The equivalent circuit with thermal resistances is shown in Figure 12.10.

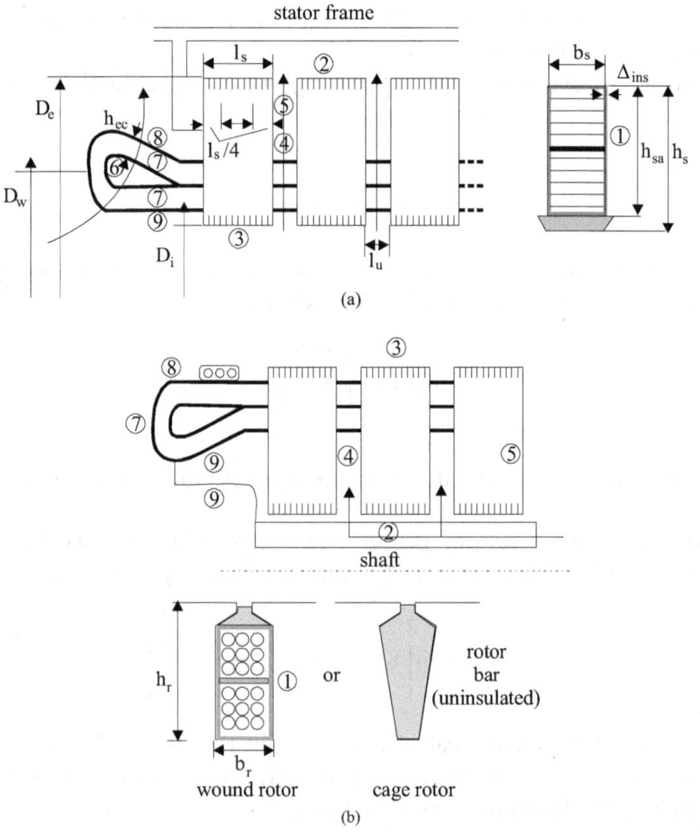

FIGURE 12.9 IM with radial ventilating channels: (a) stator winding and (b) rotor winding.

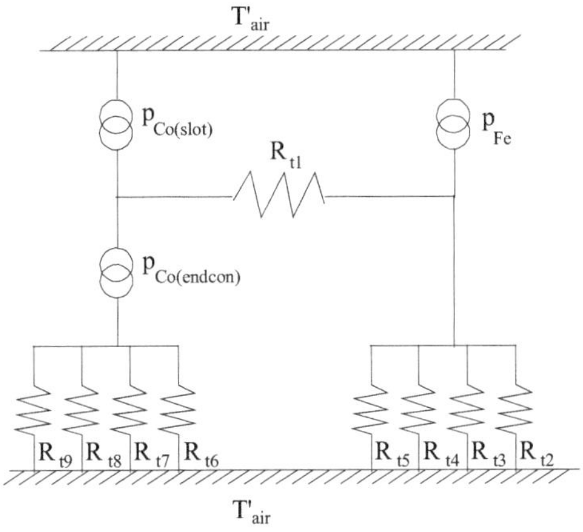

FIGURE 12.10 Equivalent thermal circuit for stator or rotor windings.

In Figure 12.10,

T'_{air} – air temperature from the entrance into the machine to the winding surface.
$P_{Co(slot)}$ – winding losses for the part situated in slots
$P_{Co(endcon)}$ – end-connection winding losses
P_{Fe} – iron losses
R_{t1} – slot insulation thermal resistance from windings in slots to core
R_{t2} – thermal resistance from core to air which is exterior (interior) to it – stator core to frame; rotor core to shaft
R_{t3} – thermal resistance from core to airgap cooling air
R_{t4} – thermal resistance from the iron core to the air in the ventilation channels
R_{t5} – thermal resistance from core to air in ventilation channel
R_{t6} – thermal resistance towards the air inside the end connections (it is ∞ for round conductor coils)
R_{t7} – thermal resistance from the frontal side of end connections to the air between neighbouring coils
R_{t8} – thermal resistance from end connections to the air above them
R_{t9} – thermal resistance from end connections to the air below them.

Approximate formulas for $R_{t1} - R_{t9}$ are

$$R_{t1} = \frac{\Delta_{ins}}{K_{ins} A_1}; \quad A_1 \approx (2h_s + b_s) n_s l_s N_s \tag{12.43}$$

n_s – number of elementary stacks ($n_s - 1$ – radial channels)
l_s – elementary stack length; N_s – stator slot number
K_{ins} – slot insulation heat transfer coefficient

$$R_{t2} = \frac{h_{cs(r)}}{K_{Fe} A_2} + \frac{1}{h_{c2} A_2}; \quad A_2 = (\pi D_e - b_f n_f) n_s l_s \tag{12.44}$$

b_f – fins width and n_f – fins number for the stator. For the rotor, $b_f = 0$ and D_e is replaced by rotor core interior diameter, $h_{cs(r)}$ – back core radial thickness, K_{Fe} – core radial thermal conductivity, and h_{c2} – thermal convection coefficient (all parameters in IS units).

$$R_{t3}=\frac{h_{sa}}{K_{Fe}A_3}+\frac{1}{h_{c3}A_3};\quad A_3=(\pi D_i-b_f N_s)n_s l_s \tag{12.45}$$

h_{c3} is the convection thermal coefficient as influenced by the air speed in the airgap in the presence of radial channels (Equation 12.14).

$$R_{t4}=\frac{\Delta_{inscon}}{K_{copper}A_4}+\frac{1}{h_{c4}A_4};\quad A_4=N_s(2h_{s0}+b_s)(n_s-1)l_u \tag{12.46}$$

l_u – axial length of ventilation channel

$$R_{t5}=\frac{2(n_s-1)(l_s/4)}{K_{Felong}A_{5s(r)}}+\frac{1}{h_{c5}A_{5s(r)}};\quad A_{5s}=2n_s\left(\frac{\pi}{4}\left(D_e^2-(D_i+2h_s)^2\right)+N_s b_{ts} h_s\right) \tag{12.47}$$

$$A_{5r}=2n_s\left(\frac{\pi}{4}\left((D_i-2h_r)^2-D_{ir}^2\right)+N_r b_{tr} h_s\right) \tag{12.48}$$

N_r – rotor slots, $b_{ts(r)}$ – stator (rotor) tooth width, and h_{c5} – convection thermal coefficient as influenced by the air speed in the radial channels.

$$\begin{aligned}R_{t6}&=\frac{\Delta_{inscon}}{K_{copper}A_6}+\frac{1}{h_{c6}A_6}\\R_{t7}&=\frac{\Delta_{inscon}}{K_{copper}A_7}+\frac{1}{h_{c7}A_7}\\R_{t8}&=\frac{\Delta_{inscon}}{K_{copper}A_8}+\frac{1}{h_{c8}A_8}\\R_{t9}&=\frac{\Delta_{inscon}}{K_{copper}A_9}+\frac{1}{h_{c9}A_9}\end{aligned} \tag{12.49}$$

R_{t6}–R_{t9} refer to winding end-connection heat transfer by thermal conduction through the electrical insulation and, by convention, through the circulating air in the machine. Areas of heat transfer A_6–A_9 depend heavily on the coils' shape and their arrangement as end connections in the stator (or rotor).

For round wire coils with insulation between phases, the situation is even more complicated but the heat flow through the end connections towards their interior or circumferentially may be neglected ($R_6 = R_7 = \infty$).

As the air temperature inside the machine was considered uniform, the stator and rotor equivalent thermal circuits as in Figure 12.10 may be treated rather independently ($p_{Fe} = 0$ in the rotor, in general). In the case that there is one stack (no radial channels), the above expressions are still valid with $n_s = 1$ and, thus, all heat transfer resistances related to radial channels are ∞ ($R_4 = R_5 = \infty$).

A thorough investigation of thermal resistances and capacitance expressions (with convection from motor frame and end space cooling) is given in [12].

12.11 A DETAILED THERMAL EQUIVALENT CIRCUIT FOR TRANSIENTS

The ultimate detailed thermal equivalent circuit of the IM should account for the three-dimensional character of the heat flow in the machine.

Although this may be done, a two-dimensional model is used. However, we may "break" the motor axially into a few segments and "thermally" connect these segments together.

To account for thermal transients, the thermal equivalent circuit should contain thermal resistances R_{ti}(°C/W), capacitors C_{ti}(J/°C), and heat sources (W) (Figure 12.11).

A detailed thermal equivalent circuit – in the radial plane – emerges from the more realistic thermal circuit of Figure 12.10 by dividing the heat sources in more components (Figure 12.12).

The stator conductor losses are divided into their in-slot and overhang (end-connection) components. The same thing could be done for the rotor (especially for wound rotors). Also, no heat transport through conduction from end connections to the coils section in slot is considered on Figure 12.12, as the axial heat flow is neglected.

Due to the machine pole symmetry, the model in Figure 12.12 is in fact one dimensional, that is, the temperatures vary radially. A kind of similar model is to be found in [4] for PM brushless and in [5] for induction motors.

In [6,13], a thermal model for three-dimensional heat flow is presented.

It is possible to augment the model with heat transfer along circumferential direction and along axial dimension to obtain a rather complete thermal equivalent circuit with hundreds of nodes.

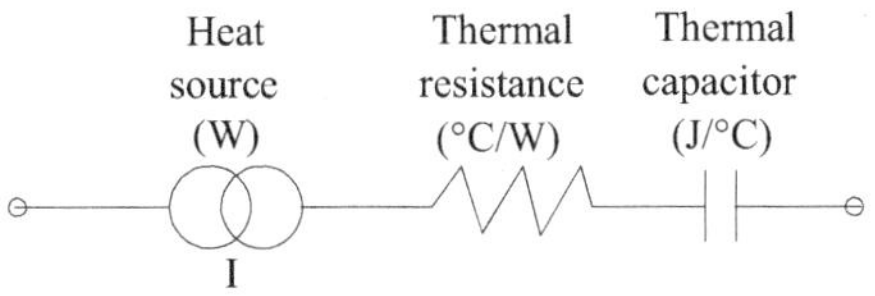

FIGURE 12.11 Thermal circuit elements with units.

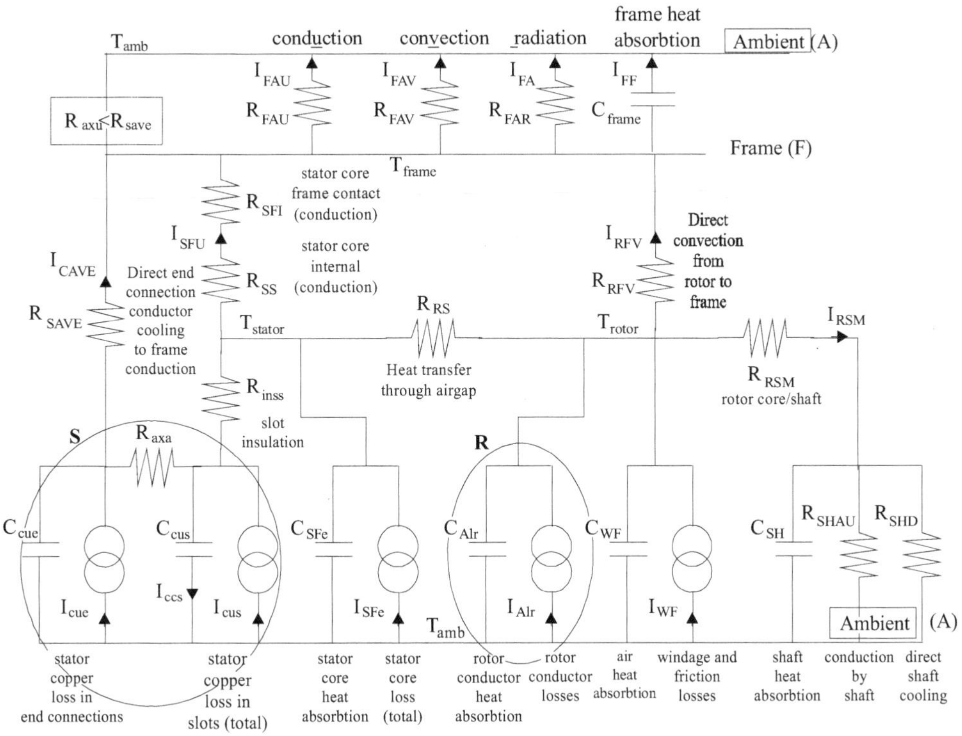

FIGURE 12.12 A more detailed thermal equivalent circuit for IMs.

12.12 THERMAL EQUIVALENT CIRCUIT IDENTIFICATION

As shown in Section 12.10, the various thermal resistances R_{ti} (or conductances $G_{ti} = 1/R_{ti}$) and thermal capacitances (C_{ti}) may be approximately calculated through analytical formulas.

As the thermal conductivities, K_i, and convection and radiation heat transfer coefficients, h_i, are dependent on various geometrical factors, cooling system, and average local temperature, at least for thermal transients, their identification through tests would be beneficial.

Once the thermal equivalent circuit structure is settled (Figure 12.12, for example), with various temperatures as unknowns, its state-space matrix equation system is

$$\frac{dX}{dt} = AX + BP \tag{12.50}$$

$$\begin{aligned} X &= \left[T_1 \ldots T_j \ldots T_n\right]^t \text{ - temperature matrix} \\ P &= \left[P_1 \ldots P_j \ldots P_m\right]^t \text{ - power loss matrix} \end{aligned} \tag{12.51}$$

P_j has to be known from the electromagnetic model. A and B are coefficient matrixes built with R_{ti} and C_{ti}.

Ideally, n temperature sensors to measure $T_1, \ldots T_n$ versus time would be needed. If it is not feasible to install so many, the model is to be simplified so that all temperatures in the model are measured.

Having experimental values of $T_i(t)$, system (12.50) may be used to determine, by an optimisation method, the parameters of the equivalent circuit. In essence, the squared error between calculated and measured (after filtering) temperatures is to be minimum over the entire time span. In Ref. [7], such a method is used and the results look good.

As some of the thermal parameters may be calculated, the method can be used to identify them from the losses and then check the heat division from its centre.

For example, it may be found that for low-power IMs at rated speed, 65% of the rotor cage losses is evacuated through airgap, 20% to the internal frame, and 15% by shaft bearing.

Also the heat produced in the stator end-connection windings for such motors is divided as 20% to the internal frame and end shields (brackets) by convection and the rest of 80% to the stator core by conduction (axially).

Consequently, based on such results, the detailed equivalent circuit should contain a conduction resistance branch from end connections to stator core as 80% of the heat goes through it (R_{axa} in Figure 12.12).

As an example of an ingenious procedure to measure the R_i and C_i parameters, or the loss distribution, we notice here the case of turning off the IM and measuring the temperature decrease in locations of interest versus time [9].

From (12.18) in the steady-state conditions,

$$p_{loss} = \frac{1}{R_t}\left(T_0 - T_a\right) \tag{12.52}$$

The temperature derivative at t = 0 (from 12.18), when the heat input is turned off, is

$$\left(\frac{dT}{dt}\right)_{t=0} = -\frac{1}{R_t C_t}\left(T_0 - T_a\right) \tag{12.53}$$

Finally,

$$p_{loss} = -C_t\left(\frac{dT}{dt}\right)_{t\to 0} \tag{12.54}$$

Measuring the temperature gradient at the moment when the motor is turned off, with C_t known, allows for the calculation of local power loss in the machine just before the machine was turned off [8,9].

This way, the radial or axial variation of losses (in the core especially) may be obtained, provided that small enough in size temperature sensors are placed in key locations. The winding average temperature is measured after being turned off by the D.C. voltage/current method of stator winding resistance.

$$R_s(T) = (R_s)_{20°C}\left[1+\frac{1}{273}(T-T_{amb})\right] \tag{12.55}$$

Typical results are shown in Figure 12.13 [9]. For the four-pole IM in question, the temperature variation along circumferential and axial directions in the back core is small, but it is notable in the stator teeth. A notable decrease of loss density with radial distance is present, as expected (Figure 12.13b).

Recent attempts to identify critical thermal parameters – such as thermal resistances of frame to air, winding to core, end winding to end-caps radiation, air cooling speed, and bearing – by D.C. and A.C. standstill tests [14], for impregnated windings [15] end-winding thermal effects for a copper cage rotor IM [16], to be used in complex thermal equivalent circuit modelling of IMs, are bringing new insights valuable to the designer of both line-start and variable speed IM drives.

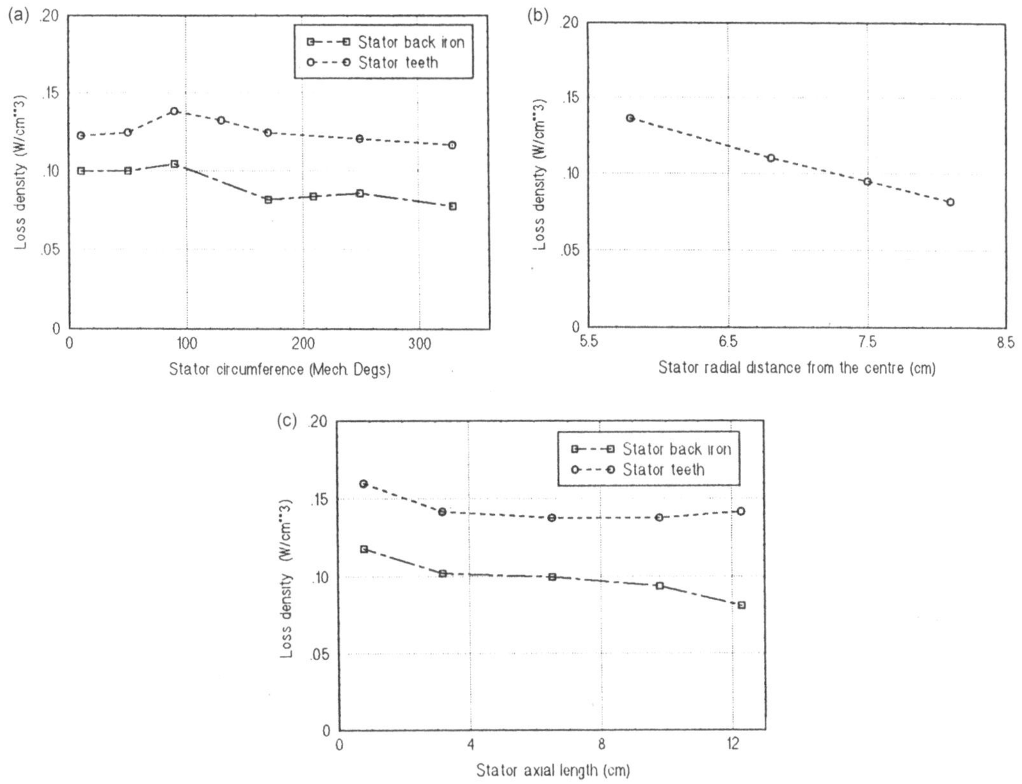

FIGURE 12.13 Iron loss density distribution (W/cm^3), (after [9]). (a) Along circumferential direction, (b) along radial direction, and (c) along axial direction.

12.13 THERMAL ANALYSIS THROUGH FEM

In theory, the three-dimensional FEM alone could lead to a fully realistic temperature distribution for a machine of any power and size, provided the localisation of heat sources and their levels are known.

The differential equation for heat flow is

$$\nabla(K\Delta T) + p_{loss} = \gamma c \frac{\partial T}{\partial t} \tag{12.56}$$

with K – local thermal conductivity (W/m°C); γ – local density (Kg/m³), p_{loss} – losses per unit volume (W/m³); c – specific heat coefficient (J/(°C·Kg)). The coefficients K, γ, and c vary throughout the machine.

Two types of boundary conditions are usually present

- Dirichlet conditions

$$T(x_b, y_b, z_b, t) = T^* \tag{12.57}$$

The ambient temperature is such a boundary around the machine.

- Neumann conditions

$$q - K_x \frac{\partial T}{\partial x} n_x - K_y \frac{\partial T}{\partial y} n_y - K_z \frac{\partial T}{\partial z} n_z = 0 \tag{12.58}$$

n_x, n_y, and n_z are the x, y, and z components of unit vector rectangular to the respective boundary surface and q – the heat flow through the surface.

As a 3D FEM would require large amounts of computation time, 2D FEM models have been built to study the temperature distribution either in the radial or axial cross-section.

The radial cross-section has a geometrical symmetry as seen in Figure 12.14.

To avoid the rotation influence on the modelling in the airgap zone, the rotor sector is replaced by a motion-independent computation sector (Figure 12.14b) [10].

After the temperature on the rotor surface is calculated and considered "frozen", the actual rotor sector is modelled.

It was found that the stator temperature varies radially and axially in a visible manner, while in the rotor, the temperature gradient is much smaller (Figure 12.14a) [10]. Similar results with 2D FEM are presented in [11,17].

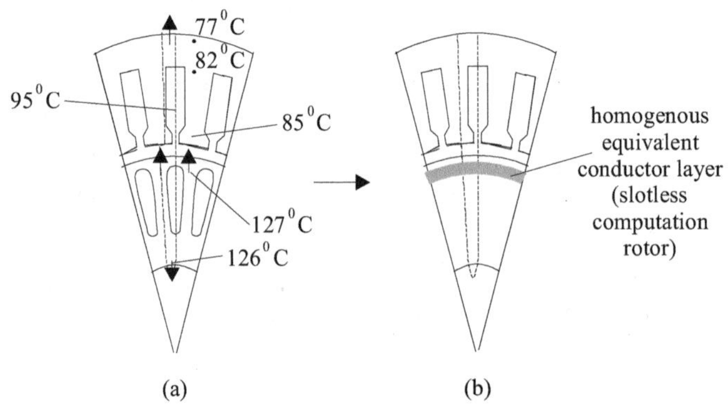

FIGURE 12.14 Computation sector (a) and its restructuring to avoid motion influence (b).

Also FEM offline imported data for IM thermal model with temperature-dependent materials properties via green transfer functions has been used for online monitoring models of IMs [18], with accent on 1° error target for hot-spot temperature.

Three-dimensional FEM may be used for a more complete IM thermal modelling, treated independently or together with the electromagnetic model for load (or speed) transients. The computation time, however, becomes large.

Still, due to the dependence of thermal resistances and capacitors and loss distribution on many "imponderable" factors, experimental methods (such as advanced calorimetric ones) are needed to validate theoretical results, especially when new machine configurations or design specifications are encountered.

12.14 SUMMARY

- Any IM design is thermally limited according to the conductor insulation class mainly: A(105°C), B(130°C), F(155°C), and H(180°C).
- The insulation life decreases very rapidly with temperature increase over the rated value (for extra 10°C, the life may be halved).
- Cooling methods, used to extract the heat from the machine and thus limit the temperature, are paramount in IM design.
- From TENV induction motors through drip-proof axial internal, radial internal to radial-axial cooling, the sophistication of cooling methods and their heat removal capacity increases steadily.
- Heat transfer takes place through thermal conduction, convection, and radiation. For IM design, convection and radiation heat transfer coefficients are usually lumped together. Radiation is notable for TENV configurations.
- The heat transfer steady state and transients may be approached either by equivalent thermal circuits (similar to R-C electric circuits) or by numerical methods (FEM).
- Thermal resistances R_{ti} (°C/W) are introduced for conduction, convection, and radiation heat transfer and thermal capacitances C_{ti} (J/°C) stand for heat absorption in the various parts of the motor.
- The heat sources p_{lossi}(W) represent "current" sources in the equivalent circuit.
- Thermal resistances decrease with cooling air speed.
- The windings have the highest temperature spots, in general.
- It has been shown that, at least in small-power IMs, about 80% of the heat (loss) in the stator coils overhangs is transmitted axially through conduction, to the winding part in slots and to the stator core.
- It has also been shown that 60%–70% of rotor losses are transferred through the airgap to the stator core.
- The axial variation of temperature depends on the relative stack length, method of cooling, and machine power. However, for powers up to hundreds of kW, unistack motors with axial external air cooling and finned frames have become recently popular for general-purpose designs.
- Information like this should be instrumental into developing adequate simplified equivalent thermal circuits. FEM 2D and 3D are adequate in getting information for designing practical equivalent circuits.
- Investigations of temperature non-uniformities in faulty IM drives keep surfacing [19] only to multiply in the near future, especially for multiphase IMs.
- Due to the thermal modelling complexity, theory and tests should go hand in hand in the future [20].
- Improved identification methods of IM thermal models with experimental validation continue to surface [21].

REFERENCES

1. S. A. Nasar (editor), *Handbook of Electric Machines*, McGraw-Hill Inc., New York, Chapter 12 (by A. J. Spisak).
2. W. H. McAdams, *Heat Transmission*, 2nd edition, McGraw-Hill Inc., New York, 1942.
3. V. T. Buyukdegirmenci, Ph. T. Krein, Induction machine characterization for short-therm or momentary stall torque, *IEEE Transactions on Industry Applications*, Vol. 51, No. 3, 2015, pp. 2237–2245.
4. J. Hondershot, T. J. E. Miller, PM brushless motor designs, OUP, 1995, Chapter 15.
5. G. Bellenda, L. Ferraris, A. Tenconi, A new simplified thermal model for induction motors for EVs applications, *Record of 7th International Conference on EMD*, IEE Conference, Publications 412, 1995, Durham, UK, pp. 11–15.
6. W. Guyglewics – Kacerka, J. Mukosiej, Thermal analysis of induction motor with axial and radial cooling ducts, *Record of ICEM*, 1992, Vol. 3, pp. 971–975.
7. G. Champenois, D. Roye, D. S. Zhu, Electrical and thermal performance predictions in inverter – fed squirrel cage induction motor drives, *EMPS*, Vol. 22, No. 3, 1994, pp. 355–369.
8. M. Benamrouche et al., Determination of iron loss distribution in inverter fed induction motors, *EMPS*, Vol. 25, No. 6, 1997, pp. 649–660.
9. H. Benamrouche et al., Determination of iron and stray load losses in induction motors using a thermometric method, *EMPS*, Vol. 26, No. 1, 1998, pp. 3–12.
10. J. Roger, G. Jinenez, The finite element application to the study of the temperature distribution inside electric rotating machines, *Record of ICEM*, 1992, Vol. 3, pp. 976–980.
11. D. Sarkar, Approximate analysis of temperature rise in an induction motor during dynamic braking, *EMPS*, Vol. 26, No. 6, 1998, pp. 585–599.
12. D. Staton, A. Boglietti, A. Cavagnino, Solving the more difficult aspects of electric motor thermal analysis in small and medium size industrial induction motors, *IEEE Transactions on Energy Conversion*, Vol. 20, No. 3, 2005, pp. 620–628.
13. A. Boglietti, A. Cavagnino, D. Staton, TEFC induction motors thermal models: A parameter sensitivity analysis, *IEEE Transactions on Industry Applications*, Vol. 41, No. 3, 2005, pp. 756–770.
14. A. Boglietti, A. Cavagnino, D. Staton, Determination of critical parameters in electric machine thermal models, *IEEE Transactions on Industry Applications*, Vol. 44, No. 4, 2008, pp. 1150–1159.
15. N. Simpson, R. Wrobel, Ph. H. Mellor, Estimation of equivalent thermal parameters of impregnated electric windings, *IEEE Transactions on Industry Applications*, Vol. 49, No. 6, 2013, pp. 2505–2515.
16. F. Ahmed, N. C. Kar, Analysis of end-winding thermal effects in a totally enclosed fan-cooled induction motor with a die cast copper rotor, *IEEE Transactions on Industry Applications*, Vol. 53, No. 3, 2017, pp. 3098–3109.
17. N. Bianchi, S. Bolognani, F. Tonel, Thermal analysis of a run-capacitor single-phase induction motor, *IEEE Transactions on Industry Applications*, Vol. 39, No. 2, 2003, pp. 457–465.
18. H. Zhang, On line thermal monitoring models for induction machines, *IEEE Transactions on Energy Conversion*, Vol. 30, No. 4, 2015, pp. 1279–1287.
19. M. Popescu, D. G. Dorrell, L. Alberti, N. Bianchi, D. A. Staton, D. Hawkins, Thermal analysis of duplex three phase induction motor under fault operation conditions, *IEEE Transactions on Industry Applications*, Vol. 49, No. 4, 2013, pp. 1523–1530.
20. G. Kylander, Temperature simulation of a 15 kW induction machine operating at variable speed, *Record of ICEM*, 2002, Vol. 3, pp. 943–947.
21. M. Rosu, P. Zhou, D. Lin, D. Ionel, M. Popescu, F. Blaabjerg, V. Rallabandi, D. Staton, *Multiphysics Simulation by Design for Electrical Machines, Power Electronics, and Drives*, IEEE Press Wiley, New York, 2018.

13 Single-Phase Induction Machines
The Basics

13.1 INTRODUCTION

Most small-power (generally below 2 kW) induction machines (IMs) have to operate with single-phase A.C. power supplies that are readily available in homes and remote rural areas. When power electronics converters are used, three-phase A.C. output is produced, and thus, three-phase induction motors may still be used.

However, for constant speed applications (the most frequent situation), the induction motors are fed directly from the available single-phase A.C. power grids. In this sense, we call them single-phase induction motors.

To be self-starting, the IM needs a travelling field at zero speed. This in turn implies the presence of **two windings** in the stator, while the rotor has a standard squirrel cage. The first winding is called the **main winding,** while the second winding (for start, especially) is called **auxiliary winding**.

Single-phase IMs may run only on the main winding once they started on two windings. A typical case of single-phase single-winding IM occurs when a three-phase IM ends up with an open phase (Chapter 7, Volume 1). The power factor and efficiency degrade while the peak torque also decreases significantly.

Thus, except for low powers (less than 1/3 kW in general), the **auxiliary winding** is also active during running conditions to improve performance.

Three main types of single-phase induction motors are in use today:

- Split-phase induction motors
- Capacitor induction motors
- Shaded-pole induction motors.

13.2 SPLIT-PHASE INDUCTION MOTORS

The split-phase induction motor has a main and an auxiliary stator winding displaced by 90° or up to 110°–120° (Figure 13.1a).

The auxiliary winding has a higher ratio between resistance and reactance, by designing it at a higher current density, to shift the auxiliary winding current $\underline{I}_a$ ahead of main winding current $\underline{I}_m$ (Figure 13.1b). A resistance which increases with temperature (PTC) may be used as "a switch" to turn off the auxiliary winding after start.

The two windings – with a 90° space displacement and a $\gamma \approx 20°–30°$ current time phase shift – produce a magnetic field in the airgap with a definite forward travelling component (from m to a). This travelling field induces voltages in the rotor cage whose currents produce a starting torque which rotates the rotor from m to a (clockwise in Figure 13.1).

Once the rotor catches speed, the starting switch is opened to disconnect the auxiliary winding, which is designed for short duty. The starting switch may be centrifugal, magnetic, or static type. The starting torque may be up to 150% rated torque, at moderate starting current, for frequent start long-running time applications. For infrequent starts and short running time, low efficiency is allowed in exchange for lower starting current with higher rotor resistance.

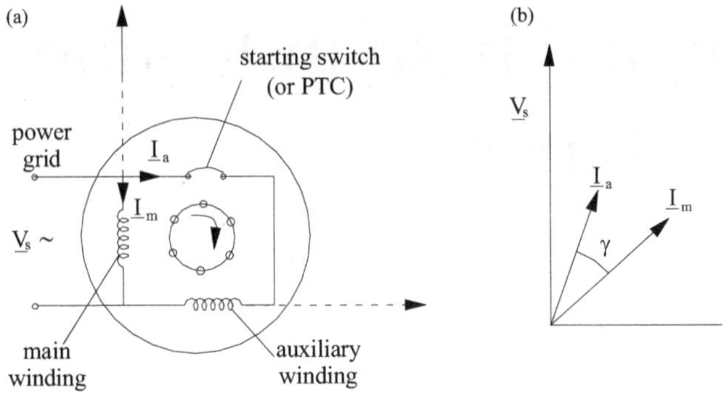

FIGURE 13.1 (a and b) The split-phase induction motor.

During running conditions, the split-phase induction motor operates on one winding only, and thus, it has a poor power factor. It is used below 1/3 kW, generally, where the motor costs are of primary concern.

13.3 CAPACITOR INDUCTION MOTORS

Connecting a capacitor in series with the auxiliary winding causes the current in that winding, $\underline{I}_a$, to lead the current in the main winding, $\underline{I}_m$, by up to 90°.

Complete symmetrisation of the two windings mmf for given slip may be performed this way.

That is, a pure travelling airgap field may be produced either at start (S = 1) or at rated load ($S = S_n$) or somewhere in between.

An improvement in starting and running torque density, efficiency, and especially in power factor is brought by the capacitor presence. Capacitor motors are of quite a few basic types:

- Capacitor-start induction motors
- Two-value capacitor induction motors
- Permanent-split capacitor induction motors
- Tapped-winding capacitor induction motors
- Split-phase capacitor induction motors
- Capacitor three-phase induction motors (for single-phase supply).

13.3.1 Capacitor-Start Induction Motors

The capacitor-start IM (Figure 13.2a) has a capacitor and a start switch in series with the auxiliary winding.

The starting capacitor produces at standstill an almost 90° phase advance of its current $\underline{I}_a$ with respect to $\underline{I}_m$ in the main winding (Figure 13.2b).

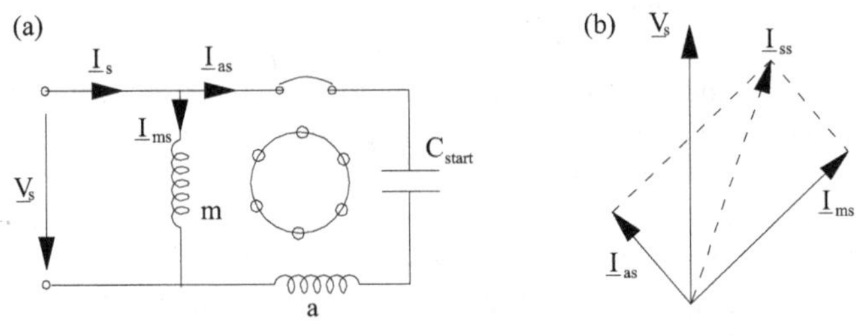

FIGURE 13.2 The capacitor-start induction motor: (a) schematics and (b) phasor diagram at zero speed.

This way, a large travelling field is produced at start. Consequently, the starting torque is large. After the motor starts, the auxiliary winding circuit is opened by the starting switch, leaving on only the main winding. The large value of the starting capacitor is not adequate for running conditions. During running conditions, the power factor and efficiency are rather low.

Capacitor-start motors are built for single or dual voltage (115 V and 230 V). For dual voltage, the main winding is built in two sections connected in series for 230 V and in parallel for 115 V.

13.3.2 The Two-Value Capacitor Induction Motor

The two-value capacitor induction motor makes use of two capacitors in parallel, one for starting, C_s, and the other for running C_n ($C_s \gg C_n$).

The starting capacitor is turned off by a starting switch, while the running capacitor remains in series with the auxiliary winding (Figure 13.3).

The two capacitors are sized to symmetrise the two windings at standstill (C_s) and, respectively, at rated speed (C_n).

High starting torque is associated with good running performance around rated speed (torque). Both power factor and efficiency are high.

The running capacitor is known to produce effects such as [1]

- 5%–30% increase of breakdown torque
- 5%–10% improvement in efficiency
- Power factor values above 90%
- Noise reduction at load
- 5%–20% increase in the locked rotor torque.

Sometimes, the auxiliary winding is made of two sections (a_1 and a_2). One section works at start with the starting capacitor C_s, while both sections work for running conditions to allow a higher voltage running (smaller) capacitor C_n (Figure 13.4).

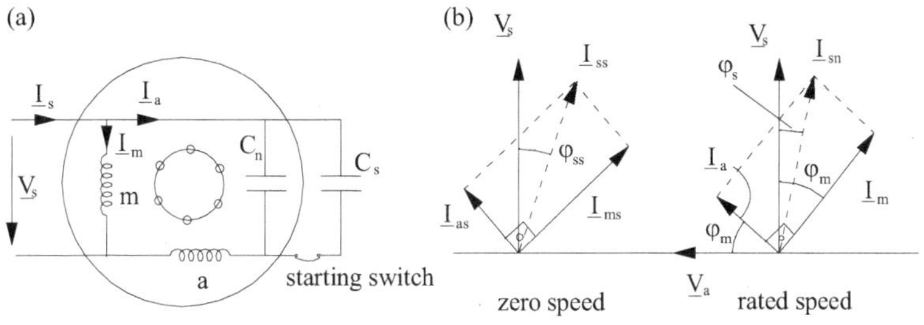

FIGURE 13.3 The two-value capacitor induction motor: (a) the schematics and (b) phasor diagrams at zero and rated speed.

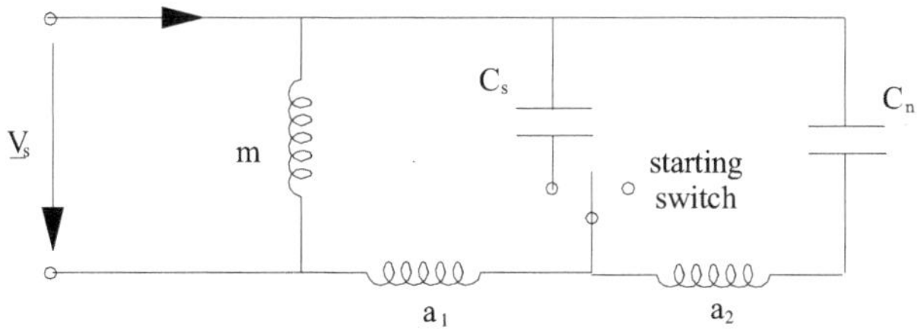

FIGURE 13.4 Two-value capacitor induction motor with two-section auxiliary winding.

13.3.3 Permanent-Split Capacitor Induction Motors

The permanent-split capacitor IMs have only one capacitor in the auxiliary winding which remains in action all the time.

As a compromise between starting and running performance, this motor has a rather low starting torque but good power factor and efficiency under load. This motor is not to be used with belt transmissions due to its low starting torque. It is, however, suitable for reversing, intermittent-duty service, and powers from 10 to 200 W. For speed reversal, the two windings are identical and the capacitor is "moved" from the auxiliary to the main winding circuit (Figure 13.5).

13.3.4 Tapped-Winding Capacitor Induction Motors

Tapped-winding capacitor induction motors are used when two or more speeds are required.

For two speeds, the T and L connections are highly representative (Figure 13.6a and b). The main winding contains two sections m_1 and m_2.

The T connection is more suitable for 230 V, while the L connection is more adequate for 115 V power grids, since this way the capacitor voltage is higher and thus its cost is lower.

The difference between high and low speed is not large unless the power of the motor is very small, because only the voltage is reduced. The locked rotor torque is necessarily low (less than the load torque at the low speed). Consequently, they are not to be used with belted drives. Unstable low-speed operation may occur as the breakdown torque decreases with voltage squared.

A rather general tapped-winding capacitor induction motor is shown in Figure 13.7 [2].

In this configuration, each coil of the single-layer auxiliary winding has a few taps corresponding to the number of speeds required.

13.3.5 Split-Phase Capacitor Induction Motors

Split-phase capacitor induction motors start as split-phase motors and then commute to permanent capacitor motors.

This way, both higher starting torque and good running performance are obtained. The auxiliary winding may also contain two sections to provide higher capacitance voltage (Figure 13.8).

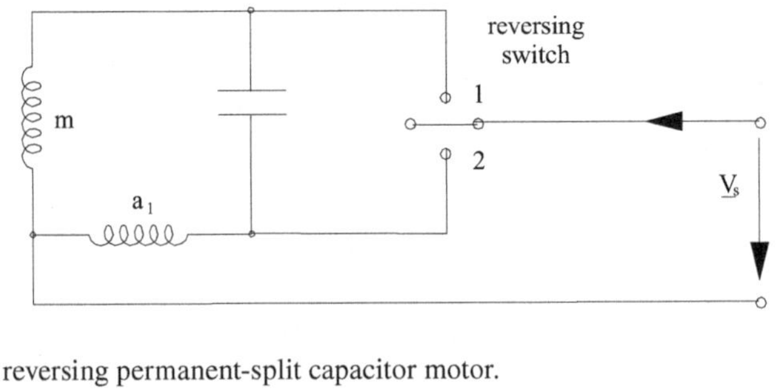

FIGURE 13.5 The reversing permanent-split capacitor motor.

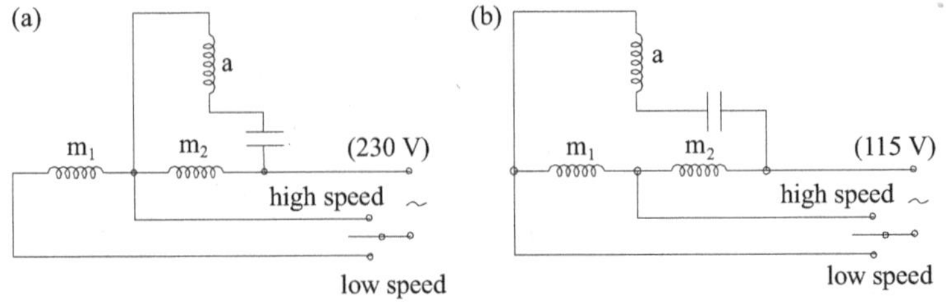

FIGURE 13.6 T (a) and L (b) connections of tapped-winding capacitor induction motors.

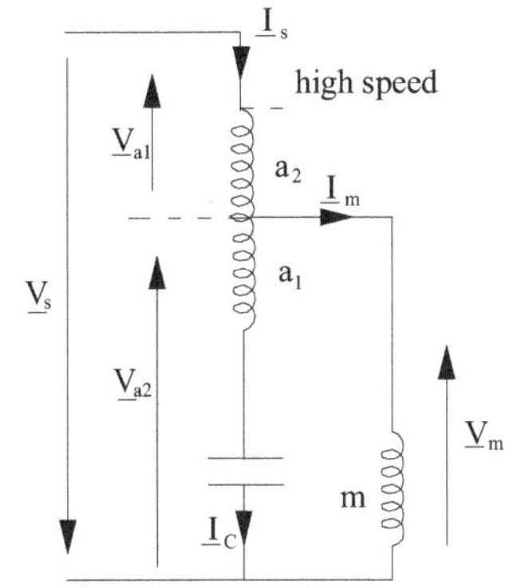

FIGURE 13.7 General tapped winding. Capacitor induction motor.

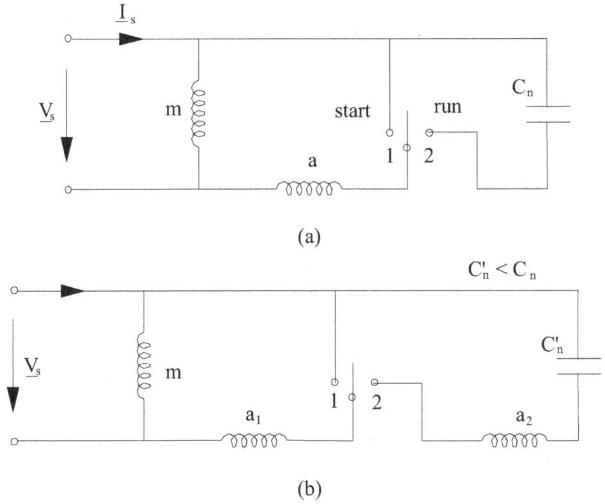

FIGURE 13.8 The split-phase capacitor induction motor (a) with single- and (b) two-section auxiliary winding.

13.3.6 Capacitor Three-Phase Induction Motors

The three-phase winding induction motors may be connected and provided with a capacitor to comply with single-phase power supplies. The typical connections, also called Steinmetz star and delta connections, are shown in Figure 13.9. As seen in Figure 13.9c, with an adequate capacitance C_n, the voltage of the three phases A, B, and C may be made symmetric for a certain value of slip ($S = 1$ or $S' = S_n$ or a value in between).

The star connection prevents the occurrence of third mmf space harmonic, and thus, its torque–speed curve does not show the large dip at 33% synchronous speed, accentuated by saturation and typical for low-power capacitor single-phase induction motors with two windings. It is not so for delta connection (Figure 13.9b) which allows for the third mmf space harmonic. However, the delta connection provides $\sqrt{3}$ times larger voltage per phase and thus, basically, 3 times more torque in comparison with the star connection, for same single-phase supply voltage.

The three-phase windings also allow for pole changing to reduce the speed in a 2:1 or other ratios (see Chapter 4, Volume 1) – Figure 13.10.

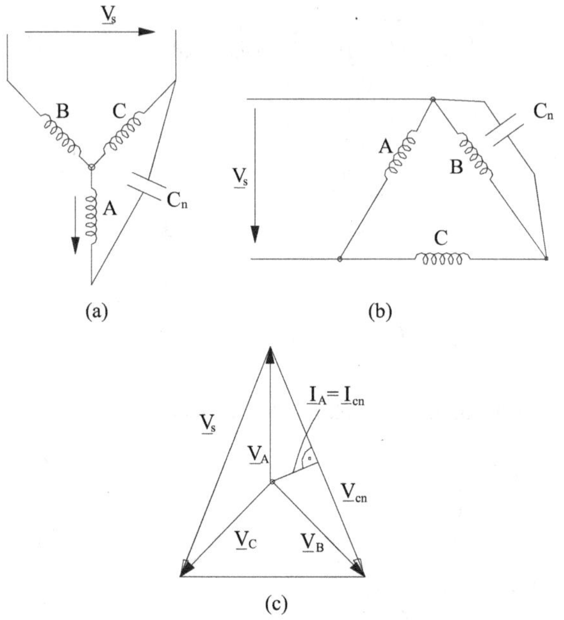

FIGURE 13.9 Three-phase winding IM with capacitors for single-phase supply: (a) star, (b) delta, and (c) phasor diagram (star).

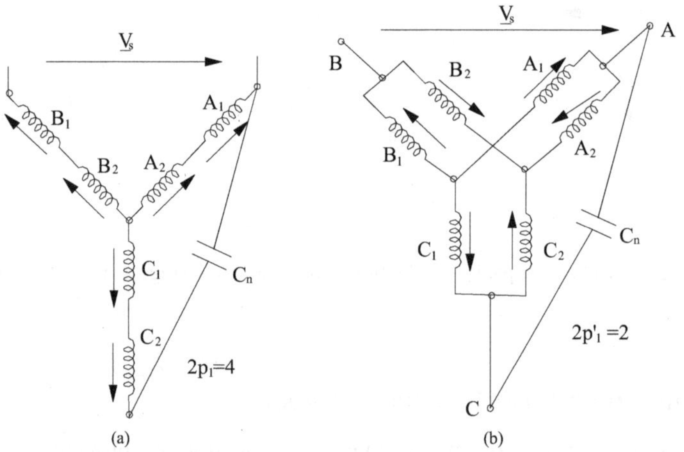

FIGURE 13.10 Pole-changing single-phase induction motor with three-phase stator winding: (a) $2p_1 = 4$ poles and (b) $2p_1' = 2$ poles.

13.3.7 Shaded-Pole Induction Motors

The shaded-pole induction motor has two (4, 6) concentrated coil stator winding, with $2p_1 = 2$, four and six poles, connected to a single-phase A.C. power grid. To provide for self-starting, a displaced short-circuited winding is located on a part of main winding poles (Figure 13.11).

The main winding current flux induces a voltage in the shaded-pole coil which in turn produces current which affects the total flux in the shaded area. So the flux in the unshaded area $\underline{\Phi}_{mm}$ is both space-wise and time-wise shifted ahead with respect to the flux in the shaded area ($\underline{\Phi}_a$).

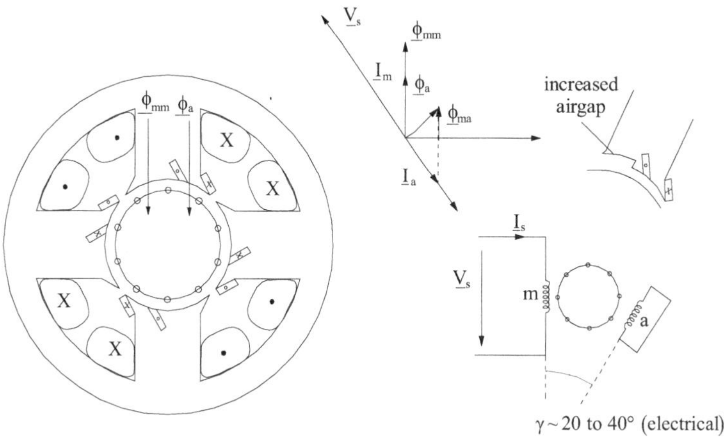

FIGURE 13.11 The four-pole shaded-pole induction motor.

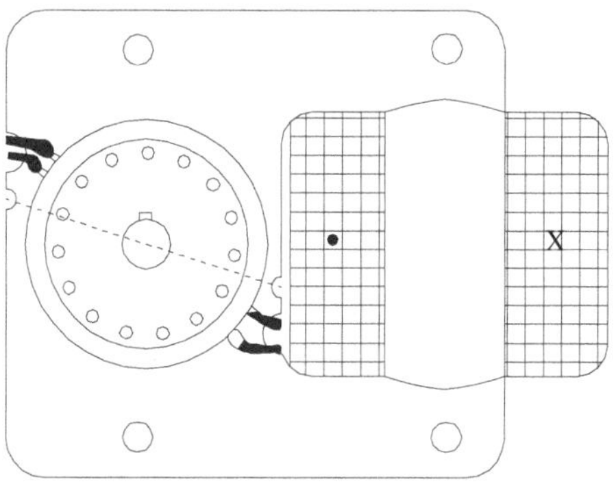

FIGURE 13.12 Two-pole C-shape shaded-pole induction motor.

These two fluxes will produce a travelling field component that will move the rotor from the unshaded to the shaded area. Some further improvement may be obtained by increasing the airgap at the entry end of stator poles (Figure 13.11). This beneficial effect is due to the third space-harmonic field reduction.

As the space distribution of airgap field is far from a sinusoid, space-harmonic parasitic asynchronous torques occur. The third harmonic is in general the largest and produces a notable dip in the torque–speed curve around 33% of synchronous speed $n_1 = f_1/p_1$.

The shaded-pole induction motor is low cost but also has low starting torque, torque density, low efficiency (less than 40%), and power factor. Due to these demerits, it is used only at low powers, between 1/25 and 1/5 kW in round frames (Figure 13.11), and less and less.

For even lower power levels (tens of mW), an unsymmetrical (C-shape) stator configuration is used (Figure 13.12). A strong increase in efficiency (say from 40% to 60% in a 100 W, two-pole such machine) may be obtained in principle by simply opening the circuit of shade-pole coils after starting by using a PTC.

13.4 THE NATURE OF STATOR-PRODUCED AIRGAP FIELD

The airgap field distribution in single-phase IMs, for zero rotor currents, depends on the distribution of the two stator windings in slots and on the amplitude ratio I_m/I_a and phase shift γ of phase currents.

When the current waveforms are not sinusoidal in time, due to magnetic saturation or due to power grid voltage time harmonics, the airgap field distribution gets even more complicated. For the time being, let us neglect the stator current time harmonics and consider zero rotor current.

Chapter 4 presented in some detail typical distributed windings for capacitor single-phase induction motors.

Let us consider here the first and the third space harmonic of stator winding mmfs

$$F_m(\theta_{es},t) = F_{1m}\cos\omega_1 t\left[\cos\theta_{es} + \frac{F_{3m}}{F_{1m}}\cos(3\theta_{es})\right] \tag{13.1}$$

$$F_a(\theta_{es},t) = -F_{1a}\cos(\omega_1 t + \gamma_i)\left[\sin\theta_{es} + \frac{F_{3a}}{F_{1a}}\sin(3\theta_{es})\right] \tag{13.2}$$

θ_{es} – stator position electrical angle.

For windings with dedicated slots and coils with same number of turns, expressions of F_{1m}, F_{3m}, F_{1a}, and F_{3a} (see Chapter 4) are

$$F_{1m\nu} = \frac{2N_m K_{Wm\nu} I_m\sqrt{2}}{\pi\nu p_1};\quad \nu = 1,3 \tag{13.3}$$

$$F_{1a\nu} = \frac{2N_a K_{Wa\nu} I_a\sqrt{2}}{\pi\nu p_1};\quad \nu = 1,3$$

The winding factors $K_{Wm\nu}$ and $K_{Wa\nu}$ depend on the relative number of slots for the main and auxiliary windings (S_m/S_A) and on the total number of stator slots S_1 (see Chapter 4).

For example, in a four-pole motor with $S_1 = 24$ stator slots, 16 slots for the main winding and 8 slots for the auxiliary winding, both of single-layer type, the coils are full pitch. Also, $q_m = 4$ slots/pole/phase for the main winding and $q_a = 2$ slots/pole/phase for the auxiliary winding. Consequently,

$$K_{Wm\nu} = \frac{\sin q_m \frac{\alpha_{c\nu}}{2}}{q_m \sin\frac{\alpha_{c\nu}}{2}} = \frac{\sin 4\cdot\frac{\pi}{24}\nu}{4\sin\frac{\pi}{24}\nu} \tag{13.4}$$

$$K_{Wa\nu} = \frac{\sin q_a \frac{\alpha_{c\nu}}{2}}{q_a \sin\frac{\alpha_{c\nu}}{2}} = \frac{\sin 2\cdot\frac{\pi}{24}\nu}{2\sin\frac{\pi}{24}\nu} \tag{13.5}$$

$F_{3m}/F_{1m} = 0.226$; $F_{3a}/F_{1a} = 0.3108$.

Let us first neglect the influence of slot openings while the magnetic saturation is considered to produce only an "increased apparent airgap".

Consequently, the airgap magnetic field $B_{go}(\theta_{er}, t)$ is

$$B_{go}(\theta_{es},t) = \frac{\mu_0}{gK_c(1+K_s)}\left[F_m(\theta_{es},t) + F_a(\theta_{es},t)\right] \tag{13.6}$$

K_s – saturation coefficient, K_c – Carter coefficient.

So, even in the absence of slot opening and saturation influence on airgap magnetic permeance, the airgap field contains a third space harmonic. The presence of slot openings produces harmonics in the airgap permeance (see Chapter 10, Section 10.3). As in single-phase IM both stator and rotor slots are semiclosed, these harmonics may be neglected in a first-order analysis.

On the other hand, magnetic saturation of slot neck leakage path produces a second-order harmonic in the airgap permeance (Chapter 10, Section 10.4). This, in turn, is small unless the currents are large (starting conditions).

The main flux path magnetic saturation tends to create a second-order airgap magnetic conductance harmonic (Chapter 10, Section 10.5)

$$\lambda_{gs}(\theta_{es}) = \frac{1}{gK_C}\left[\frac{1}{1+K_{s1}} + \frac{1}{1+K_{s2}}\sin\left(2\frac{\theta_{es}}{p_1} - \omega_{1t} - \gamma_s\right)\right] \quad (13.7)$$

K_{s1} and K_{s2} are saturation coefficients that result from the decomposition of main flux distribution in the airgap.

We should mention here that rotor eccentricity also "induces" airgap magnetic conductance harmonics (Chapter 10, Section 10.7).

For example, the static eccentricity $\varepsilon = e/g$ yields

$$\lambda_{ge}(\theta_{es}) \approx \frac{1}{g}\left(C_0 + C_1\cos\frac{\theta_{es}}{p_1}\right)$$
$$C_0 = \frac{1}{\sqrt{1-\xi^2}};C_1 = 2(C_0 - 1)/\xi \quad (13.8)$$

Now the airgap flux density $B_{g0}(\theta_{es},t)$, for zero rotor currents, becomes

$$B_{g0}(\theta_{es},t) = \mu_0\lambda_{gs}(\theta_{es})\left[F_m(\theta_{es},t) + F_a(\theta_{es},t)\right] \quad (13.9)$$

As seen from (13.9) and (13.1)–(13.2), the main flux path saturation introduces an additional third space-harmonic airgap field, besides that produced by the mmf third space harmonic. The two third harmonic field components are phase shifted with an angle γ_3 dependent on the relative ratio of the main and auxiliary winding mmf amplitudes and on their time phase lag.

The third space harmonic of airgap field is expected to produce a third-order notable asynchronous parasitic torque whose synchronism occurs around 1/3 of ideal no-load speed $n_1 = f_1/p_1$ (Figure 13.13 [3]). A rather complete solution of the airgap field distribution, for zero rotor currents (the rotor does not yet have the cage in slots), may be obtained by finite element modelling (FEM).

In face of such complex field distributions in time and space, simplified approaches have traditionally become widely accepted.

Let us suppose that magnetic saturation and mmf-produced third harmonics are neglected. That is, we return to (13.6).

13.5 THE FUNDAMENTAL MMF AND ITS ELLIPTIC WAVE

Neglecting the third harmonic in (13.1) and (13.2), the main and auxiliary winding mmfs are

$$F_m(\theta_{es},t) = F_{1m}\cos\omega_1 t \times \cos\theta_{es} \quad (13.10)$$

$$F_a(\theta_{es},t) = -F_{1a}\cos(\omega_1 t + \gamma_i)\sin\theta_{es} \quad (13.11)$$

From (13.3), F_{1a}/F_{1m} is

$$F_{1a}/F_{1m} = \left(\frac{N_m K_{wm1}}{N_a K_{wa1}}\cdot\frac{I_m}{I_a}\right)^{-1} = \left(\frac{1}{a}\frac{I_m}{I_a}\right)^{-1} \quad (13.12)$$

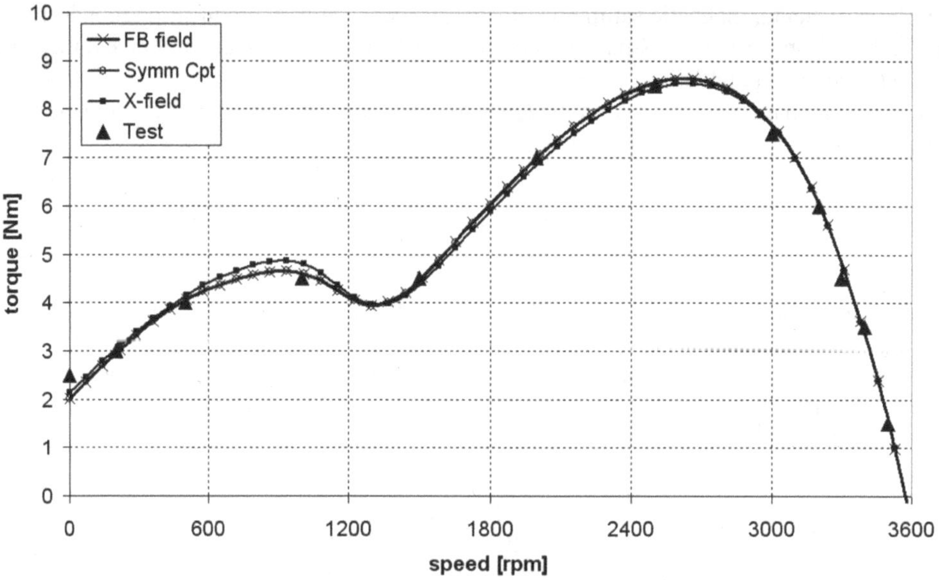

FIGURE 13.13 Typical torque/speed curve with visible third space harmonics influence around 1200 rpm (33% of 3600 rpm).

Given the sinusoidal spatial distribution of the two mmfs, their resultant space vector F(t) may be written as

$$F(t) = F_{1m} \sin\omega_1 t + jF_{1a} \sin(\omega_1 t + \gamma_i) \tag{13.13}$$

So the amplitude F(t) and its angle γ(t) are

$$F(t) = \sqrt{F_{1m}^2 \cos^2\omega_1 t + F_{1a}^2 \sin^2(\omega_1 t + \gamma_i)}$$

$$\gamma(t) = \tan^{-1}\left[-a\frac{I_a}{I_m}\frac{\cos(\omega_1 t + \gamma_i)}{\cos\omega_1 t}\right] \tag{13.14}$$

A graphical representation of (13.13) in a plane with the real axis along the main winding axis and the imaginary one along the auxiliary winding axis is shown in Figure 13.14. An elliptic hodograph is evident.

Both the amplitude F(t) and time derivative (dγ/dt) (speed) vary in time from a maximum to a minimum value.

Only for symmetrical conditions, when the two mmfs have equal amplitudes and are time shifted by $\gamma_i = 90°$, the hodograph of the resultant mmf is a circle. That is, a pure **travelling** wave ($\gamma = \omega_1 t$) is obtained. It may be demonstrated that the mmf wave speed dγ/dt is positive for $\gamma_i > 0$ and negative for $\gamma_i < 0$.

In other words, the motor speed may be reversed by changing the sign of the time phase shift angle between the currents in the two windings. One way to do it is to switch the capacitor from auxiliary to main winding (Figure 13.5).

The airgap field is proportional to the resultant mmf (13.14), so the elliptic hodograph of mmf stands valid for this case also.

When heavy magnetic saturation occurs for sinusoidal voltage supply, at zero rotor currents, the stator currents become nonsinusoidal, while the e.m.fs, and the phase flux linkages, also depart from time sinusoids.

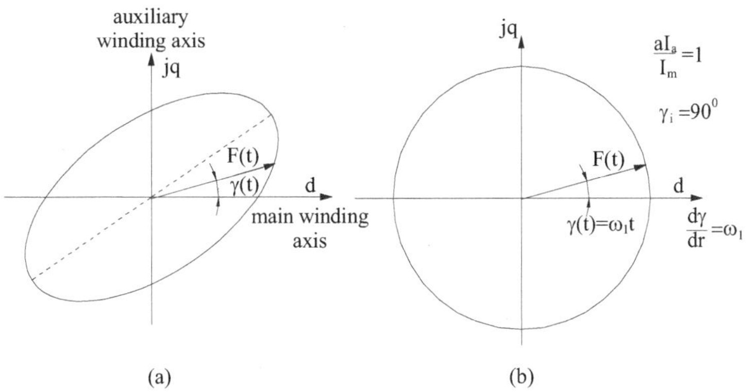

FIGURE 13.14 The hodograph of fundamental mmf in single-phase IMs (a) in general (b) for symmetrical conditions.

The presence of rotor currents further complicates the picture. It is, thus, fair to say that the computation of field distribution, currents, power, torque for the single-phase IMs is more complicated than for three-phase IMs. A magnetic saturation model will be introduced in Chapter 4, Volume 2.

13.6 FORWARD–BACKWARD MMF WAVES

The forward–backward revolving field theory is attributed to Morrill [4] and Veinott [5]. Here, only the basics are discussed.

The resultant stator mmf $F(\theta_{es},t)$ – (13.10 and 13.11) – may be decomposed in two waves

$$F(\theta_{es},t) = F_{1m}\cos\omega_1 t\cos\theta_{es} + F_{1a}\cos(\omega_1 t+\gamma_i)\sin\theta_{es}$$
$$= \frac{1}{2}F_m C_f \sin(\theta_{es} - \omega_1 t - \beta_f) + \frac{1}{2}F_m C_b \sin(\theta_{es} + \omega_1 t - \beta_b) \tag{13.15}$$

$$C_f = \sqrt{\left(1+\frac{F_{1a}}{F_{1m}}\sin\gamma_i\right)^2 + \left(\frac{F_{1a}}{F_{1m}}\right)^2 \cos\gamma_I{}^2}$$
$$\sin\beta_f = \frac{\left(1+F_{1a}\sin\gamma_i/F_{1m}\right)}{C_f} \tag{13.16}$$

$$C_b = \sqrt{\left(1-\frac{F_{1a}}{F_{1m}}\sin\gamma_i\right)^2 + \left(\frac{F_{1a}}{F_{1m}}\cos\gamma_i\right)^2}$$
$$\sin\beta_b = \frac{\left(1-F_{1a}\sin\gamma_i/F_{1m}\right)}{C_b} \tag{13.17}$$

Again, for $F_{1a}/F_{1m} = 1$ and $\gamma_i = 90°$, $C_b = 0$, and thus, the backward (b) travelling wave becomes zero (the case of circular hodograph in Figure 13.14b).

Also, the forward (f) travelling amplitude is in this case equal to $F_{1m} = F_{1a}$ (it is 3/2 F_{1m} for a three-phase symmetrical winding).

For $F_{1a} = 0$, the f.&b. waves have the same magnitude: ½ F_{1m}.

The mmf decomposition in forward and backward waves may be done even with saturation while the superposition of the forward and backward field waves is correct only in the absence of magnetic saturation.

The f.&b. decomposition is very practical as it leads to simple equivalent circuits with the slip $S_f = S$ and $S_b = 2 - S$. This is how the travelling (f.&b.)-revolving field theory of single-phase IMs evolved [4,5].

$$S_f = \frac{\omega_1 - \omega_r}{\omega_r} = S; S_b = \frac{(-\omega_1) - \omega_r}{(-\omega_1)} = 2 - S \tag{13.18}$$

For steady state, the f.&b. model is similar to the symmetrical components model. On the other hand, for the general case of transients or nonsinusoidal voltage supply, the d-q cross-field model has become standard.

13.7 THE SYMMETRICAL COMPONENTS' GENERAL MODEL

For steady state and sinusoidal currents, time phasors may be used. The unsymmetrical mmf, currents, and voltages corresponding to the two windings of single-phase IMs may be decomposed in two symmetrical systems (Figure 13.15) which are in fact the forward and backward components introduced in the previous paragraph.

It goes without saying that in general the two windings are 90° electrical degrees phase shifted spatially (unlike in Figure 13.15).

From Figure 13.15 (with $\gamma_i = 90°$),

$$\underline{A}_{a+} = j A_{m+}; \underline{A}_{a-} = -j \underline{A}_{m-} \tag{13.19}$$

The superposition principle yields

$$\begin{aligned} \underline{A}_m &= \underline{A}_{m+} + \underline{A}_{m-} \\ \underline{A}_a &= \underline{A}_{a+} + \underline{A}_{a-} \end{aligned} \tag{13.20}$$

From (13.19) and (13.20)

$$A_{m+} = \frac{1}{2}\left(\underline{A}_m - j\underline{A}_a\right) \tag{13.21}$$

$$A_{m-} = \frac{1}{2}\left(A_m + j\underline{A}_a\right) = \underline{A}_{m+}^{*} \tag{13.22}$$

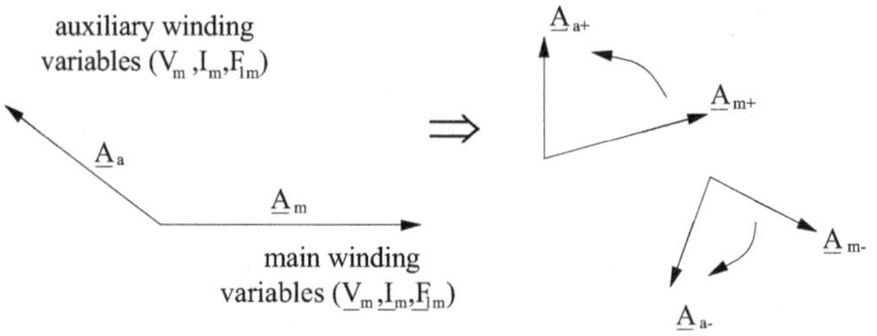

FIGURE 13.15 Symmetrical (f.&b. or + –) components of a two-phase winding motor.

Denoting by $\underline{V}_{m+}$, $\underline{V}_{m-}$, $\underline{I}_{m+}$, $\underline{I}_{m-}$ and $\underline{V}_{a+}$, $\underline{V}_{a-}$, $\underline{I}_{a+}$, $\underline{I}_{a-}$, the voltage and current components of the two winding circuits, we may write the equations of the two fictitious two-phase symmetric machines (whose travelling fields are opposite (f,b)) as

$$\begin{aligned}\underline{V}_{m+} &= \underline{Z}_{m+}\underline{I}_{m+};\ \underline{V}_{m-} = \underline{Z}_{m-}\underline{I}_{m-}\\ \underline{V}_{a+} &= \underline{Z}_{a+}\underline{I}_{a+};\ \underline{V}_{a-} = \underline{Z}_{a-}\underline{I}_{a-}\\ \underline{V}_{m} &= \underline{V}_{m+} + \underline{V}_{m-};\ \underline{V}_{a} = \underline{V}_{a+} + \underline{V}_{a-}\end{aligned} \tag{13.23}$$

The $\underline{Z}_{m+}$, $\underline{Z}_{m-}$ and $\underline{Z}_{a+}$, $\underline{Z}_{a-}$ represent the resultant forward and backward standard impedances of the fictitious IM, per phase, with the rotor cage circuit reduced to the "m", and respectively, to "a" stator winding (Figure 13.16).

We still need to solve Equations (13.23), with given slip S value and all parameters in $\underline{Z}_{m+,-}$ and $\underline{Z}_{a+,-}$ known. A relationship between the $\underline{V}_m$, $\underline{V}_a$ and the source voltage $\underline{V}_s$ is needed.

When an impedance ($\underline{Z}_a$) is connected in series with the auxiliary winding, and then both windings are connected to the power grid given voltage $\underline{V}_s$, we obtain a simple relationship

$$\underline{V}_m = \underline{V}_s;\ \underline{V}_a + (\underline{I}_{a+} + \underline{I}_{a-})\underline{Z}_a = \underline{V}_s \tag{13.24}$$

It is in general more comfortable to reduce all windings on rotor and stator to the main winding.

The reduction ratio of auxiliary to main winding is the coefficient a in Equation (13.12)

$$a = \frac{N_a}{N_m}\frac{K_{w1a}}{K_{W1m}} \tag{13.25}$$

The rotor circuit is symmetric when $a^2X_{rm} = X_{ra}$, $a^2R_{rm} = R_{ra}$.

Consequently, when

$$\begin{aligned}\Delta X_{sa} &= X_{sa} - a^2X_{sm} = 0\\ \Delta R_{sa} &= R_{sa} - a^2R_{sm} = 0\end{aligned} \tag{13.26}$$

the two equivalent circuits in Figure 13.16, both reduced to the main winding, become identical.

This situation occurs in general when both windings occupy the same number of uniform slots and the design current density for both windings is the same. When the auxiliary winding occupies 1/3 of stator periphery, its design current density may be higher to fulfill (13.26).

The voltage and current reduction formulas from the auxiliary winding to main winding are

$$\underline{V}_a' = \underline{V}_a/a;\ \underline{I}_a' = a\underline{I}_a \tag{13.27}$$

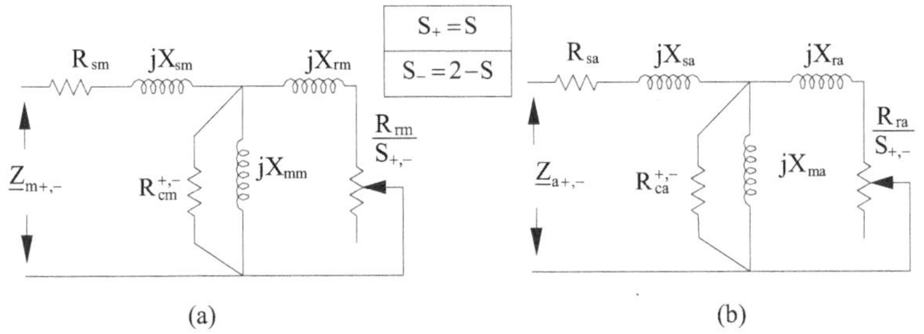

FIGURE 13.16 Equivalent symmetrical (+,–) impedances (a) reduced to the main stator winding m and (b) reduced to the auxiliary stator winding a.

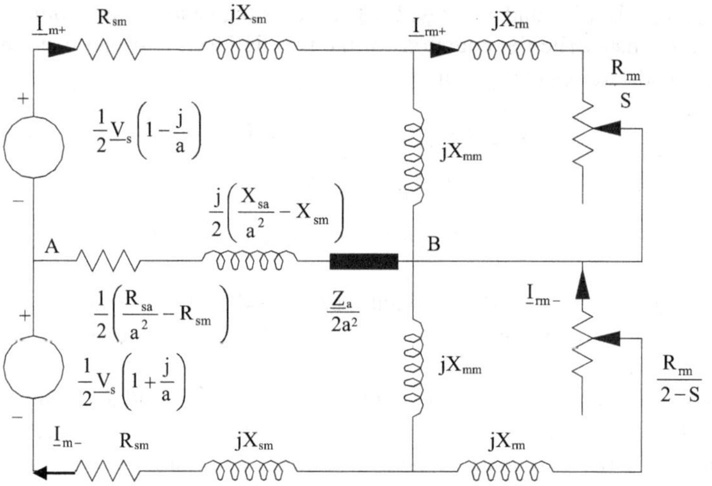

FIGURE 13.17 General equivalent circuit of split-phase and capacitor induction motor.

In such conditions, when $\Delta X_{sa} \neq 0$ and $\Delta R_{sa} \neq 0$, their values are introduced in series with the capacitor impedance $\underline{Z}_a$ after reduction to the main winding and division by 2. The division by 2 is there to conserve powers (Figure 13.17).

Note again that $\underline{Z}_a$ contains in general the capacitor. In split-phase IMs, $\underline{Z}_a = 0$. When the auxiliary winding is open, $\underline{Z}_a = \infty$. Also note that the + – voltages are now defined at the grid terminals ($V_a = V_s/a$) rather than at the winding terminals.

This is how the A to B section impedance in Figure 13.17 takes its place inside the equivalent circuit. The equivalent circuit in Figure 13.17 is fairly general. It has only two unknowns, I_{m+} and I_{m-}, and may be solved rather easily for given slip, machine parameters, and supply voltage $\underline{V}_s$.

Tapped-winding (multiple-section winding or three-phase winding) capacitor IMs may be represented directly by the equivalent circuit in Figure 13.17. Core loss resistances may be added in Figure 13.17 in parallel with the magnetisation reactance X_{mm} or with the leakage reactances (for additional losses) but the problem is how to calculate (measure) them.

Main flux path saturation may be included by making the magnetisation reactance X_{mm} a variable. The first temptation is to consider that only the forward (direct component) X_{mm} (the upper part of Figure 13.17) saturates, but, as shown earlier, the elliptic mmf makes the matters far more involved.

This is why the general equivalent circuit in Figure 13.17 is suitable when the saturation level is constant or neglected.

To complete the picture, we add here the average torque expression

$$T_e = T_{e+} + T_{e-} = \frac{2p_1}{\omega_1}\left[I_{rm+}^{\ 2} \cdot \frac{R_{rm}}{S} - \frac{I_{rm-}^{\ 2} \cdot R_{rm}}{2-S}\right] \tag{13.28}$$

For the case when the auxiliary winding is open ($Z_a = \infty$), $I_{rm+} = I_{rm-}$. In such a case for S = 1 (start), the total torque T_e is zero, as expected.

It may be argued that as the rotor currents experience two frequencies: $f_{2+} = Sf_1$ and $f_2 = (2 - S)f_1 > f_1$, the rotor cage parameters differ for the two components due to skin effects.

To account for skin effect, two fictitious cages in parallel may be considered in Figure 13.17. Their parameters are chosen to fit the rotor impedance variation from $f_2 = 0$ to $f_2 = 2f_1$ according to a certain optimisation criterion.

13.8 THE d-q MODEL

As the stator is provided with two orthogonal windings and the rotor is fully symmetric, the single-phase IM is suitable for the direct application of d-q model in stator coordinates (Figure 13.18).

The rotor will be reduced here to the main winding, while the auxiliary winding is not reduced to the main winding.

The d-q model equations in stator coordinates are straightforward

$$\begin{aligned} I_m R_{sm} - V_m &= -\frac{d\psi_m}{dt} \\ I_a R_{sa} - V_a &= -\frac{d\psi_a}{dt} \\ i_D R_r &= -\frac{d\psi_D}{dt} - \omega_r \psi_a \\ I_Q R_r &= -\frac{d\psi_Q}{dt} + \omega_r \psi_D \end{aligned} \tag{13.29}$$

The flux current relationships are

$$\begin{aligned} \psi_D &= L_{rm} I_D + L_{mm}(I_m + I_D) \\ \psi_Q &= L_{rm} I_Q + L_{am}(aI_q + I_Q)/a^2 \\ \psi_m &= L_{sm} I_m + L_{mm}(I_m + I_D) \\ \psi_a &= L_{sa} I_a + L_{am}\left(I_a + \frac{I_Q}{a}\right) \end{aligned} \tag{13.30}$$

For non-linear saturation conditions, it is better to define a, the reduction ratio, as

$$a = \sqrt{\frac{L_{am}}{L_{mm}}} \tag{13.31}$$

This aspect will be treated in a separate chapter on testing (Chapter 14, Volume 2). The magnetisation inductances L_{mm} and L_{am} along the two axes (windings) are dependent on the resultant magnetisation current I_μ

$$I_\mu(t) = \sqrt{(I_m + I_D)^2 + (aI_a + I_Q)^2} \tag{13.32}$$

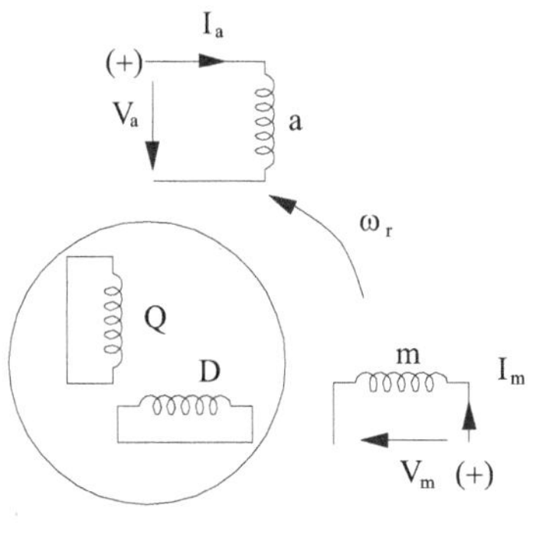

FIGURE 13.18 The d-q model of single-phase IM.

The torque and motion equations are straightforward.

$$T_e = -p_1\left(\psi_D I_Q - \psi_Q I_D\right) = p_1 L_{mm}\left(\sqrt{\frac{L_{am}}{L_{mm}}} i_a I_D - I_m I_Q\right) \quad (13.33)$$

$$\frac{J}{p_1}\frac{d\omega_r}{dt} = T_e - T_{load}\left(\omega_r, \theta_{er}, t\right) \quad (13.34)$$

Still missing are the relationships between the stator voltages V_m, V_a and the source voltage V_s. In a capacitor IM,

$$\begin{aligned} V_m(t) &= V_s(t) \\ V_a(t) &= V_s(t) - V_c(t) \end{aligned} \quad (13.35)$$

$$C\frac{dV_C(t)}{dt} = I_a(t) \quad (13.36)$$

V_C – is the capacitor voltage.

A six-order non-linear model has been obtained. The variables are: I_m, I_a, I_D, I_Q, ω_r, and V_C, while the source voltage $V_s(t)$ and the load torque T_{load} (ω_r, θ_{er}, t) constitute the inputs.

With known $L_{am}(I_\mu)$ and $L_{mm}(I_\mu)$ functions, even the saturation presence may be handled elegantly, both during transients and steady state. The implicit mathematical condition in the d-q model here is $\Delta X_{sa} = 0$ and $\Delta R_{sa} = 0$.

13.9 THE d-q MODEL OF STAR STEINMETZ CONNECTION

The three-phase winding IM, in the capacitor motor connection for single-phase supply, may be reduced simply to the d-q (m,a) model as

$$V_m + jV_a = \sqrt{\frac{2}{3}}\left(V_A(t) + V_B(t)e^{j\frac{2\pi}{3}} + V_C(t)e^{-j\frac{2\pi}{3}}\right)e^{-j\theta_0} \quad (13.37)$$

The same transformation is valid for currents, while the phase resistances and leakage inductances remain unchanged.

The voltage relationships depend on the connection of phases. For the Steinmetz star connection (Figure 13.19),

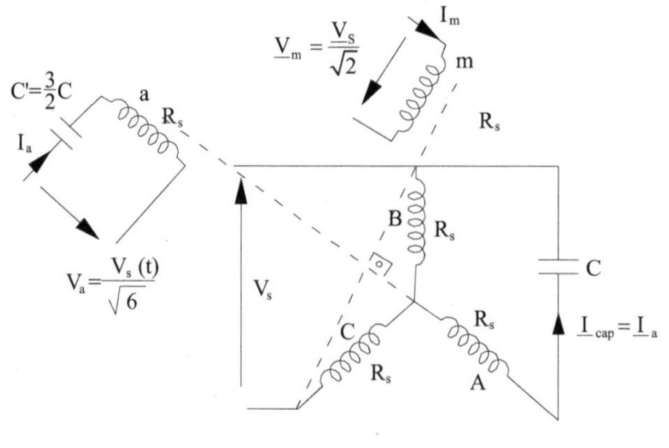

FIGURE 13.19 Three- (two-) phase equivalence.

$$\begin{aligned}
&V_B(t) - V_C(t) = V_S(t) \\
&V_A(t) + V_B(t) + V_C(t) = 0 \\
&V_B(t) - V_A(t) = V_{cap}(t) \\
&I_A(t) = I_{cap}(t) \\
&C\frac{dV_{cap}}{dt} = I_{cap}(t)
\end{aligned} \tag{13.38}$$

With $\theta_0 = -\pi/2$, d axis falls along the axis of the main winding

$$\begin{aligned}
&V_m(t) = \frac{1}{\sqrt{2}}(V_B - V_C) = \frac{V_S(t)}{\sqrt{2}} \\
&V_a(t) = +\sqrt{\frac{3}{2}}V_A(t);\ i_a(t) = \sqrt{\frac{3}{2}}i_A(t)
\end{aligned} \tag{13.39}$$

From (13.38) with (13.39)

$$\sqrt{\frac{2}{3}}V_{cap} + V_a = \frac{V_S(t)}{\sqrt{6}} \tag{13.40}$$

Notice that (as in 13.39)

$$I_a = \sqrt{\frac{3}{2}}I_{cap} \tag{13.41}$$

For the d-q model, Equation (13.40) becomes

$$\begin{aligned}
&V'_{cap} + V_a = \frac{V_S(t)}{\sqrt{6}} \\
&\frac{dV'_{cap}}{dt} = \frac{1}{C'}I_a
\end{aligned} \tag{13.42}$$

Also,

$$V'_{cap} = \frac{1}{C'}\cdot\int I_a\,dt = \sqrt{\frac{2}{3}}\frac{1}{C}\int I_{cap}\,dt = \frac{2}{3C}\int I_a\,dt = \frac{1}{\left(\frac{3}{2}C\right)}\int I_a\,dt \tag{13.43}$$

So,

$$C' = \frac{3}{2}C \tag{13.44}$$

The equivalent capacitance circuit C' in the d-q (m,a) – two phase – model is 1.5 times higher than the actual capacitance C. The source voltage to the main winding m(d) is $V_s/\sqrt{2}$ instead of V_s as it is for the usual model. For the auxiliary winding, the “virtual source” voltage is $V_s/\sqrt{6}$ (Equation 13.40). Finally, the resistance, leakage inductance, and magnetisation inductance are those of the three-phase machine per phase. Consequently, for the m,a model, stator windings are now identical.

The steady-state performance may be approached also by running transients from start (or given initial conditions) until conditions stabilise. As the state-space system of equations is solvable only via numerical methods, its solution takes time. With zero initial values for currents, capacitor voltage, and given (final) slip (speed), this time is somewhat reduced.

Alternatively, it is possible to use $j\omega_1$ instead of d/dt in Equations (13.29) and to extract the real part of torque expression, after solving the algebraic system of equations for the currents and the capacitor voltage.

This approach may be more suitable than the travelling (revolving) field model when the magnetic saturation is to be accounted for, because the resultant (elliptical) magnetic field in the machine may be accounted for directly. However, it does not exhibit the intuitive attributes of the revolving field theory. In the presence of strong computation facilities, this steady-state approach to the d-q model may revive for usage when magnetic saturation is heavy.

Finally, the reduction of the three-phase Steinmetz star connection to the m,a model performed above may also be used in the symmetrical components model (Figure 13.17).

13.10 PM-ASSISTED SPLIT-PHASE CAGE-ROTOR IMS

Cage and PM rotor IMs are vigorously proposed for low-power residential appliances in an effort to save energy by increased efficiency.

A 3-year payback time for the extra investment in such a motor (in comparison with the capacitor IM) may be considered economically advantageous.

The single flux barrier PM cage rotor (Figure 13.20a) implies strong PMs ($emf/V_s > 0.7$) as the reluctance synchronous torque component is small (low magnetic saliency $\frac{X_{qm}}{X_{dm}} < 1.5$).

In contrast, for the multiple flux barrier PM rotor IM, we may consider ferrite (or bounded NdBFe) magnets ($emf/V_s = 0.3–0.4$) used mainly to increase efficiency, power factor (smaller running capacitor), and better starting (lower PM braking torque) relying on larger reluctance synchronous torque due to higher rotor magnetic saliency $\left(\frac{X_{qm}}{X_{dm}} > 2.5\right)$.

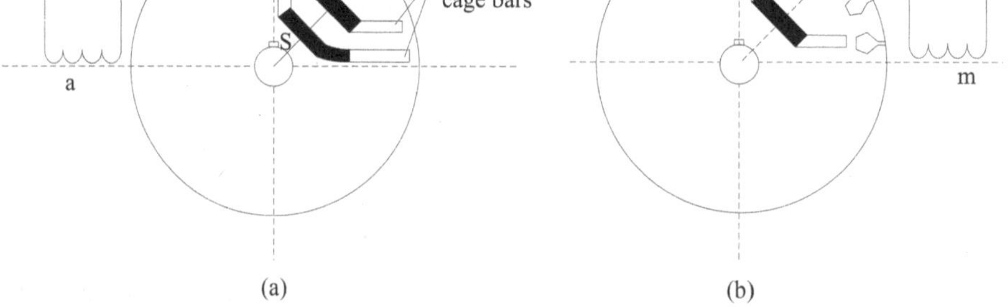

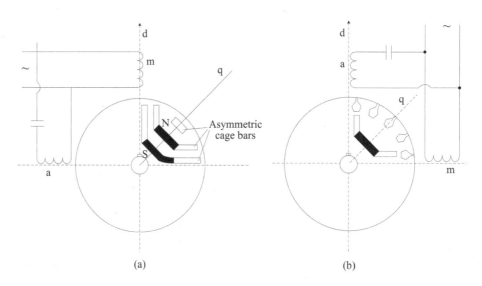

FIGURE 13.20 Four-pole cage and PM rotor capacitor IM: (a) with single flux barrier and PM rotor $\left(\frac{X_{qm}}{X_{dm}} < 1.5\right)$ and (b) with multiple flux barrier PM rotor: $\left(\frac{X_{qm}}{X_{dm}} > 2.5\right)$.

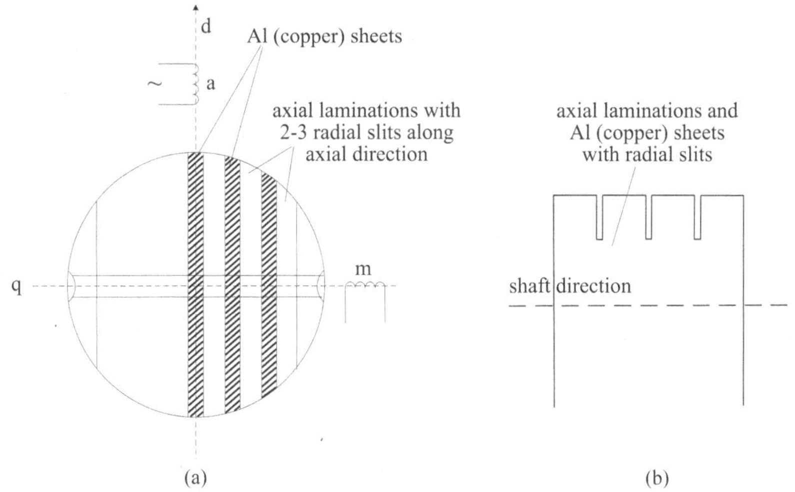

FIGURE 13.21 Cage ALA two-pole rotor (a) and slitted axial laminations and Al (copper) sheets (b).

The capacitor needed for good starting is for such motors anyway larger than for their IM in the same frame due to the PM braking torque from standstill to self-synchronisation.

The efficiency increase of a few (2%–4%) is thus payed for by PM and capacitor additional costs.

In the extreme, for two-pole motors, the PMs may be eliminated if the rotor magnetic saliency $\frac{X_{qm}}{X_{dm}} > 10$, which in principle may be obtained only with axially laminated anisotropic (ALA) non through – shaft rotors where aluminium (copper) sheets are placed between the rotor axial laminations, welded to disk-shape end rings (if any). The shaft ends are added to the structure, and in general, the fabrication of such cage ALA – rotor induction – synchronous motors is still considered involved and too expensive (Figure 13.21).

The radial slits (Figure 13.21b) are needed to reduce no-load space harmonic flux core losses in the rotor – whose flux lines go perpendicularly to lamination and Al (copper) sheet plane.

The disk-shape end rings of the sheet cage may be eliminated if the stator (rotor) stack length l_{stack}/pole pitch ≥ 1, without too large an increase in equivalent rotor resistance. Optimally designed such a cage ALA rotor RelSyn motors may qualify for premium efficiency line-start small-power applications at reasonable costs. If the cage is replaced by insulation sheets and the machine is three phase, a variable speed motor drive is obtained.

13.11 SUMMARY

- Single-phase supply induction motors have a cage rotor and two-phase stator windings: the main winding and the auxiliary winding.
- The two windings are phase shifted spatially, in general, by 90 electrical degrees.
- The auxiliary winding may have a larger resistance/reactance ratio or a series capacitance to push ahead its current phase angle with respect to that of the main winding.
- Alternatively, the auxiliary winding may be short-circuited, but in this case, its coils should occupy only a part of the main winding poles; a resistance may be added in series to increase somewhat the starting torque, while in running mode, the auxiliary winding is disconnected, for small power (below 100 W) in general. This is so also for the shaded-pole induction motor. It is not only low cost but also low performance. It is fabricated for powers below (1/5) kW and down to 10–15 W.

- The capacitor induction motor may be built in capacitor-start, dual-capacitor permanent-capacitor, tapped-winding capacitor, and split-phase capacitor configurations. Each of these types is suitable for certain applications. Except the capacitor-start IMs, all the other capacitor IMs have good efficiency and over 90% power factor but only capacitor-start, dual-capacitor, and split-phase configurations exhibit large starting torque (up to 250% rated torque).
- Speed reversal is obtained by "moving" the capacitor from the auxiliary to the main winding circuit. In essence, this move reverses the predominant travelling field speed in the airgap.
- The three-phase stator winding motor may be single-phase supply fed, adding a capacitor mainly in the so-called Steinmetz star connection. Notable derating occurs unless the three-phase motor was not designed for delta connection.
- The nature of the airgap field in capacitor IMs may be easily investigated for zero rotor currents (open rotor cage). The two stator winding mmfs and their summation contain a notable third space harmonic. So does the airgap flux. This effect is enhanced by the main flux path magnetic saturation which creates a kind of second space harmonic in the airgap magnetic conductance. Consequently, a notable third harmonic asynchronous parasitic torque occurs. A dip in the torque/speed curve around 33% of ideal no-load speed shows up, especially for low-power motors (below 1/5 kW).
- Even in the absence of magnetic saturation and slot opening effects, the stator mmf fundamental produces an elliptic hodograph. This hodograph degenerates into a circle when the two-phase mmfs are perfectly symmetric (90° (electrical) winding angle shift, 90° current phase shift angle).
- The main (m) and auxiliary (a) mmf fundamentals produce together an elliptic wave which may be decomposed into a forward (+) and a backward (−) travelling wave.
- The backward wave is zero for perfect symmetrisation of phases.
- For zero auxiliary mmf (current), the forward and backward mmf waves have equal magnitudes.
- The (+) (−) mmf waves produce, by superposition, corresponding airgap field waves (saturation neglected). The slip is S for the forward wave and 2-S for the backward wave. This is a phenomenological basis for applying symmetrical components to assess capacitor IM performance under steady state.
- The symmetrical components method leads to a fairly general equivalent circuit (Figure 13.17) which shows only two unknowns, the + and − components of main winding current, I_{m+} and I_{m-}, for given slip, supply voltage and machine parameters. The average torque expression is straightforward (13.18).
- When the auxiliary winding parameters R_{sa} and X_{sa}, reduced to the main winding R_{sa}/a^2 and X_{sa}/a^2, differ from the main winding parameters R_{sm} and X_{sm}, the equivalent circuit (Figure 13.17) contains, in series with the capacitance impedance $\underline{Z}'_a$ (reduced to the main winding: $\underline{Z}_a/a^2$) after division by 2, ΔR_{sa} and ΔX_{sa}. This fact brings a notable generality to the equivalent circuit in Figure 13.17.
- To treat the case of open auxiliary winding ($I_a = 0$), it is sufficient to put $Z_a = \infty$ in the equivalent circuit in Figure 13.17. In this case, the forward and backward main winding components are equal to each other, and thus, at zero speed (S = 1), the total torque (13.18) is zero, as expected.
- The d-q model seems ideal for treating the transients of single-phase induction motor. Cross-coupling saturation may be elegantly accounted for in the d-q model. The steady state is obtained after solving the sixth-order state system via a numerical method at given speed for zero initial variable values. Alternatively, the solution may be found in complex numbers by putting $d/dt = j\omega_1$. The latter procedure though handy on the computer, in the

presence of saturation, does not show the intuitive attributes (circuits) of the travelling (revolving) field theory of symmetrical components.

- The star Steinmetz connection may be replaced by a two winding capacitor motor provided the capacitance is increased by 50%, the source voltage is reduced $\sqrt{2}$ times for the main winding and $\sqrt{6}$ times for the auxiliary one. The two winding parameters are equal to phase parameters of the three-phase original machine. The star Steinmetz connection is lacking the third mmf space harmonic. However, the third harmonic saturation-produced asynchronous torque still holds.
- Improved, three-phase IM connections for single-phase operation are analysed in Refs. [6,7].
- Cage and PM rotor IMs are currently promoted for premium efficiency line start or variable speed drives [8,9].
- Next chapter deals with steady-state performance in detail.

REFERENCES

1. S. A. Nasar, editor, *Handbook of Electric Machines*, McGraw-Hill, New York, 1987, Chapter 6 by C. G. Veinott.
2. T. J. E. Miller, J. H. Gliemann, C. B. Rasmussen, D. M. Ionel, Analysis of a tapped-winding capacitor motor, *Record of ICEM*, 1998, Istanbul, Turkey, Vol. 1, pp. 581–585.
3. M. Popescu, C. B. Rasmussen, T. J. E. Miller, M. McGilp, Effect of MMF harmonics on single-phase induction motor performance – A unified approach, *Record of IEEE–IAS*, 2007, pp. 1164–1170.
4. W. J. Morrill, Revolving field theory of the capacitor motor, *Transactions of the American Institute of Electrical Engineers*, Vol. 48, 1929, pp. 614–632.
5. C. G. Veinott, *Theory and Design of Small Induction Motors*, McGraw-Hill, New York, 1959.
6. O. J. M. Smith, Three-phase induction generator for single-phase line, *IEEE Transactions on Energy Conversion*, Vol. 2, No. 3, 1987, pp. 382–387.
7. T. F. Chan, L. L. Lai, Single-phase operation of a three-phase induction generator using a novel line current injection method, *IEEE Transactions on Energy Conversion*, Vol. 20, No. 2, 2005, pp. 308–315.
8. T. Marcic, B. Stumberger, G. Stumberger, Comparison of induction motor and line-start IPM synchronous motor performance in a variable-speed drive, *IEEE Transactions on Industry Applications*, Vol. 48, No. 6, 2012, pp. 2341–2352.
9. L. N. Tutelea, T. Staudt, A. A. Popa, W. Hoffmann, I. Boldea, Line start 1 phase–source split phase capacitor cage–PM rotor-RelSyn motor: modeling, performance and optimal design with experiments, *IEEE Transactions on Industrial Electronics*, Vol. 65, No. 2, 2018, pp. 1772–1780.

14 Single-Phase Induction Motors *Steady State*

14.1 INTRODUCTION

Steady-state performance report in general on the no-load and on-load torque, efficiency and power factor, breakdown torque, locked-rotor torque, and torque and current versus speed.

In Chapter 13, we already introduced a quite general equivalent circuit for steady state (Figure 13.17) with basically only two unknowns – the forward and backward current components in the main windings – with space and time harmonics neglected.

This circuit portrays steady-state performance rather well for a basic assessment. On the other hand, the d–q model/cross-field model (see Chapter 13) may alternatively be used for the scope with $d/dt = j\omega_1$.

The cross-field model [1,2] has gained some popularity especially for T–L tapped and generally tapped winding multispeed capacitor motors [3,4]. We prefer here the symmetrical component model (Figure 13.19) for the split-phase and capacitor IMs as it seems more intuitive.

To start with, the auxiliary phase is open and a genuine single-phase winding IM steady-state performance is investigated.

Then, we move on to the capacitor motor to investigate both starting and running steady-state performance.

Further on, the same investigation is performed on the split-phase IM (which has an additional resistance in the auxiliary winding).

Then the general tapped winding capacitor motor symmetrical component model and performance is treated in some detail.

Steady-state modelling for space and time harmonics is given some attention.

Some numerical examples are meant to give a consolidated feeling of magnitudes.

14.2 STEADY-STATE PERFORMANCE WITH OPEN AUXILIARY WINDING

The auxiliary winding is open, in capacitor-start or split-phase IMs, after the starting process is over (Figure 14.1).

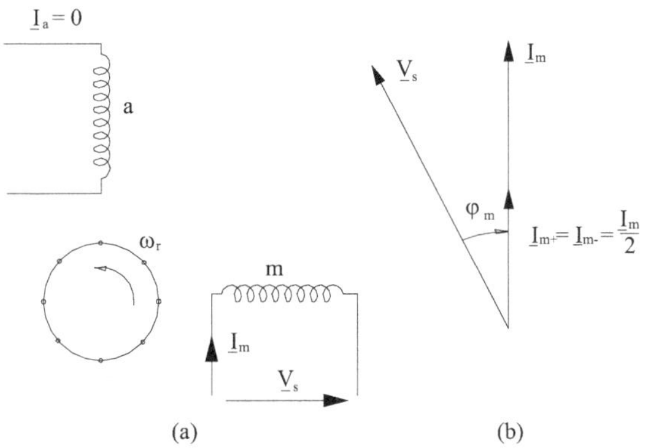

FIGURE 14.1 The case of open auxiliary winding: (a) schematics and (b) phasor diagram.

The auxiliary winding may be kept open intentionally for testing. So this particular connection is of practical interest.

According to (13.21) and (13.22), the direct and inverse main winding current components $\underline{I}_{m+}$ and $\underline{I}_{m-}$ are

$$\underline{I}_{m+,-} = \frac{1}{2}\left(\underline{I}_m \mp j\underline{I}_a\right) = \frac{1}{2}\underline{I}_m \tag{14.1}$$

The equivalent circuit in Figure 13.17 gets simplified as $\underline{Z}_a = \infty$ (Figure 14.2).

As the two current components I_{m+} and I_{m-} are equal to each other, the torque expression T_e (13.18) becomes

$$T_e = 2\frac{R_r \cdot p_1}{\omega_1}\left(\frac{I_{rm+}^2}{S} - \frac{I_{rm-}^2}{2-S}\right) \tag{14.2}$$

with

$$I_m = \frac{\underline{V}_s}{R_{sm} + jX_m + \left(\frac{\underline{Z}_{r+}}{2}\right) + \left(\frac{\underline{Z}_{r-}}{2}\right)} \tag{14.3}$$

$$\frac{\underline{Z}_{r+}}{2} = \frac{j\frac{X_{mm}}{2}\left(\frac{R_{rm}}{2S} + j\frac{X_{rm}}{2}\right)}{\frac{R_{rm}}{2S} + j\frac{(X_{mm}+X_{rm})}{2}}$$

$$\frac{Z_{r-}}{2} = \frac{j\frac{X_{mm}}{2}\left(\frac{R_{rm}}{2(2-S)} + j\frac{X_{rm}}{2}\right)}{\frac{R_{rm}}{2(2-S)} + j\frac{(X_{mm}+X_{rm})}{2}} \tag{14.4}$$

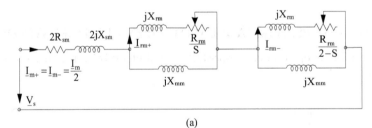

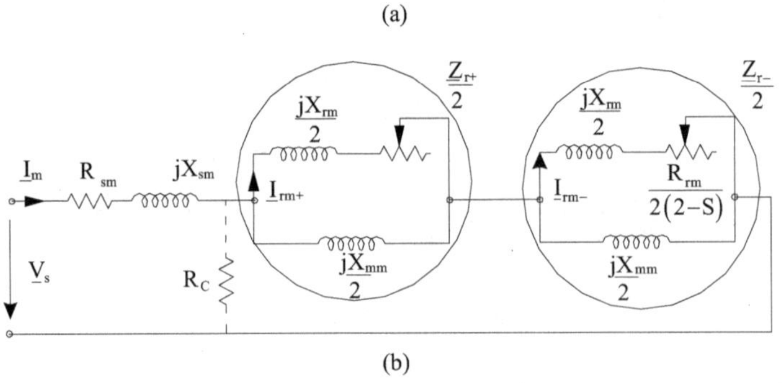

FIGURE 14.2 Equivalent circuit with open auxiliary phase: (a) with forward current $\underline{I}_{m+} = \underline{I}_{m-} = \underline{I}_m/2$ and (b) with total current $\underline{I}_m$.

It is evident that for S = 1 (standstill), though the current $\underline{I}_m$ is maximum, I_{ms}

$$I_{ms} \approx \frac{V_s}{\sqrt{(R_{sm}+R_{rm})^2+(X_{sm}+X_{rm})^2}} \tag{14.5}$$

the total torque (14.2) is zero as S = 2 – S = 1.

The torque versus slip curve is shown in Figure 14.3.

At zero slip (S = 0), the torque is already negative as the direct torque is zero and the inverse torque is negative. The machine exhibits only motoring and generating modes.

The peak torque value T_{ek} and its corresponding slip (speed) S_K are mainly dependent on stator and rotor resistances R_{sm} and R_{rm} and leakage reactances X_{sm} and X_{rm}.

As in reality, this configuration works for split-phase and capacitor-start IMs, the steady-state performance of interest relate to slip values $S \le S_K$.

The loss breakdown in the machine is shown in Figure 14.4.

Winding losses p_{cos} and p_{cor} occur both in the stator and the rotor. Core losses occur mainly in the stator, p_{Fes}.

Additional losses occur in iron and windings and both in the stator and the rotor p_{adds} and p_{addr}.

The additional losses are similar to those occurring in the three-phase IM. They occur on the rotor and stator surface and as tooth-flux pulsation in iron, in the rotor cage and the rotor laminations, due to transverse interbar currents between the rotor bars.

The efficiency η is, thus,

$$\eta = \frac{P_2}{P_1} = \frac{P_1 - \sum p}{P_1} \tag{14.6}$$

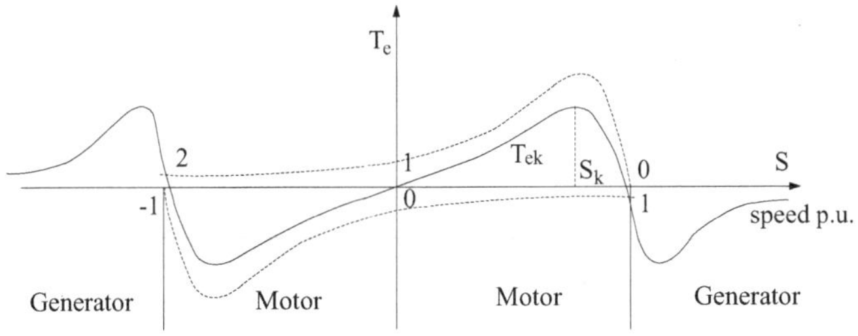

FIGURE 14.3 Torque/speed curve for open auxiliary winding (single-phase winding operation).

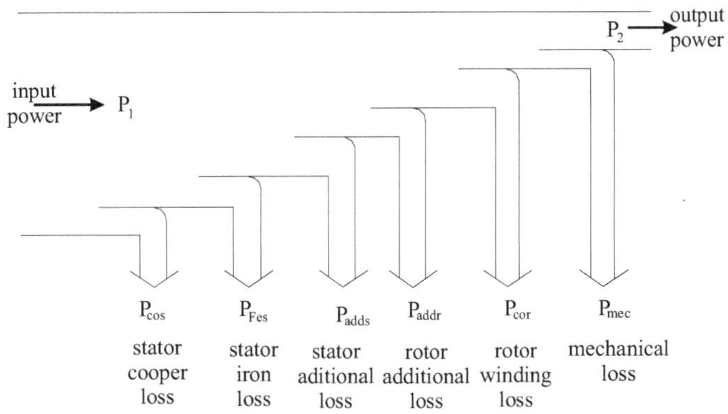

FIGURE 14.4 Loss breakdown in the single-phase IM.

$$\sum p = p_{cos} + p_{Fes} + p_{cor} + p_{adds} + p_{addr} + p_{mec} \tag{14.7}$$

A precise assessment by computation of efficiency (of losses, in fact) constitutes a formidable task due to complexity of the problem caused by magnetic saturation, slot openings, skin effect in the rotor cage, transverse (interbar) rotor currents, etc.

In a first approximation, only the additional losses $p_{add} = p_{adds} + p_{addr}$ and the mechanical losses p_{mec} may be assigned a percentage value of rated output power. Also, the rotor skin effect may be neglected in a first approximation, and the interbar resistance may be lumped into the rotor resistance.

Moreover, the fundamental core loss may be calculated by simply "planting" a core resistance R_r in parallel, as shown in Figure 14.2b.

A numerical example follows.

Example 14.1 One Phase IM Performance with the Auxiliary Phase Open – A Case Study

Let us consider a permanent capacitor (C = 8 μF) six-pole IM fed at 230 V, 50 Hz, whose rated speed n_n = 940 rpm. The main and auxiliary winding and rotor parameters are: R_{sm} = 34 Ω, R_{sa} = 150 Ω, a = 1.73, X_{sm} = 35.9 Ω, $X_{sa} = a^2X_{sm}$, X_{rm} = 29.32 Ω, R_{rm} = 23.25 Ω, and X_{mm} = 249 Ω.

The core loss is p_{Fes} = 20 W and the mechanical loss p_{mec} = 20 W; the additional losses p_{add} = 5 W.

With the auxiliary phase open, calculate

a) The main winding current and power factor at n = 940 rpm
b) The electromagnetic torque at n = 940 rpm
c) The mechanical power, input power, total losses, and efficiency at 940 rpm
d) The current and torque at zero slip (S = 0).

Solution

a) The no-load ideal speed n_1 is

$$n_1 = \frac{f_1}{p_1} = \frac{50}{3} = 16.66 \text{ rps} = 1000 \text{ rpm}$$

The value of rated slip S_n is

$$S_n = \frac{n_1 - n_n}{n_1} = \frac{1000 - 940}{1000} = 0.06$$

$$2 - S_n = 2 - 0.06 = 1.94!$$

The direct and inverse component rotor impedances $Z_{r+}/2$, $Z_{r-}/2$ (14.3)–(14.4) are

$$\frac{Z_{r+}}{2} = \frac{1}{2}\frac{j249(23.25/0.06 + j29.32)}{23.25/0.06 + j(29.32 + 249)} \approx 52.7 + j87$$

$$\frac{Z_{r-}}{2} = \frac{1}{2}\frac{j249(23.25/1.94 + j29.32)}{23.25/1.94 + j(29.32 + 249)} \approx 4.8 + j13.5$$

The total impedance at S = 0.06 in the circuit of Figure 14.2b $\underline{Z}_{11}$ is

$$\underline{Z}_{11} = R_{sm} + jX_{sm} + \frac{Z_{r+}}{2} + \frac{Z_{r-}}{2} = 92.8 + j137$$

Now the RMS stator current $\underline{I}_m$, (14.3), is

$$|I_m| = \left|\frac{V_s}{\underline{Z}_{11}}\right| = \frac{230}{165.19} \approx 1.392\ A$$

The power factor $\cos\varphi_{11}$ is

$$\cos\varphi_{11} = \frac{\text{Real}(\underline{Z}_{11})}{|\underline{Z}_{11}|} = \frac{92.8}{165.19} \approx 0.5618!$$

b) The torque T_e is

$$T_e = \frac{p_1}{\omega_1} I_m^2 \left[\text{Re}(\underline{Z}_{r+}) - \text{Re}(\underline{Z}_{r-})\right] = \frac{3}{314} \cdot 1.392^2[52.7 - 4.8] = 0.8867\ \text{Nm} \tag{14.8}$$

Notice that the inverse component torque is less than 10% of direct torque component.

c) The iron and additional losses have been neglected so far. They should have occurred in the equivalent circuit but they did not.

So,

$$p_{cos} + p_{cor} = P_1 - T_e \cdot 2\pi n = I_m^2\,\text{Re}\left[\underline{Z}_{11}\right] - T_e \cdot 2\pi n_1$$

$$= 1.392^2 \cdot 92.8 - 0.8867 \cdot 2\pi \cdot 16.6 = 87.1\ \text{W}$$

The mechanical power P_{mec} is

$$P_{mec} = T_e \cdot 2\pi n - p_{mec} = 0.8867 \cdot 2\pi \cdot 16.666(1 - 0.06) - 20 = 85.20\ \text{W}$$

If we now add the core and additional losses to the input power, the efficiency η is

$$\eta = \frac{85.2}{87.1 + 85.2 + 2 + 20 + 5} = 0.4275$$

The input power P_1 is

$$P_1 = \frac{P_{mec}}{\eta} = \frac{85.2}{0.4975} = 199.30\ \text{W}$$

The power factor, including core and mechanical losses, is

$$\cos\varphi = P_1/(V_s I_m) = 199.30/(230 \cdot 1.392) = 0.6225!$$

d) The current at zero slip is approximately

$$(I_m)_{S=0} \approx \left|\frac{\underline{V}_s}{R_{sm} + j\left(X_{sm} + \dfrac{X_{mm}}{2}\right) + \dfrac{(\underline{Z}_{r-})_{S=0}}{2}}\right| \approx \left|\frac{230}{34 + j(35.9 + 249) + 4.8 + j13.5}\right| = 0.764\ \text{A}$$

Only the inverse torque is not zero

$$(T_e)_{S=0} = -(I_m)_{s=0}^2\,\text{Re}\frac{(\underline{Z}_{r-})_{S=0}}{2} \cdot \frac{P_1}{\omega_1} \approx -0.764^2 \cdot 4.8 \cdot \frac{3}{314} = -0.0268\ \text{Nm}$$

Negative torque means that the machine has to be shaft-driven to maintain the zero slip conditions.

In fact, the negative mechanical power equals half the rotor winding losses as $S_b = 2$ for $S = 0$.

So the rotor winding loss $(p_{cor})_{s=0}$ is

$$(p_{cor})_{S=0} = 2\times(T_e)_{S=0}\cdot 2\pi n_1 = 2\times 0.0268\times 2\pi\cdot 16.6 = 5.604\text{ W}$$

The total input power $(P_1)_{S=0}$ is

$$P_{10} = \left(R_{sm} + \text{Re}\left(\underline{Z}_b\right)_{s=0}\right)\left(I_m\right)_{s=0}^2 \approx (34+4.8)\cdot 0.764^2 = 22.647\text{ W}$$

The stator winding losses $p_{cos} = 34\times 0.764^2 = 19.845$ W.

Consequently, half of rotor winding losses at $S = 0$ ($2.8\text{ W} = P_{10} - p_{cos}$) are supplied from the power source while the other half is supplied from the shaft power.

The frequency of rotor currents at $S = 0$ is $f_{2b} = S_b f_1 = 2f_1$.

The steady-state performance of IM at $S = 0$ with an open auxiliary phase may be used for loss segregation and some parameters' estimation.

14.3 THE SPLIT PHASE AND THE CAPACITOR IM: CURRENTS AND TORQUE

For the capacitor IM, the equivalent circuit of Figure 13.17 remains as it is with

$$\underline{Z}_a = -j/\omega_1 C \tag{14.9}$$

as shown in Figure 14.5.

For the split-phase IM, the same circuit is valid but with $C = \infty$ and $R_{sa}/a^2 - R_{sm} > 0$.

Also for a capacitor motor with a permanent capacitor in general: $R_{sa}a^2 - R_{sm} = 0$ and $X_{sa}a^2 - X_{sm} = 0$; that is, it uses about the same copper weight in both, main and auxiliary windings.

Notice that all variables are reduced to the main winding (see the two voltage components). This means that the main and auxiliary actual currents $\underline{I}_m$ and $\underline{I}_a$ are now

$$\underline{I}_m = \underline{I}_{m+} + \underline{I}_{m-};\underline{I}_a = j\left[\underline{I}_{m+} - \underline{I}_{m-}\right]/a \tag{14.10}$$

Let us denote

$$\underline{Z}_+ = R_{sm} + jX_{sm} + \frac{JX_{mm}\left(jX_{rm} + R_{rm}/S\right)}{R_{rm}/S + j(X_{mm} + X_{rm})} \tag{14.11}$$

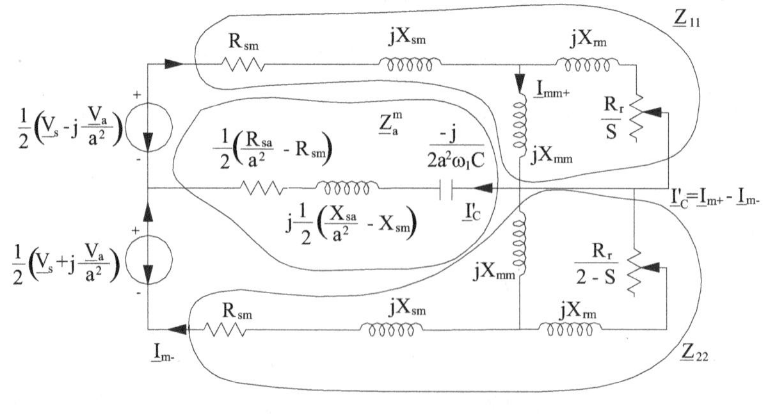

FIGURE 14.5 Equivalent circuit for the capacitor IM ($V_a = V_s$) in the motor mode.

$$\underline{Z}_a{}^m = \Delta R_{sm} + j\Delta X_{sm} - \frac{j}{2a^2}X_c; X_c = \frac{1}{\omega_1 C} \tag{14.12}$$

$$\Delta R_{sm} = \frac{1}{2}\left(R_{sa}/a^2 - R_{sm}\right); \Delta X_{sm} = \frac{1}{2}\left(X_{sa}/a^2 - X_{sm}\right) \tag{14.13}$$

$$\underline{Z}_- = R_{sm} + jX_{sm} + \frac{jX_{mm}\left(jX_{rm} + R_{rm}/(2-S)\right)}{R_{rm}/(2-S) + j\left(X_{rm} + X_{mm}\right)} \tag{14.14}$$

The solution of the equivalent circuit in Figure 14.5, with (14.11)–(14.14), leads to matrix equations

$$\begin{vmatrix} \frac{\underline{V}_s(1-j/a)}{2} \\ \frac{\underline{V}_s(1+j/a)}{2} \end{vmatrix} = \begin{vmatrix} \underline{Z}_+ + \underline{Z}_a^m & -\underline{Z}_a^m \\ -\underline{Z}_a^m & \underline{Z}_- + \underline{Z}_a^m \end{vmatrix} \cdot \begin{vmatrix} \underline{I}_{m+} \\ \underline{I}_{m-} \end{vmatrix} \tag{14.15}$$

with the straightforward solution

$$\underline{I}_{m+} = \frac{\underline{V}_s}{2}\frac{(1-j/a)\left(\underline{Z}_- + 2\underline{Z}_a^m\right)}{\underline{Z}_+ \cdot \underline{Z}_- + \underline{Z}_a^m\left(\underline{Z}_+ + \underline{Z}_-\right)} \tag{14.16}$$

$$\underline{I}_{m-} = \frac{\underline{V}_s}{2}\frac{(1+j/a)\left(\underline{Z}_+ + 2\underline{Z}_a^m\right)}{\underline{Z}_+ \cdot \underline{Z}_- + \underline{Z}_a^m\left(\underline{Z}_+ + \underline{Z}_-\right)} \tag{14.17}$$

The electromagnetic torque components T_{e+} and T_{e-} retain the expressions (14.8) under the form

$$T_{e+} = \frac{2p_1}{\omega_1}I_{m+}{}^2\left[\mathrm{Re}\left(\underline{Z}_+\right) - R_{sm}\right] \tag{14.18}$$

$$T_{e-} = \frac{2p_1}{\omega_1}I_{m-}{}^2\left[\mathrm{Re}\left(\underline{Z}_-\right) - R_{sm}\right] \tag{14.19}$$

The factor 2 in (14.18) and (14.19) is due to the fact that the equivalent circuit in Figure 14.5 is per phase and it refers to a two-phase winding model.

Apparently, the steady-state performance problem is solved once we have the rather handy Equations (14.10)–(14.19) at our disposal.

We now take a numerical example to get a feeling of magnitudes.

Example 14.2 Capacitor Motor Case

Let us consider again the capacitor motor from Example 14.1; this time, with the auxiliary winding on and R_{sa} changed to $R_{sa} = a^2 R_{sm}$, $C' = 4\mu F$ and calculate

a) The starting current and torque with $C_s = 8\mu F$
b) For rated slip $S_n = 0.06$: supply current and torque
c) The main phase I_m and auxiliary phase current I_a and the power factor at rated speed ($S_n = 0.06$)
d) The phasor diagram.

Solution

a) For S = 1, from (14.10) to (14.13),

$$\left(\underline{Z}_{+}\right)_{S=1} = \left(\underline{Z}_{-}\right)_{S=1} = \underline{Z}_{sc} \approx R_{sm} + R_{rm} + j\left(X_{sm} + X_{rm}\right)$$
$$= (34 + 23.25) + j(35.9 + 29.32)$$
$$= 57.5 + j65.22$$

Note that $\Delta R_{sm} = \frac{R_{sa}}{a^2} - R_{sm} = 0$ and $\Delta X_{sm} = \frac{X_{sa}}{a^2} - X_{sm} = 0$ and thus

$$\underline{Z}_a{}^m = \frac{-jX_c}{2a^2} = \frac{-j}{2\cdot 1.73^2 \cdot 2\pi \cdot 50 \times 8 \cdot 10^{-6}} \approx -j66.34$$

From (14.16) and (14.17)

$$\left(\underline{I}_{m+}\right)_{S=1} = \frac{230}{2}\frac{\left[(1 - j/1.73)(57.5 + j65.22) - 2j66.7\right]}{(57.5 + j65.22)^2 - j66.7\cdot(57.5 + j65.22)\times 2}$$
$$= 1.444 - j1.47$$

$$\left(\underline{I}_{m-}\right)_{S=1} = \frac{230}{2}\frac{\left[(1 + j/1.73)(57.5 + j65.22) - 2j\cdot 66.7\right]}{(57.5 + j65.22)^2 - j66.7\cdot(57.5 + j65.22)\times 2}$$
$$= 0.3 - j0.5$$

Consequently, the direct and inverse torques are (14.17 and 14.18)

$$\left(T_{e+}\right)_{S=1} = \frac{2\times 3}{314}\cdot 2.06^2(57.5 - 34) = 1.9066 \text{ Nm}$$

$$\left(T_{e-}\right)_{S=1} = \frac{-2\times 3}{314}\cdot 0.583^2(57.5 - 34) = -0.1526 \text{ Nm}$$

The inverse torque T_{e-} is much smaller than the direct torque T_{e+}. This indicates that we are close to symmetrisation through adequate a (turn ratio) and C.

The main and auxiliary winding currents $\underline{I}_m$ and $\underline{I}_a$ are (14.10)

$$\underline{I}_m = \underline{I}_{m+} + \underline{I}_{m-} = 1.444 - j1.47 + 0.30 - j0.50 \approx 1.744 - j1.97$$

$$\underline{I}_a = \frac{j\left(I_{m+} - I_{m-}\right)}{a} = \frac{j(1.44 - j1.47 - 0.3 + j0.5)}{1.73} = j0.6613 + 0.56$$

$$\underline{I}_s = \underline{I}_m + \underline{I}_a = 1.744 - j1.97 + j0.6693 + 0.56 = 2.304 - j1.308$$

At the start, the source current is lagging the source voltage (which is a real variable here) despite the influence of the capacitance.

The capacitance voltage,

$$\left(V_c\right)_{S=1} = \frac{\left(I_a\right)_{S=1}}{\omega C_s} = \frac{0.866}{314\cdot 8\cdot 10^{-8}} = 344.75 \text{ V}$$

b) For rated speed (slip) – S_n = 0.06, the same computation routine is followed. As the capacitance is reduced twice

$$\underline{Z}_a^m = -j133.4$$

$$\left(\underline{Z}_+\right)_{S=0.06} = R_{sm} + jX_{sm} + \frac{jX_{mm}\left(\frac{R_{rm}}{S} + jX_{rm}\right)}{\frac{R_{rm}}{S} + j\left(X_{mm} + X_{rm}\right)}$$

$$= 34 + j23.25 + \frac{j249\left(\frac{23.25}{0.06} + j29.32\right)}{\frac{23.25}{0.06} + j(249 + 29.32)} = 139.4 + j197.75$$

$$\left(\underline{Z}_-\right)_{S=0.06} = 340 + j23.25 + \frac{j249\left(\frac{23.25}{1.94} + j29.32\right)}{\frac{23.25}{1.94} + j(249 + 29.32)} = 46.6 + j50.25\Omega$$

$$\left(\underline{I}_{m+}\right)_{S_n=0.06} = \frac{230}{2}\frac{\left[(1 - j/1.73)(46.6 + j50.2) - 2j133.4\right]}{(139.4 + j197)(46.6 + j50.25) - j133.4\cdot\left(139.4 + 46.66 + j(197 + 50.25)\right)}$$

$$= 0.525 - j0.794$$

$$\left(\underline{I}_{m-}\right)_{S_n=0.06} = \frac{230}{2}\frac{\left[(1 + j/1.73)(139 + j197.5) - 2j133.4\right]}{(139.4 + j197)(46.6 + j50.25) - j133.4\cdot(139.4 + 46.66 + j(197 + 50.25))}$$

$$= 0.1016 + j0.0134$$

$$\left(T_{e+}\right)_{S=0.06} = \frac{2\times 3}{314}\times(0.9518)^2(139.4 - 34) = 1.8245 \text{ Nm}$$

$$\left(T_{e-}\right)_{S=0.06} = -\frac{2\times 3}{314}\times(0.1025)^2(46.6 - 34) = -2.53\times 10^{-3} \text{ Nm}$$

Both the inverse current and torque components are very small. This is a good sign that the machine is almost symmetric. Perfect symmetry is obtained for $\underline{I}_{m-} = 0$.

The main and auxiliary winding currents are

$$I_m = I_{m+} + I_{m-} = 0.525 - j0.794 + 0.1016 + j0.0134 = 0.6266 - j0.7806$$

$$I_a = j\frac{\left(I_{m+} - I_{m-}\right)}{a} = \frac{j(0.525 - j0.794 - 0.1016 - j0.0134)}{1.73} = j0.274 + 0.466$$

The phase shift between the two currents is almost 90°.

Now the source current $(\underline{I}_s)_{s=0.06}$ is

$$\left(I_s\right)_{S=0.06} = 0.6266 - j0.7806 + j0.274 + 0.466$$

$$= 1.0926 - j0.5066$$

The power factor of the motor is

$$\cos\varphi_s = \frac{\text{Re}\left(\underline{I}_s\right)}{\left|\underline{I}_s\right|} = \frac{1.0926}{1.2043} = 0.9072!$$

The power factor is good. The 4 μF capacitance has brought the motor close to symmetry providing a good power factor.

The speed torque curve may be calculated, for steady state, point by point as illustrated above (Figure 14.6).

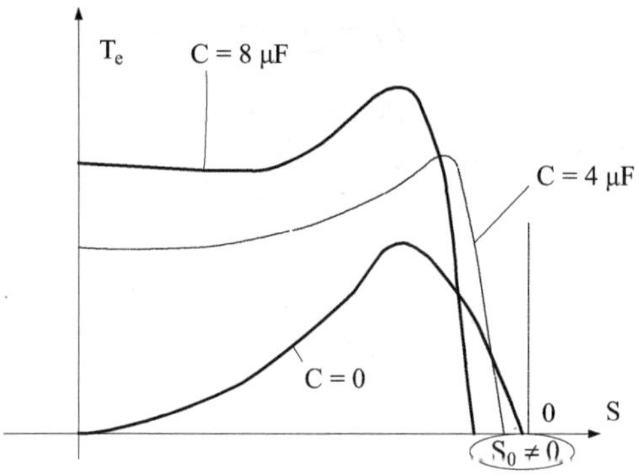

FIGURE 14.6 Torque–speed curve of capacitor motors from Example 14.2 (qualitative).

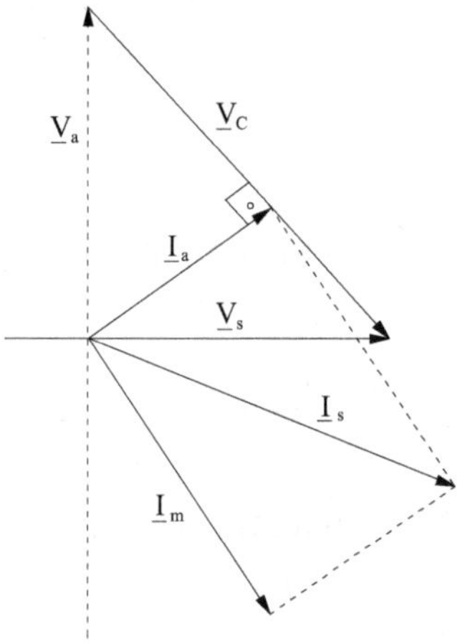

FIGURE 14.7 Phasor diagram for S = 0.06, C = 4 μF, and the data of Example 14.2.

A large capacitor at start leads to higher starting torque due to symmetrisation tendency at S = 1. At low slip values, however, there is too much inverse torque (Figure 14.6) as the motor departures from symmetrisation conditions.

A proper capacitor at rated load would again lead the symmetrisation conditions (zero inverse current), and thus, good performance is obtained.

This is how the dual-capacitor IM evolved. The ratio of the two capacitances is only 2/1 in our case, but as power goes up in the KW range, this ratio may go up to 6/1.

The phasor diagram at S = 0.06 is shown in Figure 14.7.

As the equivalent circuit (Figure 14.5) and the current and torque component expressions (14.16)–(14.19) remain unchanged, the case of split-phase IM (without a capacitor, but with an additional resistance: $\Delta R_{sm} \neq 0$ by design or for real) may be treated the same way as the capacitor IM, except for the content of $\underline{Z}_a^m$.

14.4 SYMMETRISATION CONDITIONS

By symmetrisation we mean zero inverse current component I_{m-} for a given slip. This requires by necessity a capacitance C dependent on slip. From (14.17) with $\underline{I}_{m-} = 0$, we obtain

$$(1 + j/a)\underline{Z}_+ + 2\underline{Z}_a^m = 0; \underline{Z}_+ = R_+ + jX_+ \tag{14.20}$$

with

$$\underline{Z}_a^m = -j\frac{1}{2\omega Ca^2} + \frac{1}{2}\left(\frac{R_{sa}}{a^2} - R_{sm}\right) + \frac{1}{2}j\left(X_{sa}/a^2 - X_{sm}\right) = R_a^m + jX_a^m \tag{14.21}$$

The direct total impedance $\underline{Z}_+$ may be written as

$$\underline{Z}_+ = Z_+e^{j\varphi_+};\varphi_+ = \tan^{-1}\left(X_+/R_+\right) \tag{14.22}$$

With (14.21 and 14.22), Equation (14.20) yields two symmetrisation conditions

$$R_+ - \frac{X_+}{a} = -2R_a^m = -\left(R_{sa}/a^2 - R_{sm}\right) \tag{14.23}$$

$$\frac{R_+}{a} + X_+ = +\frac{1}{\omega Ca^2} - \left(X_{sa}/a^2 - X_{sm}\right) \tag{14.24}$$

The two unknowns of Equations (14.23) and (14.24) are

- C and a for the capacitor IM
- R_{sa} and a for the split-phase IM.

The eventual additional resistance in the auxiliary winding of the split-phase IM is lumped into R_{sa}.

For the permanent capacitor motor, in general, $\Delta R_{sm} = R_{sa}/a^2 - R_{sm} = 0$ and $\Delta X_{sm} = X_{sa}/a^2 - X_{sm} = 0$. So we obtain,

$$a = \frac{X_+}{R_+} = \tan\varphi_+ \tag{14.25}$$

$$X_c = \frac{1}{\omega C} = \frac{X_+}{\cos^2\varphi_+} = Z_+ \cdot a\sqrt{a^2+1} \tag{14.26}$$

φ_+ is the direct component power factor angle.

A few remarks are in order

- The turn ratio a for symmetrisation is equal to the tangent of the direct component of power factor angle. This angle varies with slip for a given motor.
- So, in dual capacitor motor, only, it may be possible to match a and C both for start and running conditions when and only when the value of turn ratio a changes from low to high speed running when the capacitor value is changed. This change of a means a dual section main or auxiliary winding.

For the split-phase motor, $C = \infty$ and $\Delta R_{sm} \neq 0$ and $\Delta X_{sm} \neq 0$, and thus, from (14.23) to (14.24), even if

$$X_{sa}/a^2 < X_{sm} \tag{14.27}$$

we obtain

$$a = \frac{-R_+ \pm \sqrt{R_+^2 - 4(X_+ - X_{sm})X_{sa}}}{X_+ - X_{sm}} \tag{14.28}$$

It is now clear that as $X_+ > X_{sm}$ there is no *a* to symmetrise the split-phase IM, as expected.

Choosing a and R_{sa} should thus follow other optimisation criterion.

For the dual-capacitor motor IM in Example 14.2, the values of a and C for symmetrisation, according to (14.25) and (14.26) are

$$(a)_{S=1} = \frac{(X_+)_{S=1}}{(R_+)_{S=1}} = \frac{65.22}{57.5} = 1.134$$

$$(C)_{S=1} = \frac{1}{314\cdot\left(\sqrt{57.5^2+65.22^2}\right)\times 1.134\sqrt{1+1.134^2}} = 21.36\times 10^{-6}\text{F}$$

$$(a)_{S=0.06} = \frac{(X_+)_{S=0.06}}{(R_+)_{S=0.06}} = \frac{197.25}{139.4} = 1.415$$

$$(C)_{S=0.06} = \frac{1}{314\left(\sqrt{139.25^2+197.25^2}\right)\cdot 1.415\sqrt{1+1.415^2}} = 5.378\cdot 10^{-6}\text{F}$$

The actual values of the capacitor were 8 μF for starting (S = 1) and 4 μF for rated slip (S = 0.06). Also the ratio of equivalent turns ratio was a = 1.73 in Example 14.2.

It is now evident that with this value of a (a = 1.73), no capacitance C can symmetrise the machine perfectly for S = 1 or S = 0.06. Once the machine is made, finding a capacitor to symmetrise the machine is difficult. In practice, in general, coming close to symmetrisation conditions for rated load is recommendable, while other criteria for a good design have to be considered.

14.5 STARTING TORQUE AND CURRENT INQUIRES

For starting conditions (S = 1) $\underline{Z}_+ = \underline{Z}_- = \underline{Z}_{sc}$, so (14.16) and (14.17) become

$$\underline{I}_{m+} = \frac{\underline{V}_s}{2\underline{Z}_{sc}}\left[1 - \frac{j}{a}\frac{\underline{Z}_{sc}}{\underline{Z}_{sc} + 2\underline{Z}_a^m}\right] \tag{14.29}$$

$$\underline{I}_{m-} = \frac{\underline{V}_s}{2\underline{Z}_{sc}}\left[1 + \frac{j}{a}\frac{\underline{Z}_{sc}}{\underline{Z}_{sc} + 2\underline{Z}_a^m}\right] \tag{14.30}$$

The first components of + – currents in (14.29) and (14.30) correspond to the case when the auxiliary winding is open $\left(\left|\underline{Z}_a^m\right| = \infty\right)$.

We may now write

$$\underline{Z}_{sc} = Z_{sc}e^{j\varphi_{sc}};\ \underline{Z}_a^m = Z_a^m e^{j\varphi_a};\ K_a = \frac{2\left|\underline{Z}_a^m\right|}{\left|\underline{Z}_{sc}\right|} \tag{14.31}$$

This way, (14.29) may be written as

$$\underline{I}_{m+} = \underline{I}_{sc}(1+\alpha+j\beta)/2 \tag{14.32}$$

$$\underline{I}_{m-} = \underline{I}_{sc}(1-\alpha-j\beta)/2 \tag{14.33}$$

with

$$\alpha + j\beta = \frac{-j}{a\left(1+K_a\exp(j\gamma)\right)} \tag{14.34}$$

and

$$\gamma = \varphi_{sc} - \varphi_a$$

Let us denote the starting torque of two-phase symmetrical IM by T_{esc}.

$$T_{esc} = 2\frac{p_1}{\omega_1} \cdot I_{sc}^2 \operatorname{Re}(Z_{r+})_{S=1} \tag{14.35}$$

The p.u. torque at start t_{es} is

$$t_{es} = \frac{(T_e)_{S=1}}{T_{esc}} = \frac{|\underline{I}_{m+}|^2 - |\underline{I}_{m-}|^2}{|\underline{I}_{sc}|^2} \tag{14.36}$$

Making use of (14.32) and (14.33), t_{es} becomes

$$t_{es} = \alpha = \frac{K_a \sin\gamma}{a\left(1 + K_a^2 + 2K_a \cos\gamma\right)} \tag{14.37}$$

$$\beta = -(1 + K_a \cos\gamma)/\left(1 + K_a^2 + 2K_a \cos\gamma\right)/a \tag{14.38}$$

A few comments are in order

- The relative starting torque depends essentially on the ratio K_a and on the angle γ.
- We should note that for the capacitor motor with windings using the same quantity of copper ($\Delta R_{sa} = 0$ $\Delta X_{sa} = 0$),

$$(K_a)_C = \frac{1}{\omega C a^2 |\underline{Z}_{sc}|} \tag{14.39}$$

Also for this case, $\varphi_a = -\pi/2$ and thus,

$$(\gamma)_c = \varphi_{sc} + \pi/2 \tag{14.40}$$

- On the other hand, for the split-phase capacitor motor with, say, $\Delta X_{sa} = 0$, $C = \infty$.

$$(K_a)_{\Delta R_{sa}} = \frac{\Delta R_{sa}}{|\underline{Z}_{sc}|} = \frac{\left(R_{sa} - a^2 R_{sm}\right)}{a^2 |\underline{Z}_{sc}|} \tag{14.41}$$

Also, $\varphi_a = 0$ and $\gamma = \varphi_{sc}$.

For the capacitor motor, with (14.39) and (14.40), the expression (14.37) becomes

$$t_{es} = \frac{\cos\varphi_{sc}}{a\left[a^2 Z_{sc} \cdot \omega_1 C + \dfrac{1}{\omega_1 C Z_{sc} a^2} - 2\sin\varphi_{sc}\right]} \tag{14.42}$$

The maximum torque versus the turns ratio *a* is obtained for

$$a_K = \sqrt{\frac{\sin\varphi_{sc}}{Z_{sc}\omega C}}$$
$$(t_{es})_{max} = \frac{\sqrt{Z_{sc}\omega_1 C \sin\varphi_{sc}}}{\cos\varphi_{sc}} \tag{14.43}$$

For the split-phase motor (with ΔR_{sa} replacing $1/\omega C$) and $\gamma = \varphi_{sc}$, no such maximum starting torque occurs.

The maximum starting torque conditions are not enough for an optimum starting.

At least the starting current also has to be considered.

From (14.32) and (14.33)

$$\underline{I}_m = (\underline{I}_{m+} + \underline{I}_{m-}) = \underline{I}_{sc} \tag{14.44}$$

$$\underline{I}_a = j\frac{(\underline{I}_{m+} - \underline{I}_{m-})}{a} = j\frac{(\alpha + j\beta)}{a}\underline{I}_{sc} \tag{14.45}$$

The source current $\underline{I}_s$ is

$$\underline{I}_s = \left[1 + \frac{(-\beta + j\alpha)}{a}\right]\underline{I}_{sc}$$
$$\frac{\underline{I}_s}{\underline{I}_{sc}} = \sqrt{\left(1 - \frac{\beta}{a}\right)^2 + \frac{\alpha^2}{a^2}} \tag{14.46}$$

Finally, for the capacitor motor

$$\frac{\underline{I}_s}{\underline{I}_{sc}} = \sqrt{\left[1 + \frac{t_{es}}{a}\frac{(\omega C Z_{sc} a^2 - \sin\varphi_{sc})}{\cos\varphi_{sc}}\right]^2 + \frac{t_{es}^2}{a^2}} \tag{14.47}$$

The p.u. starting torque and current expressions reflect the strong influence of capacitance C, the short-circuit impedance Z_{sc}, its phase angle φ_{sc}, and the turns ratio a. As expected, a higher rotor resistance would lead to higher starting torque.

In general, $1.5 < a < 2.0$ seems to correspond to most optimisation criteria. From (14.42), it follows that in this case, $Z_{sc}\omega C < 1$.

For the split-phase induction motor with $\gamma = \varphi_{sc}$ from (14.38)

$$\beta = \frac{-(1 + K_a\cos\varphi_{sc})}{a(1 + K_a^2 + 2K_a\cos\varphi_{sc})} \tag{14.48}$$

with α (t_{es}) from (14.37)

$$\beta = \frac{-(1 + K_a\cos\varphi_{sc})}{K_a\sin\varphi_{sc}}\cdot t_{es} \tag{14.49}$$

With K_a from (14.41), the starting current (14.46) becomes

$$\frac{I_s}{I_{sc}} = \sqrt{\left[1 + \frac{t_{es}}{a}\frac{\left(\frac{|Z_{sc}|a^2}{(R_{sa} - a^2R_{sm})} + \cos\varphi_{sc}\right)}{\sin\varphi_{sc}}\right]^2 + \left(\frac{t_{es}}{a}\right)^2} \tag{14.50}$$

Expressions (14.47) and (14.50) are quite similar except for the role of φ_{sc}, the short-circuit impedance phase angle. They allow to investigate the existence of a minimum I_s/I_{sc} for given starting torque for a certain turns ratio a_{ki}.

Especially for the capacitor start and split-phase IM, the losses in the auxiliary winding are quite important as, in general, the design current density in this winding (which is to be turned off after start) is to be large to save costs.

Let us notice that for the capacitor motor, the maximum starting torque conditions (14.43) lead to a simplified expression of I_s/I_{sc} (14.47)

$$\left(\frac{I_s}{I_{sc}}\right)_{t_{es}\max} = \sqrt{1 + \left(\frac{(t_{es})_{max}}{a_K}\right)^2} = \sqrt{1 + \frac{(Z_{sc}\omega C)^2}{\cos^2\varphi_{sc}}} \tag{14.51}$$

For $Z_{sc}\omega C < 1$, required to secure $a_K > 1$, both the maximum starting torque (14.43) and the starting current are reduced.

For an existing motor, only the capacitance C may be changed. Then directly (14.43) and (14.47) may be put to work to investigate the capacitor influence on starting torque and starting losses (squared starting current).

As expected, there is a capacitor value that causes minimum starting losses (minimum of $(I_s/I_{sc})^2$), but that value also does not lead to maximum starting torque.

The above inquires around starting torque and current show clearly that this issue is rather complex, and application-dependent design optimisation is required to reach a good practical solution (see Ref. [5] for more details).

14.6 TYPICAL MOTOR CHARACTERISTICS

To evaluate capacitor or split-phase motor steady-state performance, the general practice rests on a few widely recognised characteristics.

- Torque (T_e) versus speed (slip)
- Efficiency (η) versus speed
- Power factor ($\cos\varphi_s$) versus speed
- Source current (I_s) versus speed
- Main winding current (I_m) versus speed
- Auxiliary winding current (I_a) versus speed
- Capacitor voltage (V_c) versus speed
- Auxiliary winding voltage (V_a) versus speed.

All these characteristics may be calculated via the circuit model in Figure 14.5 and Equations (14.16–14.19) with (14.11)–(14.14), (14.44) and (14.45), for given motor parameters, capacitance C, and speed (slip).

Such typical characteristics for a 300 W, two-pole, 50 Hz permanent capacitor motor are shown in Figure 14.8.

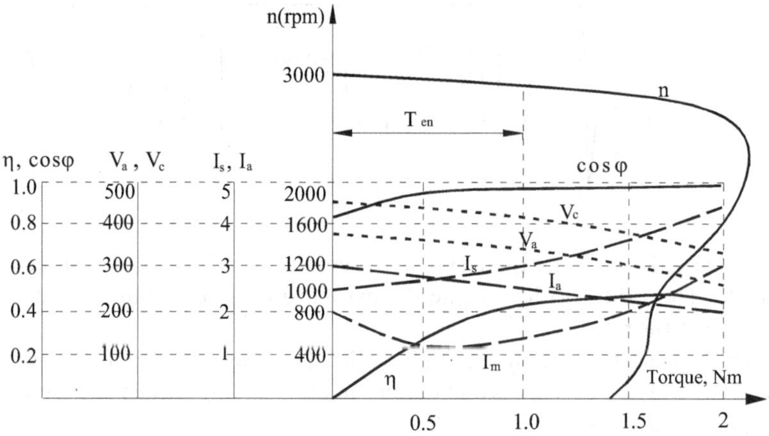

FIGURE 14.8 Typical steady-state characteristics of a 300 W, two pole, 50 Hz.

The characteristics shown in Figure 14.8 prompt remarks such as

- The torque is zero for a speed smaller than $n_1 = f_1/p_1 = 50$ rps (3000 rpm).
- The efficiency is only moderate as a rather large rotor resistance has been considered to secure good starting torque for moderate starting current.
- The power factor is high all over the torque (speed) range.
- The capacitor and auxiliary winding voltages (V_c and V_a) decrease with torque increase (above breakdown torque speed).
- Notice that data in Figure 14.8 correspond to possible measurements above breakdown torque speed and not from start. Only the torque–speed curve covers the entire speed range (from start).
- From zero torque to rated torque (about 1 Nm), the source current increases only a little while I_a decreases and I_m has a minimum below rated torque.
- The characteristics on Figure 14.8 are typical for a permanent capacitor IM where the designer has to trade efficiency for starting torque.

14.7 NON-ORTHOGONAL STATOR WINDINGS

It was shown that for the split-phase motor, with a more resistive auxiliary winding, there is no way to even come close to symmetry conditions with orthogonal windings.

This is where the non-orthogonal windings may come into play.

As the time phase angle shift $(\varphi_a - \varphi_m) < 90°$, placing the windings spatially at an angle $\xi > 90°$ intuitively leads to the idea that more forward travelling field might be produced [6].

Let us write the expressions of two such mmfs

$$F_m\left(\theta_{es}, t\right) = F_{m1}\cos\theta_{es} \cdot \sin\omega_1 t \tag{14.52}$$

$$F_a\left(\theta_{es}, t\right) = F_{a1}\cos\left(\theta_{es} + \xi\right) \cdot \sin\left(\omega_1 t + \left(\varphi_a - \varphi_m\right)\right) \tag{14.53}$$

To simplify mathematics, let us suppose that $F_{m1} = F_{a1}$ and that $(\xi > 90°)$,

$$\xi + \left(\varphi_a - \varphi_m\right) = 180° \tag{14.54}$$

$$F(\theta_{es},t) = F_m(\theta_{es},t) + F_a(\theta_{es},t)$$
$$= \frac{F_{m1}}{2}\left[\sin(\omega_1 t - \theta_{es}) + \sin(\omega_1 t - \theta_{es} + 180 - 2\xi)\right] \quad (14.55)$$

For $\xi = 90°$, the symmetry condition of capacitor motor is obtained. This corresponds to the optimum case.

For $\xi > 90°$, still, a good direct travelling wave can be obtained. Its amplitude is affected by $\xi \neq 90°$.

It is well understood that even with a capacitor motor, such an idea may be beneficial in the sense of saving some capacitance. But it has to be remembered that placing the two stator windings non-orthogonally means the using in common of all slots by the two phases.

For a split-phase IM condition, (14.54) may not be easily met in practice as $\varphi_a - \varphi_m = 30°–35°$ and $\xi = 145°–150°$ is a rather large value.

To investigate quantitatively the potential of non-orthogonal windings, a transformation of coordinates (variables) is required (Figure 14.9).

The m & a non-orthogonal (real) windings are gradually transformed into orthogonal α–β windings. The conservation of mmfs and apparent power leads to

$$\underline{I}_\alpha = \underline{I}_m + a\underline{I}_a \cos\xi$$
$$\underline{I}_\beta = a\underline{I}_a \sin\xi$$
$$\underline{V}_\alpha = \underline{V}_m \quad (14.56)$$
$$\underline{V}_\beta = \frac{V_a/a - V_m \cos\xi}{\sin\xi}$$

The total stator copper losses for the m,a and α & β model are

$$P_{m\&a} = R_{sm}I_m^{\ 2} + R_a I_a^{\ 2} \quad (14.57)$$

$$P_{\alpha\&\beta} = R_{sm}I_\alpha^{\ 2} + \frac{R_a}{a^2}I_\beta^{\ 2}$$
$$= R_{sm}I_a^{\ 2} + R_a I_a^{\ 2} + R_{sm}\cdot 2I_a \cdot a \cdot I_m \cos\xi \cos(\varphi_m - \varphi_a) \quad (14.58)$$

φ_m and φ_a are the phase angles of $\underline{I}_m$ and $\underline{I}_a$. It is now evident that only with $\varphi_m - \varphi_a = 90°$ (the capacitor motor), the a & m and α & β are fully equivalent in terms of losses.

As long as ξ is close to 90° (up to 100°–110°) and $\varphi_m - \varphi_a = 70°–80°$, the two models are practically equivalent.

The additional impedance in the β circuit is obtained from

$$\underline{Z}_a I_a^{\ 2} = Z_{a\beta} I_\beta^{\ 2} \quad (14.59)$$

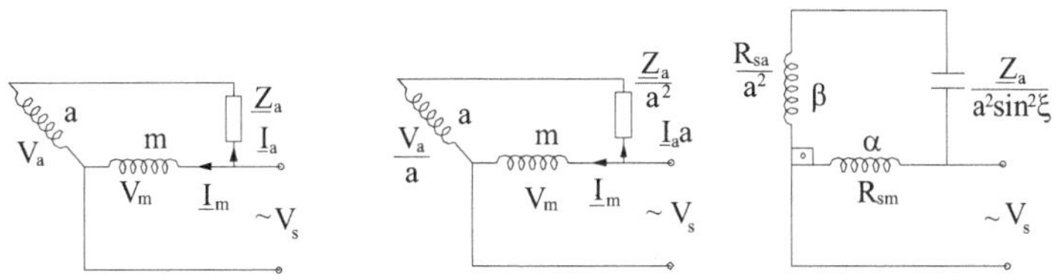

FIGURE 14.9 Evolutive equivalent schematics of non-orthogonal winding IM.

With (14.56), (14.59) becomes

$$\underline{Z}_{a\beta} = \frac{\underline{Z}_a}{a^2 \sin^2 \xi} \tag{14.60}$$

as shown in Figure 14.9.

The relationship between the main and auxiliary terminal voltages is

$$\begin{aligned} \underline{V}_s &= \underline{V}_a + \underline{Z}_a \underline{I}_a = \underline{V}_{a0} + (R_{sa} + jX_{sa})\underline{I}_a + \underline{Z}_a \underline{I}_a \\ \underline{V}_s &= \underline{V}_m = \underline{V}_{m0} + \underline{I}_m (R_{sm} + jX_{sm}) \end{aligned} \tag{14.61}$$

To consider the general case of stator windings when $R_{sa}/a^2 \neq R_{sm}$ and $X_{sa}/a^2 \neq X_{sm}$, we have introduced the total e.m.f.s $\underline{V}_{a0}$ and $\underline{V}_{m0}$ in (14.61).

Also, we introduce a coefficient m which is unity if $R_{sa}/a^2 \neq R_{sm}$ and $X_{sa}/a^2 \neq X_{sm}$ and zero, otherwise,

$$\begin{aligned} \underline{V}_s &= \underline{V}_{a0} + \left[\underline{Z}_a + (R_{sa} + jX_{sa})(1-m)\right]\underline{I}_a \\ \underline{V}_s &= \underline{V}_m = \underline{V}_{m0} + (R_{sm} + jX_{sm})(1-m)\underline{I}_m \end{aligned} \tag{14.62}$$

We now apply the symmetrical components method to V_α and V_β

$$\begin{aligned} &\begin{vmatrix} \underline{V}_+ \\ \underline{V}_- \end{vmatrix} = \frac{1}{2}\begin{vmatrix} 1-j \\ 1+j \end{vmatrix}\begin{vmatrix} \underline{V}_\alpha \\ \underline{V}_\beta \end{vmatrix} \\ &\underline{V}_\alpha = \underline{V}_+ + \underline{V}_- \\ &\underline{V}_\beta = j(\underline{V}_+ - \underline{V}_-) \end{aligned} \tag{14.63}$$

With (14.56), (14.60), and (14.62), Equation (14.63) yields only two equations [6]:

$$\begin{aligned} \underline{V}_s &= a\left[j(\underline{V}_+ - \underline{V}_-)\sin\xi + (\underline{V}_+ + \underline{V}_-)\cos\xi\right] - \frac{\left[\underline{Z}_a + (R_{sa} + jX_{sa})(1-m)\right]}{ja\sin\xi}(\underline{I}_+ - \underline{I}_-) \\ \underline{V}_s &= \underline{V}_+ + \underline{V}_- + (R_{sm} + jX_{sm})\left[\underline{I}_+ + \underline{I}_- - \frac{\cos\xi}{\sin\xi} j(\underline{I}_+ - \underline{I}_-)\right] \end{aligned} \tag{14.64}$$

The + – component equations are

$$\underline{V}_1 = \underline{Z}_+ \underline{I}_+ ; \underline{V}_2 = \underline{Z}_- \underline{I}_- \tag{14.65}$$

with

$$\underline{Z}_+ = (R_{sm} + jX_{sm})m + \frac{jX_{mm}\left(\frac{R_{rm}}{S} + jX_{rm}\right)}{\frac{R_{rm}}{S} + j(X_{mm} + X_{rm})} \tag{14.66}$$

$$\underline{Z}_- = (R_{sm} + jX_{sm})m + \frac{jX_{mm}\left(\frac{R_{rm}}{2-S} + jX_{rm}\right)}{\frac{R_{rm}}{2-S} + j(X_{mm} + X_{rm})} \tag{14.67}$$

Notice again the presence of factor m (with values of m = 1 and 0 as explained above).

The electromagnetic torque components are

$$\begin{aligned} T_{e+} &= 2\left[\mathrm{Re}\left(\underline{V}_+\underline{I}_+^*\right) - R_{sm}\left(\underline{I}_+\right)^2 m\right]\frac{p_1}{\omega_1} \\ T_{e-} &= 2\left[\mathrm{Re}\left(\underline{V}_-\underline{I}_-^*\right) - R_{sm}\left(\underline{I}_-\right)^2 m\right]\frac{p_1}{\omega_1} \\ T_e &= T_{e+} + T_{e-} \end{aligned} \tag{14.68}$$

Making use of (14.65), $\underline{I}_+$ and $\underline{I}_-$ are eliminated from (14.64) to yield

$$\underline{V}_+ = \frac{\underline{a}_0}{\underline{a}_1}\underline{V}_s \tag{14.69}$$

$$\underline{V}_- = -\frac{V_s - \underline{V}_+\left[1 - (R_{sm} + jX_{sm})(1-m)\left(\frac{1 - j\frac{\cos\xi}{\sin\xi}}{\underline{Z}_+}\right)\right]}{1 + \frac{(R_{sm} + jX_{sm})}{\underline{Z}_-}\left(1 + j\frac{\cos\xi}{\sin\xi}\right)} \tag{14.70}$$

$$\underline{C}_2 = -aj\sin\xi + a\cos\xi + \frac{\left[\underline{Z}_a + (R_{sa} + jX_{sa})(1-m)\right]}{ja\underline{Z}_- \cdot \sin\xi} \tag{14.71}$$

$$\underline{a}_0 = 1 - \frac{\underline{C}_2}{1 + \frac{(R_{sm} + jX_{sm})}{\underline{Z}_-}(1-m)\left(1 + j\frac{\cos\xi}{\sin\xi}\right)} \tag{14.72}$$

$$\begin{aligned} a_1 &= aj\sin\xi + a\cos\xi - \frac{\underline{Z}_a + (R_{sa} + jX_{sa})(1-m)}{ja\,Z_+\sin\xi} \\ &-\left(1 - \underline{a}_0\right)\left[+1 + \frac{(R_{sm} + jX_{sm})}{\underline{Z}_+}(1-m)\left(1 - j\frac{\cos\xi}{\sin\xi}\right)\right] \end{aligned} \tag{14.73}$$

14.8 SYMMETRISATION CONDITIONS FOR NON-ORTHOGONAL WINDINGS

Let us consider the case of m = 1 ($R_{sa} = a^2R_{sm}$, $X_{sa} = a^2X_{sm}$), which is most likely to be used for symmetrical conditions.

In essence,

$$V_- = 0 \tag{14.74}$$

$$\underline{V}_+ = \underline{V}_\alpha = \underline{V}_m = \underline{V}_s \tag{14.75}$$

Further on, from (14.69)

$$\underline{a}_0 = \underline{a}_1 \tag{14.76}$$

With m = 1, from (14.71) to (14.73),

$$\underline{C}_2 = -aj\sin\xi + a\cos\xi + \frac{\underline{Z}_a}{ja\underline{Z}_- \cdot \sin\xi} \tag{14.77}$$

$$\underline{a}_0 = 1 - \underline{C}_2 \tag{14.78}$$

$$\underline{a}_1 = aj\sin\xi + a\cos\xi - \frac{\underline{Z}_a}{ja\underline{Z}_+ \cdot \sin\xi} + \underline{a}_0 - 1 \tag{14.79}$$

Finally, from (14.79) with (14.76)

$$\underline{Z}_a = aZ_+ \sin\xi\left[j(a\cos\xi - 1) - a\sin\xi\right] \tag{14.80}$$

For the capacitor motor, with orthogonal windings ($\xi = \pi/2$)

$$\underline{Z}_a = -a\underline{Z}_+[a + j] = \frac{-j}{\omega C} \tag{14.81}$$

Condition (14.81) is identical to (14.26) derived for orthogonal windings.

For a capacitor motor, separating the real and imaginary parts of (14.80) yields [6]

$$a = \frac{X_+}{R_+\sin\xi + X_+\cos\xi} = \frac{\sin\varphi_+}{\sin(\varphi_+ + \xi)} \tag{14.82}$$

$$\frac{1}{\omega C} = a^2 X_+ \sin\xi + aR_+ \sin\xi(1 - a\cos\xi) \tag{14.83}$$

with

$$\tan\varphi_+ = X_+/R_+ \tag{14.84}$$

The results in Table 14.1 are fairly general [6].

When the winding displacement angle ξ increases from 60° to 120° (for the direct component power factor angle $\varphi_+ = 45°$), both the turns ratio a and the capacitance reactance steadily increase.

That is to say, a smaller capacitance is required for say $\xi = 110°$ than for $\xi = 90°$, but, as expected, the capacitance voltage V_c and the auxiliary winding voltage are higher.

TABLE 14.1
Symmetrisation Conditions for $\varphi_+ = 45°$

ξ	60°	70°	80°	90°	100°	110°	120°
a	0.732	0.78	0.863	1.00	1.232	1.232	2.73
$\frac{1}{\omega C X_+}$	0.732	1.109	1.456	2.00	2.967	4.944	12.046!

Example 14.3 Performance: Orthogonal versus Non-orthogonal Windings

Let us consider a permanent capacitor IM with the parameters $R_{sm} = 32\ \Omega$, $R_{sa} = 32\ \Omega$, $R_{rm} = 35\ \Omega$, $L_{rm} = 0.2$ H, $L_{sm} = L_{sa} = 0.1$ H, $L_{mm} = 2$ H, $p_1 = 1$, $a = 1$, $V_s = 220$ V, $\omega_1 = 314$ rad/sec, $\xi = 90°$ (or 110°), $C = 5\ \mu F$, and R_{core} (parallel resistance) = 10,000 Ω.

Let us find the steady-state performance versus slip for orthogonal ($\xi = 90°$) and non-orthogonal ($\xi = 110°$) placement of windings.

Based on the above theory, a C++ code was written to solve this problem, and the results are shown in Figures 14.10 and 14.11.

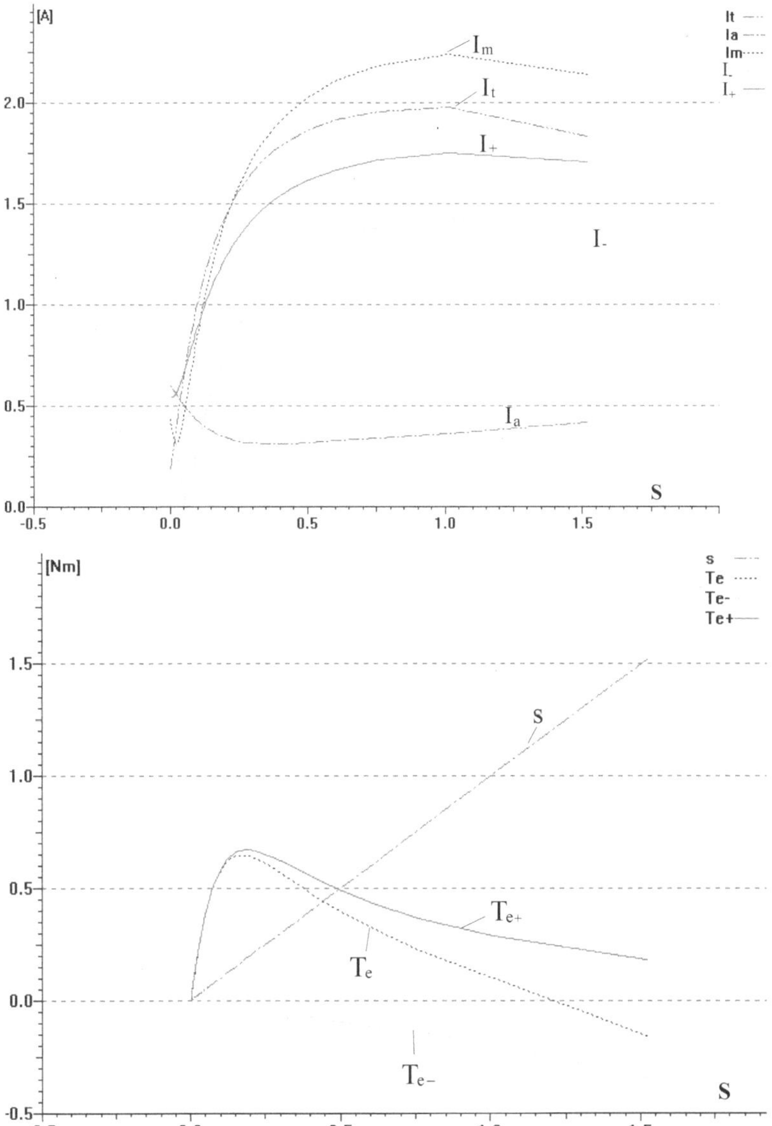

FIGURE 14.10 Performance with orthogonal windings ($\xi = 90°$): I_t – total current, I_a – auxiliary current, I_m – main current, and I_+, I_- – positive and negative sequence currents. T_e, T_{e-}, T_{e+} – total, positive, and negative sequence torques, $S(T_e)$ – slip, ITA_t – efficiency, and $cosFI_t$ – power factor. V_c, V_a, V_+, V_-, V_m – capacitor, auxiliary, +,– sequence and main winding voltage versus slip.

(Continued)

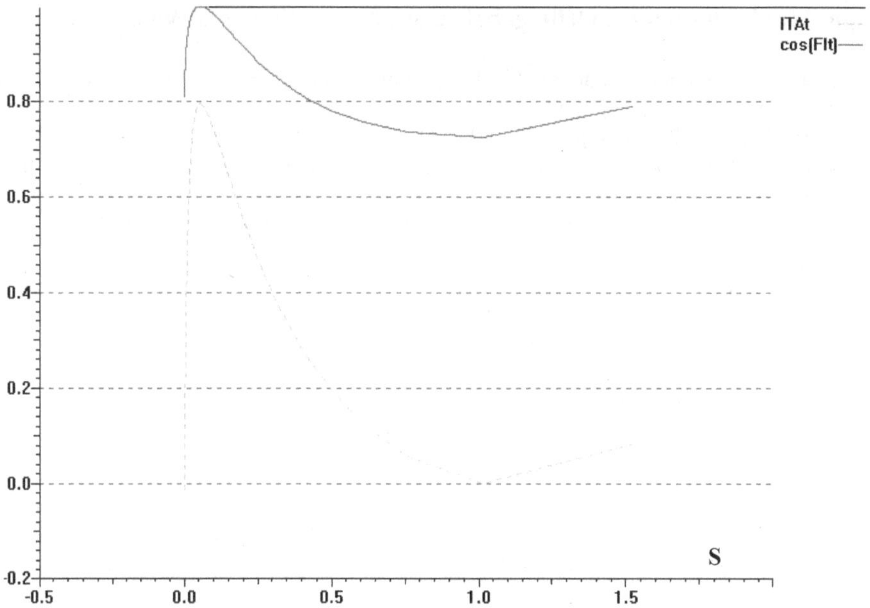

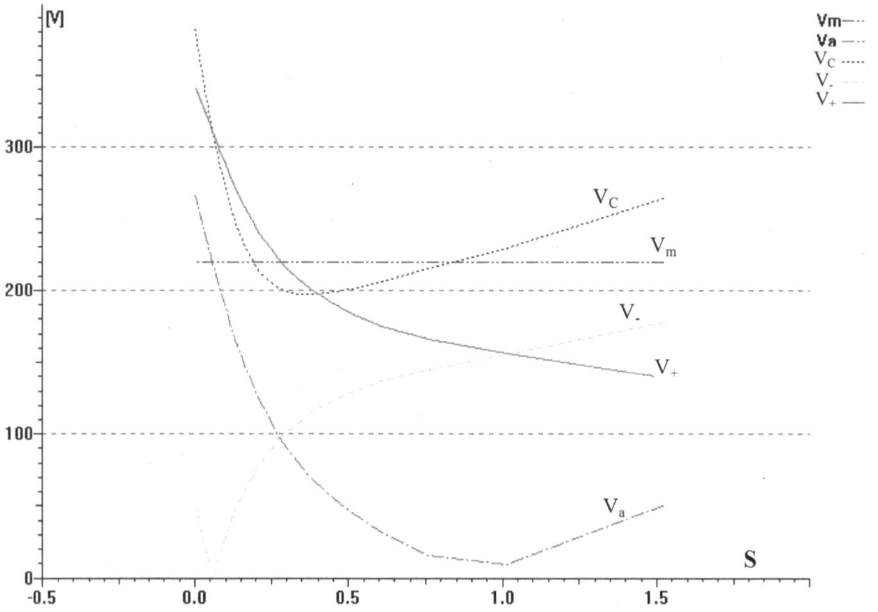

FIGURE 14.10 (CONTINUED) Performance with orthogonal windings ($\xi = 90°$): I_t – total current, I_a – auxiliary current, I_m – main current, and I_+, I_- – positive and negative sequence currents. T_e, T_{e-}, T_{e+} – total, positive, and negative sequence torques, $S(T_e)$ – slip, ITA_t – efficiency, and $cosFI_t$ – power factor. V_c, V_a, V_+, V_-, V_m – capacitor, auxiliary, +,– sequence and main winding voltage versus slip.

The performance illustrated in Figures 14.10 and 14.11 leads to remarks such as

- The torque/speed curves are close to each other but more starting torque is obtained for the non-orthogonal windings motor.
- The symmetrisation (zero I_- and V_-) slip is larger for the non-orthogonal winding motor.

- The power factor of the non-orthogonal winding motor is larger at 20% slip but smaller at slip values below 10%.
- The efficiency curves are very close.
- The capacitor voltage $\underline{V}_c$ is larger for the non-orthogonal winding motor.

It is in general believed that non-orthogonal windings make a more notable difference in split/phase rather than in capacitor IMs.

The mechanical characteristics ($T_e(S)$) for two different capacitors ($C = 5 \cdot 10^{-6}$ F and $C_a = 20 \cdot 10^{-6}$ F) are shown qualitatively in Figure 14.12 for orthogonal windings.

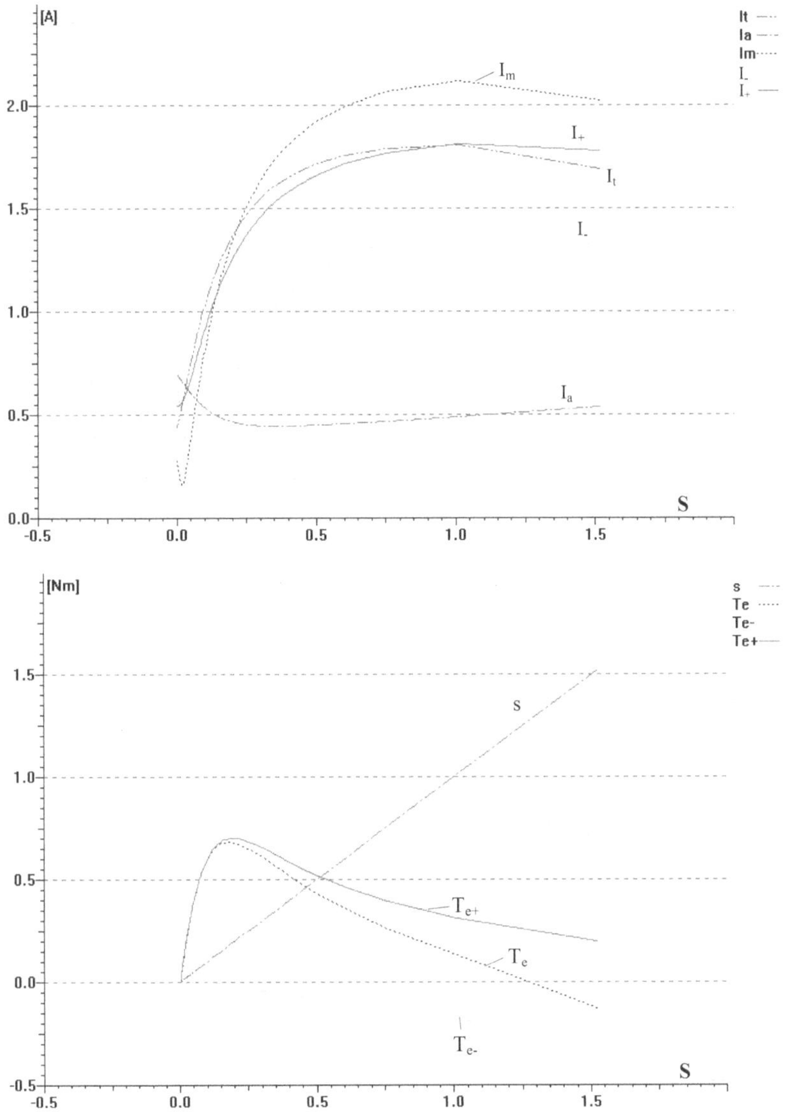

FIGURE 14.11 Performance with non-orthogonal windings (ξ = 110°)°): I_t – total current, I_a – auxiliary current, I_m – main current, I_+, I_- – positive and negative sequence currents. T_e, T_{e-}, T_{e+} – total, positive, and negative sequence torques, $S(T_e)$ – slip, ITA_t – efficiency, $cosFI_t$ – power factor. V_c, V_a, V_+, V_-, V_m – capacitor, auxiliary, +,– sequence, and main winding voltage versus slip.

(Continued)

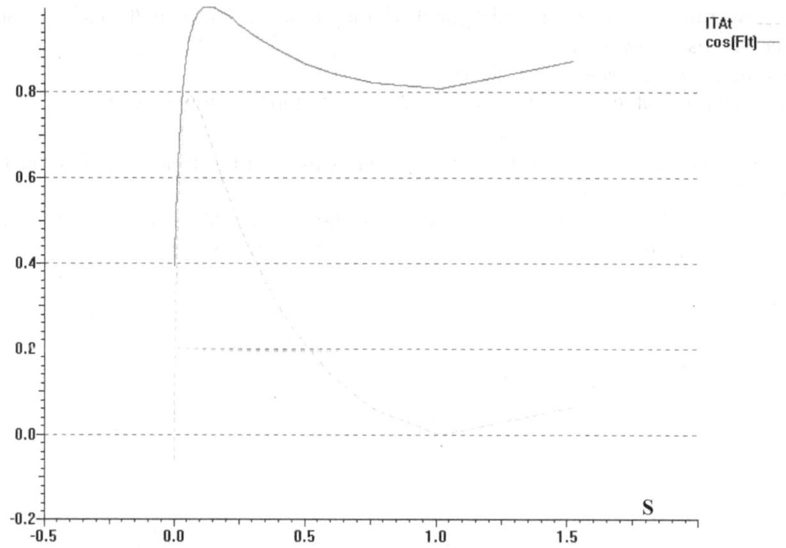

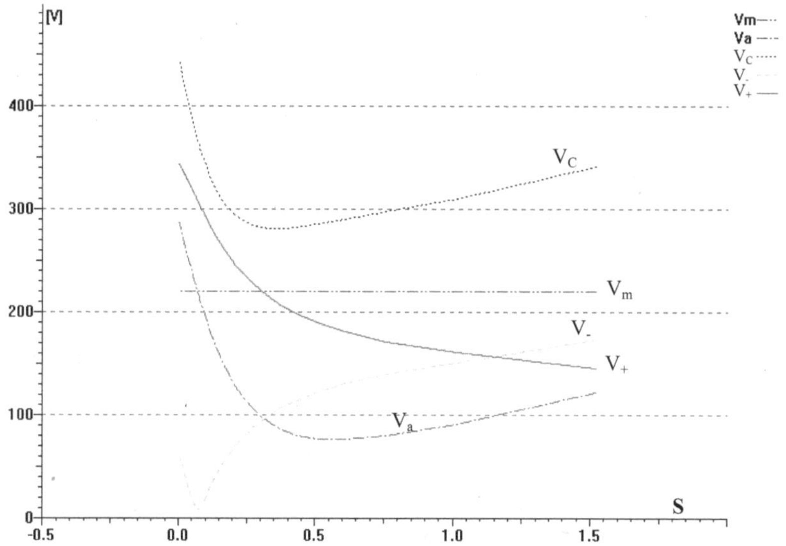

FIGURE 14.11 (CONTINUED) Performance with non-orthogonal windings ($\xi = 110°$)°): I_t – total current, I_a – auxiliary current, I_m – main current, I_+, I_- – positive and negative sequence currents. T_e, T_{e-}, T_{e+} – total, positive, and negative sequence torques, $S(T_e)$ – slip, ITA_t – efficiency, $cosFI_t$ – power factor. V_c, V_a, V_+, V_-, V_m – capacitor, auxiliary, +,– sequence, and main winding voltage versus slip.

As expected, the larger capacitor produces larger torque at start, but it is not adequate at low slip values because it produces too large an inverse torque component.

The same theory can be used to investigate various designs in the pursuit of chosen optimisation criterion such as minimum auxiliary winding loss during starting.

The split-phase motor may be investigated as a particular case of capacitor motor with either $\underline{Z}_a = 0$ and increased R_{sa} value or with the original (real) R_{sa} and $\underline{Z}_a = R_a$. Above a certain slip, the auxiliary phase is turned off which means $|\underline{Z}_a| \approx \infty$.

A general circuit theory to handle two or three stator windings located at random angles has been developed in [5,7].

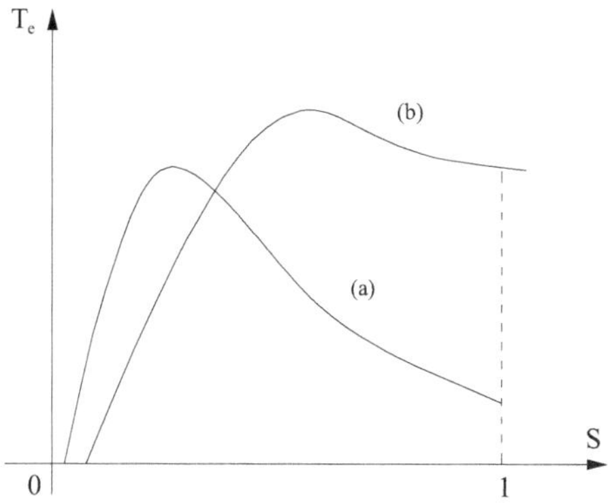

FIGURE 14.12 Torque/slip curves (a) $C = 5 \cdot 10^{-6}$ F and (b) $C_a = 20 \cdot 10^{-6}$ F.

14.9 MMF SPACE HARMONIC PARASITIC TORQUES

So far, only the fundamental of mmf space distribution along rotor periphery has been considered for the steady-state performance assessment.

In reality, the placement of stator conductors in slots leads to space mmf harmonics. The main space harmonic is of the third order. Notice that the current is still considered sinusoidal in time.

From the theory of three-phase IMs, we remember that the slip S_ν for the νth space harmonics is

$$S_\nu = 1 \mp \nu(1 - s) \tag{14.85}$$

For a single-phase IM (the auxiliary phase is open), the space harmonic mmfs are stationary, so they decompose into direct and inverse travelling components for each harmonic order.

Consequently, the equivalent circuit in Figure 14.2 may be adjusted by adding rotor equivalent circuits for each harmonic (both direct and inverse waves) (Figure 14.13).

Once the space harmonic parameters $X_{rm\nu}$ and $X_{mn\nu}$ are known, the equivalent circuit shown in Figure 14.13 allows for the computation of the additional (parasitic) asynchronous torques.

In the case when the reaction of rotor current may be neglected (say for the third harmonic), their equivalent circuits shown in Figure 14.13 are reduced to the presence of their magnetisation reactances $X_{mm\nu}$. These reactances are not difficult to calculate as they correspond to the differential leakage reactances already defined in detail in Chapter 6, Section 6.1.

The presence of slot openings and magnetic saturation enhances the influence of mmf space harmonics on additional losses in the rotor bars and parasitic torques.

Too large a capacitor would trigger deep magnetic saturation which produces a large third-order space harmonic. Consequently, a deep saddle in the torque/slip curve, around 33% of ideal no-load speed, occurs (Figure 14.14). A thorough modelling of split-phase capacitor IM for steady-state via forward/backward revolving field, symmetrical components, and cross-field methods is amply documented in Ref. [8].

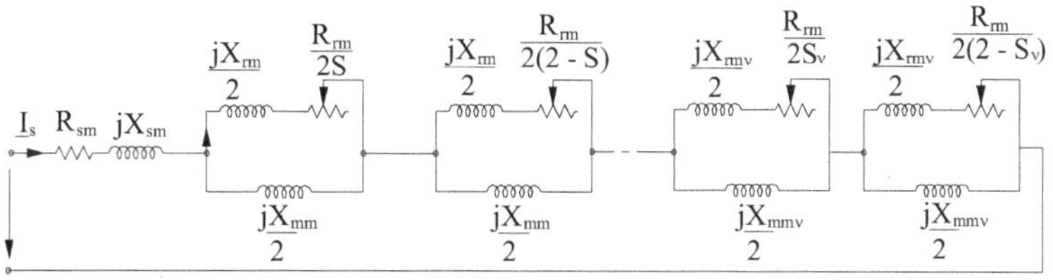

FIGURE 14.13 Equivalent circuit of single-phase IM with mmf space harmonics included.

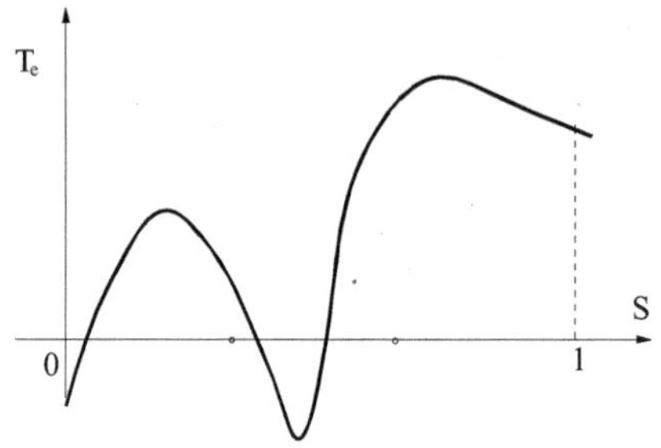

FIGURE 14.14 Grossly oversized capacitor IM torque–speed curve.

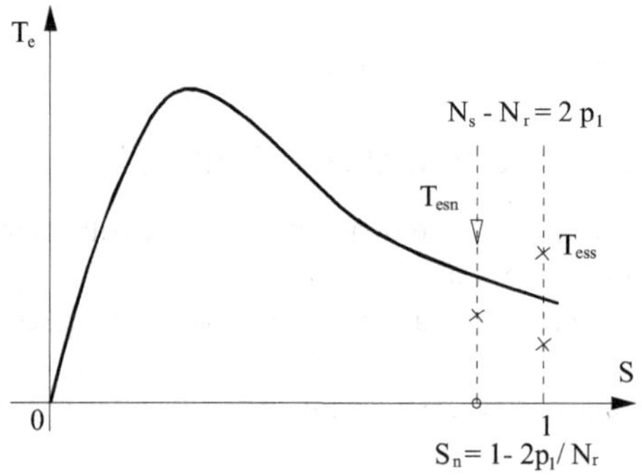

FIGURE 14.15 Synchronous parasitic torques.

Besides asynchronous parasitic torques, synchronous parasitic torques may occur as they do for three-phase IMs. The most important speeds at which such torques may occur are zero and $2f_1/N_r$ (N_r-rotor slots).

When observing the conditions

$$N_s \neq N_r \tag{14.86}$$

$$|N_r - N_s| \neq 2p_1 \tag{14.87}$$

the synchronous torques are notably reduced (Figure 14.15).

The reduction of parasitic synchronous torques, especially at zero speed, is required to secure a safe start and reduce vibration, noise, and additional losses.

14.10 TORQUE PULSATIONS

The speed of the elliptical travelling field, present in the airgap of single-phase or capacitor IMs, varies in time in contrast to the symmetrical IM where the speed of the circular field is constant.

As expected, torque pulsations occur, even during steady state.

The equivalent circuits for steady state are helpful to calculate only the time average torque.

Besides the fundamental mmf-produced torque pulsations, the space harmonics cause similar torque pulsations.

The d–q models for transients, covered in Chapter 1, Volume 2, will deal with the computation of torque pulsations during steady state also.

Notable radial uncompensated forces occur, as in three-phase IMs, unless the condition (14.88) is met

$$N_s \pm p_1 \neq N_r \tag{14.88}$$

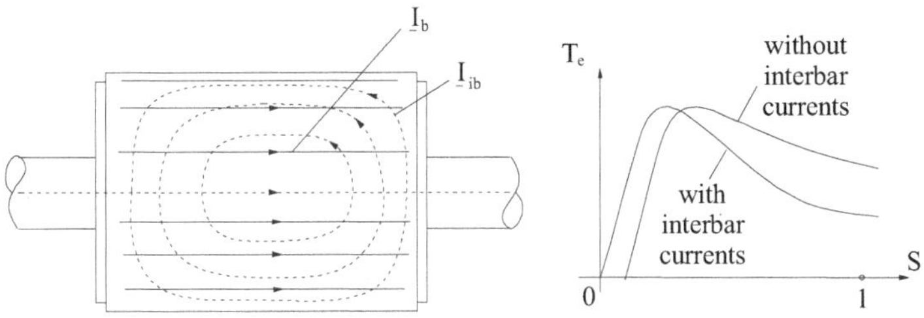

FIGURE 14.16 Interbar rotor currents and their torque ($2p_1 = 2$ poles).

14.11 INTERBAR ROTOR CURRENTS

In most single-phase IMs, the rotor cage bars are not insulated, and thus, rotor currents occur between bars through the rotor iron core. Their influence is more important in two-pole IMs where they span over half the periphery for half a period (Figure 14.16).

The interbar current ($\underline{I}_{ib}$) mmf axis is displaced by 90° (electrical) with respect to the rotor bar current ($\underline{I}_b$) mmf.

This is why the interbar currents are called transverse currents.

The interbar currents depend on the rotor skewing, besides the contact resistance between rotor bars and rotor iron and on rotor slip frequency.

The torque/speed curve may be influenced notably (as in Figure 14.16) for two-pole IMs.

The slot harmonics may further increase the interbar currents as the frequency of their produced rotor currents is much higher than for the fundamental ($f_2 = Sf_1$).

Special thermal treatments of the rotor may increase the bar-core rotor resistance to reduce the interbar currents and, thus, increase the torque for large slip values.

The influence of rotor skewing and rotor stack length/pole ratio on interbar current losses is as for the three-phase IMs (see Chapter 11, Section 11.12). Longer stacks tend to accentuate the interbar current effects. Also a large number of rotor slots ($N_r >> N_s$) lead to large interbar currents.

14.12 VOLTAGE HARMONICS EFFECTS

Voltage time harmonics occur in a single-phase IM either when the motor is fed through a static power converter or when the power source is polluted with voltage (and current) time harmonics from other static power converters acting on the same power grid.

Both odd and even time harmonics may occur if there are imbalances within the static power converters.

High-efficiency low rotor resistance single-phase capacitor IMs are more sensitive to voltage time harmonics than low-efficiency single-phase induction motors with open auxiliary phase in running conditions.

For some voltage time harmonics, frequency resonance conditions may be met in the capacitor IM, and thus, time harmonics high currents occur. They in turn lead to marked efficiency reduction.

As the frequency of time voltage harmonics increases, so does the skin effect. Consequently, the leakage inductances decrease and the rotor resistance increases markedly. A kind of leakage flux "saturation" occurs. Moreover, main flux path saturation also occurs.

So the investigation of voltage time harmonics influence presupposes known, variable parameters in the single-phase capacitor IM.

Such parameters may be calculated or measured. Correctly measured parameters are the safest way, given the non-linearities introduced by magnetic saturation and skin effect.

Once these parameters are known, for each voltage time harmonic, equivalent forward and backward circuits for the main and auxiliary winding may be defined.

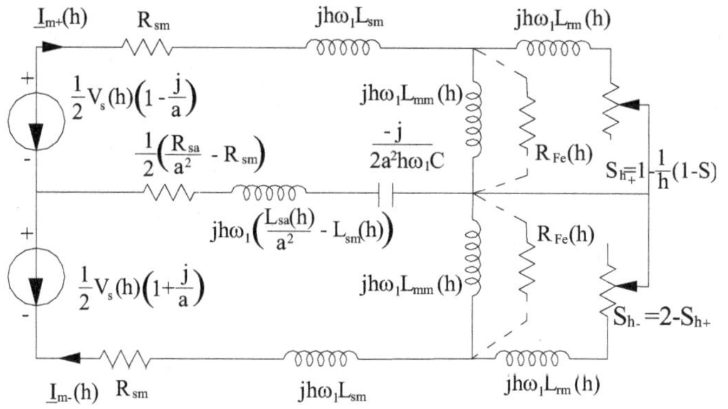

FIGURE 14.17 Time harmonics voltage equivalent circuit of capacitor IMs.

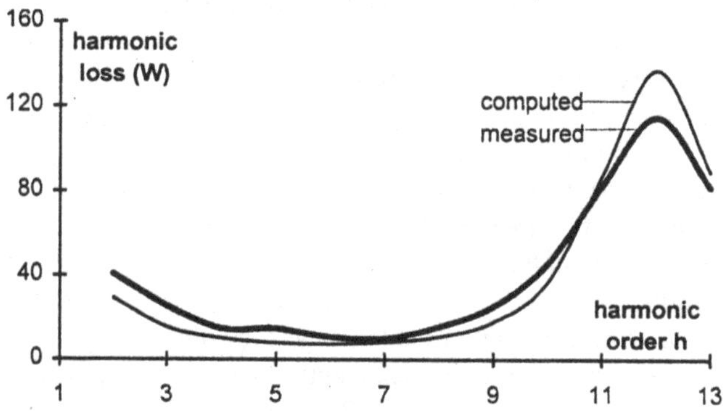

FIGURE 14.18 Harmonic losses at rated load and 10% time voltage harmonics.

For a capacitor IM, the equivalent circuit of Figure 14.5 may be simply generalised for the time harmonic of order h (Figure 14.17).

The method of solution is similar to that described in Section 14.3.

The main path flux saturation may be considered to be related only to the forward (+) component of the fundamental. X_{mm} in the upper part of Figure 14.5 that depends on the magnetisation current I_{mm+}. Results obtained in a similar way (concerning harmonic losses), for distinct harmonic orders, are shown in Figure 14.18, for a 1.5 kW, 12.5 μF capacitor single-phase induction motor [9].

Resonance conditions are met somewhere between the 11th and 13th voltage time harmonics. An efficiency reduction of 5% has been measured for this situation [9].

It is generally argued that the computation of total losses, including stray load losses, is very difficult to perform as visible in Figure 14.18. Experimental investigation is imperative, especially when the power decreases and the number of poles increases and severe interbar rotor currents may occur.

The neglection of space harmonics in the equivalent circuit as shown in Figures 14.5 and 14.17 makes the latter particularly adequate only from zero to breakdown torque slip.

14.13 THE DOUBLY TAPPED WINDING CAPACITOR IM

As already alluded to in Chapter 13, tapped windings are used with capacitor multispeed IMs or with dual-capacitor IMs (Figure 13.4) or with split-phase capacitor motors (Figure 13.8).

In the L or T connections (Figure 13.6), the tapping occurs in the main winding. For dual-capacitor or split-phase capacitor IMs (Figures 13.4 and 13.8), the tapping is placed in the auxiliary winding. Here we are presenting a fairly general case (Figure 14.19).

The two taps are t_a, for the auxiliary winding, and t_m for the main winding. In practice, only one tap is used at any time.

When only the tapping on the main winding is used, $t_a = 0$ and $0 < t_m < 0.5$ in general. On the contrary, when the tapping is done in the auxiliary winding, $0 < t_a < 0.5$ with $t_m = 0$ [10].

As it is evident from Figure 14.19, the main winding is present with a single circuit even for $t_m \neq 0$ (main winding tapping). Meanwhile, the tapped auxiliary winding is present with both sections aux 1 and aux 2.

Consequently, between the two sections, there will be a mutual inductance.

Denoting by X, the total main self-inductance of the auxiliary winding, $X_1 = t_a^2 X$ represents aux 1 and $X_2 = (1 - t_a^2)X$ refers to aux 2. The mutual inductance X_{12} is evidently $X_{12} = Xt_a(1 - t_a)$. The coupling between the two auxiliary winding sections may be represented as in Figure 14.20.

Besides the main flux inductances, the leakage inductance components and the resistances are added. What is missing in Figure 14.20 is rotor and main windings e.m.f.s in the auxiliary winding sections and the main winding.

As part of main winding is open (Figure 14.19) when tapping is applied, the equivalent circuit is much simpler (no coupling occurs).

Let us denote as forward and backward $\underline{Z}_f$ and $\underline{Z}_b$, the rotor equivalent circuits

$$\underline{Z}_f = \frac{1}{2} \frac{jX_{mm}(R_{rm}/S + jX_{rm})}{R_{rm}/S + j(X_{mm} + X_{rm})} \tag{14.89}$$

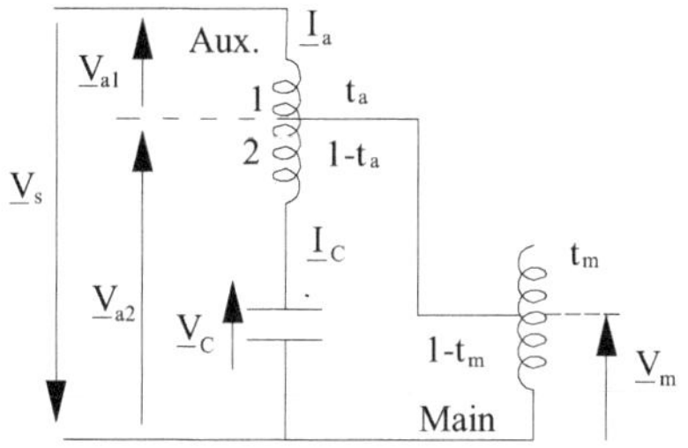

FIGURE 14.19 Doubly tapped winding capacitor IM.

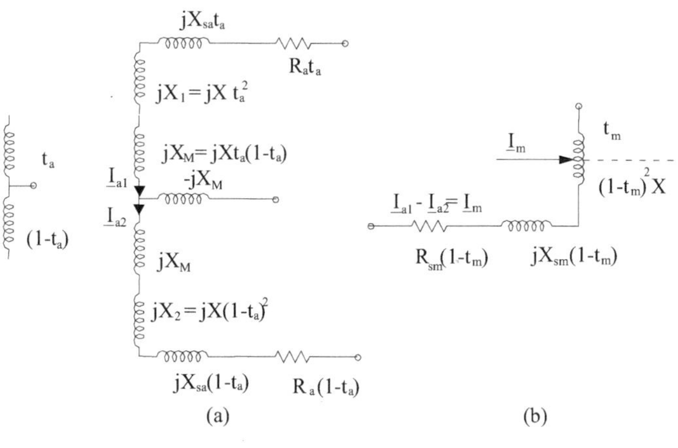

FIGURE 14.20 The tapped auxiliary (a) and main (b) winding equivalent (noncoupled) circuit.

$$\underline{Z}_b = \frac{1}{2}\frac{jX_{mm}\left(R_{rm}/(2-S)+jX_{rm}\right)}{R_{rm}/(2-S)+j\left(X_{mm}+X_{rm}\right)} \tag{14.90}$$

We may now apply the revolving field theory (Morrill's theory) to this general case, observing that each of the two sections of the auxiliary winding and the main winding are producing self-induced forward and backward e.m.f.s $\underline{E}_{fa1}$, $\underline{E}_{ba1}$, $\underline{E}_{fa2}$, $\underline{E}_{ba2}$, $\underline{E}_{fm}$, and $\underline{E}_{bm}$. The main winding tapping changes the actual turn ratio to $a/(1-t_m)$; also $\underline{Z}_f$ and $\underline{Z}_b$ to $\underline{Z}_f(1-t_m)^2$ and $\underline{Z}_b(1-t_m)^2$.

$$\begin{aligned}
\underline{E}_{fa1} &= t_a^2 a^2 \cdot \underline{Z}_f \underline{I}_{a1}\\
\underline{E}_{ba1} &= t_a^2 a^2 \cdot \underline{Z}_b \underline{I}_{a1}\\
\underline{E}_{fa2} &= (1-t_a)^2 a^2 \cdot \underline{Z}_f \underline{I}_{a2}\\
\underline{E}_{ba2} &= (1-t_a)^2 a^2 \cdot \underline{Z}_b \underline{I}_{a2}\\
\underline{E}_{fm} &= \underline{Z}_f \cdot (1-t_m)^2 \underline{I}_m\\
\underline{E}_{bm} &= \underline{Z}_b \cdot (1-t_m)^2 \underline{I}_m
\end{aligned} \tag{14.91}$$

Besides these self-induced e.m.f.s, there are motion-induced e.m.f.s from one axis (m axis) to the other (orthogonal) (a axis), produced both by the forward and backward components; in all, there are four such e.m.f.s in the main winding and two for each of the two auxiliary windings: $\underline{E}_{fa1m}$, $\underline{E}_{fa2m}$, $\underline{E}_{ba1m}$, $\underline{E}_{ba2m}$, $\underline{E}_{fma1}$, $\underline{E}_{bma1}$, $\underline{E}_{fma2}$, and $\underline{E}_{bma2}$.

$$\begin{aligned}
\underline{E}_{fa1m} &= -jat_a \underline{Z}_f \underline{I}_{a1}(1-t_m)\\
\underline{E}_{ba1m} &= jat_a \underline{Z}_b \underline{I}_{a1}(1-t_m)\\
\underline{E}_{fa2m} &= -ja(1-t_a)\underline{Z}_f \underline{I}_{a2}(1-t_m)\\
\underline{E}_{ba2m} &= -ja(1-t_a)\underline{Z}_b \underline{I}_{a2}(1-t_m)
\end{aligned} \tag{14.92}$$

Now we may easily construct the equivalent circuit model of the entire machine (Figure 14.21).

It is only a matter of algebraic manipulations to solve the equivalent circuit in Figure 14.21. After eliminating $\underline{V}_m$, $\underline{V}_{a1}$, $\underline{V}_{a2}$, and $\underline{I}_m$, two equations are obtained. They have $\underline{I}_{a1}$ and $\underline{I}_{a2}$ as variables

$$\begin{aligned}
\underline{Z}_{11}\underline{I}_{a1} + \underline{Z}_{12}\underline{I}_{a2} &= \underline{V}_s\\
\underline{Z}_{21}\underline{I}_{a1} + \underline{Z}_{22}\underline{I}_{a2} &= 0
\end{aligned} \tag{14.93}$$

with

$$\underline{Z}_{11} = (1-t_m)(R_{sm}+jX_{sm}) + (\underline{Z}_f+\underline{Z}_b)(1-t_m)^2 + t_a(R_{sa}+jX_{sa}) + t_a^2 a^2(\underline{Z}_f+\underline{Z}_b)$$

$$\underline{Z}_{22} = -(1-t_m)(R_{sm}+jX_{sm}) - (\underline{Z}_f+\underline{Z}_b)(1-t_m)^2 - (1-t_a)(R_{sa}+jX_{sa}) - (1-t_a)^2 a^2(\underline{Z}_f+\underline{Z}_b) - \underline{Z}_c$$

$$\underline{Z}_{12} = -ja(1-t_m)(\underline{Z}_f-\underline{Z}_b) - (1-t_m)(R_{sm}+jX_{sm}) - (\underline{Z}_f+\underline{Z}_b)(1-t_m)^2 + jt_a(1-t_a)a^2X_{mm}$$

$$\underline{Z}_{21} = -ja(1-t_m)(\underline{Z}_f-\underline{Z}_b) + (1-t_m)(R_{sm}+jX_{sm}) + (\underline{Z}_f+\underline{Z}_b)(1-t_m)^2 - jt_a(1-t_a)a^2X_{mm}$$

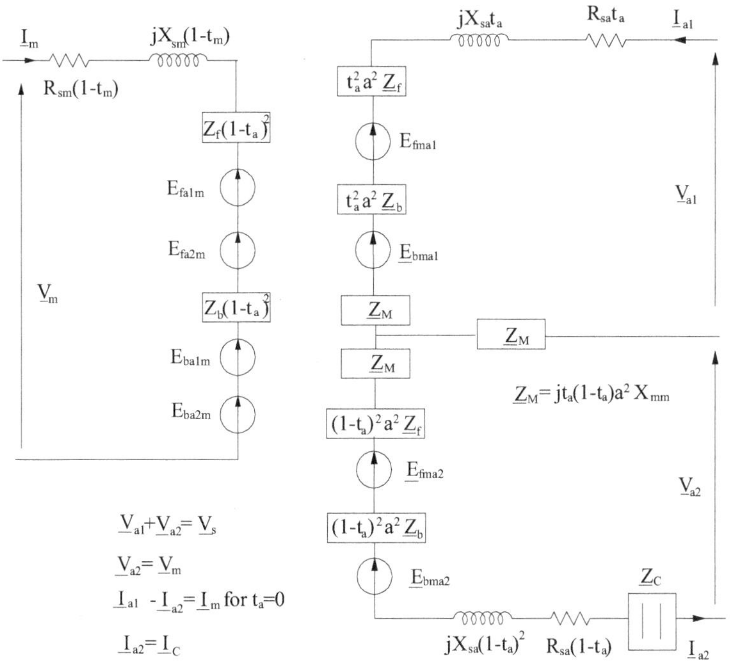

FIGURE 14.21 The equivalent circuit of doubly tapped winding capacitor induction motor.

For no tapping in the main winding ($t_m = 0$), the results in Ref. [10] are obtained. Also with no tapping at all ($t_m = t_a = 0$), (14.91–14.94) degenerate into the results obtained for the capacitor motor ((14.16)–(14.17) with (14.10); $I_{a2} = I_a$, $V_m = V_{a2} = V_s$).

The equivalent circuit in Figure 14.20, for no tapping ($t_m = t_a = 0$), degenerates into the standard revolving theory circuit of the capacitor IM (Figure 14.22).

The torque may be calculated from the power balance

$$T_e = \Big[\mathrm{Re}\big(\underline{V}_m \underline{I}_m^*\big) - R_{sm}(1-t_m)I_m^{\,2} + \mathrm{Re}\big(\underline{V}_{a1}\underline{I}_{a1}^*\big) + \mathrm{Re}\big(V_{a2}I_{a2}^*\big) - t_a R_{sa} I_{a1}^{\,2} - (1-t_a)R_{sa}I_{a2}^{\,2}\Big]\cdot p_1/\omega_1 \tag{14.94}$$

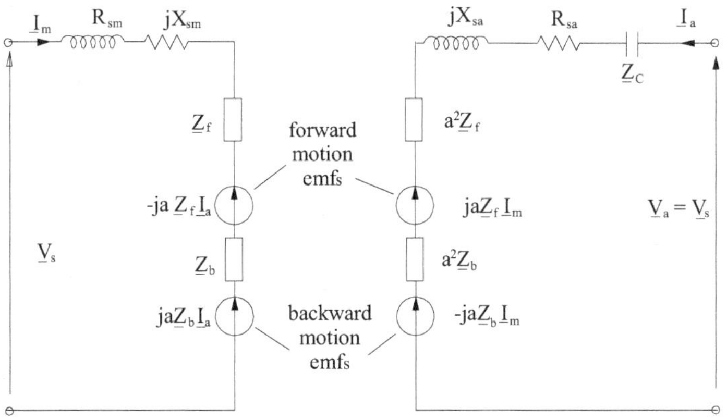

FIGURE 14.22 The revolving theory equivalent circuit of the capacitor motor.

14.13.1 Symmetrisation Conditions

The general symmetrisation may be determined by zeroing the backward (inverse) current components. However, based on geometrical properties, it is easy to see that, for orthogonal windings, orthogonal voltages in the main and auxiliary winding sections, with same power factor angle in both windings, correspond to symmetry conditions (Figure 14.23).

The two currents I_{a2} and I_m are 90° phase shifted as the voltages are. On the other hand, the current I_{a1} in the first section of the auxiliary winding a_1 is highly reactive and it serves basically for voltage reduction. This way some speed reduction is produced.

When no tapping is performed $t_a = 0$, $V_{a1} = 0$, and thus, $\dot{V}_{a2} = V_m = V_s$, and simpler symmetrisation conditions are met (Figure 14.24).

This case corresponds to highest $\underline{V}_{a2} - \underline{V}_c$ voltage; that is, for highest auxiliary active winding voltage. Consequently, the highest speed is expected.

14.13.2 Note on Magnetic Saturation and Steady-State Losses

As mentioned previously (in Chapter 13), magnetic saturation produces a flattening of the airgap flux density distribution. That is, a third harmonic e.m.f. and, thus, a third harmonic current may occur. The elliptic character of the magnetic field in the airgap makes the accounting of magnetic saturation even more complicated. Analytical solutions to this problem tend to use an equivalent sinusoidal current to handle magnetic saturation. It seems, however, that only FEM could offer a rather complete solution to magnetic saturation.

This appears even more so if we need to calculate the iron losses. On the other hand, experiments could be useful to assess magnetic saturation and fundamental and additional losses both in the core and the rotor cage or losses caused by the interbar rotor iron currents. More on these complicated issues is presented in Chapter 2 and Chapter 11 from Volume 2.

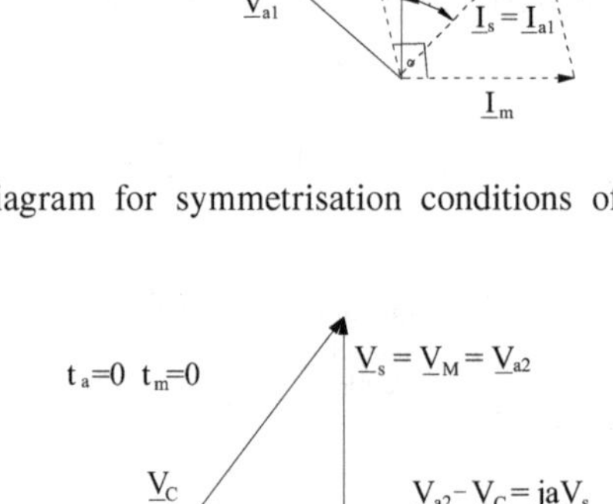

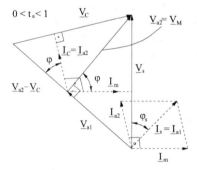

FIGURE 14.23 The phasor diagram for symmetrisation conditions of the generally tapped winding capacitor IM.

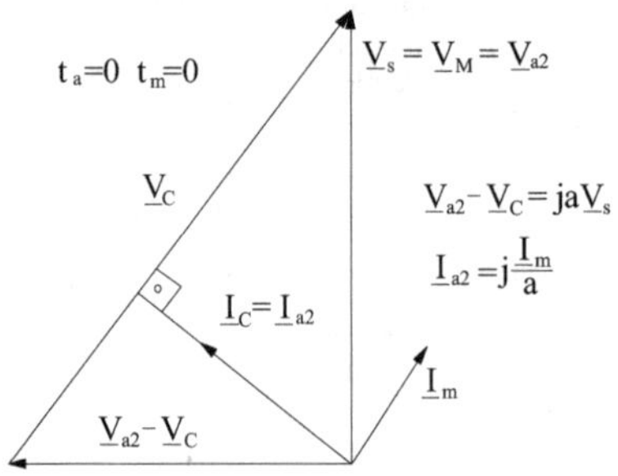

FIGURE 14.24 The phasor diagram for symmetrisation conditions with no tapping.

14.13.3 2/4 Pole Split-Phase Capacitor Induction Motor (after Ref. [11])

Household appliances such as refrigerator compressors require small speed operation (half speed) at 25% of rated power. Designing the machine for two-pole mode for rated torque T_{en} and checking its performance for four-pole mode at $T_{en}/4$, while providing good transition torque seems a practical way to save energy in variable capacity home compressor drives.

When operation at rated torque is short, it is possible to use only the main winding connected to the grid. It is a known fact that for 2p = 2 pole operation, a 2p′ = 4 pole winding has a strong third space harmonic which has to be reduced (Figure 14.25).

A compensation winding C is used for the scope (Figure 14.26).

The placement of windings M (main), A (auxiliary), and C (compensation) in slots for 2p = 2 is shown in Figure 14.27.

The compensation winding not only reduces the airgap flux density harmonics in 2p = 2 mode but also keeps the latter large to secure almost 4 times more torque at high speeds (2p = 2) – Figure 14.28.

Further improvements in torque ratio are obtained if the two main winding coils are placed one pole pitch apart (for the 2p = 4 configuration) because in the two-pole connection, the third airgap flux density harmonic is much smaller for the main winding; consequently, the compensation winding brings about even more additional torque.

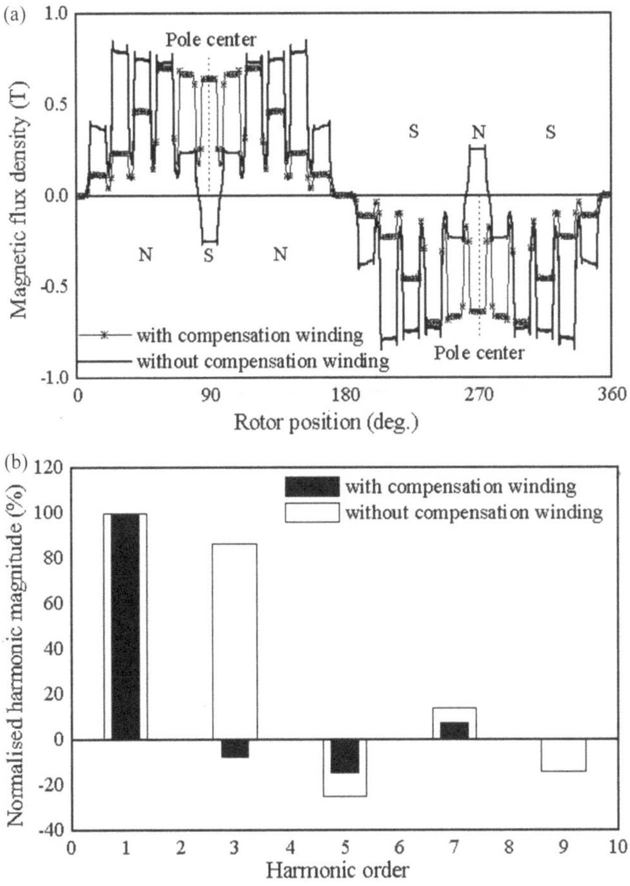

FIGURE 14.25 Airgap flux density distribution for 2p = 2 (with a four-pole winding in 2p = 2 connection) with and without compensation winding (Figure 14.26) (after Ref. [11]). (a) Magnetic flux density distribution [11]. (b) Discrete Fourier transform (DFT) of the magnetic flux density distribution.

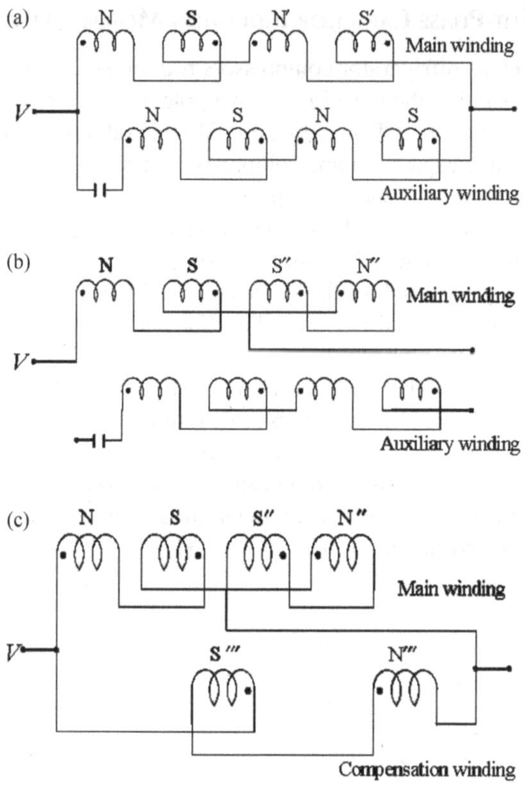

FIGURE 14.26 Winding pattern: (a) four-pole operation, (b) main winding only for two-pole connection, and (c) main plus compensation winding for two-pole connection.

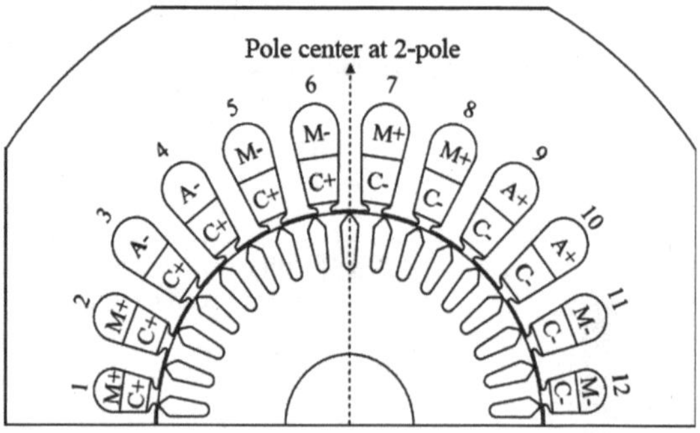

FIGURE 14.27 Four-pole windings (M and A) in 2p = 2 poles connection with compensation winding C.

Good efficiency (70%) in the 2p = 4 connection has been reported at 70 W rated power [11].

For 2p = 2 connection mode, there is still additional copper losses (in the middle of the poles of main winding) which hamper the efficiency, but still the transition from 2p = 4 to 2p = 2 mode shows large available torque.

A fairly complete treatment of single-phase IM transients including space harmonics is developed in Ref. [12].

A rather complete 2D and 3D FEM steady-state and transient analysis of capacitor IM is available in Ref. [13]. It seems that carefully designed 2D FEM suffices. Even the latter required one computation day for one operation point.

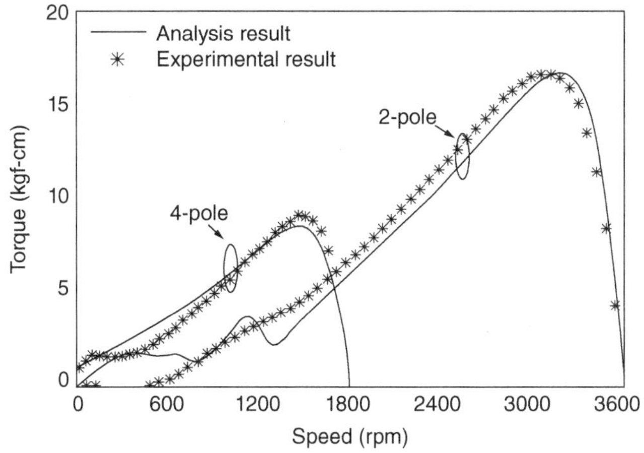

FIGURE 14.28 Torque/speed curves for 2p = 2 and 2p = 4 (after Ref. [11]).

14.14 SUMMARY

- The steady-state performance of single-phase IM is traditionally approached by the revolving field theory or the + – symmetrical components model. In essence, the voltages and currents in the two phases are decomposed into + and – components for which the slip is S and 2-S. Superposition is applied. Consequently, magnetic saturation cannot be accounted for unless the backward component is considered not to contribute to the level of magnetic saturation in the machine.
- On the other hand, the d–q (or cross-field) model (in stator coordinates), with $d/dt = j\omega_1$ may be used for steady state. In this case, keeping the current sinusoidal, magnetic saturation may be iteratively considered. This model has gained some popularity with T–L or generally tapped winding multispeed capacitor IMs.
- The revolving theory (or + – or fb) model leads, for the single-phase IM with open auxiliary winding, to a very intuitive equivalent circuit composed of + and – equivalent circuits in series. The torque is made of + and – components and is zero at zero speed (S = 1) as T_+ and T_- are equal to each other. At zero slip (S = 0), $T_+ = 0$ but $T_- \neq 0$ (and negative). Also at S = 0, with the auxiliary phase open, half of the rotor cage losses are covered electrically from stator and half mechanically from the driving motor.
- The split-phase and capacitor IM travelling field (+ –) model yields a special equivalent circuit with two unknowns I_{m+} and I_{m-} (the main winding current components). The symmetrisation conditions are met when $I_{m-} = 0$. As the speed increases above breakdown torque speed, so do the auxiliary winding and capacitor voltages for the permanent capacitor IM. Capacitor IMs have a good power factor and a moderate efficiency in general.
- Besides the configuration with orthogonal windings, it is feasible to use non-orthogonal windings especially with split-phase IMs in order to increase the output. It has been shown that symmetrisation conditions correspond to

$$\xi + (\varphi_a - \varphi_m) = 180°$$

where ξ is the electrical spatial angle of the two windings and $\varphi_a - \varphi_m$ is the time phase angle shift between $\underline{I}_a$ and $\underline{I}_m$.

The above conditions cannot be met in split-phase IMs, but they may be reached in capacitor motors. Apparently less capacitance is needed for $\xi > 90°$, and slightly more starting torque may be produced this way. However, building non-orthogonal windings means implicitly to use most or all slots for both main and auxiliary windings.

- The split-phase IM may be considered as a particular case of the capacitor motor with $Z_a = R_{aux}$. Above a certain speed, the auxiliary phase is opened ($R_{aux} = \infty$).
- Space harmonics occur due to the winding placement in slots. The third harmonic and the first-slot harmonic are especially important. The equivalent circuit may be augmented with the space harmonics contribution (Figure 14.13). Asynchronous, parasitic, forward, and backward (+, –) torques occur.
- For the permanent capacitor IM, the symmetrisation conditions ($I_{m-} = 0$) yield rather simple expressions for the turns ratio "a" between auxiliary and main winding and the capacitor reactance X_c

$$a = \tan\varphi_+; X_c = \frac{X_+}{\cos^2\varphi_+}$$

where φ_+ is the power factor angle for the maximum starting torque + component and X_+ is the + component equivalent reactance.

- Starting torque is a major performance index. For a capacitor motor, it is obtained for a turn ratio a_K

$$a_K = \sqrt{\frac{\sin\varphi_{sc}}{Z_{sc}\omega C}}$$

  Unfortunately, the maximum starting torque condition does not provide, in general, for overall optimum performance.

  As for an existing motor, only the capacitance C may be changed, care must be exercised when the turns ratio a is chosen.
- Starting current is heavily dependent on the turns ratio a and the capacitor C. A large capacitor is needed to produce sufficient starting torque, but starting current limitation has also to be observed.
- In general, for good starting performance of a given motor, a much larger capacitor C_s is needed in comparison with the condition of good running performance C ($C_s >> C$).
- Torque, main and auxiliary currents, I_m and I_a, auxiliary winding voltage V_a, capacitor voltage V_c, efficiency, and power factor versus speed constitute standard steady-state performance characteristics.
- The third harmonics may produce, with a small number of slots/pole, a large depth in the torque/speed curve around 33% of ideal no-load speed ($n_1 = f_1/p_1$). Magnetic saturation tends to accentuate this phenomenon.
- When $N_s = N_r$, synchronous parasitic torques occur at zero speed. For $N_s - N_r = 2p_1$, such torques occur at $S = 1 - 2p_1/N_r$. So, in general

$$N_s \neq N_r; N_s - N_r \neq 2p_1$$

to eliminate some very important synchronous parasitic torques.

- To avoid notable uncompensated radial forces, $N_s \pm p_1 \neq N_r$, as for three-phase IMs.
- As the rotor bars are not insulated from rotor core, interbar currents occur. In small ($2p_1 = 2$) motors, they notably influence the torque/speed curves acting as an increased

rotor resistance. That is, the torque is reduced at low slips and increased at high slip values when the bar-core contact resistance is increased.

- Voltage time harmonics originating from static power converters may induce notable additional losses in high-efficiency capacitor IMs, especially around electrical resonance conditions, when the efficiency may be reduced by 3%–5%. Such resonance voltage time harmonics have to be filtered out.
- Winding tapping is traditionally used to reduce speed for various applications. A fairly general revolving field model for dual winding tapping was developed. Symmetrisation conditions are illustrated (Figures 14.23 and 14.24). They show the orthogonality of main and auxiliary winding voltages and currents for orthogonally placed stator windings.
- Magnetic saturation is approachable through the + (forward) magnetisation inductance functional $L_{mm}(I_{mm+})$, but it seems to lead to large errors, especially when the backward current component is notable [14]. A more complete treatment of saturation is performed in Chapter 2, Volume 2, on transients.
- Practical 2/4 pole changing split-phase IM has been proposed recently [11].

REFERENCES

1. P. H. Trickey, Capacitor motor performance calculations by the cross field theory, AIEE *Transacions*, Vol. 75, February 1957, pp. 1547–1552.
2. C. G. Veinott, *Theory and Design of Small Induction Motors*, McGraw-Hill, New York, 1959.
3. B. S. Guru, Performance equations and calculations on T and L connected tapped-winding capacitor motors by cross field theory, *Electric Machines & Electromechanics*, Vol. 1, No. 4, 1977, pp. 315–336.
4. B. S. Guru, Performance equations and calculations on L - and T - connected multispeed capacitor motors by cross field theory, *Electric Machines & Electromechanics, IBID*, Vol. 2, 1977, pp. 37–48.
5. J. Stepina, *The Single Phase Induction Motors*, paragraph 3.7 (book), Springer Verlag, Berlin, Germany, 1982 (in German).
6. I. Boldea, T. Dumitrescu, S. A. Nasar, Steady state unified treatment of capacitor A.C. motors, *IEEE Transactions*, Vol. EC-14, No. 3, 1999, pp. 557–582.
7. S. Williamson, A. C. Smith, A unified approach to the analysis of single phase induction motors, *IEEE Transactions*, Vol. IA-35, No. 4, 1999, pp. 837–843.
8. M. Popescu, C. B. Rasmussen, T. J. E. Miller, M. McGilp, Effect of MMF harmonics on single-phase induction motor performance – A unified approach, *Record of IEEE-IAS*, 2007, pp. 1164–1170.
9. D. Lin, T. Batan, E. F. Fuchs, W. M. Grady, Harmonic losses of single phase induction motors under nonsinusoidal voltages, *IEEE Transactions*, Vol. EC-11, No. 2, 1996, pp. 273–282.
10. T. J. E. Miller, J. H. Gliemann, C. B. Rasmussen, D. M. Ionel, Analysis of tapped winding capacitor motor, *Record of ICEM*, 1998, Istanbul, Turkey, Vol. 2, pp. 581–585.
11. H. Nam, S. K. Jung, G. H. Kang, J. P. Hong, T. U. Jung, S. M. Beck, Design of pole change single-phase IM for household appliances, *IEEE Transactions*, Vol. IA-40, No. 3, 2004, pp. 780–788.
12. H. A. Toliyat, N. Sargolzaei, A comprehensive method for transient modeling of single phase induction motors including the space harmonics, *EMPS Journal*, Vol. 26, No. 3, 1998, pp. 221–234.
13. J. Bacher, F. Waldhart, A. Muetze, 3-D FEM calculation of electromagnetic properties of single-phase induction machines, *IEEE Transactions*, Vol. EC-31, No. 1, 2015, pp. 142–149.
14. C. B. Rasmussen, T. J. E. Miller, Revolving-field polygon technique for performance prediction of single phase induction motors, Record of IEEE-IAS-2000, Annual meeting.

Index

T